Atomic Force Microscopy/Scanning Tunneling Microscopy

Atomic Force Microscopy/Scanning Tunneling Microscopy

Edited by

Samuel H. Cohen
Mona T. Bray and
Marcia L. Lightbody
U.S. Army Natick Research, Development and Engineering Center
Natick, Massachusetts

Plenum Press • New York and London

Library of Congress Cataloging in Publication Data

Atomic force microscopy/scanning tunneling microscopy / edited by Samuel H. Cohen, Mona T. Bray and Marcia L. Lightbody.
p. cm.
Includes bibliographical references and index.
ISBN 0-306-44890-4
1. Atomic force microscopy. 2. Scanning tunneling microscopy. I. Cohen, Samuel H. II. Bray, Mona T. III. Lightbody, Marcia L. IV. U.S. Army Natick Research, Development and Engineering Center Atomic Force Microscopy/Scanning Tunneling Microscopy (AFM/STM) Symposium (1st: 1993: Natick, Mass.)
QH212.A78A86 1994
502.'.8'2—dc20 94-48894
CIP

Proceedings of the First U.S. Army Natick Research, Development and Engineering Center Atomic Force Microscopy/Scanning Tunneling Microscopy (AFM/STM) Symposium, held June 8–10, 1993, in Natick, Massachusetts

ISBN 0-306-44890-4

Printed in the United States of America

PREFACE

The first U.S. Army Natick Research, Development and Engineering Center Atomic Force/Scanning Tunneling Microscopy (AFM/STM) Symposium was held on June 8-10, 1993 in Natick, Massachusetts. This book represents the compilation of the papers presented at the meeting.

The purpose of this symposium was to provide a forum where scientists from a number of diverse fields could interact with one another and exchange ideas. The various topics included application of AFM/STM in material sciences, polymers, physics, biology and biotechnology, along with recent developments including new probe microscopies and frontiers in this exciting area.

The meeting's format was designed to encourage communication between members of the general scientific community and those individuals who are at the cutting edge of AFM, STM and other probe microscopies. It immediately became clear that this conference enabled interdisciplinary interactions among researchers from academia, industry and government, and set the tone for future collaborations.

Expert scientists from diverse scientific areas including physics, chemistry, biology, materials science and electronics were invited to participate in the symposium. The agenda of the meeting was divided into three major sessions. In the first session, Biological Nanostructure, topics ranged from AFM of DNA to STM imaging of the biomolecule tubulin and bacterial luciferase to the AFM of starch polymer double helices to AFM imaging of food surfaces. The second group of papers presented in the session on Nanostructure of Materials included papers on STM studies of fullerenes, the use of AFM in metallography, STM and AFM studies of conducting polymer films and AFM studies of nanoclusters on mica. The third session, Methodologies and Techniques, featured papers on molecular imaging and surface forces, use of STM to image in an electrochemical environment, photon tunneling microscopy, and mechanical techniques for sharpening STM tips. The papers presented over the three-day period represented a wide range of specialties and there was extensive cross fertilization of ideas.

Organizing such a symposium and editing a book that contains the papers presented are demanding tasks. Review of all the papers for reformatting to a uniform style required a great deal of time and a greater amount of patience. The results, however, justify the tireless efforts that went into the planning. There were many individuals who helped us in this regard. Special thanks and acknowledgments go to Ms. Georgette Nedder who expedited mailings and other administrative functions during the initial planning of the symposium and to Mr. Raymond Andreotti who, time after time, came to our rescue when we were having difficulties with our data entry. Thanks also go to Dr. Donald Robertson who assisted in the execution of a number of tasks and to Ms. Deborah Cobban for typing assistance.

Those who chaired the daily sessions - Dr. Wilbert Bowers, Dr. Heidi Schreuder-Gibson and Mrs. Margaret Auerbach - were outstanding in keeping the talks on schedule. Team efforts by Natick personnel were vital to the success of this symposium. Notable service was provided by Protocol, Audiovisual, Records Management, Communications Security, Motor Pool, and Facilities Engineering. Finally, Mr. Frank Yee, Mr. David Cameron, Mr. Thomas Cook and Ms. April Doyle, all from reproduction, deserve our sincere thanks for their prompt and professional response to all our printing requests.

Samuel H. Cohen
Mona T. Bray
Marcia L. Lightbody

May 1994

CONTENTS

NANOSTRUCTURE OF MATERIALS

METHODOLOGIES AND TECHNIQUES

KEYNOTE ADDRESS:

MATERIALS RESEARCH INSTRUMENTATION DEVELOPMENT: A NEW PARADIGM

J. H. Hopps

Division of Materials Research
National Science Foundation
Washington, DC 20550

Abstract: Graduate and postgraduate education in science and engineering in the U.S. has long been recognized as outstanding. In materials science and engineering, graduate and postgraduate education is not only outstanding, but as a result of its intrinsic multidisciplinary nature, it is having a positive pervasive impact upon the culture within major research universities in the U.S. Although generally excellent, this paper focuses on areas in which new initiatives are being taken to more broadly enhance education and training in materials science and engineering. We briefly discuss the issues of training modes, undergraduate materials education and research, vocational and "shop floor" level training, as well as some fundamental infrastructure issues that will impact our ability to broadly enhance materials science and engineering education in the U.S.

INTRODUCTION

Instrumentation and equipment are keys to and major drivers for innovative research and development. Instrumentation and equipment determine what experimental research can be viably pursued and the level of productivity that can be sustained in these pursuits. For example, they determine our ability to synthesize and to process advanced materials, as well as our ability to monitor, and therefore to control, such synthesis and processing in both the laboratory and manufacturing environment.

The U.S. Army Natick Research, Development and Engineering Center has long pioneered the use of state-of-the-art instrumentation to assist in carrying out fundamental research for the benefit of our nation's soldiers. For example, when Natick first opened in 1954, it possessed one of the first RCA transmission electron microscopes. Over the intervening years, Natick has obtained scanning or transmission electron microscopes along with other diverse scientific equipment for use in a broad spectrum of research endeavors. In 1993 Natick purchased an atomic force/scanning tunneling microscope, further proof of the priority the organization places in providing instrumentation for use in aiding research.

Atomic Force Microscopy/Scanning Tunneling Microscopy
Edited by S.H. Cohen *et al.*, Plenum Press, New York, 1994

INFRASTRUCTURE PERSPECTIVES

A major challenge facing the Foundation is the wider deployment of advanced research instrumentation and equipment. Wider deployment is needed to quicken the pace of scientific and technological breakthroughs, for example, in the synthesis and the processing, and ultimately in the manufacturing, of advanced materials. Wider deployment for research would also have a profound effect on the research-based component of both graduate and undergraduate education. More generally, the wider deployment of advanced research instrumentation and equipment could enhance the educational opportunities and experiences at a broader range of institutions. This potential is particularly significant with the increasing recognition of the need to strengthen technical education programs at the baccalaureate level, as well as within the curricula of two-year community colleges and technical schools.

There are many impediments to the wider deployment of advanced research instrumentation and equipment. These include potentially spiraling costs, both capital and operating, as well as design complexities, which make reliability low, maintainability difficult, and virtually necessitate the kind of general facilities support infrastructure that can only be maintained in our larger research universities. These impediments are tied, to a large extent, to the way we develop equipment. Performance has been virtually the sole driver of instrumentation and equipment design, particularly in the limited competitiveness of today's commercial market for research apparatus; a market in which the U.S. plays no significant role in many of its sectors. Another factor is that the design of much of our advanced research instrumentation and equipment tends to be of the over-the-wall type. That is, the designs consist predominately of the repackaging of instruments developed by bench scientists. In these instances, the manufacturing engineering value-added is minimal.

There is much to be gained by the reversal of the current advanced research instrumentation and equipment development approach, which unchecked, will likely continue to lead us along the ever-increasing cost spiral viewed as inevitable by many. Design, development and manufacturing processes for advanced research instrumentation and equipment can be structured to yield products that are optimized over a much larger set of simultaneous objectives. Such a set might include enhanced performance (e.g., accuracy, resolution, capacity, throughput, etc.), friendlier user interface, flexible operation and applicability, higher reliability, improved maintainability, as well as lower cost (both capital and operating). Most of these attributes relate directly to the general goal of wider deployment. Such instrumentation and equipment, designed for, and deployed at, the research level, can provide critical baseline design experiences for commercial applications.

The NSF is taking the initiative in stimulating the development of the next generation of advanced research instrumentation and equipment according to a new paradigm. This paradigm consists of the close collaboration between bench scientists who are specialists in particular instrumentation areas and experts in the area of advanced manufacturing technologies working together within the framework of concurrent engineering. It is this approach that will maximize the manufacturing engineering value-added. This capability to manufacture must become an up-front consideration. The development involves, of course, more than just fabrication. Critical aspects of the overall activity include research and development of new materials, processes and components. A collaborative concurrent engineering framework provides a systematic and rational method for establishing research and development priorities in support of advanced research instrumentation and equipment development.

The approach that is outlined requires not only teamwork and interdisciplinarity within the university, as well as partnerships with industry, it will also require teamwork and interdisciplinarity within the Foundation. Neither the universities nor NSF can do business as usual in the development of advanced research instrumentation and equipment.

ATOMIC FORCE MICROSCOPY FOCUS

It is within this context that the National Science Foundation is embarking upon a program to develop high performance scanning tip microscopes. The impetus for this Foundation initiative is the April 1993 document "Atomic Imaging and Manipulation for Advanced Materials", which is the Report of the NSF Panel on Atomic Resolution Microscopy. The document is from an NSF-sponsored workshop held in Washington, DC on May 25-28, 1992. According to the Report:

> We now stand at the birth of entirely new technologies for observing and manipulating individual atoms and molecules ...Driving the resolution of scanning tip and electron microscopies to ultimate atomic-scale limits (1 Å and below) and making them widely available in an effective way should be a national goal in science and technology.

The scope of the report was broader than that of this specific conference. It included scanning tip microscopies and atomic manipulation, as well as electron microscopies. Among the various scanning tip techniques addressed in the report were scanning tunneling microscopy, scanning tunneling spectroscopy, atomic force microscopy, field-ion microscopy, and ballistic electron microscopy. Electron microscopies considered included atomic resolution chemical analysis, atomic resolution tomography, electromagnetic field imaging, cryogenic microscopy, and in-situ microscopy. The general finding of the report was that important new science will become accessible as a result of the development of the atomic-resolution microscopy of materials. Among the anticipated major new findings were chemical identification at the atomic level and manipulation at the atomic level. The Panel concluded that major contributions to the development of advanced materials for industrial purposes would result from high performance electron microscopies and scanning tip techniques. It recommended "The U.S. must sustain a significant and effective program in this area," and "Substantial program in these areas will strengthen U.S. competitiveness."

The panel recognized the fact that much progress in materials science has been made by investigators who have had access to user friendly, low cost, fast turnaround, compact instruments.

Specific recommendations were made for scanning tip microscopies in the areas of component and procedure development. Component development focused on new tips, specialized stages, flexible atomic manipulation apparatus and the incorporation of electromagnetic spectroscopy. The procedure development focus included low or high temperatures in ultrahigh vacuum, a range of controlled atmospheres with variable temperature and other environmental controls, as well as flexible atomic manipulation.

A focus for research activity was also provided by the Panel. This included recommendations for the advancement of our state of knowledge in the area of tunneling, contact and imaging processes. Underlying all of the recommendations of the Panel were driving intellectual themes, which included, among others, atom manipulation, surface modification, liquid-phase and vapor-phase crystal growth, and polymer structure.

CONCLUSIONS

A formal NSF response to the report has not yet been developed. A viable framework for implementing such a program already is, nonetheless, in place or emerging. Current programs within the Foundation that are relevant to such an activity include the National Nanofabrication Users Facility Network being supported by the Engineering, the Mathematical and Physical Sciences, and the Biological Sciences Directorates. Also of significance are new focus areas in the Division of Materials Research such as molecular manufacturing and the advanced processing of materials. The existing program base of the Division, which includes individual investigators, centers and facilities, currently supports virtually all of the driving intellectual themes.

The formal response that is under development consists of a collaborative effort by the Division of Materials Research of the Directorate for Mathematical and Physical Sciences and the Division of Design and Manufacturing Systems of the Directorate for Engineering in pursuing the development of scanning tip microscopy instrumentation within the framework of concurrent engineering. It is anticipated that the impacts of this concurrent engineering approach will be (1) design optimizations, within the limits of current technology, relative to a "full set" of goals, (2) systematic and traceable identification of component and subsystem development priorities, and (3) systematic and traceable identification of research priorities. Such an approach has been able to evolve, in part, as a result of innovative leadership within upper management at NSF which has facilitated increased cross-directorate and cross-divisional activities, increased emphasis upon interdisciplinary activities, and the emergence of the virtual organization and programs concept within NSF for agile response to new opportunities.

BIOLOGICAL NANOSTRUCTURE

SCANNING FORCE MICROSCOPY ON LIVING VIRUS-INFECTED CELLS

W. Häberle,[1] D.P.E. Smith,[1] J.K.H. Hörber,[2] and C.P. Czerny[3]

[1]IBM Research Division, Munich Physics Group
Schellingstrasse 4, 8000 Munich 40, Germany
[2]EMBL
Meyerhofstrasse 1, 6900 Heidelberg, Germany
[3]Tierärztliche Fakultät der Ludwig-Maximillians-Universität
Veterinärstrasse 13, 8000 Munich 22, Germany

Abstract: Using scanning force microscopy (SFM), it is possible to follow processes on the membrane of single living cells. Monkey kidney cells were imaged under normal growth conditions with a resolution of approximately 10 nm. Upon adding a suspension of pox virus, pronounced changes in the cell membrane were observed. Image sequences of these processes on the cell surface can be monitored over time. The cell membrane was studied to determine how it is affected by changes in temperature. The end of the micropipette holding the cell was also examined by SFM.

INTRODUCTION

The samples that have been studied with scanning force microscopy (SFM) have evolved from inorganic to organic and, more recently, from nonliving to living. The first surfaces studied were those such as graphite[1] and boron nitride.[2]

It was immediately recognized that SFM is an ideal tool for studying organic films[3]. Rapid progress was then made in applying this technique to biologically relevant material, principally by the group of Hansma. Notable examples in this direction were the imaging of amino acid crystals,[4] fibrinogen, polyalanine,[5] lipid-protein membranes,[6] and the membrane protein bacteriorhodopsin.[7]

The next stage in the evolution of the instrument was to examine whole cells under the SFM. The first examples were human erythrocytes and lymphocytes, *E. Coli*[8] and the archaebacterium *Halobacterium halobium*.[9] For these experiments the samples were dehydrated and hence could no longer be accurately described as living. In another study, red and white blood cells were fixed with glutaraldehyde to increase their rigidity and imaged in a buffer solution.[9]

A completely new approach is fixing cells by the suction of a micropipette and

Atomic Force Microscopy/Scanning Tunneling Microscopy
Edited by S.H. Cohen *et al.*, Plenum Press, New York, 1994

imaging them in a normal growth medium. This allows the cells to be kept alive for several days.[10,11] With this development, the dynamics of cell physiology could be studied much as it is done with optical microscopy, but with the benefit of resolution that can be as high as 10 nm. The first experiments of this nature, performed on living erythrocytes, demonstrated the ability to observe molecular marker attachment and cell shrinkage in high salt concentrations.[10] Another study, described in this paper, shows the infection of a monkey kidney cell by a pox virus. Since SFM locally measures ultrasmall forces, it is thus also possible to study the mechanics of cellular processes. We demonstrate this ability by cooling a cell to 11° C and observing that it stiffens. In another example of how SFM may impact the understanding of cell physiology, we show SFM images taken of the tips of patch-clamp pipettes used for holding the cell.

EXPERIMENTAL METHODS

The SFM used in our studies had to be designed with some special considerations in mind. We chose to capture whole cells on the end of a micropipette since this method would not distort the cell membrane as much as attaching the cell to a flat substrate. For example, in the experiments of Henderson et al.,[12] in which glial cells were plated to glass cover slips, actin filaments of the cytoskeleton were observed to dominate the SFM images. We believe that holding cells with a micropipette is a more suitable way to observe the membrane surface rather than submembranal structures. The approach additionally offers the possibility to manipulate the cell, for example, by injecting drugs or genetic material.

The micropipettes, similar to those used for the patch clamp technique,[13] are made by drawing a pipette with an outside diameter of 0.8 mm to a diameter of just 2-4 μm. The pipette is then mounted on a piezoelectric tube scanner, which in turn is secured to an XYZ manipulator. A small, flexible Teflon™ tube allows a vacuum to be drawn on the pipette to capture cells in suspension. The remaining cells are then flushed from the chamber. Using the XYZ manipulator, the cell is brought in close proximity to the force sensing cantilever. The piezoelectric tube is then used to scan the cell membrane over the pyramidal tip built onto the end of the cantilever.

In order to position the cell properly with respect to the cantilever of the SFM, it was important to be able to observe simultaneously the pyramid on the lever and the cell attached to the pipette. The kinds of cells we used (human red blood cells, monkey kidney cells, and human tumor cells) were between 5-20 μm in diameter, requiring an optical microscope with a magnification of 300X to 600X. To achieve this magnification, and at the same time to keep the necessary working distance between the objective and mechanical components required for SFM, we used an inverse optical microscope. In this configuration, the objective is below the liquid-filled chamber holding the cell and light is directed downward through the chamber from above (Figure 1). The lever is fixed with respect to the chamber and the optical components are adjusted to detect angular deflections of the lever of the order of 1 μrad.

With our most recent setup, the end of a single-mode optic fiber is brought as close as possible to the lever using a similar manipulator as for the pipette (Figure 2). The last several millimeters of the fiber's protective jacket is removed. The minimum distance is determined by the diameter of the fiber cladding and the geometry of the lever. We normally use fiber for 633 nm light, which has a nominal cladding diameter of 125 m and levers which are 200 μm long. Holding the fiber at an angle of 45° with respect to the lever means that we can safely bring the fiber core to within about 150 μm of the desired spot on the lever. The 4 μm diameter core has a numerical aperture of 0.1 and the light

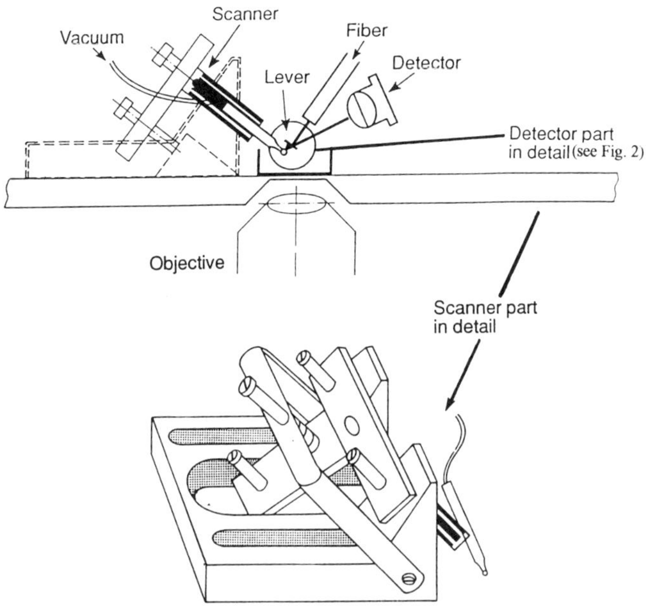

Figure 1. Scanning force microscope incorporating a cantilever, a micropipette attached to piezoelectric scanner, and a fiber optic detection scheme. (a) View of all components in a chamber open at the top. (b) Detail of XYZ translator used to adjust pipette and optic fiber with respect to cantilever.

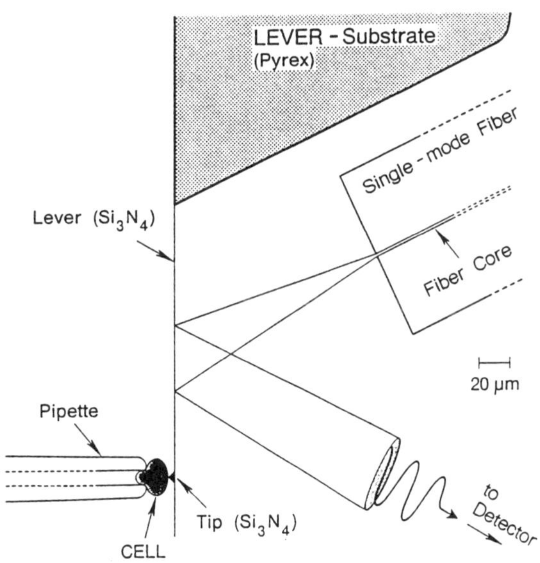

Figure 2. (Figure 1 insert) Schematic of optical detection system using single-mode optic fiber and bicell photodetector.

emerging from the fiber therefore expands with an apex angle of 6. For the geometry given above, the smallest spot size achievable is 50 μm, approximately the size of the triangular region at the end of the cantilever (Figure 3).

Due to the construction of the mechanical pieces holding the pipette and fiber positioner, the closest we can bring the position-sensitive (bicell) detector (PSD) to the lever is approximately 2 cm. This implies a minimum spot size of 2.1 mm, within the 3 mm x 3 mm borders of the PSD. Using a 2 mW HeNe laser we believe that under normal operating conditions the displacement sensitivity is below 0.1 Å with a signal-to-noise ratio of 10 and a bandwidth of 1 Hz, comparable to the sensitivity of other optical detection techniques.

The advantage of our method is that there are no lenses and there is no air/liquid or air/solid interface across which the incoming light beam must travel. At the present time the outgoing light travels across a liquid/air interface on route to the PSD. In future versions we plan to use a light pipe to eliminate this interface also, so we have no longer any air interface.

Figure 3. HeNe laser reflecting off a silicon nitride lever. A pipette holding a cell has been brought to the pyramidal tip at the end of the lever.

We now describe three methods for building an SFM based on the principles discussed above, i.e., incorporating a micropipette, optical lever detection, and an inverse optical microscope. With a few changes it is also possible to use a normal optical microscope.

1. Our original apparatus uses a small bath of 50-100 μL to hold the cell, lever, and the end of the optic fiber (Figure 2). The pipette and the optic fiber are placed at an angle of 45° from one another. This apparatus is described by Häberle et al.[10,11] It offers a very good view of the cell through the inverse microscope. The first version used a tunnel sensor

to detect the lever deflection, but this technique proved to be unreliable for the long time-scale measurements common in biological experiments. The high sensitivity of the tunnel detector also caused the system to have a rather small dynamic range when operating in a "variable force" mode. The incorporation of the fiber-optic detection scheme greatly increased the ease of use of the system. The short optical path length, the result of using no lenses or mirrors, gave the instrument a very high mechanical stability. One disadvantage of this kind of sample chamber is that it is difficult and time-consuming to change the solution in the bath.

2. A new system we have recently built uses a sample chamber open at two sides. The liquid is held by surface tension between two horizontal parallel plates (Figure 4). The lever

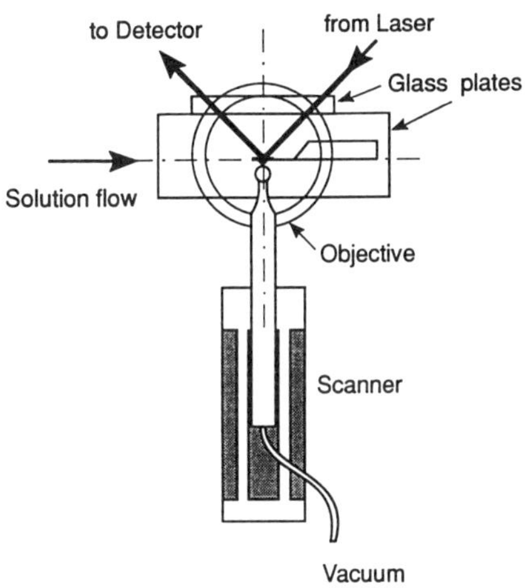

Figure 4. Design of the open chamber, which allows good flow of solution.

is mounted perpendicular to the plane of the microscope objective, allowing very fine adjustment of the cell to the lever pyramid. The open sides allow for a good flow of liquid through the chamber so that changing the growth solution or introducing chemical or biological additives is faster and more efficient. On the other hand, the large open surface area of the chamber results in a high rate of evaporation, which necessitates a constant influx of liquid to replenish the chamber. Since the top surface is closed, the optical path length of the detection laser is rather long, about 20 cm in total. This fact demands a very stable mechanical construction in order to avoid a high level of vibrational noise. The total weight of this construction is rather high (over 800 g) compared to the first setup (less than 100 g).

3. We have plans for another construction which consists of a completely enclosed chamber (Figure 5). This should ensure better temperature control and eliminate problems

of evaporation. The bath will also stay free of unwanted contamination. However, it will be more difficult to bring the pipette into the chamber and to position the cell for scanning.

STUDYING THE SURFACE OF THE MICROPIPETTE

Using the second SFM construction shown in Figure 4, it is very easy to take an SFM image of the end of the micropipette itself. Hence, one can examine the condition of the pipette before a cell or a cell membrane has been attached to it. This ability may be extremely important to better understand patch-clamp technology.[13] In this technique, one measures the resistance between the electrolyte inside and outside of a pipette opening that is covered by a membrane. The high resistance of the patch, 10-100 GΩ for a well-sealed pipette 10 to 1μm in diameter, makes it possible to measure currents through even a single ion channel. These studies have led to a greatly improved understanding of ion transport through biological membranes.

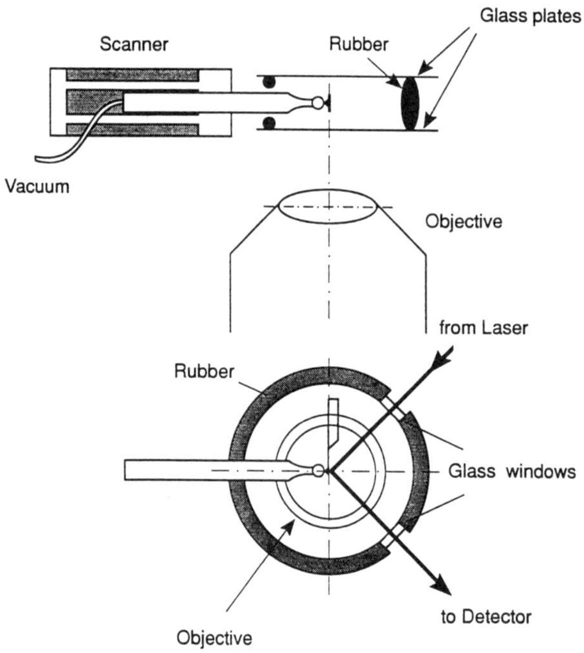

Figure 5. Design of the proposed chamber, which would be completely closed on all sides.

The pipettes are produced by heating the middle of small glass capillaries and pulling the ends apart with well-defined forces. The final diameter at the tip is determined by the parameters of this process. The tips are then heat polished, which significantly reduces the surface roughness arising from the glass breaking at the end of the pulling process.

Until now, the sealing surface of pipettes has only been investigated with scanning electron microscopy (SEM), making it necessary to coat the pipettes with metal. Coating, however, distorts the size of the central hole and the wall thickness, in addition to change the surface roughness. SFM, in contrast, not only gives higher resolution but also allows the surface to be investigated under normal operating conditions. In particular, a saltwater solution can cause important changes in the borosilicate and aluminum silicate glasses commonly used. The ability to study the molecular structure of the pipette tip will be important in determining the forces involved in the glass/membrane seal.

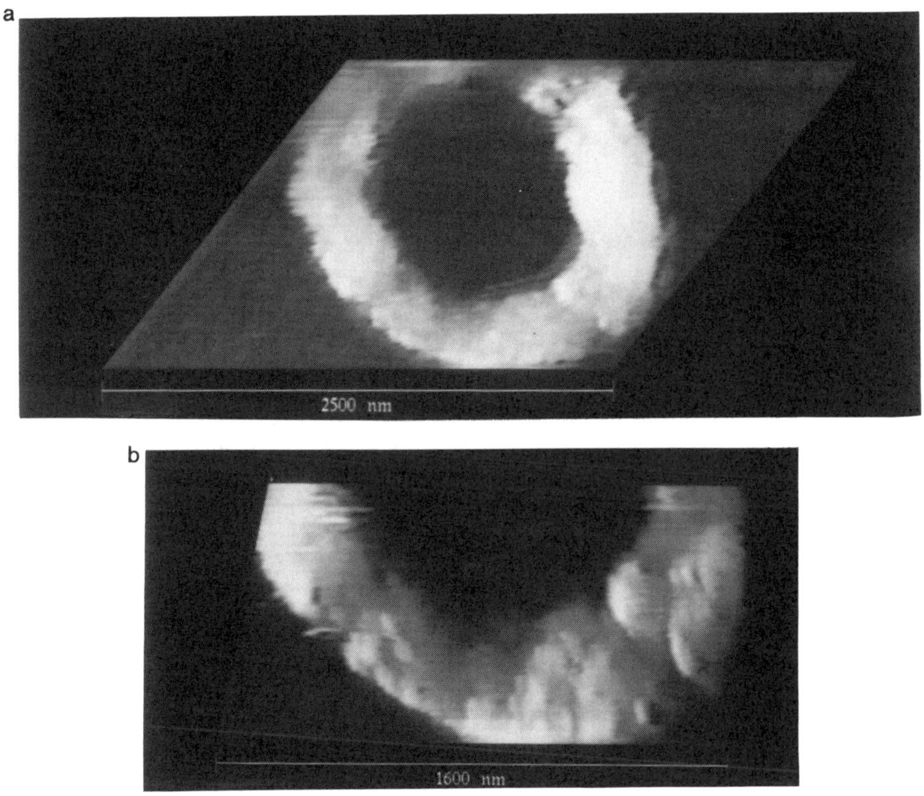

Figure 6. Tip of a patch clamp pipette before flame polishing: (a) SFM view of almost the entire circular opening; (b) higher magnification of the lower right corner

The 3-μm diameter patch-clamp pipette shown in the SFM images in Figure 6 (a) and (b) was imaged in a 150 mM KCl solution. This pipette was not heat polished after the pulling procedure. The SFM images show grooves that might be the result of the pulling process or of subsequent contamination. These topographical features might make it impossible to achieve a GΩ membrane patch clamp. Such inspection of pipettes *in situ* immediately before use could prove to be extremely valuable.

SURFACE PROCESSES ON VIRUS-INFECTED CELLS

We have found that cells can be continuously and stably scanned for several days in growth solution with forces of around 1 nN.[11,14] In agreement with SEM images, we observe that the surface of the monkey kidney epithelial cell is characterized by many finger-like cytoplasmic extensions known as microvilli. The cell surface displays micron-scale changes on time scales of several hours and 10-nm-scale changes on the order of several minutes. The former is likely to be indicative of changes in the plasma membrane while the latter may have to do with alterations in the cytoskeleton.

The vaccinia virus was added to the sample chamber as a suspension in growth solution. This virus was chosen because, with a diameter of 200-300 nm, it is one of the largest known viruses. The SFM image shown in Figure 7 is of a cultured monkey kidney cell in standard growth solution that had been infected by the virus 19 hours earlier.[11,14]

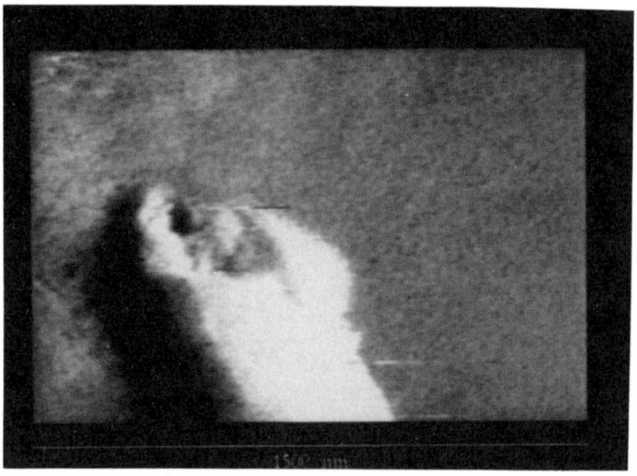

Figure 7. SFM image 19 hours after virus infection: probable exocytosis of a progeny vaccinia virus from the end of a microvillus on the surface of a monkey kidney cell.

This image shows what we believe to be the process of exocytosis of a progeny virus leaving the cell at the end of a microvillus. The roughly spherical protrusion at the end of the finger-like structure is 200-300 nm in diameter, a size that would confirm its identification as a vaccinia virus. Using high-voltage electron microscopy (EM), Stokes[15] observed that 20 hours after infection with vaccinia virus the entire cytoplasm, including the cell periphery, of the monkey kidney cell contained an abundance of virus particles. He found that one mechanism of viral release was the penetration of microvilli cores by individual viruses (Figure 8). He often observed that a virus sat in a slight depression at the microvilli tip. The SFM image in Figure 7 is consistent with these electron microscopy (EM) findings.

It is interesting to note that EM has the ability to examine large areas of a cell frozen at a certain moment in time while SFM can examine a relatively small area continuously for very long time scales. For instance, a series of SFM images like Figure 7 shows the evolution of a single microvillus over time; the EM image in Figure 8 shows several microvilli captured at a single moment but at different stages of development.

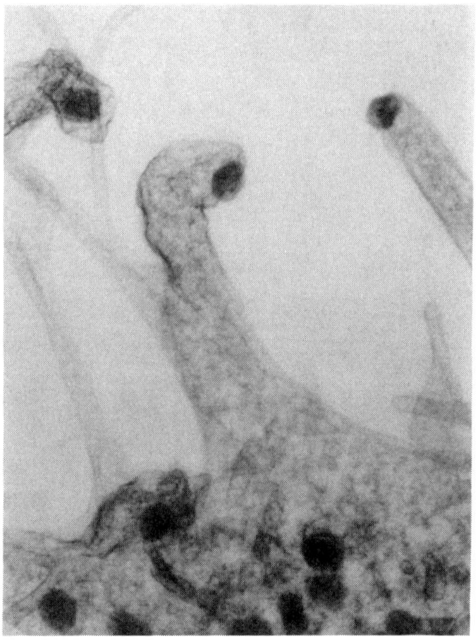

Figure 8. Electron micrograph, taken 20 hours after infection, showing vaccinia viruses at different stages of penetration along microvilli of a monkey kidney cell. From Stokes[15].

TEMPERATURE-CONTROLLED SURFACE PROCESSES

Biological systems are exquisitely sensitive to changes in temperature; the characteristics of living cells can be dramatically altered over the range of 4° C to 40° C. The metabolic rate can be reduced to practically zero near 0° C, and the mechanical properties of the cell can become considerably more rigid. Both these changes are significant for SFM. Firstly, since the time for a complete SFM scan is typically several seconds, many normal metabolic processes must be slowed down considerably for them to be recorded at all. For example, we have never been able to follow clearly the process of viral endocytosis, and we believe this may be because it is too rapid for SFM at room temperature to follow explicitly. Secondly, because SFM depends upon detecting the deflection of a cantilever with a finite stiffness, increasing the stiffness of the cell should make it possible to resolve more delicate cellular structures.

We have conducted preliminary experiments which address the second point. The bath containing a monkey kidney cell attached to a micro pipette was cooled to 11° C. The stiffness of the cell was noticeably higher than in the earlier room temperature experiments. This fact not only improved the SFM imaging but also made holding the cell in the pipette much easier. The contrast of the SFM images was significantly higher, and finer spatial details could be resolved. Figure 9 demonstrates how both the contrast and the resolution decrease as the temperature is increased from 12° C to 28° C. We believe that by cooling the sample to near 10° C it will be possible to image very soft cells such as tumor cells and white blood cells *in vivo*, which until now has proved impossible.

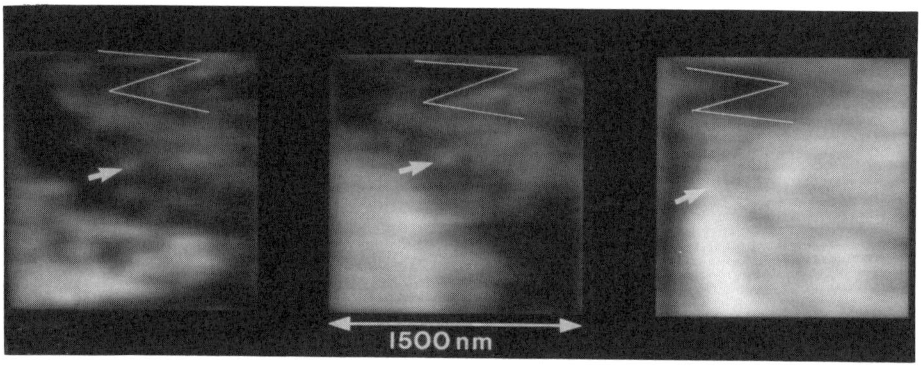

Figure 9. Changes in the surface of monkey kidney cell when the temperature of the bath is changed from 12° C to 16° C to 28° C. Arrows denote the position of the same structures in each image. At lower temperatures the cell is noticeably stiffer and the images have more contrast and show finer details.

CONCLUSION

With its currently attainable resolution of approximately 10 nm, SFM gives access to processes such as antibody binding, infection of cells by viruses, and the dynamics of actin filaments. When the additional step of refrigerating the samples is taken, we believe it may be possible to clarify many details regarding immune reactions, the formation of cell pores, endo- and exocytosis, and other dynamic interactions between the cytoplasm, the plasma membrane, and external agents.

ACKNOWLEDGMENT

The authors thank the Max Plank Institute for Medical Research in Heidelberg for its support in building the second SFM setup described here.

REFERENCES

1. G. Binnig, C. H. Gerber, E. Stoll, T.R. Albrecht, and C.F. Quate, Atomic resolution with atomic force microscope, *Europhys. Lett.* 3:12, 1281-1286 (1987).
2. T.R. Albrecht and C.F. Quate, Atomic resolution imaging of a nonconductor by atomic force microscopy, *J. Appl. Phys.* 62:2599-2602 (1987).
3. O. Marti, H.O. Ribi, B. Drake, T.R. Albrecht, C.F. Quate, and P.K. Hansma, Atomic force microscopy of an organic monolayer, *Science* 239:50-52 (1988).
4. S. Gould, O. Marti, B. Drake, L. Hellemans, C.E. Bracker, P.K Hansma, N.L. Keder, M.M. Eddy, and G.D. Stucky, Molecular resolution images of amino acid crystals with the atomic force microscope, *Nature* 332:332-334 (1988).
5. B. Drake, C.B. Prater, A.L. Weisenhorn, S.A.C. Gould, T.R. Albrecht, C.F. Quate, D.S. Channell, H.G. Hansma, and P.K. Hansma, Imaging crystals, polymers, and processes in water with the atomic force microscope, *Science* 243:1586-1589 (1989).
6. M. Egger, F. Ohnesorge, A.L. Weisenhorn, S.P. Heyn, B. Drake, C.B. Prater, S.A.C. Gould, P.K. Hansma, and H.E. Gaub, Wet lipid-protein membranes imaged at submolecular resolution by atomic force microscopy, *J. Struct. Biol.* 103:89-94 (1990).
7. H. Butt, K.H. Downing, and P.K. Hansma, Imaging the membrane protein bacteriorhodopsin with the atomic force microscope, *Biophys. J.* 58:1473-1480 (1990a).
8. S.A.C. Gould, B. Drake, C.B. Prater, A.L. Weisenhorn, S. Manne, H.G. Hansma, and P.K. Hansma, From atoms to integrated circuit chips, blood cells, and bacteria with the atomic force microscope, *J. Vac. Sci. Technol.* A8:369-373 (1990).
9. H.J. Butt, E.K. Wolff, S.A.C. Gould, B.D. Northern, C.M. Peterson, and P.K. Hansma, Imaging cells with the atomic force microscope, *J. Struct. Biol.* 105:54-61 (1990b).
10. W. Häberle, J. K. H. Hörber, and G. Binnig, Force microscopy on living cells. *J. Vac. Sci. Technol.* B9:1210-1213 (1991).
11. W. Häberle, J.K. Hörber, F. Ohnesorge, D.P.E. Smith, and G. Binnig, In situ investigation of single living cells infected by viruses, *Ultramicroscopy*, 42-44: 1161-1167 (1992).
12. E. Henderson, P.G. Haydon, and D.S. Sakaguchi, Actin filament dynamics in living glial cells imaged by atomic force microscopy, *Science* 257:1944-1946 (1992).
13. B. Sakmann and E. Neher, "Single-Channel Recording," Plenum Press, New York, 37-51 (1983).
14. J.K.H. Hörber, W. Häberle, F. Ohnesorge, G. Binnig, H.G. Liebich, C.P. Czerny, H. Mahnel, and A. Mayr, Investigation of living cells in the nanometer regime with the scanning force microscope, *Scanning Microscopy* 6:919-930 (1992).
15. G.V. Stokes, High-voltage electron microscope study of the release of vaccinia virus from whole cells, *J. Virology* 18:636-643 (1976).

SCANNING PROBE MICROSCOPY IMAGING AND CHARACTERIZATION OF BIOLOGICAL STRUCTURES FROM BIOMOLECULES TO LIVING CELLS

Vincent B. Pizziconi, Darren L. Page, Catherine T. Connolly, and Pamela A. Diamond

Chemical, Bio and Materials Engineering Department
Arizona State University
Box 876006
Tempe, Arizona 85287-6006

Abstract: Much progress has been made on the imaging and characterization of biological structures using scanning probe microscopy (SPM) since the first images were reported on DNA almost a decade ago. SPM has now demonstrated its utility in imaging a diverse group of biological structures ranging from biomolecular and macromolecular structures to supramolecular structures and even living biological cells. It is now apparent that SPM has the potential to yield unique insight into biological structure and function. However, more work is needed to better understand probe-sample interactions, better define sample preparatory and SPM operational conditions, and develop independent image interpretation methods before the full potential of SPM imaging of biomolecular structures can be realized. Progress made in these areas will also help delineate the specific utility of the SPM methods (STM and AFM) for imaging biologicals. This is particularly true for larger biological structures, such as cells, which pose unique challenges and opportunities for SPM. This paper reviews the progress made to date on the SPM imaging and characterization of biomolecular structures from macromolecules to living cells and their self-assemblies.

INTRODUCTION

Much of what we have learned about physical systems has depended greatly upon the tools that were available, with major advances made particularly from the introduction of new techniques. This has certainly been true of our understanding of biological systems where much of what we have learned about living systems has resulted from the innovation and advances in microscopy techniques. Indeed, the first living cells were seen with the light microscope over three centuries ago, which eventually led to the notion of cells as fundamental units of living systems credited to Schleiden and Schwann in 1838. It was

Atomic Force Microscopy/Scanning Tunneling Microscopy
Edited by S.H. Cohen *et al.*, Plenum Press, New York, 1994

approximately a century later that the advent of the much higher resolution electron microscope in 1939, which bridged the gap between light microscopy, X-ray diffraction and birefringent microscopy, led to a much deeper understanding of cell biology and constituent macromolecules. While significant advances in imaging biological structures continue to be made in optical (e.g., laser confocal) and electron microscopy (e.g., environmental EM) techniques, this last decade has witnessed another major microscopy innovation with the advent of scanning probe microscopy (SPM), which perhaps holds the greatest potential to date in furthering our understanding and exploitation of biological systems.

Relative to the evolution of the optical and electron microscopy, SPM systems are still very early in their development and far from being considered a mature technology. Even so, the rapidly growing number of studies reported in the literature on diverse synthetic and natural material surfaces attest to the widespread utility and promise of SPM as a unique surface characterization tool. As a class, SPM generically represents any method that precisely controls the movement of either a probe or sample in three dimensions by piezoelectric elements as a means to obtain probe-to-surface distances that result in physical interactions that are transducible to high resolution surface topography. It presently includes scanning tunneling microscopy (STM) and atomic force microscopy (AFM) methods which are commercially available. Other SPM techniques currently under development include scanning ion, magnetic force, photon scanning tunneling, and thermal profiler microscopes. The extent and diversity of SPM development and applications are impressive for such a relatively short time period since SPM's inception and these are summarized elsewhere.[1]

The scanning elements of STM and AFM, depicted in Figure 1 and Figure 2 respectively, distinguish the basic operating characteristics between these SPM-related but distinctly different microscopy methods. Briefly, STM exploits electron tunneling phenomena, which is favored to occur when a conductive probe is placed in very close proximity to a conductive surface separated by only a thin insulating gap whereas AFM is based on attractive/repulsive forces occurring between a cantilever-mounted probe and a sample and is not restricted to conductive probe/sample systems.

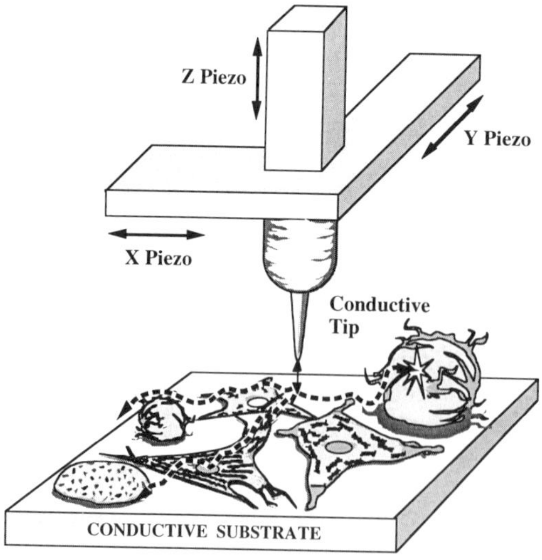

Figure 1. Schematic of STM operation

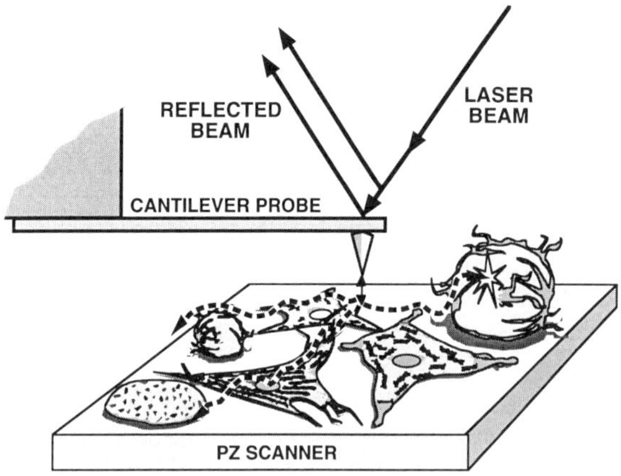

Figure 2. Schematic of AFM operation

A particularly attractive generic feature of these SPM systems is their ability to obtain both two- and three-dimensional topographical features of material surfaces at extremely high resolution in either vacuum, air, or liquid environments. It is this latter ability of SPM systems, i.e., the imaging of biological structure directly in a liquid environment, that is most distinguishing from other high resolution microscopies used for imaging biological materials. Thus, one important advantage of SPM is its elimination of the need for special preparatory techniques (e.g., critical point drying, sample coating) or extreme environments (e.g., high vacuum) required of conventional high resolution microscopy that ultimately limits utility in imaging native biological structures and characterizing function. An additional advantage for AFM imaging in liquid environments is the elimination of capillary forces. Clearly, the excitement of SPM to biomedical researches is its unique capability to image biomolecular surface structure at high resolution in the native state, its potential to directly assess native biological structure-property relations, as well as to study biological dynamical events in real time, in situ. More detailed descriptions on the physics of STM[2-7] and AFM[7-8] are available.

These SPM researchers are currently assessing the potential of SPM in providing a deeper understanding of native biological structures and their interactions in natural and engineered artificial environments. If SPM can afford additional insight, it will advance our ability to exploit more fully biological materials for applications in molecular medicine, molecular and cellular bioengineering and biotechnology. Clearly, SPM, by virtue of its ability to operate in an aqueous environment, has the potential to provide such insight into these applications. As such, it has provided our group a powerful new method to help provide a rational basis for the design of biomolecular-based materials and devices. Potential areas of study include biomolecular structures that exhibit self-assembly and molecular recognition, which are attractive properties that are not presently achievable by synthetically engineered materials alone. Specific biological structures under study by these investigators include extracellular matrix (ECM) proteins, ECM cell binding sequences, and living cells presently being considered in the design of hybrid microstructures,[9] such as whole cell biosensors.[10] Prior to discussing these results, an attempt is made herein to provide a more global perspective of the SPM biological efforts to date as many SPM researchers are making significant strides in the imaging and characterization of biological structures.

Biological Applications of STM and AFM

The number of publications reporting the utility of SPM for characterizing biological structures has grown dramatically over this last decade even though its potential was initially envisioned to be most suitable for imaging rigid, nonbiological material structures just a little over a decade ago.[11-12] Notwithstanding the initial success of SPM on the characterization of synthetic structures, early SPM researchers quickly moved ahead with their initial pioneering work on imaging the first biological structure, i.e., DNA, using STM. The research potential of SPM for imaging biological structures was clearly expanded and enhanced shortly thereafter with the advent of AFM in 1986. Although still early in the application stage of imaging biologicals, STM and AFM have now been widely used to image and characterize a variety of complex structures as witnessed by the diverse studies on biological structures summarized herein along with the studies reported at this timely atomic force/scanning tunneling microscopy symposium.

The range of biological structures studied by SPM now spans from biomolecules to living cells. It includes a wide diversity of basic molecular structures (amino acids and biominerals),[13-15] macromolecular structures (polypeptides, nucleic acids, polysaccharides, carbohydrates, and proteins),[16-52] and supramolecular structures (membrane protein bilayers, fragments, filaments, and vesicles)[52-90] to intact biological cells (procaryote/ eucaryote)[59,70,71,92-106] and even organized tissue.[107] It also has been used on a more limited basis with some success to explore its potential for real time studies of biological structures including DNA replication,[23] virus/cell interaction,[93] and cell activation.[96,101] Tables 1, 2, 3, and 4 exemplify the diversity of SPM research on biological structures according to the various biological subgroups listed. (*Note:* reference numbers without underscoring in Tables denote STM studies; underscored reference numbers denote AFM studies.)

Table 1. Biomolecular structures studied by SPM

•Amino acids	**• Biomineral**
Glycine[13-14]	Hydroxylapatite[15]
Lysine[13]	
Tryptophan[14]	
Leucine[14]	
Methionine[14]	

Table 2. Macromolecular structures studied by SPM

• Polypeptides	**• Nucleic acids/ Polynucleotides**
Valinomycin[16]	poly (dA-dT)[20]
Telopeptides of collagen[19]	poly (rA)[21]
poly (γ-benzene L glutamate)[16,17,18]	poly (rU)[21]
	tRNA[13]
• Polysaccharides	DNA[13,16,22-31]
Xanthan[16]	Plasmid DNA[32-34]
Glycogen[39]	Lambda-DNA[36,37]
	ds-RNA[37,38]
• Carbohydrates	RNApolymerase DNA complex[32]
Beta-cyclodextrin[17]	DNA-cytochrome c copolymer[35]

Table 2. *(cont'd)*

- **Proteins/Enzymes**
  - Albumin[14]
  - Vicilin[16]
  - Gramicidine[42]
  - Chymotrypsinogen[40]
  - Collagen type 1[19,44]
  - Laminin[10]
  - Phosphorylase[39,46-48]
  - Phosphorylase b[45]
  - IgG-BSA complex [50]/IgG [51]
  - Zein[16]
  - Fibrinogen[14,45]
  - Lysozyme[14,40]
  - α-macroglobulin[41]
  - HMW wheat subunit[17]
  - Fibronectin[10,42]
  - Phosphorylase kinase[39,45-48]
  - Reverse transcriptase[49]
  - von Willebrand Factor[52]

Table 3. Supramolecular assemblies studied by SPM

- **Cell Membrane Proteins (isolated)**
  - HPI protein [52]
  - F-actin [57,58-59]
  - Phage T4 polyheads[61]
  - Lipoprotein ice nucleator [63]
  - Membrane-bound ferritin[65]
  - Nicotinic acetylcholine receptor channel [54,55,56]
  - Chloroplasts[60]
  - Bacteriophage T7 [62]
  - Bacteriorhodopsin [64,65]
  - Cholera toxin[66]
- **Cell Membrane Fragments**
  - Purplemembrane[63,64,67,68,69,70]
  - HPI monolayer[75,77]
  - TE 671neural membrane[78]
  - Human erythrocyte membrane [80]
  - Membrane gap junctions [85]
- **Cell Membrane Fragments** *(Continued)*
  - Bacterialsheaths/sheathhoops[71,72,73,73,74,74,75]
  - Oocyte cytoskeleton[78]
    - Membrane fragment-liver ascities cells[79]
    - Lipid-protein LB Structures[57,64,78,81-84]
- **Cell Cytoskeletal and Motile Structures**
  - Bacterial flagella[86]
  - Microtubules[87-89]
  - Soybean PE nonbilayer lipid tubes[78]
  - Intermediate filaments [90]
- **Cell Membrane Vesicles**
  - Rod cell of bovine retina[84]
- **Hard tissue**
  - Dentin and enamel[107]

Table 4. Biological cells imaged by scanning probe microscopy

	State		SPM Probe	
Cell Type	Dry	Hydrated	STM	AFM
MAMMALIAN CELLS				
Red blood cells[70, 91,92-95]	•	•	•	•
White blood cells[70,94,97,104]	•	•	•	•
Platelets[59,91,96]		•	•	•
Endothelial cells[98]		•		•
Sperm[106]		•		•
Cultured Cells				
RBL-2H3[91, 101]	•	•	•	•
WEHI-3[91]		•	•	
CHO fibroblast[102]		•	•	
T-24 cancer cell (bladder)[102]		•	•	
Lung carcinoma 131 line[103]		•		•
Kidney cell MA-104[93]		•	•	
AMPHIBIAN CELLS				
Glial cells[99]		•		•

(Continued)

Table 4. *(cont'd)*

Cell Type	State: Dry	State: Hydrated	SPM Probe: STM	SPM Probe: AFM
AVIAN CELLS				
Chick bone cell[100]		•		•
PLANT CELLS				
L. subcostata [94]		•		•
N. odorata [94]		•		•
MICROBIOLOGICAL				
Mycobacterium [105]	•		•	•
Yeast[105]	•		•	•
Bacillus coagulans E38-66[93]		•		•
E. coli[71]	•			•
Halobacterium[94]	•			•

Inspection of these summary tables indicates that SPM studies have been conducted on a broad array of biological constituents. It is also evident that both STM and AFM have been approximately equally utilized for study of biological materials. This outcome may appear somewhat surprising in view of the poor conductivity associated with biological materials, which is thought to limit the utility of STM for biologicals, at least for larger (thicker) biological samples. In fact, there does exist a clear bias toward a specific SPM method within a given biological group. For example, it can be seen that the predominant SPM method for imaging smaller biomolecular structures (polypeptides, proteins, DNA) is STM while both STM and AFM are about equally utilized for imaging the more intermediate, supramolecular (cell complexes) biological structures, whereas AFM is clearly the preferred choice for imaging cellular structures. This outcome is consistent with the view that small biopolymer structures are considered to be thin enough to be considered part of the tunneling barrier height (insulative gap) and still expect tunneling to occur. Also, AFM is the preferred SPM by many researchers for imaging the larger biological structures due to its inherent ability to image relatively thick, poorly conducting and nonconducting materials.

In spite of the demonstrated utility of STM for the direct imaging of biologicals, there remains a concern over STM resolution and reproducibility.[108] This concern has been recently addressed,[109] which indicates that STM can yield reproducible results of at least macromolecular structures with high spatial resolution. However, it still remains unclear how tunneling, or more generally, how charge-flow occurs particularly in the larger biological structures. But the fact remains that many STM studies have been conducted successfully on a wide variety of biological structures, both coated and uncoated, including living cells by these investigators and others (and other apparent poor conductors such as synthetic polymers[110]) even though the specific charge-flow mechanism remains presently unknown.

Thus, our current view is that both STM and AFM continue to offer a high potential to image and characterize biomolecular materials in their native and possible functional state at unprecendented resolution. If so, it is quite possible that STM and AFM may eventually provide independent and/or complementary topographical and potentially functional information on these complex living biological structures.

EXPERIMENTAL METHODS AND MATERIALS

The focus of these studies was to assess the utility of STM for imaging a wide range of uncoated, and fully hydrated biological structures from macromolecules to living cells. All experiments were conducted at ambient temperature and pressure using uncoated, native biological stuctures that were immobilized to HOPG graphite substrates by natural or synthetic fixation methods and imaged immersed in aqueous buffer solutions.

SPM Studies of ECM Nanostructure

One phase of our SPM activities has been to image and characterize native laminin and fibronectin structure, two ECM proteins involved in cellular self-assembly that are currently being evaluated as molecular linkages between living cells and synthetic microstructures. Fresh graphite surfaces were created prior to assembly of the custom microwell by carefully cleaving the exposed graphite surface layer with scotch tape. Approximately 100 μL of Tris buffer solution containing 15 μm/mL of ECM protein was placed into the custom HOPG-ZYB graphite well assembly which was affixed to the scanning base. ECM proteins were allowed to adsorb onto the freshly cleaved HOPG substrate maintained for approximately 30 minutes at room temperature prior to initiating imaging. STM experiments were carried out at room temperature and under ambient pressure conditions using an Angstrom TAK 3.0 STM microscope. Apiezon-coated platinum-iridium and tungsten STM probes were then manually advanced to within close proximity of the surface, followed by automatic approach until a constant tunneling current of 1 nanoampere was achieved.

SPM of Living Biological Cells

The second phase of our ongoing SPM activities has focused on the feasibility of imaging uncoated, living biological cells using STM. Several diverse types of mammalian cells were studied. These include RBC's, platelets, and immune cells derived from culture. Cell preparations varied depending on the specific separation technique required and on the cells ability to anchor to the graphite substrate. Cell imaging was performed directly in buffered aqueous solutions contained in custom, 7 mm I.D. microwells (~100 μL) temporarily sealed against HOPG-ZYB graphite substrates with apiezon grease. Freshly collected bovine platelets were prepared according to the method of Mustard.[111] These inherently surface-active cells were added to freshly cleaved substrate surfaces and allowed to naturally adhere to the untreated graphite surface prior to study with STM. However, STM microwells were pretreated with either silane or poly-L-lysine prior to the addition of fresh RBC's that required some form of immobilization. RBC's were then allowed to remain on substrates for ~2 hours before microwell buffer was replaced by fresh phosphate-buffered saline (PBS) to remove nonadherent cells. Anchorage-dependent WEHI-3 cells, a feeder cell line used to propagate mast cells (RBL-2H3), were cultured directly onto freshly cleaved graphite surfaces, both with and without fibronectin pretreatment. Fibronectin pretreatment of graphite wells were used to hasten cell adhesion and to increase cell spreading. This not only resulted in better immobilization but also induced a flattened cell morphology, thus making them more amenable for scanning within the maximun z-distance of approximately 5 μm. Fibronectin (100 μg/mL) was applied in minimal volumes to coat the well bottoms. The wells were air dried under a laminar flow hood for one hour before cell suspensions were added. STM pilot data were obtained on an Angstrom TAK 3.0 using insulated (apiezon wax) platinum-iridium and tungsten probes. STM images were obtained using a tip bias voltage ranging from -0.2V to + 0.3V. Set point currents ranged between 0.44 to 1.0 nanoamperes. Scan rates varied between 0.56 to 2.19 lines/s.

RESULTS

Images of uncoated, native and living biological structures were successfully obtained at both low and high resolution in all macromolecular and cellular experiments using the STM apparatus independent of their size, thickness, and aqueous medium.

ECM Nanostructure

High quality STM images of laminin and fibronectin were obtained which correlated well with the reported molecular structure and dimensions obtained from electron microscopy studies.[9] Figure 3 shows a cruciform structure of laminin, which is remarkably similar to that obtained from previous TEM studies but provides much greater molecular detail of this extracellular matrix molecule in this 3-D mode. The latter is made particularly evident from the cross-sectional plot constructed from the 2-D nanograph shown in Figure 4. In this figure, high contrast areas can be seen at the intersection of the long and short arms with this fully hydrated laminin molecule located. Similarly, the characteristic "cotter pin" structure of fibronectin can be clearly seen in Figure 5 along with three-dimensional detail of its molecular structure.

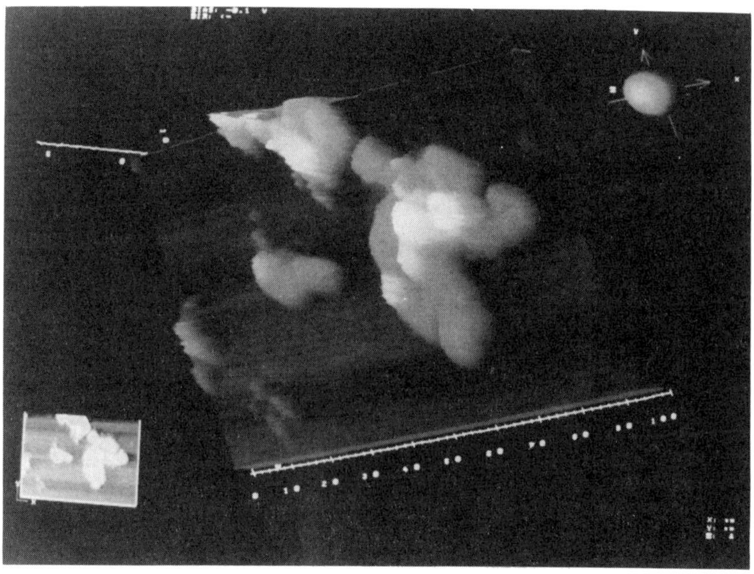

Figure 3. STM image of laminin.

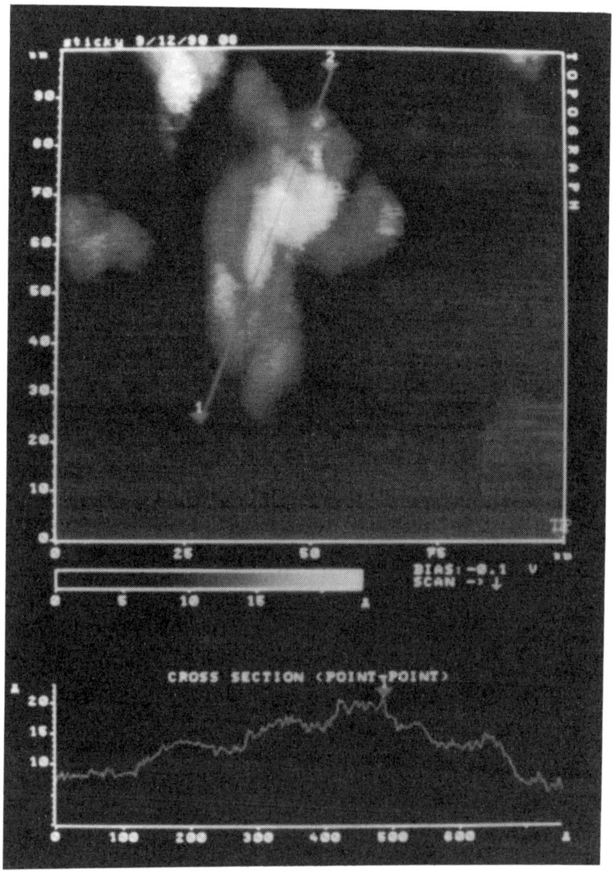

Figure 4. 3-D image of laminin.

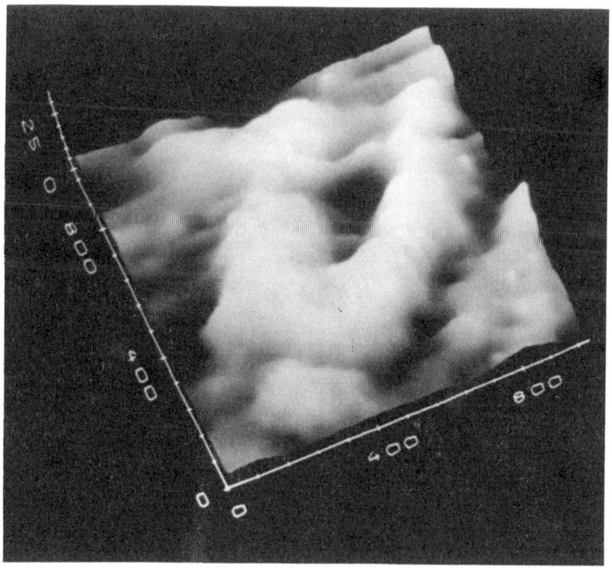

Figure 5. STM of fibronectin

STM of Living Cells

STM images of several types of living whole cells were obtained as single cell entities and as cell aggregates. Figure 6 is a 2-D image of a platelet aggregate. Figure 7 is a higher resolution 3-D image of the platelet aggregate exposing several cell-cell contact areas. Figure 8 is a 2-D global depiction of an RBC aggregate (rouleaux), while Figure 9 depicts an RBC cell-cell contact region, and Figure 10 depicts a low resolution image of an RBC showing its well known biconcave geometry. Figure 11 is a 2-D higher resolution image

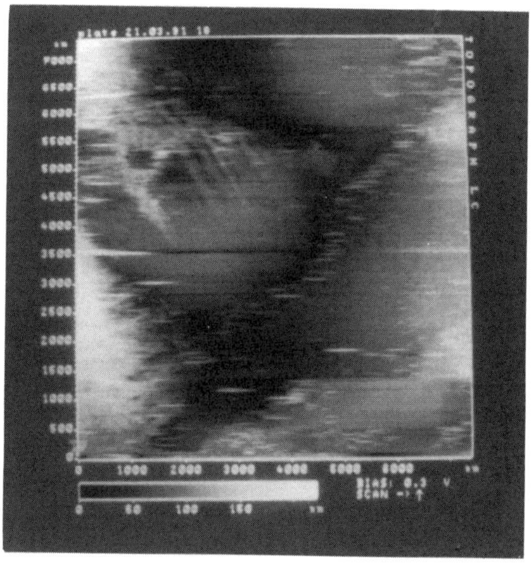

Figure 6. 2-D Image of platelet aggregate.

of a human RBC surface that is analogous to the "tire tread" region showing many surface extensions. Figure 12 is a 3-D image of an even higher resolution of the RBC surface depicting macromolecular size surface extensions. Figure 13 is a 2-D STM image of a much larger, living WEHI-3 cell depicting its attachment processes that formed during culturing directly onto the graphite surface. Figure 14 and Figure 15 are a 2-D and 3-D close-up view of a WEHI-3 foot process apparatus, respectively. Figure 16 depicts WEHI-3 cell outgrowth processes propagating along the HOPG surface.

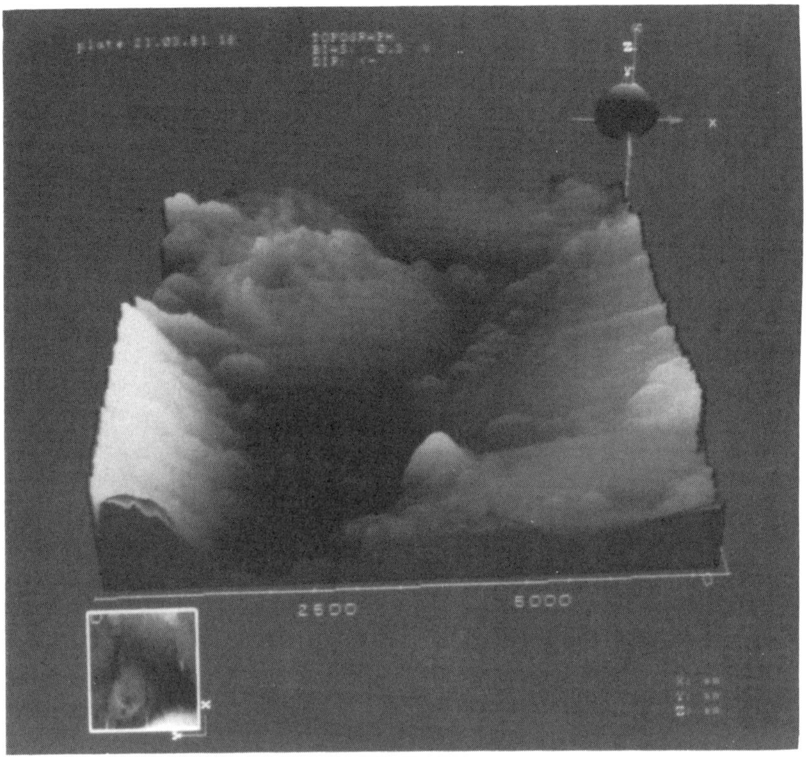

Figure 7. 3-D image of platelet cell-cell contact regions.

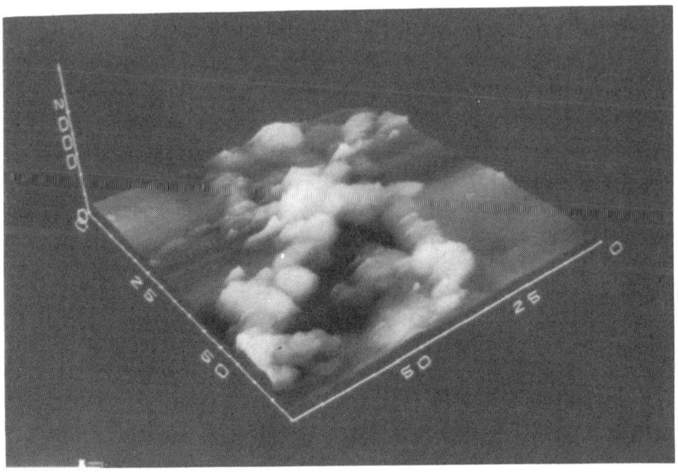

Figure 8. STM image of RBC rouleaux.

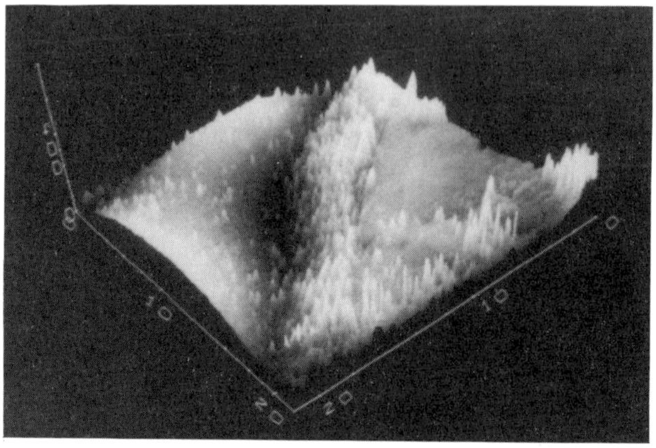

Figure 9. STM image of RBC cell-cell contact region.

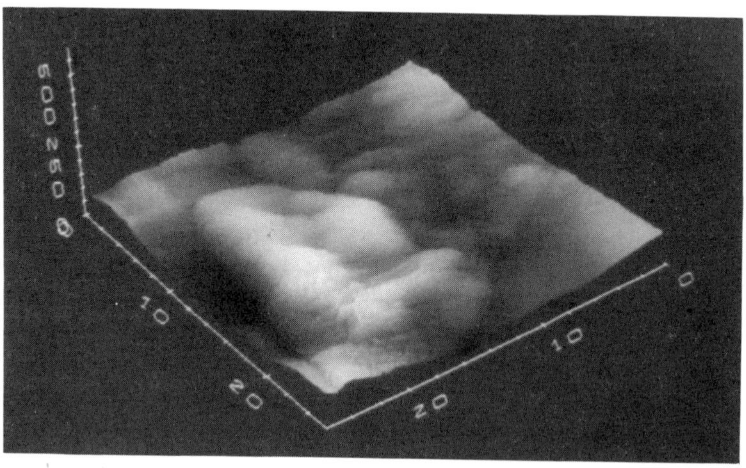

Figure 10. STM image of biconcave RBC geometry.

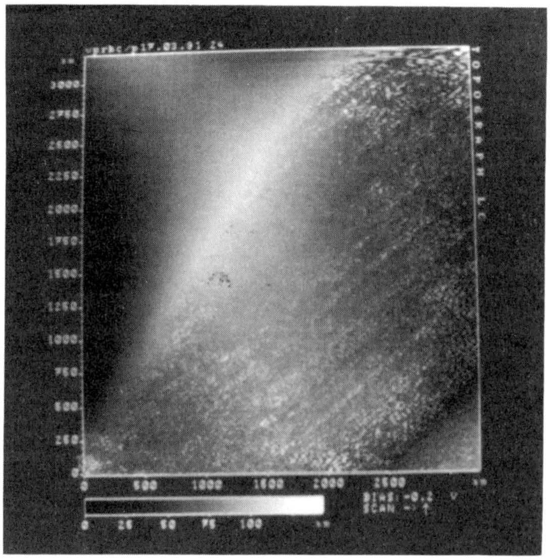

Figure 11. 2-D image of RBC surface.

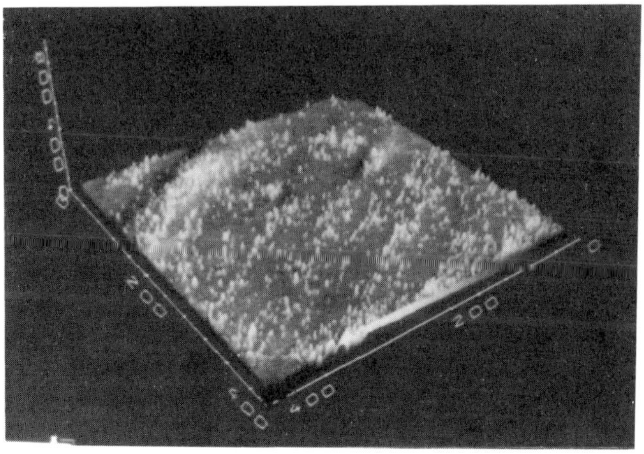

Figure 12. High resolution 3-D image of RBC surface.

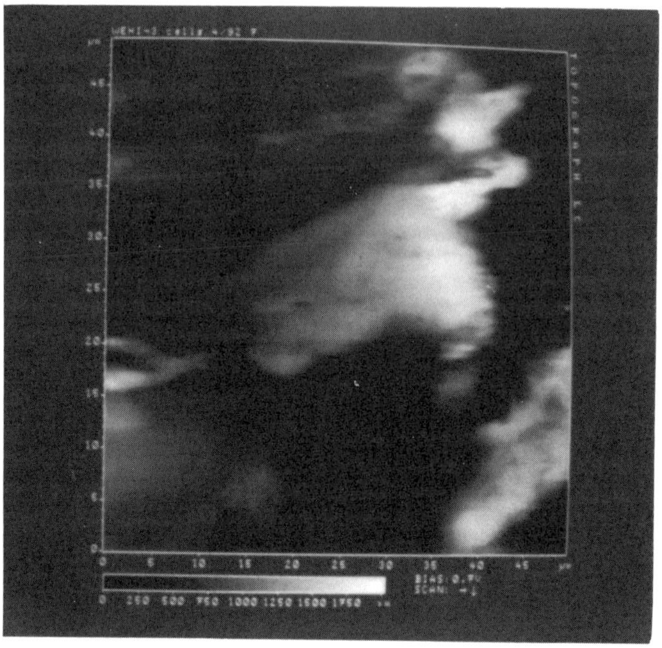

Figure 13. 3-D STM image of living WEHI-3 cell.

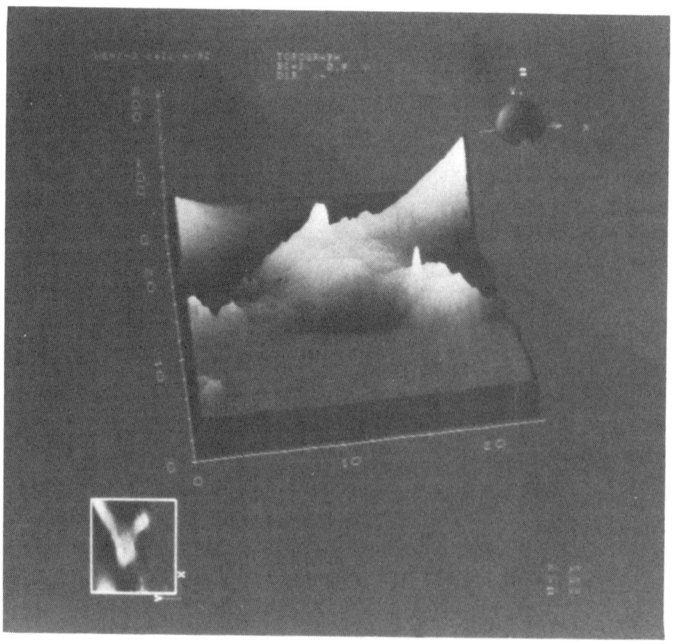

Figure 14. 2-D STM of WEHI-3 foot process.

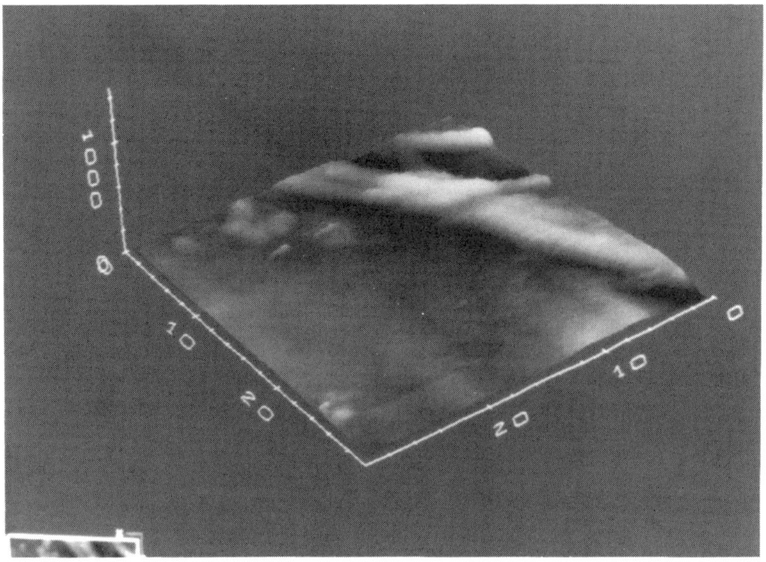

Figure 15. 3-D STM image of WEHI-3 foot process.

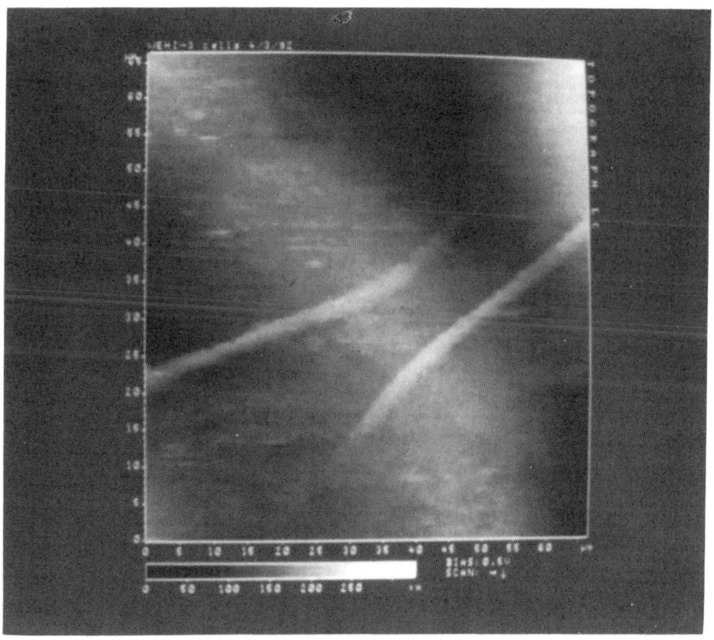

Figure 16. STM image of WEHI-3 cell extension.

DISCUSSION

ECM Nanostructure

The STM images of individual ECM proteins were consistent in their overall planar molecular dimensions and structure with those reported by transmission electron micrographs (TEM) of ECM proteins but offer much greater molecular detail of these adhesive proteins, particularly in the 3-D mode. For example, in addition to obtaining a cruciform-like structure remarkably similar to that obtained by TEM, the three-dimensional STM image of laminin shown in Figure 4 reveals significant topological variances as indicated by variations in color intensity seen in this laminin STM nanograph. Interestingly, the regions of lightest color intensity (highest z) correspond to one of the major cell binding domains of laminin.[9] Similarly, Figure 5 is a three-dimensional STM nanograph of fibronectin, which portrays the characteristic "cotter pin"-like structure again remarkably similar to reported TEM studies,[9] as well as having apparent topological variances that may be associated with potential binding consequences.

There was, however, a noted discrepancy in the absolute height and width dimensions between these STM results and reported TEM results for these ECM structures that have been commonly observed by other researchers who have conducted STM experiments on uncoated biological structures.[112,113] In this study, the STM results indicate that the ECM protein structures are larger in diameter but lower in thickness when compared to reported TEM measurements. Since the molecular diameters of the individual chains appear to be significantly larger in the STM nanographs, these results suggest that the macromolecule may have undergone deformation from local tip forces. This is an important issue to resolve before the full potential of STM is realized in characterizing biological structures. Also, closer inspection of Figure 4 suggests that the long arm (A chain) may have folded back onto itself probably as a result of the rastering probe. If so, this could explain what appears to be the globular portion of the A chain (cell binding region) residing on or near the other cell binding region located at the intersection of the A chain and the B chains, thereby accounting for the bimodal surface protrusion in this region.

Also, part of the success of obtaining these high resolution images was due to the natural adherence of these ECM proteins to the HOPG graphite surface probably due to nonspecific hydrophobic forces associated with these anchorage-related matrix proteins. This was true for both individual and molecular aggregates of laminin and fibronectin which were seen in all tunneling experiments. The variation in deposited protein was expected since no attempt was made to control macromolecular aggregation nor natural self-association mechanisms associated with these ECM proteins. However, ECM protein self-association was unlikely to occur under these experimental conditions whereby the ECM concentrations used (10 μg/mL) were well below the reported critical self-assembly levels of 0.1 mg/mL.[114]

These STM results on ECM protein morphology clearly demonstrate the potential of SPM for imaging biomolecular structures in their fully hydrated state at high molecular resolution. The STM nanographs of laminin and fibronectin, which were consistent with TEM images but additionally yielded important 3-D structural aspects, demonstrate the potential of this valuable tool for gaining insight into complex biomolecular structure-function heuristics. Continued progress in this area will be essential for understanding extracellular matrix formation and function in natural cell assembly processes and for successfully exploiting hybrid microstructure design.

Living Cells

Both low and high resolution topological images of uncoated, living whole cells and cell aggregates were obtained from the STM apparatus as shown in Figure 6 through Figure 16. The overall dimensions of the cells imaged corresponded to their dimensions obtained from optical measurements and reported values. The direct measurement of the cell height or thickness (z contrast data) yielded thickness values ranging from tenths of μm to several μm, which is consistent with cell thickness in order of magnitude but clearly far in excess of tunneling barrier height dimensions. It is also noteworthy that tunneling conditions were observed to be much more erratic (i.e., large fluctuation of tunneling current) in all cell experiments, independent of cell type, compared to the tunneling responses obtained when imaging macromolecules. Further, scanning conditions varied markedly depending on whether imaging was performed on groups of cells (decreased scan rate and increased gain) or on a single cell (higher scan rate and lower gain).

Many interesting cell surface features were evident in the STM images, including surface protrusions, cell-cell contact interfaces, and cell-surface adhesion complexes formed in situ in their natural, functional state. More specifically, the two-dimensional image and three-dimensional high resolution image obtained from one of the RBC experiments shown in Figure 11 and Figure 12, respectively, reveal the presence of cell-surface protrusions of relative high density, which is consistent with cell surface receptor magnitude. This preliminary result suggests that it may be possible to characterize a variety of important cell surface structures at high resolution in their functional state. This includes the potential to study molecular cell recognition events associated with cell receptor-ligand binding phenomena. In addition, the cell-cell contact interfaces, clearly seen in the images of cell aggregates shown in Figure 7 and Figure 9, indicate that it may be possible to study cell-cell adhesion processes in situ. Also, the images of cell-substrate extension processes (filopodia) obtained in situ from anchorage-dependent cells (WEHI-3) confirmed their immediate purpose of stabilizing these cells during the STM run and, more importantly, may provide new approaches in understanding cell adhesion phenomena.

In addition, the cell immobilization techniques used in this study, either natural cell processes or binding structures (i.e., anchorage-dependent cell types or use of ECM linkers) or synthetic approaches (i.e., poly-l-lysine), depending on the cell type studied, were found to be acceptable in meeting STM immobilization requirements. In comparison, the naturally anchored cells were found to be considerably more stable than suspension cells immobilized with synthetic linkers. However, both of these approaches failed (as depicted in cartoon form in Figures 1 and 2) when the cell type, in its immobilized state, exceeded the maximum z-dimension allowable (typically 4-5 μm for most SPM scopes). Thus, although it is possible to obtain images from very large cellular structures by attaining a flattened cell morphology via adhesive and spreading processes as was achieved in this study, the design of SPM systems having a larger z-dimension working range to accommodate a wide range of cellular structures would be of great value for these applications. Alternative approaches to overcome this limitation include affixing a single cell on an appropriate attachment and positioning the probe to scan over a portion of its surface, but this may limit the scope of cell studies otherwise possible.

These living cell experiments indicate that STM can provide unique insight into several areas of importance in molecular cell biology and cellular bioengineering, including cell-cell recognition and adhesion, cell-substrate adhesion, cell signalling, and cell growth and differentiation, among others. However, as biological cells are very flexible structures composed of dynamic surface components, there remain several important issues that need to be overcome before the full potential of SPM for characterizing cellular systems in the living state at high resolution can be realized.

PROBLEMS AND PROSPECTS OF SPM FOR BIOLOGICAL IMAGING

Presently, SPM can easily attain molecular resolution of biological structures with either STM or AFM. However, the prospect of obtaining higher, atomic resolution of three-dimensional surface structure of biological materials directly in their liquid, fully functional state remains compelling. In order to obtain this goal a much better understanding of both STM and AFM imaging mechanisms will be required. In the former case, the imaging mechanism associated with STM for biological materials remains unknown and needs to be delineated. Several researchers have proposed tunneling-related mechanisms,[115-118] but other conduction mechanisms associated with charge-flow through thin insulating films are also possible[9] and have gained experimental support.[119] Although the exact charge-flow mechanism remains unclear, the results of this and other STM studies cited clearly indicate that STM can indeed image a wide array of biological structures including living biological cells. In the latter case, the surface area over which the force is operating during imaging with AFM also remains unknown.

In addition to poorly understood probe-sample interactions, the soft physical nature of biological materials significantly hampers image resolution due to the substantial physical forces induced by the SPM probes. These forces, although on the order of nanonewtons, can be substantial when applied over the extremely small surface areas typically utilized during scanning probe experiments, particularly during high resolution imaging. The experimental results by many researchers verify that these induced forces are capable of (a) deforming the sample resulting in variance in sample dimensions as reported herein for the ECM proteins, (b) bending, twisting, or stretching these typically very soft and flexible macromolecules resulting in a distortion of the structure as indicated by streaky or cloudy images, or (c) removing the sample completely from the substrate with even the possibility of its reattachment to the scanning tip itself. This outcome becomes exacerbated when attempting to image biological cells in the living state due to the dynamic reorganization of membrane cell surface components. Based upon direct measurements of lateral diffusion coefficients of selected membrane components ranging from 0.1-0.2 x 10^{-8} cm^2/s,[120] the present capability of SPM to image molecular and certainly any atomic features of interest at normal scan rates can easily be exceeded. However, atomic resolution may be within reach if biological systems can be made to act more like rigid structures and/or be able to operate SPM with lower tip-induced forces.

Several sample-based approaches have been attempted to improve image resolution of biologicals. For macromolecular and supramolecular structures, these included the use of sample coating procedures, and freeze-fracture replication as a means to improve sample rigidity, as well as constructing Langmuir-Blodgett film substrates as a means to improve immobilization and control sample orientation.[108] For the larger cellular structures, earlier attempts to make cells more rigid used conventional fixation procedures with images obtained primarily in air. More recent attempts utilized cryogenic temperature systems. Low temperature SPM approaches take advantage of the glass transition temperature at which the stiffness of biological structures has been estimated to increase by four orders of magnitude.[95] Reported efforts to achieve this end employed the use of liquid nitrogen and sample cold stages to image cell membrane fragments, but problems with contamination and thermal drift limited the utility of this approach. This outcome is expected to improve with the use of liquid helium.[121]

Another interesting approach to obtain better resolution of living biological cells using AFM has recently been reported that took advantage of natural cell activation processes to induce a more rigid, dynamic cell state due to cytoskeletal assembly processes resulting in cell stiffening.[101] In this study, images obtained on unactivated RBL-2H3 mast cells depicted only a featureless, smeared appearance whereby stimulated mast cells exhibited their underlying cytoskeletal network. These results indicate that methods which can alter biological physical properties in favor of improved imaging without losing integrity are highly desirable. In this case, cell activation sequences may be followed in real time and clearly demonstrate the potential utility of SPM imaging biological structures in their native and fully functional state. Interestingly, these investigators also found that SPM images of critical point dried mast cells (whether fixed or living) markedly differed from the SPM images obtained from otherwise identical, noncritical point dried mast cells in all cases, strongly suggesting that microscopy methods that are capable of imaging and characterizing native biological structures may also be inherently more meaningful.

Finally, a potentially significant AFM instrument advancement has been made using tapping mode AFM to minimize frictional forces resulting in much improved resolution of biological structures in liquids.[122]

CONCLUSIONS

SPM is clearly in its infancy but is progressing very fast relative to the 300 years of development for the optical microscope and approximately 50 years for the development of the electron microscope. This is particularly true for its development in biological applications where SPM is rapidly becoming the tool of choice for characterizing native biological structure and function. However, more work is needed in several critical areas before the full potential of SPM of biologicals can be realized. This includes (a) the delineation of the constitutive properties of biological materials in terms of mechanical properties, conduction mechanisms, and surface interactions that will assist in (b) the development of reproducible methods for controlled sample deposition and immobilization, (c) a better understanding of tip-sample interactions, (d) the identification of optimal experimental conditions for high resolution images of native macromolecular and cellular structures, (e) the development of environmental cells that allow online control of experimental conditions in situ, and (f) the adoption of existing microscopy techniques and, where needed, the development of customized SPM techniques for image interpretation and characterization. If these factors can be adequately addressed, it may be quite possible to achieve atomic resolution of biological structures including cell surfaces of both procaryotic and eucaryotic cells. Thus, STM is expected to greatly augment other established microscopies in establishing molecular heuristics that may lead to a better understanding of biological organization and function. This outcome will certainly assist those researchers interested in the design of smart materials and devices. In this light, SPM is a potentially valuable tool for gaining insight into complex biological structure-function heuristics essential for the rational design of hybrid molecular devices.

ACKNOWLEDGMENT

Support of this work was made in part by NIH Division of Research Resources under grant number BRSG 2 SO4 RR07112.

REFERENCES

1. S.R.Snyder and H.S.White, Scanning tunneling microscopy, atomic force microscopy, and related techniques, *Anal.Chem.*, 64(12):116R-133R (1992).

2. J.A. Golovchenko, The tunneling microscope: a new look at the atomic world, *Science*, 232:48-53 (1986).
3. M.E. Welland and M.E. Taylor, Scanning Tunneling Microscopy, *in*: "Modern Microscopies: Techniques and Applications," Duke, P.J., and Michette, A.G., eds., Plenum Press, New York N.Y. (1990).
4. R.J. Behm, N. Garcia, H. Rohrer, "Scanning Tunneling Microscopy and Related Methods" Academic Publishers, Dordrecht, The Netherlands (1990).
5. C.J. Chen, "Introduction to Scanning Tunneling Microscopy," Oxford University Press, Oxford (1993).
6. D.A. Bonnell, "Scanning Tunnelling Microscopy and Spectroscopy," VCH Publishers, Inc., New York (1993).
7. O. Marti and M. Amrein, "STM and SFM in Biology," Acad. Press, Inc., San Diego (1993).
8. D. Sarid, "Scanning Force Microscopy: With Applications to Electric, Magnetic, and Atomic Forces" Oxford University Press, New York, NY (1991).
9. V.B. Pizziconi and D.L. Page, Elucidation of macromolecular assemblies: use of scanning tunneling microscopy for molecular bioengineering of cellular self-assemblies in molecular device design, *in*: Macromolecular assemblies in polymeric systems, Stroeve, P., Balazs, A.C., eds., ACS Symposum Series 493:256-277 (1992).
10. V.B.Pizziconi and D.L.Page, Image and characterization of macromolecular interface structures for whole cell biosensors", *in*: Synthetic Microstructures in Biological Research, Schnur, J. and Peckerar, M., eds., Plenum Press, New York, NY (1992).
11. G.H. Binnig, H. Rohrer., Ch.Gerber, E.Wiebel., Surface studies by scanning tunneling microscopy, *Phys. Rev. Lett.* 49:57-61 (1982).
12. G. Binnig and H. Rohrer, Scanning tunneling microscopy, *Surface Sci.* 126:236-244 (1983).
13. J. Mou, W. Sun, J. Yan, W.S. Yang, C. Liu, Z. Zhai, Q. Xu, and Y. Xie, Underwater scanning tunneling microscopy of organic and biological molecules, *J. Vac. Sci. Technol.* B 9 (3):1566-1569, (1991).
14. L. Feng, J.D. Andrade, and C.Z. Hu, Scanning tunneling microscopy of proteins on graphite surfaces, *Scanning Microscopy* 3(2): 399-410 (1989).
15. L. Siperko and W. Landis, Atomic scale imaging of hydroxyapatite and brushite in air by force microscopy, *Appl. Phys. Lett.* 61(21):2610-2612 (1992).
16. V.J. Morris, T.J. McMaster, A.P. Gunning, J. Mingins, A.S. Tatham, and E. Mitchell, Scanning tunneling microscopy of biological macromolecules, *Polymer Preprints*, 33(1):737-738 (1992).
17. M.J. Miles, T. McMaster, H.J. Carr, A.S. Tatham, P.R. Shewry, J.M. Field, P.S. Belton, D. Jeenes, B. Hanley, M. Whittam, P. Cairns, V.J. Morris, and N. Lambert, Scanning tunneling microscopy of biomolecules, *J. Vac. Sci. Technol.* A8 (1): 698-702 (1990).
18. T.J. McMaster, H. Carr, M.J. Miles. P. Cairns, and V.J. Morris, Polypeptide structures imaged by the scanning tunneling microscope, *J. Vac. Sci. Tech.*, A8(1): 648-651 (1990).
19. H. Snellman, L.J. Pelliniemi, R. Penttinen, R. Laiho, Application of scanning tunneling microscopy for imaging of CNBr-peptides of type I collagen, *J. Vac. Sci. Technol.* A 8 (1):692-694 (1990).
20. R. Coratger, A. Chahboun, F. Ajustron, J. Beauvillian, M. Erard, and F. Amalric, Scanning tunneling microscopy of a liquid crystalline phase of poly((dA-dT)_(dA-dT)) induced by a histone H1 peptide, *Ultramicroscopy* 34: 141-147 (1990).
21. P.G. Arscott and V.A. Bloomfield, Scanning tunneling microscopy of nucleic acids and polynucleotides, *Ultramicroscopy* 33:127-131 (1990).
22. M.G. Youngquist, R.J. Driscoll, T.R. Coley, W.A. Goddard, and J.D. Baldeschweiler, Scanning tunneling microscopy of DNA: Atom-resolved imaging, general observations and possible contrast mechanism, *J. Vac. Sci. Technol.* B 9 (2):1304-1309 (1991).
23. G. DeStasio, D. Rioux, G. Margaritondo, D. Mercanti, L. Trasatti, and C. Moore, Scanning tunneling microscopy of deoxyribonucleic acid during replication, *J. Vac. Sci.Technol.* A 9 (4):2319-2321 (1991).
24. J.A. DeRose, S.M. Lindsay, L.A. Nagahara, P.I. Oden, T. Thundat, and R.L. Rill, Electrochemical deposition of nucleic acid polymers for scanning probe microscopy, *J. Vac. Sci. Technol.* B 9 (2):1166-1170 (1991).
25. C. Bustamante and D. Dunlap, Application of scanning tunneling microscopy to structural biology, *Seminars in Cell Biology* 2:179-185 (1990).
26. M. Salmeron, T. Beebe, J. Odriozola, T. Wilson, D.F. Ogletree, and W. Siekhaus, Imaging of biomolecules with the scanning tunneling microscope: Problems and prospects, *J. Vac. Sci. Technol.*. A 8 (1): 635-641 (1990).

27. L.A. Nagahara, T. Thundat, P.I. Oden, S.M. Lindsay, and R.L. Rill, Electrochemical deposition of molecular adsorbates for in situ scanning probe microscopy, *Ultramicroscopy* 33:107-116 (1990).
28. L.A. Bottomley, J.N. Haseltine, D.P. Allison, R.J. Warmack, T. Thundat, R.A. Sachleben, G.M. Brown, R.P. Woychik, K. Bruce Jacobson, and T.L. Ferrell, Scanning tunneling microscopy of DNA:The chemical modification of gold surfaces for immobilization of DNA, *J. Vac.Sci.Technol.* A 10 (4):591-595 (1992).
29. T. Thundat, L.A. Nagahara, P. Oden, and S.M. Lindsay, Direct observation of bioelectrochemical processes by scanning tunneling microscopy, *J. Vac. Sci Technol.* A 8 (1): 645-647 (1990).
30. C. Bendixen, F. Besenbacher, E. Laegsgaard, I. Stensgaard, B. Thomsen, and O.Westergaard, Deoxyribonucleic acid structures visualized by scanning tunneling microscopy, *J. Vac. Sci. Technol.* A8(1):703-705 (1990).
31. D.P. Allison, J.R. Thompson, K.B. Jacobson, R.J. Warmack, and T.L. Ferrell, Scanning tunneling microscopy and spectroscopy of plasmid DNA, *Scanning Microscopy* 4 (3):517-522 (1990).
32. C. Bustamante and J. Vesenka, Imaging of DNA molecules and DNA-RNA polymerase complexes by scanning force microscopy, *Polymer Preprints*, 33(1): 743-744 (1992).
33. T. Thundat, R.J. Warmack, D.P. Allison, L.A. Bottomley, A.J. Lourenco, and T.L. Ferrell, Atomic force microscopy of deoxyribonucleic acid strands adsorbed on mica: The effect of humidity on apparent width and image contrast, *J. Vac. Sci. Technol.* A 10(4):630-635 (1992).
34. H.G. Hansma, J. Vesenka, C. Siegerist, G. Kelderman, H. Morrett, R.L. Sinsheimer, V. Elings, C. Bustamante, P.K. Hansma, Reproducible imaging and dissection of plasmid DNA under liquid with the atomic force microscope, *Science*, 256:1180-1184 (1992).
35. J. Yang, K Takeyasu and Z. Shao, Atomic force microscopy of DNA molecules, *FES* 10933, 30 (2):173-176 (1992).
36. Y.L.Lyubchenko, A.A.Gall, LS.Shlyakhtenko, R.E.Harrington, B.L.Jacobs, P.I.Oden, and S. M.Lindsay, Atomic force microscopy imaging of double stranded DNA and RNA, *J. of Biomolec.Struct.Dynam.* 10(3): 589-606 (1992).
37. Y.L. Lyubchenko, B.L. Jacobs, and S.M. Lindsay, Atomic force microscopy of reovirus dsRNA: a routine technique for length measurements, *Nucleic Acids Res.* 20(15):3983-3986 (1993).
38. Y. Lyubchenko, L. Shlyakhtenko, R. Harrington, P. Oden, and S. Lindsay, Atomic force microscopy of long DNA: imaging in air and under water, *Proc. Natl. Acad. Sci. USA*, 90: 2137-2140 (1993).
39. R.D. Edstrom, M.A. Miller, V.B. Elings, X. Yang, R. Yang, G. Lee, and D.F. Evans, Scanning tunneling microscopy and atomic force microscopy visualization of the components of the skeletal muscle glycogenolytic complex, *J. Vac. Sci. Technol.* B 9 (2):1248-1252 (1991).
40. L. Haggerty, B.A. Watson, M.A. Barteau, and A.M. Lenhoff, Ordered arrays of proteins on graphite observed by scanning tunneling microscopy, *J. Vac. Sci. Technol.* B 9(2): 1219-1222 (1991).
41. H. Arakawa, K. Umemura, and A. Ikai, Protein images obtained by STM, AFM, and TEM, *Nature* 358:171-173 (1992).
42. R. Emch, X. Clivaz, C. Taylor-Denes, P. Vaudaux, and P. Descouts, Scanning tunneling microscopy for studying the biomaterial-biological tissue interface, *J. Vac. Sci. Technol.* A 8 (1):655-658 (1990).
43. R. Wigren, H. Elwing, R. Andsson, S. Welin, and I. Lundstrom, Structure of adsorbed fibrinogen obtained by scanning force microscopy, *FEBS* 280(2): 225-228 (1991).
44. E.A.G. Chernoff, Atomic force microscope images of collagen fibers, *J. Vac. Sci. Technol.* A 10(4):596-599 (1992).
45. V. B. Elings, R. D. Edstrom, M. H. Meinke, X. Yang, R. Yang, and D. F. Evans, Direct observations of enzymes and their complexes by scanning tunneling microscopy, *J. Vac. Sci. Technol.* A8 (1): 652-654 (1990).
46. R.D. Edstrom, M.H.Meinke, X.Yang, R.Y. and D. F. Evans, Scanning tunneling microscopy of the enzymes of muscle glycogenolysis, *Ultramicroscopy* 33:99-106 (1990).
47. G. Lee, D.F. Evans, V. Elings, and R.D. Edstrom, Observation of phosphorylase kinase and phosphorylase b at solid-liquid interfaces by scanning tunneling microscopy, *J. Vac. Technol. B* 9 (2):1236-1241 (1991).
48. R.D. Edstrom, M.H. Meinke, X. Yang, R. Yang, V. Elings, and D.F. Evans, Direct visualization of phosphorylase-phosphorylase kinase complexes by scanning tunneling and atomic force microscopy, *Biophys. J.* 58:1437-1448 (1990).
49. H. Lewerenz, H. Jungblu S. A. Campbell, M. Giersig, and D. Muller, Direct observation of reverse transcriptases by scanning tunneling microscopy, *Aids Res. Human Retrovirus.* 8(9):1663-1667 (1992).
50. C.H. Olk, J. Heremans, P.S. Lee, D. Dziedzic, and N.E. Sargent, IgG antibody and antibody-antigen

complex imaging by scanning tunneling microscopy, *J. Vac. Sci. Technol.* B 9 (2):1268-1271 (1991).

51. J.N. Lin, B. Drake, A.S. Lea, P.K. Hansma, and J.D. Andrade, Direct observation of immunoglobulin adsorption dynamics using the atomic force microscope, *Langmuir* 6(2):241-248 (1990).
52. S.T. Eppell, F.R. Zypman, and R.E. Marchant, Probing the resolution limits and tip interactions of AFM in the study of von Willebrand Factor, *Trans. Soc. Biomater.*, 19:210 (1993).
53. R. Guckenberger, W. Wiegrabe, A. Hillebrand, T. Hartmann, Z. Wang, and W. Baumeister, Scanning tunneling microscopy of a hydrated bacterial surface protein, *Ultramicroscopy* 31:327-332 (1989).
54. R. Emch, X. Clivaz, C. Taylor-Denes, P. Vaudaux, and P. Descouts, Scanning tunneling microscopy for studying the biomaterial-biological tissue interface, *J. Vac. Sci. Technol.* A 8 (1):655-658 (1990).
55. R. Lal and L. Yu, Atomic force microscopy of cloned nicotinic acetylcholine receptor expressed in Xenopus oocytes, *Proc. Natl. Acad. Sci . USA* 90:.7280-7284 (1993).
56. J.K.H. Horber, F.M. Schuler, V. Witzemann, K.H. Schroter, H. Muller, and J.P. Ruppersberg, Imaging of cell membrane proteins with a scanning tunneling microscope, *J. Vac,Sci.Technol.*. B 9 (2):1214-1217 (1991).
57. J. Masai, T. Shibata, S. Kondo, and S. Ishiwata, Scanning tunneling microscopy of actin filament, *J. Vac. Sci. Technol.* B 9 (2): 1177-1179 (1991).
58. A.L. Weisenhorn, B. Drake, C.B. Prater, S.A.C. Gould, P.K. Hansma, F. Ohnesorge, M. Egger, S.-P.Heyn, and H.E.Gaub, Immobilized proteins in buffer imaged at molecular resolution by atomic force microscopy, *Biophys. J.* 58: 1251-1258 (1990).
59. M. Radmacher, R.W. Tillmann, M. Fritz, H.E. Gaub, From molecules to cells: imaging soft samples with the atomic force microscope, *Science* 257:1900-1905 (1992).
60. B. Mainsbridge and T. Thundat, Scanning tunneling microscopy of chloroplasts, *J. Vac. Sci. Technol.* B9(2):1259-1262 (1991).
61. A. Stemmer, A. Hefti, U. Aebi, and A. Engel, Scanning tunneling and transmission electron microscopy on identical areas of biological specimens, *Ultramicroscopy* 30: 263-280 (1989).
62. R. Garcia, D. Keller, J. Panitz, D. Bear, C. Bustamante, Imaging of metal-coated biological samples by scanning tunneling microscopy, *Ultramicroscopy* 27:367-374 (1989).
63. K. Yeung, E. Wolf, and J. Duman, A scanning tunneling microscopy study of an insect lipoprotein ice nucleator, *J. Vac. Sci. Technol.* B9 (2):1197-1201 (1991).
64. H.-J. Butt, C.B. Prater, and P.K. Hansma, Imaging purple membranes dry and in water with the atomic force microscope, *J. Vac. Sci. Technol. B* 9(2):193-1196 (1991).
65. H-J. Butt, K. Downing, and P. Hansma, Imaging the membrane protein bacteriorhodopsin with the atomic force microscope, *Biophys. J.* 58:1473-1480 (1990).
66. J. Yang, L.K. Tamm, T.W. Tillack, and Z. Shao, New approach for atomic force microscopy of membrane proteins; the imaging of cholera toxin, *J. Mol. Biol.* 229: 286-290 (1993).
67. K.A. Fisher, K.C. Yanagimoto, S.L. Whitfield, R.E. Thomson, M.G.L. Gustafsson, and J. Clarke, Scanning tunneling microscopy of planar biomembranes, *Ultramicroscopy* 33:117-126 (1990).
68. C.B. Prater, M.R. Wilson, and J. Garnaes, Atomic force microscopy of biological samples at low temperature, *J. Vac. Sci. Technol.* B 9 (2):989-991 (1991).
69. A.A. Kononenko, E.P. Lukashev, V.I. Panov, and E. Fedorov, Scanning tunneling microscopy of functionally active membrane fragments of halobacteria containing bacteriorhodopsin, Dokl. Akad. Nauk. SSSR, 315 (5):1252-1255 (1990).
70. S.A.C .Gould, B. Drake, C.B. Prater, A.L. Weisenhorn, S. Manne, H.G. Hansms, J. Massie, M. Longmire, V. Elings, B.D. Northern, B. Mukergee, C.M. Peterson, W. Stoeckenius, T.R. Albrecht, and C.F. Quate, From atoms to integrated circuit chips, blood cells, and bacteria with the atomic force microscope, *J. Vac. Sci.Technol.* A 8 (1):369-373 (1990).
71. B.L. Blackford, M.O. Watanabe, D.C. Dahn, M.H. Jericho, G. Southam, and T.J. Beveridge, The imaging of a complete biological structure with the scanning tunneling microscope,*Ultramicroscopy* 27:427-432 (1989).
72. B.L.Blackford and M.H. Jericho, A metallic replica/anchoring technique for scanning tunneling microscope or atomic force microscope imaging of large biological structures, *J.Vac. Sci. Technol. B* 9 (2):1253-1258 (1991).
73. G. Southam, M. Firtel, B.L. Blackford, M.H. Jericho, W. Xu, P.J. Mulhern, and T.J. Beveridge, Transmission electron microscopy, scanning tunneling microscopy, and atomic force microscopy of the cell envelope layers of the archaeobacterium methanospirillum hungatei GP1, *J. Bacteriol.* 175(7):1946-1955 (1993).
74. B.L. Blackford, M.H. Jericho, P.J. Mulhern, and C. Frame, Scanning tunneling microscope imaging of

hoops from the cell sheath of the bacteria methanospirillum hungatei and atomic force microscope imaging of complete sheathes, *J.Vac.Sci.Technol.B* 9(2):1242:1247 (1991).

75. M.H. Jericho, B.L. Blackford, D.C. Dahn, C. Frame, and D. Maclean, Scanning tunneling microscopy imaging of uncoated biological material, *J. Vac, Sci. Technol.* A 8(1):661-666 (1990).
76. M. Amrein, Z. Wang, and R. Guckenberger, Comparative study of a regular protein layer by scanning tunneling microscopy and transmission electron microscopy, *J. Vac. Sci. Technol. B* 9 (2):1276-1281 (1991).
77. W. Wiegrabe, M. Nonnenmacher, R. Guckenberger, and O.Wolter, Atomic force microscopy of a hydrated bacterial surface protein, *J. Microscopy* 163(1):79-84 (1991).
78. J. Ruppersberg, J.K. Heinrich Horber, C. Gerber, and G. Binnig, Imaging of cell membraneous and cytoskeletal structures with a scanning tunneling microscope, *FEBS Letters* 257(2):460-464 (1989).
79. S. Ya-Xian, J. Yue-Kan, X. San-Duo, Y. Jun-En, and L. Ke-Chun, The surface structure of artificial and natural membranes as studied by scanning tunneling microscopy, *J.Vac. Sci.Technol.* A 8(1):695-697 (1990).
80. M. Gaczynska, M. Chwialkowski, W. Olejniczak, S. Woljczuk, and G. Bartosz, Scanning tunneling microscopy of human erythrocyte membranes, *Biochem. Biophys. Res.Comm.*, 181(2):600-603 (1991).
81. S. Singh and D.J. Keller, Atomic force microscopy of supported planar membrane bilayers, *Biophys. J.* 60:1401-1410 (1991).
82. M. Egger, F. Ohnesorge, A.L. Weisenhorn, S.P. Heyn, B. Drake, C.B. Prater, S.A.C. Gould, P.K. Hansma, and H.E. Gaub, Wet lipid-protein membranes imaged at submolecular resolution by atomic force microscopy, *J. Struct. Biol.* 103: 89-94 (1990).
83. C. Luo, C. Zhu, L. Ruan, G. Huang, C. Dai, Z. Cheng, C. Bai, Y. Su, S. Xu, K. Lin, and J.D. Baldeschweiler, Scanning tunneling microscopy of the phosphatidylcholine bilayers, *J. Vac. Sci..Technol.* A 8 (1):684-686 (1990).
84. M. Dreschler, K.P. Hofmann, H.J. Cantow, Imaging of membrane vesicles by TEM and AFM, *Polymer Preprints* 33(1):739-740 (1992).
85. J.H. Hoh, R. Ratneshwar, S.A. John, J.P. Revel, and F. Arnsdorf, Atomic force microscopy and dissection of gap functions, Science 253: 1405-1408 (1991).
86. N. Nakagiri, H. Fujisaki, and S. Aizawa, Scanning tunneling microscopy of bacterial flagella, *J. Vac. Sci. Technol.* B 9 (2):1202:1205 (1991).
87. Y. Simic'-Krstic', M. Keley, C. Schneiker, M. Krasovich, R. McCuskey, D. Koruga, and S. Hammeroff, Direct observation of microtubules with the scanning tunneling microscope, *FASEB J.* 3:2184-2188 (1989).
88. Y. Simic'-Krstic', M. Voelker, M. Andjelkovic, M. Trifunovic, S. Hammeroff, and D. Koruga (in press).
89. S. Hammeroff, Y. Simic'-Krstic', L. Vernetti, Y.C. Lee, D. Sarid, and R. McCuskey, Scanning tunneling microscopy of cytoskeletal proteins and intermediates: microtubules and intermediate filaments, *J. Vac. Soc. Technol.* A 8(1):687-691 (1990).
90. L. Vernetti, D. Satid, A.J. Gandolphi, R.B. Nagle, S. Hammeroff, R. McCuskey, and A.E. Cress, The topographical structure of cyclokeratin intermediate filaments using scanning tunneling microscopy, *Nanobiology* 1:379 (1992).
91. V.B. Pizziconi and D.L. Page, Scanning probe microscopy of living biological cells, *Polymer Preprints* 33(1):747-748 (1992).
92. W. Haberle, J.K.H. Horber, and G. Binnig, Force microscopy on living cells, *J. Vac. Sci.Technol.* B 9 (2):1210-1213 (1991).
93. J.K.H. Horber, W. Haberle, F. Ohnesorge, G. Binnig, H.G. Liebich, C.P. Czerny, H. Mahnel, and A. Mayr, Investigation of living cells in the nanomater regime with the scanning force microscope, Scanning Microscopy 6 (4):919-930 (1992).
94. H.-J. Butt, E.K. Wolff, S.A.C. Gould, B. Dixon Northern, C.M. Peterson, and P.K. Hansma, Imaging cells with the atomic force microscope, *J. Struct. Biol.* 105: 54-61 (1990).
95. J. Yang, L.K. Tamm, A.P. Somlyo, Z. Shao, Promises and Problems of biological atomic force microscopy, *J. Microscopy* 171(3):183-198 (1993).
96. M. Fritz, M. Radmacher, and H. E. Gaub, In vitro activation of human platelets triggered and probed by atomic force microscopy, *Exp. Cell Res.* 205:187-190 (1993).
97. J.W. Dai, Y.K. Jiao, Q. Dong, Y.X. Su, K.C. Lin, J. He, G.Y. Shang, and J.E. Yao, The surface structure of natural membrane of macrophages in water as studied by the scanning tunneling microscope, *J. Vac. Sci. Technol.* B 9 (2):1184-1188 (1991).

98. K. Barbee, Peter F. Davies, and Ratneshwar Lal, Shear-stress induced reorganization of the surface topography of living endothelial cells imaged by atomic force microscopy, *Circ. Res.* 74(1):163-171 (1994).

99. E. Henderson, P.G. Haydon, D.S. Sakaguchi, Actin filament dynamics in living glial cells imaged by atomic force microscopy, *Science* 257:1944-1946 (1992).

100. USA Microscopy and Analysis, January, 1994

101. L. Chang, T. Kious, M. Yorgancioglu, D. Keller, and J. Pfeiffer, Cytoskeletal of living, unstained cells imaged by scanning force microscopy, *Biophys. J.* 64:1282-1286 (1993).

102. E. Ito, T. Takahashi, K. Hama, T. Yohioka, W. Mizutani, H. Shimizu, and M. Ono, An approach to imaging of living cell surface topography by scanning tunneling microscopy, *Biochem. & Biophys. Res.Comm.* 177(2): 636-643 (1991).

103. S. Kasas, V. Gotzos, and M.R. Celio, Observation of living cells using the atomic force microscope, *Biophys. J.* 64:539-544 (1993).

104. C.A.J. Putman, B.G. deGrooth, P.K. Hansma, Niek F. van Hulst, and Jan Greve, Immunogold labels: cell-surface markers in atomic force microscopy, *Ultramicroscopy* 48: 177-182 (1993).

105. A.A. Garcia and C. Pettigrew, Imaging BCG cells and yeast cell walls using scanning probe microscopy, *Polymer Preprints* 33(1):745-746 (1992).

106. T. Tomie, H. Shimizu, T. Majima, M. Yamada, T. Kanayama, H. Kondo, M. Yano, and M. Ono, Three-dimensional readout of flash x-ray images of living sperm in water by atomic force microscopy, *Science*, 252:691-693 (1991).

107. H. Hatano, T. Fujihara, S. Kurata, and H. Negishi, Scanning tunneling microscopic observation of dental structures of a permanent tooth, *Anal. Lett.*, 23(1): 47-55 (1990).

108. J.A.N. Zasadzinski and P.K. Hansma, Scanning tunneling microscopy and atomic force microscopy of biological surfaces, *Ann. N.Y. Acad. Sci.* 589:476-491 (1990).

109. T.W. Jing, A.M. Jeffrey, J.A. DeRose, Y.L. Lyubchenko, L.S. Shlyakhtenko, R.E. Harrington, E. Appella, J. Larsen, A. Vaught, D.Rekesh, F-X. Lu, and S.M. Lindsay, Structure of hydrated oligonucleotides studied by in-situ scanning tunneling microscopy, *Proc. Natl. Acad. Soc.* (in press).

110. D.H. Reneker, R. Patil, and S.J. Kim, Morphology and surface interactions of polymer molecules observed with scanning tunneling and atomic force microscopy, *Polymer Preprints* 33(1):790-791 (1992).

111. J. F. Mustard, D. W. Perry, M.G. Arlie, and M.A. Packman, Preparation of suspensions of washed platelets from humans, *Brit. J. Hemat.* 22:193 (1972).

112. Z. Wang, T. Hartman, W. Baumeister, and R. Guckenberger, Thickness determination of biological samples with a z-calibrated scanneling tunneling microscope, *Proc.Natl.Acad. Sci.* 87:9343-9347 (1990).

113. R. Garcia, Comments on direct visualization of protein complexes by scanning tunneling microscopy, *Biophys. J.* 60:738 (1991).

114. P.D.Yurencho, E.C.Tsilibary, A.S. Charonis, and H. Furthmayr, Laminin polymerization in vitro: evidence for a two step assembly with domain specificity *J. Biol. Chem.* 260(12):7636-7644 (1985).

115. J.-Y. Yuan and Z. Shao, Simple model of image formation by scanning tunneling microscopy of non-conducting materials, *Ultramicroscopy* 34:223-226 (1990).

116. R. Garcia and N. Garcia, Electron conductance in organic chains: why are STM experiments possible on bare biological samples?, *Chem. Phys. Lett.* 173 (1): 44-50 (1990).

117. P. Dietz and K.-H. Herrmann, Scanning tunneling microscopy of thin organic films on conducting substrates, *Surf. Sci.* 232:339-345 (1990).

118. S.M. Lindsay, O.F. Sankey, Y. Li, C. Herbst, and A. Ruprecht, Pressure and resonance effects in scanning tunneling microscopy of molecular absorbates, *J.Phys.Chem.* 94:4655-4660 (1990).

119. S.L.Tang, A.J. McGhie, and A.Suna, Molecular-resolution imaging of insulating macromolecules with the scanning tunneling microscope via a nontunneling, electric-field induced mechanism, *Phys. Rev. B* 47 (7) 3850-3856 (1993).

120. M. de Brabander, R. Nuydens, A. Ishihara, B. Holifield, K. Jacobson, and H. Geerts, Lateral diffusion and retrograde movements of individual cell surface components on single motile cells observed with nanovid microcsopy, *J. Cell Biol.* 112(1):111-124 (1991).

121. C.B. Prater, M.R. Wilson, and J. Garnaes, Atomic force microscopy of biological samples at low temperature, *J.Vac.Sci.Technol. B* 9 (2): 989-991 (1991).

122. P.K. Hansma, J.P. Cleveland, M. Radmacher, D.A. Walters, P.E. Hillner, M. Bezanilla, M. Fritz, D. Vie, H.G. Hansma, C.B. Prater, J. Massie, L. F ukunaga, J. Gurley, and V. Elings, Tapping mode atomic force microscopy in liquids, *Appl.Phys.Lett.* 64(13):1738-1740 (1994).

RESOLUTION AND LIMITATIONS IN BIOLOGICAL APPLICATIONS OF ATOMIC FORCE MICROSCOPY

Jie Yang, Lukas K. Tamm, and Zhifeng Shao

Department of Molecular Physiology and Biological Physics
University of Virginia, Box 449
Charlottesville, VA 22908

Abstract: Atomic force microscopy (AFM) has been applied to image DNA and a membrane protein: cholera toxin. By use of the Kleinschmidt method, DNA molecules were picked up on carbon-coated mica surfaces and imaged by AFM in air and in organic solvents. The resolution was found to be closely related to the adhesion force and a resolution of 3 - 6 nm was routinely obtained when the adhesion force was below 3 nN. The role of the adhesion force, the tip condition and the specimen preparation on resolution and imaging quality will be discussed. Polymerized diacetylene phosphatidylcholine (DAPC) bilayers provide a relatively stable matrix for studying membrane proteins. When cholera toxin (complete or B-subunit oligomer) was bound to mixed bilayers of DAPC and the receptor glycolipid G_{M1}, the subunit structure was well resolved by AFM in buffer, without crystallization. The resolution was better than 2 nm with excellent reproducibility for a probe force of 0.3 - 0.5 nN. These results show that individual biomacromolecules under native conditions can be imaged by AFM with high spatial resolution.

INTRODUCTION

The atomic force microscope[2] (AFM) has been applied to the study of the surface topology of various materials.[26,27] Because of its abilities to acquire high resolution surface structures of nonconductive specimens and to operate in aqueous solutions, this microscope is extremely promising for determining the structures of biological molecules under native conditions. Although the AFM tip is normally quite large,[1,30] it is actually quite routine to obtain "atomic" resolution on crystalline specimens, as shown in Figure 1, where the cubic lattice structure on [100] plane of a NaCl crystal and atomic steps of 0.2-0.3 nm height are clearly seen. However, on noncrystalline specimens, the resolution becomes limited by the extent of the AFM tip. Furthermore, for hydrated biological materials, such as membrane proteins, the situation becomes considerably more complex.

Atomic Force Microscopy/Scanning Tunneling Microscopy
Edited by S.H. Cohen *et al.*, Plenum Press, New York, 1994

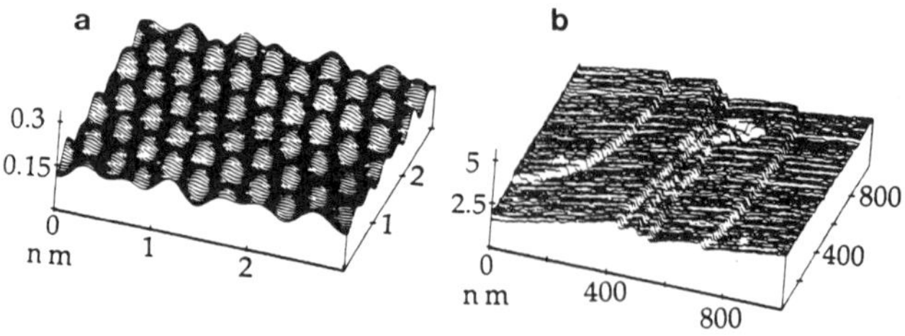

Figure 1. These are AFM images of a NaCl crystal obtained in air using a cantilever with a spring constant of 0.06 N/m. Both images were obtained on a freshly cleaved surface (the [100] plane). (a) was obtained at a scanning speed of 39 Hz. The image shown was a surface plot after Fourier averaging to emphasize the periodic atomic structure. The distance between nearest neighbors corresponding well to the distance between ions of the same polarity. We attribute the corrugation to the negative ion Cl^-, due to its larger van der Waals radius. (b) was the result of a larger area scanning (1 μm^2). The steps have heights between 2 to 3 Å, indicating atomic single layers. These steps were very stable. But scanning over a prolonged period of time (2 to 3 hours) showed drifts of the edges at some area of these steps, possibly due to some kind of surface reorganization since the humidity was about 30% which may facilitate surface atomic movement.

Experimental results have shown that, for randomly distributed soft biological macromolecules and cells on a solid support, resolutions are in the range from a few nm to sub µm. For example, 3-60 nm resolution was obtained by AFM on DNA molecules in air and organic solvents[3,10-12,17-19,29,31-33,36] and about 30 nm in aqueous solutions.[19] When operated in solution, 1-2 nm resolution was achieved on the a membrane-bound protein cholera toxin, on a solid support and without Fourier averaging.[34]

At present, it appears that the limiting factors for AFM of biological specimens are quite different for operating in air and in solution. However, it is clear that the applied probe force is a critical in both cases. In solution, 0.1 nN is achievable. Even at this force, the pressure on the specimen would be about 3000 atm if we assume that the contact area had a radius of 1 nm This pressure is very large for biological specimens to withstand. Since high resolution images can be obtained on such materials, the total probe force must be balanced by a much larger area, by which the resolution is not determined upon. The most likely explanation is that small protrusions on a rather blunt tip actually determine the lateral resolution.[35] This argument is indirectly supported by the fact that even fully hydrated cells can withstand the pressure of a probe to some extent (few nN probe force) without apparent damage.[5,8,9,13,22,28] If this is indeed the case, then, a super sharp tip can only be used with caution due to the build-up of pressure in the area of contact. It may be actually a better choice to have a smooth blunt tip with sharp small protrusions of sub-nm dimension. Secondly, the probe force is likely to induce elastic or even plastic deformation on biological materials because of their intrinsic softness when fully or even partially hydrated, which presents a serious limitation in high resolution AFM imaging, although the extent of such deformation is quite difficult to study by any means, and no definitive data are available for a realistic estimate.

In this paper, we show recent results from our laboratory that could illustrate the current status of biological AFM. Applications on DNA, supported planar lipid bilayers, and a membrane protein are presented. Promises and limitations on resolution on soft biological specimens are discussed.

AFM OF DNA IN AIR

Using a simplified Kleinschmidt method,[31,33] we have imaged DNA molecules with excellent reproducibility under various conditions in order to find the factors that would affect the achievable resolution. To separate effects of different factors, we studied a number of DNA specimens using the commercial Si_3N_4 tips under a controlled humidity. We found that the lateral resolution was closely related to the adhesion force, rather than the humidity. Especially, in several cases, even though the same tip was used to image the same molecule, the resolution was quite different for different adhesion forces. Here the adhesion force is defined by the hysteresis of the tip, when retracting from the specimen surface. Unfortunately, the adhesion force cannot be adjusted at will, so that to quantify the relationship between the adhesion force and the resolution, we must rely on a large number collection of images. Figure 2 is a log-log plot of the lateral resolution vs the adhesion force, based on measurements over more than six months. The lateral resolution (R) was shown to relate to the adhesion force (F) as $R \propto \sqrt{F}$. For adhesion forces larger than 30 nN, DNA images were no longer obtainable. One intuitive explanation of this result is that the adhesion force is proportional to the "contact" area, that dominates the resolution, such as the main bump on the tip. Therefore the resolution should be proportional to the square root of this area: $\propto \sqrt{F}$. Although the details of this relationship cannot be determined by these measurements alone, it is plausible to assume that some sort of contamination could build up around the "bump" to increase the "contact" area.

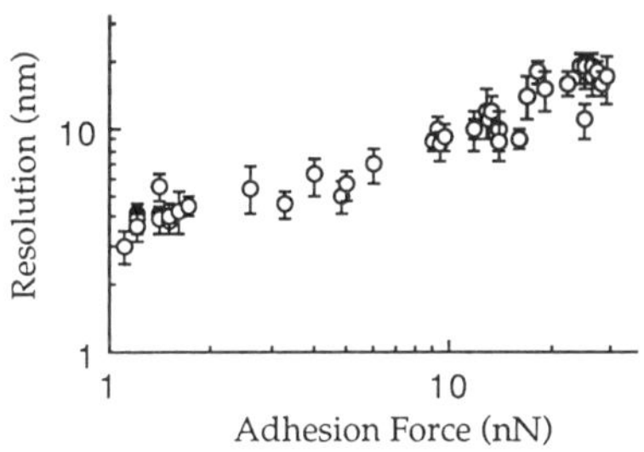

Figure 2. The log-log plot of the adhesion force (F) vs the lateral resolution (R) in the imaging of DNA molecules. Each datum point in this plot was obtained under one adhesion force averaged along the DNA molecule. The slope in this plot is the power law relation with which the lateral resolution is dependent on the adhesion force. A least-squares fit of the data gives a power law of 0,48, leading to the condusion of $R \propto \sqrt{F}$ discussed in the text.

Studies of other factors under otherwise optimal conditions also show results consistent with the assumption of relating adhesion force to tip contamination. For example, with tips showing an adhesion force < 3 nN at humidities below 40%, we studied the effects on imaging due to humidity changes. We found that as long as the tip is kept engaged while changing the humidity, the image quality and resolution as well as the adhesion force remain almost the same. But if the tip were disengaged at high humidities (> 45%), a reengagement usually showed a larger adhesion force, possibly due to the backmigration of contaminants at higher humidities when the tip is exposed to air. Another example is to apply a larger probe force externally. The result showed the resolution was unaffected by the applied force, until it became so large that the DNA molecules were damaged.

Under the assumption that the adhesion force is primarily determined by contamination, some sort of tip cleaning (conditioning) should be possible, at least in principle. Indeed, we found that scanning a carbon-coated substrate alone often caused a reduction of the adhesion force for a given tip. If the adhesion force remained small (~ 2 nN) for a prolonged scan of more than 2 min with a scan area of 10 μm at slow speeds, usually such a tip can image a DNA specimen with high resolution if the specimen exchange was done quickly (< 1 min). If a tip were left in air for a longer period of time after cleaning, it normally did not show high resolution images of DNA and the adhesion force on the DNA specimen also became large. At humidities larger than 45%, most often this cleaning procedure was not effective, that no matter how fast the specimen exchange was completed (usually at least 20 s), the adhesion force always became larger again on the DNA specimen, although it was small on the cleaning carbon film substrate for a very long period (> 5 min) and stable. At lower humidities, a tip can normally be cleaned several times. Figure 3 shows two examples of this cleaning experiments. Figure 3 (a) is an image of a DNA molecule obtained with a cleaned tip, the adhesion force was about 2 nN with a typical resolution about 4 nm. Figure 3 (b) is the same molecule imaged by the same tip later when somehow the tip was contaminated. When this tip was cleaned again, it obtained high resolution images of DNA as shown in Figure 3 (c), with an adhesion force again about 2 nN. Although it remains possible that the scanning on the carbon film caused some structural change that resulted in some sort of a supertip, all results seem to favor the notion of contamination. Further study is definitely needed, in order to shed some light on this issue.

SUPPORTED SYNTHETIC LIPID BILAYERS

In our study of supported phospholipid bilayers prepared by using a Langmuir trough, we found that several regular phospholipids, such as dimyristoyl-, dipalmitoyl-, distearoyl-, and palmitoyl-oleoyl-phosphatidylcholine as well as dimyristoyl-, dipalmitoyl-, and palmitoyl-oleoyl-phosphatidylethanolamine, showed relatively low success rates (< 40%) of ordered lipid bilayer transfer on to a freshly cleaved mica, as judged from low resolution AFM scans. Higher resolution scanning did not yield any reproducible images in our hands, although the height measurements at the edge of defects of successfully transferred bilayers were consistent with the expected values of single bilayers of the different lipids. It is possible that the hydrophilic interactions between the headgroups and the mica surface are not sufficient to maintain a stable adhesion during lipid transfer and/or AFM scanning. In contrast, we found that it was much easier to prepare stable bilayers with polymerizable phospholipids. These diacetylene lipid bilayers were polymerized by UV light after their transfer on to the mica surface. The success rate with diacetylene phosphotidylcholine (DAPC) bilayers (both DAPC-25 and DAPC-23, where 25 and 23 represent the acyl chain lengths, respectively) was much higher (> 95%). The height measured at the edge of these lipid bilayers, 72±9 Å for DAPC-25 and 66±6 Å for DAPC-23, respectively, were consistent with expected values from neutron diffraction or molecular modeling.

Higher resolution images of these DAPC bilayers showed a highly reproducible ridged structure of about 1 nm periodicity, consistent with studies of other DAPC lipids by electron microscopy and X-ray diffraction.[16-24] Figure 4 is a high resolution image of a DAPC-23 bilayer after low pass filtering to remove scanning lines. The ridge structure is easily identifiable. The modulation along the ridge direction is not periodic and cannot be used to identify individual headgroup with any degree of certainty, due to the limited resolution and the low contrast. To obtain high resolution images of these bilayers we used an optimal probe force of 0.5 - 1 nN, which was larger than the lowest possible probe force

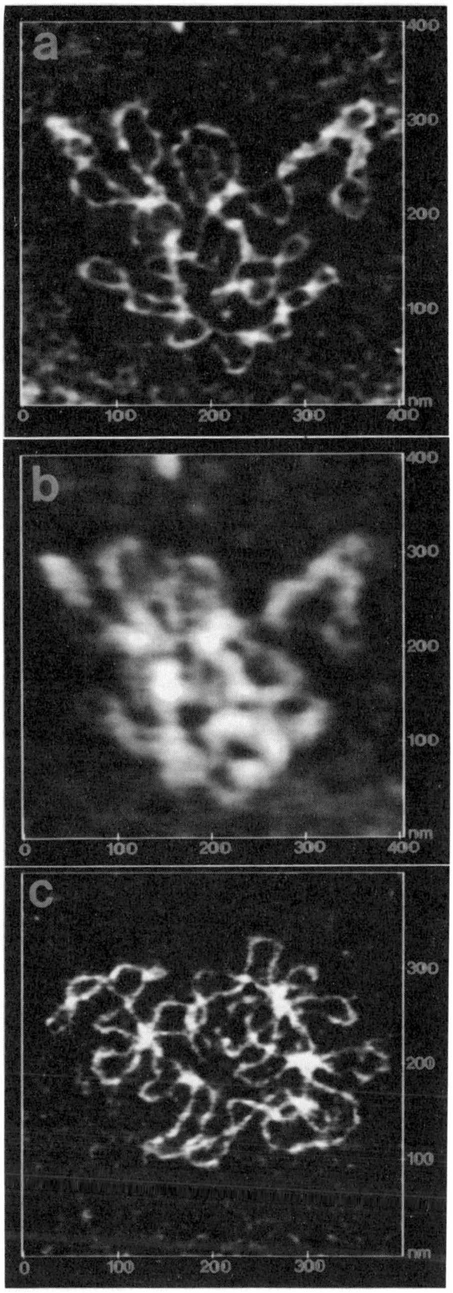

Figure 3. AFM images of DNA molecules obtained with the same tip (k = 0.06 N/m) on the same specimen. (a) and (b) were also the same DNA molecule. All images were obtained with a scanning speed of 5.79 Hz. (a) was imaged after cleaning the tip by the procedure described in the text. The adhesion force was about 2 nN; (b) was the same molecule, obtained later (~ 1 hour). The adhesion force for (b) was about 25 nN. In this case, the adhesion force was suddenly increased during a normal search of other DNA molecules. This increase of the adhesion force was irreversible on the same specimen, so that when the tip was brought back to the position where the image (a) was taken, we obtained image (b). Then, the tip was disengaged and the tip was allowed to scan a carbon film on mica. (c) was obtained after cleaning the tip. The adhesion force became about 2 nN again, and the resolution also recovered.

of the instrument for stable imaging. The fact that a certain minimum probe force was required for obtaining high resolution images may be indirect evidence for the effect of the hydration force that must be overcome to probe underlying structures.

These studies show that the lateral cross-linking of the polymerizable DAPCs increased the mechanical strength of the supported bilayers so that they could withstand the impact of the probe. Another interesting result is that the vertical compression was quite small in these images. This may be an indication that the probe force was balanced by a larger area than the resolution would indicate.

MEMBRANE-BOUND CHOLERA TOXIN IN AQUEOUS SOLUTION

Our studies of the DAPCs showed that polymerizable phospholipid bilayers should be a suitable substrate for the study of membrane proteins, due to the improved reproducibility and much larger surface coverage when compared with isolated native membrane fragments.[4,14] The method was applied to study the membrane-bound protein cholera toxin. This protein has a mass of 85 kD, consists of 5 B-subunits for its binding to the membrane surface receptor, ganglioside glycolipid G_{M1} and a toxic A subunit which blocks the hydrolysis of GTP by the α-subunit of trimeric G-proteins.[7,15] After incubation of a DAPC bilayer with 25% mole G_{M1} with cholera toxin B-subunit oligomers, excess proteins were removed by washing and the lipids were cross-linked by UV light. The specimens were imaged by AFM in either deionized water or in a low salt buffer. It was found that the probe force must be lower than 1 nN for successful imaging of individual toxin molecules. For closely but randomly packed toxin molecules, the pentameric structure of the B-subunit oligomers was well resolved, with a lateral resolution between 1-2 nm, and the measured size was consistent with the molecular dimension determined by electron microscopy of 2-d crystals.[20,23,25] It was found that a minimum probe force was also required (~ 0.3 nN) for stable imaging. Figure 5 is an AFM image of closely packed cholera toxin B-subunit oligomers on DAPC-23 bilayer. The pentameric structure can be seen on many individual molecules. In this special case, the resolution achieved by AFM is comparable to that achieved on 2-d crystals by electron microscopy.[20,23,25]

We also imaged the complete cholera toxin. The A-subunit appeared to fill the central hole of the B-subunit rings and actually protrude out of the membrane in these preparations. At larger probe forces, the damage to the specimen became extensive. A tip-induced aggregation of toxin molecules on the bilayer was observed, indicating the removal of the toxin molecules from the bilayer surface by the probe. This phenomenon together with the requirement for a minimum probe force for high resolution images limits the probe force to a narrow operation range. Although the compression with the bilayer alone was not obvious, we observed a vertical compression of up to 50% with cholera toxin. In view of this quite large deformation and although it appeared to be elastic, further improvement in lateral resolution does not seem optimistic. One may have to improve upon the mechanical strength of the specimen, or to use 2-d crystallized preparations in order to improve the spatial resolution. Again, commercial tips were used for these studies.

CONCLUSIONS

Just a few years after the invention of AFM, its use to visualize DNA in air has already achieved a resolution close to the native molecular dimension. But, the control of the imaging condition has not been reliably solved. At present, higher spatial resolution to resolve helical structure and the goal of imaging individual basis has not been

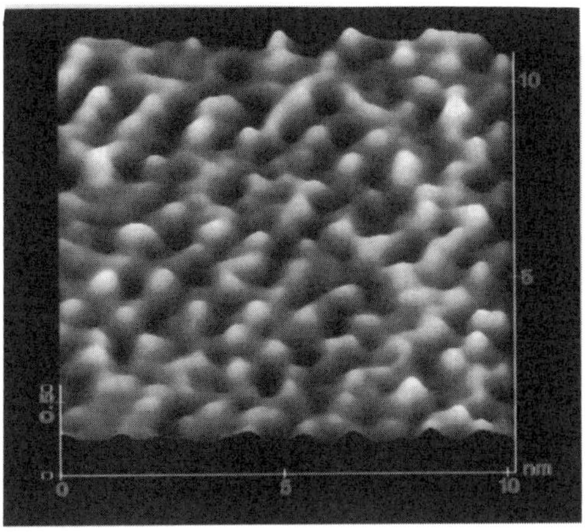

Figure 4. DAPC-23 bilayer surface on freshly cleaved mica by AFM in water. This is an image (60° surface plot) after low pass filtering to remove scanning lines. The ridge structure can be easily identified. The spacing between ridges in this case is ~ 1 nm. The irregularity of corrugations along the ridge makes it difficult to associate each bump to an individual headgroup. We have observed structures of the same character for almost every specimen prepared. Therefore, it is a plausible conclusion that the polymerizable DAPC bilayers after chemical crosslinking have a periodic ridge structure of about 1 nm periodicity. The direction of the ridge may correspond to the direction of polymerization. The probe force was about 0.6 nN with a scanning speed of 39 Hz.

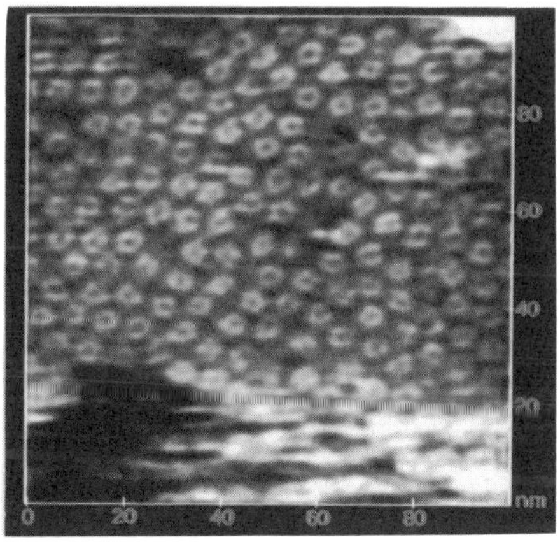

Figure 5. Cholera toxin B-subunit oligomers on DAPC-23:GM1 (4:1) bilayers. This image was obtained in a low salt buffer (10 mM TrisCl, pH 7.5, 10 mM NaCl, 1 mM EDTA). A clear boundary is seen, which may be the edge of the underlying bilayer. The pentameric subunit structure can be clearly seen on many molecules. The probe force was 0.4 nN with a scanning speed of 8.68 Hz. The same image usually remained stable for only several repeated scans, afterwards, the image quality gradually degraded. The resolution achieved here is 1-2 nm depending on definition. Since the vertical compression is up to 50% and at least 0.3 nN probe force was required for stable imaging, further improvement on resolution is unlikely, unless the mechanical strength and the stability of the specimen are substantially improved.

achieved. In this case, the STM might be a better choice for its intrinsically higher resolution. However, for DNA conformation and DNA-protein interaction, the AFM should be more useful due to their larger molecular dimensions.

Before the AFM can be used routinely in a biological laboratory, we must solve the problem of tip contamination. Possible solutions may include laser ablation and different tip materials, or even clean room operation. For fully hydrated specimens, a theoretical understanding of the role of hydration force in AFM imaging is useful to justify future instrumental development to achieve nm resolution on various specimens. These development may include higher force sensitivity, larger formats for 2-d crystal specimens etc. It appears that solid supported specimens are more suitable for high resolution structural studies by the AFM. The resolution of about 1 nm achieved on randomly packed toxin molecules by AFM in aqueous solutions is quite encouraging for the usefulness of AFM in biological research. Since the AFM directly obtains the surface structure, the information should complement well with electron microscopy and X-ray diffraction.

A unique ability of AFM is to image biological materials at native condition at nm resolution which could be used to study conformational changes of some well prepared specimens. 2-d crystals can also be imaged by AFM to further improve the spatial resolution. However, the soft nature of biomacromolecules is an intrinsic property, which is a serious limitation for AFM imaging due to the inevitable probe force, which probably will not be much reduced, particularly at room temperature where thermal fluctuation is considerable. One approach to circumvent this problem is to lower the temperature, since the Young's modulus of typical biopolymers can increase up to 104 fold at liquid nitrogen temperature.[6] Recent success of a low temperature AFM operated under ambient pressure indicated that the technology is almost ready to make a decisive step in this direction.[21] It is quite possible that sub-nm resolution could be achieved on some biological specimens at cryogenic temperatures in the near future.

ACKNOWLEDGMENTS

We thank Drs. A.P. Somlyo, A.V. Somlyo and J. Mou for useful discussions and Dr. S. Tatulian for the preparation of some LB films. The work presented is supported by grants from US Army Research Office, Whitaker Foundation, Jeffress Memorial Trust, National Institutes of Health and the National Science Foundation.

REFERENCES

1. T.R. Albrecht, S. Akamine, T.E. Carver, and C.F. Quate, Microfabrication of cantilever styli for the atomic force microscope, *J. Vac. Sci. Technol.* A8, 3386-3396 (1990).
2. G. Binnig, C.F. Quate, and C.H. Gerber, Atomic Force Microscope, *Phys. Rev. Lett.* 56, 930-933 (1986).
3. C. Bustamante, J. Vesenka, C.L. Tang, W. Rees, M. Guthod, and R. Keller, Circular DNA molecules imaged in air by scanning force microscopy, *Biochemistry* 31, 22-26 (1992).
4. H.-J. Butt, K.H. Downing, and P.K. Hansma, Imaging the membrane protein bacteriorhodopsin with the atomic force microscope, *Biophys. J.* 58, 1473-1480 (1990).
5. H.-J. Butt, E.K. Wolff, S.A.C. Gould, B.D. Northern, C.M. Peterson, and P.K. Hansma, Imaging cells with the atomic force microscope, *J. Struct. Biol.* 105, 54-61(1990).
6. K.L. Dorrington, The theory of viscoelasticity in biomaterials, *in*: "The Mechanical Properties of Biological Materials," Cambridge University Press, Cambridge (1979).
7. D.M. Gill, Mechanism of action of cholera toxin, *Adv. Cyc. Nucleo. Res.* 8, 85-118 (1977).
8. W. Haberle, J.K.H. Horber, and G. Binnig, Force microscopy of living cells, *J. Vac. Sci. Techno.* 9, 1210-1213 (1991).

9. W. Haberle, J.K.H. Horber, F. Ohnesorge, D.P.E. Smith, G. Binnig, In situ investigation of single living cells infected by viruses, *Ultramicroscopy* 42-44, 1161-1167 (1992).
10. H.G. Hansma, R.L. Sinsheimer, M.-Q. Li, and P.K. Hansma, Atomic force microscopy of single- and double-stranded DNA, *Nucleic. Acids Res.* 20, 3585-3590 (1992).
11. H.G. Hansma, J. Vesenka, C. Siegerist, G. Kelderman, H. Morrett, P.L. Sinsheimer, V. Elings, C. Bustamante, P.K. Hansma, Reproducible imaging and dissection of plasmid DNA under liquid with atomic force microscopy, *Science* 256, 1180-1184 (1992).
12. E. Henderson, Imaging and nanodissection of individual supercoiled plasmid by atomic force microscopy, *Nucleic Acids Res.* 20, 445-447 (1992).
13. E. Henderson, P.G. Hayelon, and D.S. Sakaguchi, Actin filament dynamics in living glial cells imaged by atomic force microscopy, *Science* 257, 1944-1946 (1992).
14. J.H. Hoh, R. Lal, S.A. John, J.-P. Revel, and M.F. Arnsdorf, Atomic force microscopy and dissection of gap junctions, *Science* 253, 1405-1408 (1991).
15. J. Holmgren, Actions of cholera toxin and the prevention and treatment of cholera, *Nature* 292, 413-417 (1981).
16. D. Johnston, S. Sanghera, M. Pons, and D. Chapman, Phospholipid polymers-sythesis and spectral characteristics, *Biochim. Biophys. Acta* 602, 57-69 (1980).
17. Y.L. Lyubchenko, A.A. Gall, L.S. Shlyakhtenko, R.E. Harrington, and S.M. Lindsay, Atomic force microscopy imaging of large double stranded DNA molecules, *Biophys. J.* 61, A149 (1992).
18. Y.L. Lyubchenko, B. L. Jacobs, and S.M. Lindsay, Atomic force microscopy of reovirus dsRNA: a routine technique for length measurements, *Nucleic Acids Res.* 20, 3983-3986 (1992).
19. Y.L. Lyubchenko, P.I. Oden, D. Lampner, S.M. Lindsay and K.A. Dunker, Atomic force microscopy of DNA and bacteriophage in air, water and propanol: the role of adhesion forces, *Nucl. Acids Res.* in press.
20. G. Mosser, and A. Brisson, Structural analysis of two-dimensional arrays of cholera toxin B-subunit, *J. Electron Micro. Tech.* 18, 387-394 (1991).
21. J. Mou, J. Yang, and Z. Shao, An optical detection low temperature atomic force microscope at ambient pressure for biological research, *Rev. Sci. Instrum.* in press.
22. M. Radmacher, R.W. Tillmann, M. Fritz, and H.F. Gaub, From molecules to cells: imaging soft samples with the atomic force microscope, *Science* 257, 1900-1905 (1992).
23. R.A. Reed, J. Mattai, and G.G. Shipley, Interaction of cholera toxin with ganglioside G_{M1} receptors in supported lipid monolayers, *Biochemistry* 26, 824-832 (1987).
24. D.G. Rhodes, A. Xu, and R. Bittman, Structure of polymerizable lipid bilayers V: synthesis, bilayer structure and proterties of diacetylenic ether and ester lipids, *Biochim. Biophys. Acta* 1128, 93-104 (1992).
25. H.O. Ribi, D.S. Ludwig, K.L. Mercer, G.K. Schoolnik, R.D. Kornberg, Three-dimensional structure of cholera toxin penetrating a lipid membrane, *Science* 239, 1272-1276 (1988).
26. D. Rugar, and P. Hansma, Atomic force microscopy, *Physics Today* 43, 23-30 (1990).
27. D. Sarid, "Scanning Force Microscopy," Oxford University Press, Oxford, New York (1990).
28. T. Tomie, H. Shimizu, T. Majima, M. Yamada, T. Kanayama, H. Kondo, M. Yano, and M. Ono, Three-dimensional readout of flash X-ray images of living sperm in water by atomic force microscopy, *Science* 252, 691-693 (1991).
29. J. Vesenka, M. Guthold, C.L. Tang, D. Keller, E. Delaine, and C. Bustamante, Substrate preparation for reliable imaging of DNA molecules with the scanning force microscope, *Ultrmicroscopy* 42-44, 1243-1249 (1992).
30. H. Ximen, and P.E. Russell, Microfabrication of AFM tips using focused ion and electron beam techniques, *Ultramicroscopy* 42-44, 1526-1532 (1992).
31. J. Yang and Z. Shao, The effect of probe force on the resolution of atomic force microscopy of DNA. Ultramicroscopy, in press.
32. J. Yang, A.V. Somlyo, M.K. Reedy, K. Takeyasu, L.K. Tamm, M. Allietta, T.W. Tillack, and Z. Shao, Biological applications of AFM, *in*: "Proc. 50th EMSA Annual Meeting," Boston, MA., 1138-1139 (1992).
33. J. Yang, K. Takeyasu and Z. Shao, Atomic force microscopy of DNA molecules, *FEBS Lett.* 301, 173-176 (1992).
34. J. Yang, L.K. Tamm, T.W. Tillack, Z. Shao, New Approach for Atomic Force Microscopy of Membrane Proteins: the Imaging of Cholera Toxin, *J. Mol. Biol.* 229, 286-290 (1993).
35. J.A.N. Zasadzinski, C.A. Helm, M.L. Longo, A.L. Weisenhorn, S.A.C. Gould, and P.K. Hansma, Atomic force microscopy of hydrated phosphatidylethanolamine, *Biophys. J.* 59, 755-760 (1991).
36. F. Zenhausern, M. Adriah, B.T. Heggeler-Bordier, R. Emch, M. Jobin, M. Taborelli, P. Descouts, Imaging of DNA by scanning force microscopy, *J. Struct. Biol.* 108, 69-73 (1992).

SCANNING TUNNELING MICROSCOPY IMAGING OF BIOMOLECULES: *I.* TUBULIN IN MICROTUBULES AND MONOLAYERS *II.* BACTERIAL LUCIFERASE--A MODEL SYSTEM FOR ANESTHESIA

Stuart R. Hameroff,[1] Djuro Koruga,[2] J. Simic-Krstic,[2] M. Trifunovic,[2] and M. Voelker[3]

[1]Advanced Biotechnology Laboratory, Department of Anesthesiology
University of Arizona Health Sciences Center
Tucson, Arizona 85724
[2]Molecular Machines Research Center, University of Belgrade
27 Marta 80
1100 Belgrade, Yugoslavia
[3]Optical Sciences Center
University of Arizona
Tucson, Arizona 85721

Abstract: We used scanning tunneling microscopy (STM) to image two types of biomolecules using both graphite and gold substrates: 1) tubulin - the component protein of cytoskeletal microtubules - in two different forms: cylindrical microtubules and in zinc-induced tubulin monolayer sheets. In both cases STM images show periodic structures whose dimensions correlate with tubulin monomers and dimers. The shapes of the dimers differ in the two cases, apparently because of different surface charge distributions. 2) bacterial luciferase, a photoemitting enzyme whose function is inhibited by general anesthetic gases proportional to their anesthetic potency. We imaged bacterial luciferase in its "activated" form (in the presence of NADH) and in ethanol, which acts as an anesthetic.

INTRODUCTION

STM has successfully imaged atomic surfaces of numerous metals and semiconductors in vacuum, air, water, and ionic solution, and potential technological applications abound.[3,4,17,18] Application of STM to biological structures may be limited by poor electrical conductivity of biomolecules, their coating with an adsorbate layer,[23] their flexible elasticity,[9] and poor localization and stability for STM approach. Further, STM artifacts from graphite, a commonly used substrate for biomolecules, have resembled periodic filaments and been mistaken for DNA.[7] These obstacles have been circumvented by imaging carbon-casted replicas of biomembranes,[37] coating biological materials with a conductor,[32] using gold or other substrates or using STM-related atomic force microscopy in which deflection of a piezo-driven cantilever reveals surface structure.[27] In particular, atomic force microscopy has dynamically imaged the thrombin-catalyzed assembly of fibrinogen into fibrin.[10] STM itself, however, may have the greatest resolution capabilities, and reports of successful STM imaging of uncoated biological materials suggest that electron current somehow flows through or around biomolecules to provide useful imaging in some circumstances.[4,5,9,19,23,34]

We have used STM to image biomolecules in two general areas: 1) cytoskeletal microtubules and zinc-induced monolayer sheets of microtubule subunit protein tubulin, 2) the photoemitting enzyme bacterial luciferase, its activated form, and in the presence of ethanol.

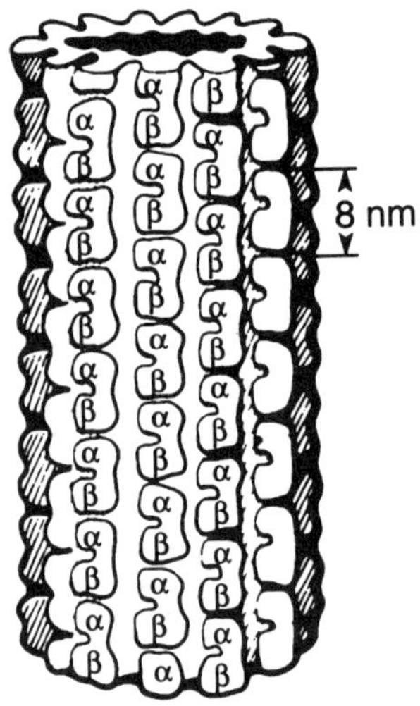

Figure 1. Cylindrical structure of microtubule showing 13 protofilament chains of α and β tubulin monomers. Lattice arrangement of α/β tubulin dimers show skewed hexagonal packing arrangement. Optical diffraction from Amos and Klug, 1974.[2]

I. STM OF CYTOSKELETAL MICROTUBULES AND ZINC-INDUCED TUBULIN SHEETS

The cytoskeleton structurally and dynamically organizes living eukaryotic cells. The most important components of the cytoskeleton are microtubules (MT), 25 nanometer ("nm") diameter cylindrical polymers whose walls are composed of 13 longitudinal protofilaments which are each chains of 8 nm by 4 nm by 4 nm subunit proteins:

"tubulin"[11] (Figure 1). Each tubulin subunit is a dimer weighing 100 kD and consisting of two 50 kD monomers: α and Β tubulin. Under proper conditions including precise concentrations of divalent cations Ca^{++} and Mg^{++}, tubulin dimers self-assemble into cylindrical microtubules.[11] The arrangement of tubulin subunits in microtubules has been observed by electron microscopy[25,26], x-ray diffraction, optical diffraction[2] and STM.[15,29] In the presence of excess zinc (Zn^{++}), tubulin dimers self-assemble into flat monolayer sheets rather than cylindrical microtubules.[22] The structure of zinc induced tubulin sheets has been analyzed by electron microscopy and optical diffraction.[31] In this study we have used STM to image tubulin in two forms: 1) within microtubules, and 2) within zinc-induced monolayer sheets.

METHODS

2-(N-Morpholino) ethanesulfonic acid (MES), ethylene bis (oxyethylenenitrilo tetraacetic acid (EGTA), and GTP were obtained from Sigma Chemical Co., (St. Louis, MO.). For both microtubules and zinc-induced sheets, tubulin was isolated from pig brain tissue by three cycles of temperature dependent assembly and disassembly.[30] Tubulin was incubated in reassembly buffer (0.1 M MES, 1 mM EGTA, 1 mM GTP, 0.5 mM $MgCl_2$, pH 6.5) containing 4 M glycerol at 37°C. For tubulin sheets, 0.25 mM $ZnSO_4$ was added to the buffer. Samples from the reassembly reaction were taken after 30 min and diluted fivefold in reassembly buffer containing 0.1% glutaraldehyde. After 10 min, one drop of sample was smeared onto a chip of freshly cleaved, highly oriented pyrolytic graphite (HOPG).

For Zn^{++} - tubulin, a gold substrate was also used. For STM, tubulin samples on HOPG or gold were placed under atmospheric conditions at room temperature on the sample stage of a Nanoscope II STM (Digital Instruments, Santa Barbara, Calif.). Mechanically sharpened, 80/20 platinum/iridium tips were negatively biased with respect to the sample. STM operating conditions for microtubule tubulin were 66 mV and 0.4 nA in a constant current mode and for zinc-tubulin 61 mV and 1nA in a constant height mode. Both used high and lowpass filtering.

For electron microscopy, drops of tubulin polymerized in solution with and without zinc were applied to carbon-coated 400 mesh copper grids and stained with 1% uranyl acetate in aqueous solution. Samples were then examined in a Philips 300 transmission electron microscope at 60 kV using a 30-μ objective aperture.

RESULTS

Electron microscopy of the microtubule tubulin preparation (without zinc) verified presence of microtubules (Figure 2). STM images of the HOPG substrate with reassembly buffer showed no periodic structures. STM images of microtubule preparations produced images consistent with microtubule structure and dimensions, with identifiable longitudinal striations representing protofilaments. Figure 3 shows an image of several rows of about four "peanut shaped," 8 nm x 4 nm structures, which appear to represent tubulin dimer subunits within four protofilaments of a microtubule. Figure 4 shows electron microscopy of zinc-induced "ribbon-like" sheets of tubulin. Figure 5 illustrates an STM image of zinc-induced tubulin sheets deposited on gold. Zn^{++} - tubulin was often folded and twisted as previously reported[22] and multilayered structures are evident. The major features in these images are periodic domains which we interpret as arrays of tubulin subunits. STM images of zinc-induced sheets are shown in increasing orders of magnification in Figures 5 to 7. Longitudinal striations approximately 5 nm wide are subdivided into periodic domains spaced 4 nm apart. These appear to represent tubulin subunit monomers; however, their "double-crescent" shape differs from the "peanut"-shape observed with tubulin in microtubules.

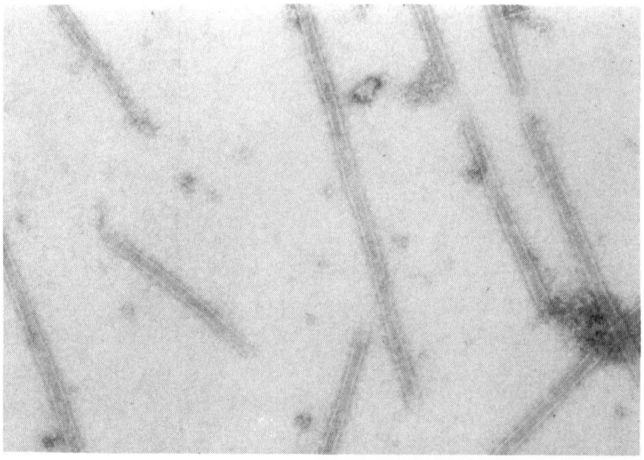

Figure 2. Electron micrograph of microtubules. Width of field is 1.2 μm.

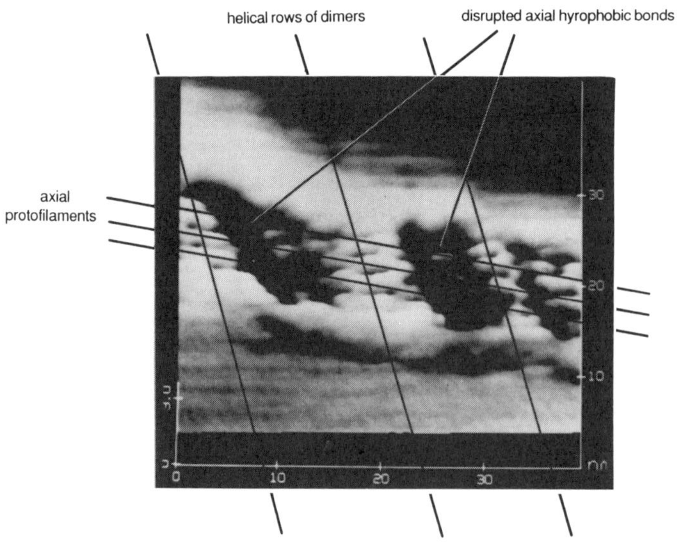

Figure 3. STM image of cylindrical microtubule embedded in glycerol. Three rows of 8 nm long tubulin dimers are seen. The dimers appear separated along axial protofilaments.

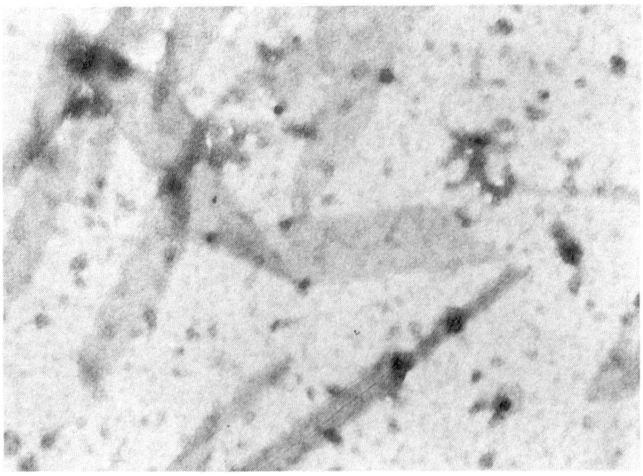

Figure 4. Electron micrograph of Zn^{++} - induced sheets of tubulin. Some twisting and folding are evident. Width of field is 1.2 μm.

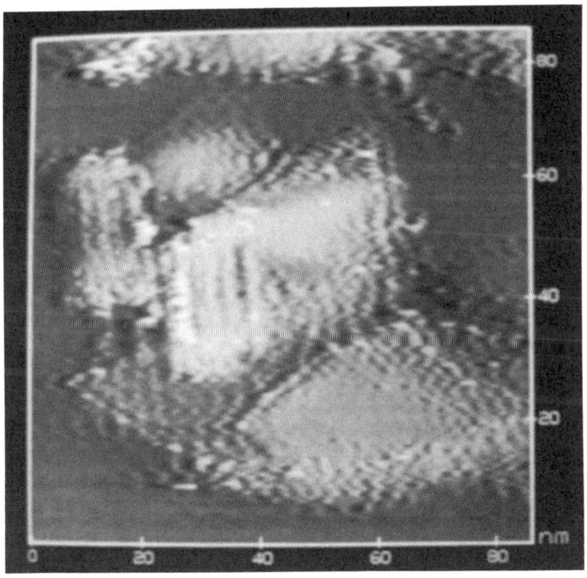

Figure 5. "Low-resolution" STM image of tubulin in zinc-induced sheet. Tubulin subunits show a periodical structure and some twisting and folding of sheets are evident. Scale shown in nm.

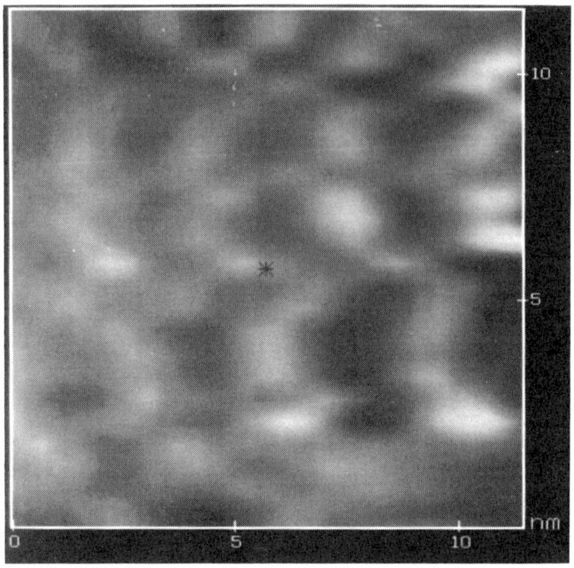

Figure 6. Higher resolution STM image of tubulin in zinc-induced sheet. Periodic 4 nm "double-crescent" configurations are seen.

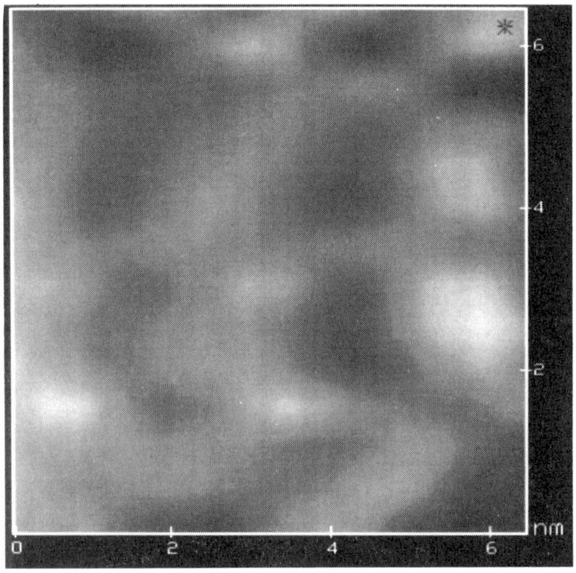

Figure 7. Highest magnification of STM image of tubulin in zinc-induced sheet. "Double-crescent" structure of 4 nm is evident.

DISCUSSION

Tubulin subunits polymerized in microtubules demonstrate a cylindrical surface of hexagonal packing and cubic packing in the axial direction of protofilaments.[20] From results of electron microscopy,[25,26] x-ray diffraction,[2] and symmetry theory[20,21] we conclude that the hexoctahedral pattern of tubulin subunits is based on the *m3m* [-*Oh*(6/4)] crystallographic group with symmetric formula $3L_4$ $4L_3$ $6L_2$ 9PC (three axes of four-fold symmetry, four axes of three-fold symmetry, six axes of two-fold symmetry, nine symmetry planes and a center of symmetry). The primary structure of the tubulin subunit is known to have about 2.5 times more negative charges than positive charges. The shape of the tubulin dimer hybrid molecular orbital is represented in Figure 8 (a).

Since the tubulin dimer requires GTP in its structure and since also in an aqueous environment there is one exchangeable GTP site on each of the α and β monomers, under the influence of water the energy state of the dimer will be characterized by the energy potential:

$$GTP^{4-} + H_2O \rightarrow GDP^{-3} + HPO_4^{2-} + H^+ \ (0.49 \text{ eV})$$

Thus two monomer subunits will share an energy potential of H^+ (0.49 eV) and the shape of molecular orbitals based on tubulin amino acids (negative: aspartic and glutamic acids, and positive: lysine, arginine and histidine) will depend on the chemical environment according to definite symmetry properties.[20,21] Under the influence of GTP, the four-fold symmetry of the *m3m* symmetry group should be replaced by the *4mm* symmetry subgroup. Consequently the surface of the tubulin dimer will have ∞m symmetry (Curie symmetry) and antisymmetry $\infty/m'm(13)$. The shape of the tubulin dimer molecular orbital under the thermodynamic influence of GTP mediated through $[H^+]$ should therefore have symmetric and antisymmetric components as shown in Figure 8 (b), and the structure of the charge density of tubulin dimers in microtubules should be as shown in Figure 8(c). The STM image shown in Figure 3 resembles Figure 8 (c) implying that the biomolecular STM image shows a hybrid of molecular orbitals based on crystallographic properties and on electronic tunneling and protonic conductivity.

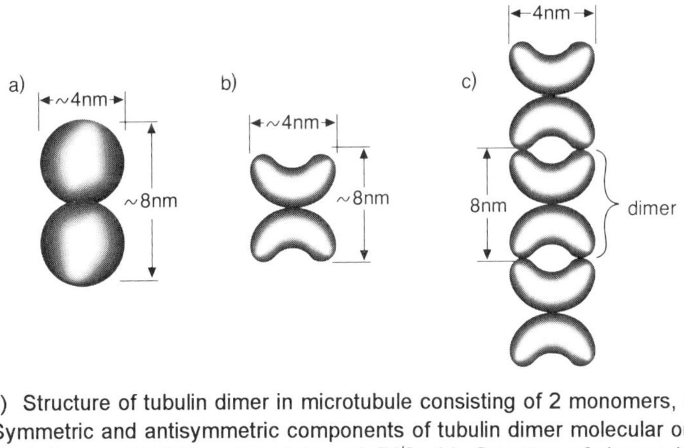

Figure 8. (a) Structure of tubulin dimer in microtubule consisting of 2 monomers, α and β tubulin. (b) Symmetric and antisymmetric components of tubulin dimer molecular orbital under thermodynamic influence of GTP mediated through $[H^+]$. (c) Structure of charge density of tubular dimers in microtubules.

The site of GTP binding is affected by tubulin's interaction with metal ions such as Mg^{++} or Zn^{++}. The zinc-induced tubulin sheets show the influence of GTP-Zn^{++}, whereas the microtubules show the influence of GTP-Mg^{++}. Figures 9 (a) and 9 (c) show how, in the zinc-induced sheets, both symmetric and antisymmetric components split into two parts, with the antisymmetric component changing into a symmetric one. The predicted shape of the tubulin dimer in Figure 9 (c) is a result of the influence of GTP and zinc on the elastic properties of tubulin dimers. The twisted shape of each dimer under the influence of zinc is as shown in Figure 9 (b) and matches images reconstructed from electron microscopic observation of Zn^{++} - tubulin[31] as shown in Figure 10.

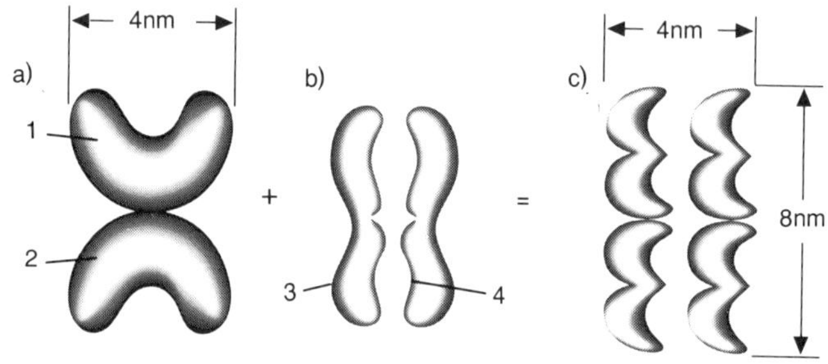

Figure 9. (a) Symmetric and antisymmetric components of tubulin. (b) Twisting of Zn++ tubulin. (c) Further reduction into symmetric and antisymmetric components due to Zn++ and GTP.

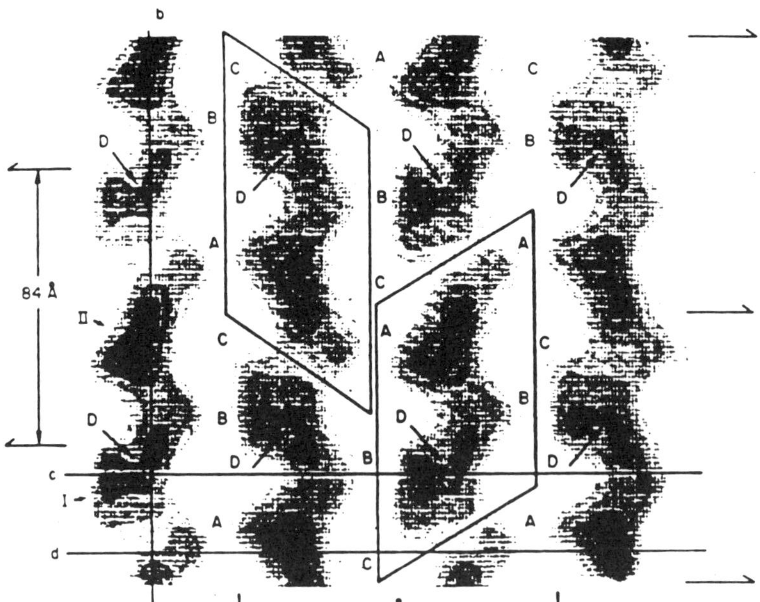

Figure 10. Projection of a planar zinc-tubulin sheet by an inverse Fourier transform based on electron microscopy data.[31] Compare with Figure 9 (c).

II. STM OF BACTERIAL LUCIFERASE--A MODEL SYSTEM FOR ANESTHESIA

Light emission by bioluminescent bacteria and fireflies is inhibited by low levels of general anesthetics.[1,8,36] Concentrations required to inhibit this bioluminescence correlate remarkably well with those needed to induce general anesthesia in animals and humans,[12,14] suggesting a common mechanism of action. This suggestion is further enhanced by the observations that both anesthetic induction in animals and inhibition of bioluminescence are reversed by pressure.[13] Isolation of the luminescent enzyme luciferase and demonstration that anesthetics inhibit luciferase luminescence in the absence of a membrane environment have been used to argue that the site of anesthetic action in the brain is protein, rather than lipid. It has been further argued that anesthetics cause a conformational change in luciferase, or prevent conformational change necessary for the protein's function. In the case of bacterial luciferase, NADH promotes activation resulting in light emission. We have used STM to image bacterial luciferase and attempt to visualize effects related to activation and anesthetic effect.

METHODS

Bacterial luciferase ("BL") from Vibrio harveyi and NADH were obtained from Sigma (St. Louis, Mo) and prepared in 50 mM phosphate buffer, pH 7.0. Four sample solutions were studied: 1) BL 0.4nM, 2) BL 0.4nM + NADH 0.2nM, 3) BL 0.4nM + ethanol 1.5M and 4) BL 0.4nM, NADH 0.2mM and ethanol 1.5M (1.5 M ethanol is the EC_{50} for *V. harveyi* BL photoemission inhibition[5]). Control solutions without BL were also studied. A drop of each solution was smeared onto a gold substrate and imaged at room temperature and atmospheric pressure with a Nanoscope II STM (Digital Instruments, Santa Barbara, CA). STM conditions are shown in the images.

RESULTS

Control images on gold (no BL) showed no structures. Figure 11 shows BL alone (solution 1). In each, a pair of structures approximately 10 nm x 5 nm with several peaks is observed. Figure 12 shows BL and NADH (solution 2). A trio of 6 nm x 2 nm structures (note different scale bars) is observed. Figure 13 shows BL and ethanol (solution 3). A 10 nm long structure without peaks is seen. Figure 14 shows BL, NADH and ethanol (solution 4). Curved structure with total length of about 10 nm is observed. Figures shown are representative of numerous, similar images in each category.

DISCUSSION

Our STM images of BL are consistent with known dimensions of the enzyme (no crystallographic data are available). NADH "activation" appears to cause contraction of BL. In the presence of ethanol, BL appears "softened." This may be denaturation, an effect on the tunneling imaging process, or a conformational change in the protein. Ethanol and NADH appear to cause a partial contraction of BL which may represent denaturing, prevention of NADH contraction or some combination.

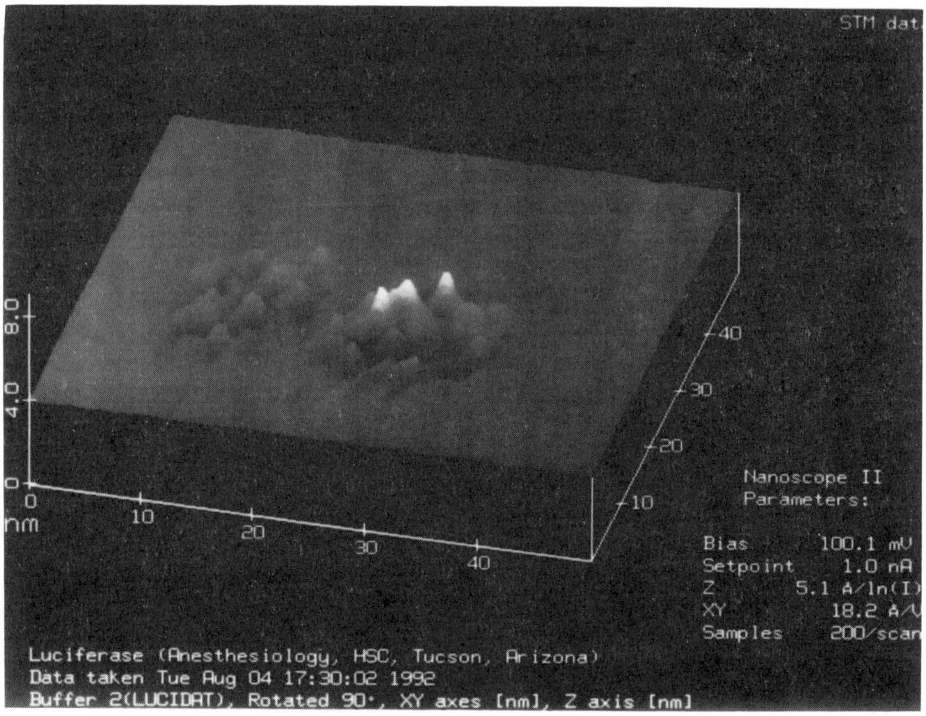

Figure 11. STM of bacterial luciferase on gold substrate (either 2 proteins or double image).

SUMMARY

We have used STM to image the surface topography of microtubules, zinc-induced tubulin monolayer sheets, and bacterial luciferase. STM images of microtubules show rows of tubulin subunit dimers. STM images of zinc-induced tubulin sheets show 5nm wide longitudinal striations comprised of 4nm periodic structures which match tubulin monomer subunits. Comparing these results with electron microscopy[31], STM images have greater detail. Comparison between STM images of tubulin in protofilaments and zinc-induced monolayer sheets suggests different conformation and/or charge distribution for these two polymeric forms of tubulin.

STM imaging of bacterial luciferase, its activation by NADH, and effects of ethanol (as a model anesthetic) suggest that NADH contracts bacterial luciferase. Ethanol may prevent the contraction, but the effects we observed may be due to denaturing or direct effects on the tunneling process.

Like other proteins, tubulin and luciferase are insulators and the mechanism of tunneling permitting resolution is unclear. This question has been evident in biomolecular studies by STM and was recently reviewed.[6] Proposed mechanisms for conduction and STM imaging in proteins include: a) surface conduction (electrons travel across protein surface via hydrated salts from buffer), b) resonant state tunneling effects,[24] c) through bond tunneling (electrons traverse the protein by electron dense bonds). Other contrast mechanisms have been postulated which include tunneling through thin specimens, tip-

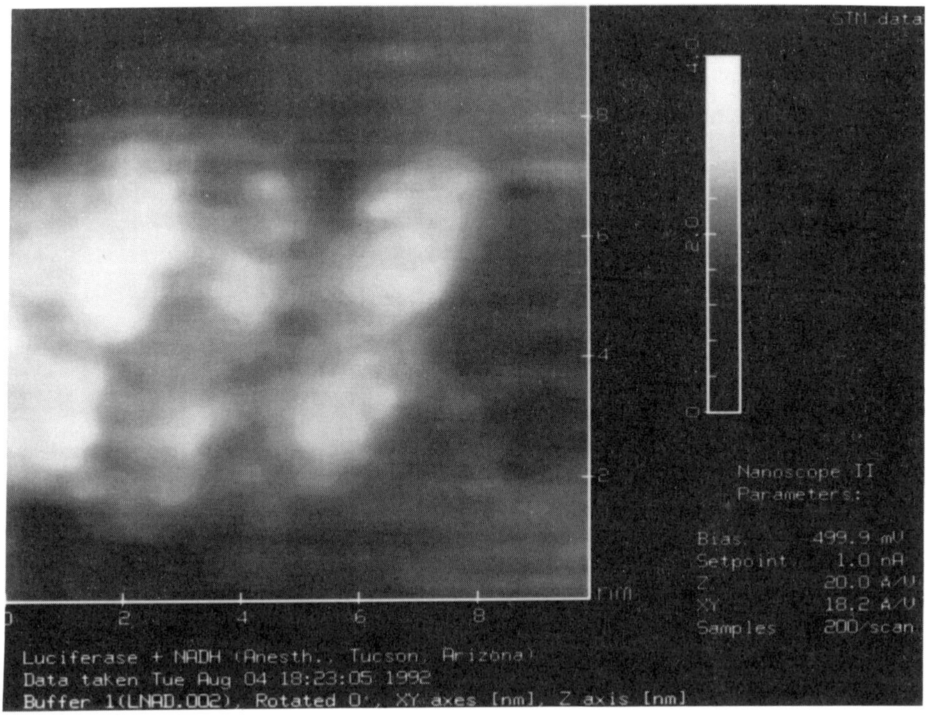

Figure 12. STM of bacterial luciferase and NADH. Contracted length, triple image compared to protein alone.

sample elastic interactions and coating of the sample by graphite from the substrate. The latter is not so surprising when it is appreciated that tip-graphite interactions can produce periodic structures, some of which have been mistaken for DNA.[7] In some studies, imaging of biomolecules has been obtained "negatively," in which an image is produced as a "hole" in the graphite. This is thought to represent insulation causing the tip to bury itself into the sample, shielding it from the graphite. Imperfect tips can also influence images obtained. Multiple tips (for example, a tip with two or more "tips") can produce multiple images. Our images of bacterial luciferase, in which pairs or trios of the protein were obtained may represent a multiple tip effect. Tips can also move and/or alter the sample during scanning; these effects can cause disturbing artifacts but have also been proposed to permit (when reproducibly controllable) future modification and manipulation of biomolecules.

In the present study, changing conditions may also alter image quality. For example, the ethanol images of luciferase often have a distinct "softened" quality that may be due to the effects of the ethanol on the tunneling process rather than on the protein itself. However, this effect could also be relevant to anesthetic mechanism. Anesthetics have been shown to inhibit electron mobility,[16] and such inhibition, for example by affecting intra-protein resonant electron states linked to conformational dynamics, may be important for anesthetic effects.

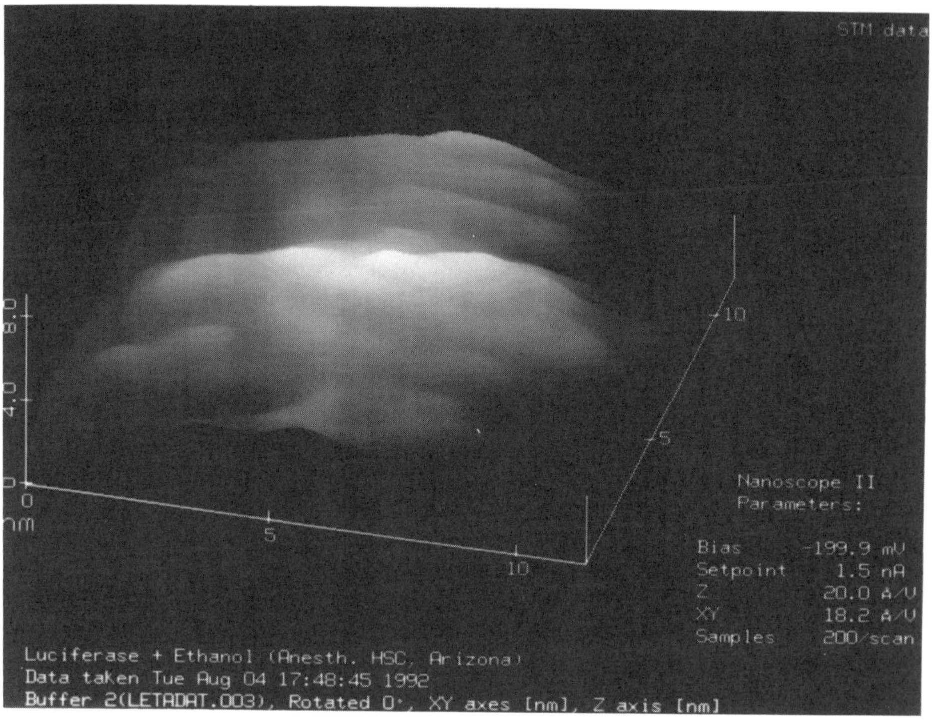

Figure 13. STM of bacterial luciferase and ethanol. Elongated, smoothed image of proteins is seen.

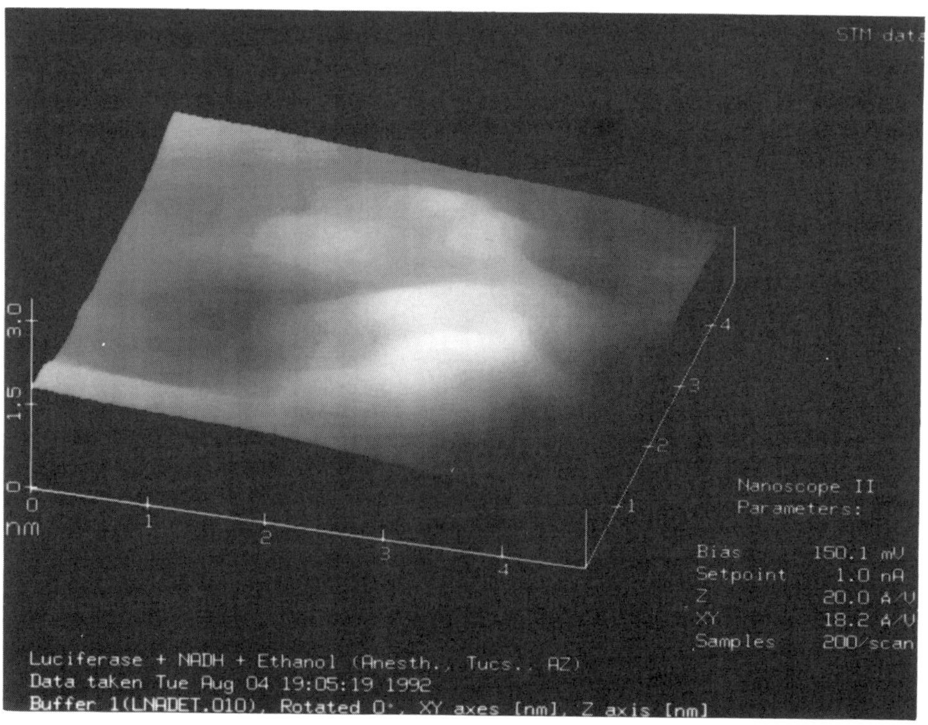

Figure 14. STM of bacterial luciferase, NADH, and ethanol shows curved structures with total length of about 10 nm.

ACKNOWLEDGMENTS

The authors are indebted to Lawrence Vernetti, Virgil Eling, Jay Gandolfi, and Dror Sarid for collaboration and to Richard Hofstad and Daphne Anthony for technical support. S.R. Hameroff is partially supported by a National Science Foundation grant.

REFERENCES

1. G. Adey, B. Wardley-Smith, D. White, Mechanism of inhibition of bacterial luciferase by anaesthetic, *Life Sci.*, 17:1849-1854 (1975).
2. L.A. Amos, A. Klug, Arrangement of subunits in flagellar microtubules, *J. Cell Sci.*, 14:523-550 (1974).
3. M. Amrein, R. Durr, A. Stasiak, H. Gross, G. Travaglini, Scanning tunneling microscopy of uncoated rec-DNA complexes, *Science*, 243:1709-1710 (1989).
4. A.M. Baro, R. Miranda, J. Alaman, N. Garcia, G. Binning, H. Rohrer, C. Berger, J.L. Carraseosa, Determination of surface topography of biological specimens at high resolution by scanning tunneling microscopy, *Nature*, 315:253-255 (1985).
5. T.P. Beebe, T.E. Wilson, F. Ogletree, J.E. Katz, R. Balhorn, M.B. Salmeron, W.J. Siekhaus, Direct observation of native DNA structures with the scanning tunneling microscope, Science, 243:370-372 (1989).
6. B.L. Blackford, M.H. Jericho, P.J. Mulhern, A review of scanning tunneling microscope atomic force microscope imaging of large biological structures: problems and prospects, *Scanning Microscopy*, 5:970-918 (1991).
7. C.R. Clemmer, T.P. Beebe, Jr., Graphite: a mimic for DNA and other biomolecules in scanning tunneling microscope studies, *Science*, 251:640-642 (1991).
8. S. Curry, W.R. Lieb, N.P. Franks, Effects of general anesthetics on the bacterial luciferase enzyme from Vibrio harveyi: an anesthetic target site with differential sensitivity, Biochem., 29:4641-4652 (1990).
9. D.C. Dahn, M.O. Watanabe, B.L. Blackford, M.H. Jericho, T.J. Beveridge. Scanning tunneling microscopy imaging of biological structures, *J. Vac. Sci. Technol. A.*, 6:548-552 (1988).
10. B. Drake, C.B. Prater, A.L. Weisenhorn, S.A.C. Gould, T.R. Albrecht, C.F. Quate, D.S. Cannell, H.G. Hansma, P.K. Hansma, Imaging crystals, polymers, and processes in water with the atomic force microscope, *Science*, 243:1586-1589 (1989).
11. P. Dustin, "Microtubules," Springer-Verlag, Berlin (1984).
12. N.P. Franks, W.R. Lieb, Do general anaesthetics act by competitive binding to specific receptors? *Nature*, 310:599-601 (1984).
13. N.P. Franks, W.R. Lieb, Mapping of general anaesthetic target sites provides a molecular basis for cutoff effects, *Nature*, 316:349-351 (1985).
14. N.P. Franks, W.R. Lieb, Molecular mechanisms of general anaesthesia, *Nature*, 300:487-493 (1982).
15. S. Hameroff, Y. Simic-Krstic, L. Vernetti, Y.C. Lee, S. Dror. J. Weidmann, V. Elings, K. Kjoller, R. McCuskey Scanning tunneling microscopy of cytoskeletal proteins: microtubules and intermediate filaments, *J. Vac. Sci. Technol.*, A8:687-691(1990).
16. S. R. Hameroff, R.C. Watt, Do anesthetics act by altering electron mobility? *Anesth. Analg.*, 62:936-940 (1983).
17. P.K. Hansma, V.B. Elings, O. Marti, C.E. Bracker, Scanning tunneling microscopy and atomic force microscopy: application to biology and technology, *Science*, 242:209-216 (1988).
18. P. K. Hansma, J. Tersoff, Scanning tunneling microscopy, *J. Appl. Phys.*, 61:R1-R23 (1987).
19. J.K.H. Horber, C.A. Lang, T.W. Hansch, W.M. Heckl, H. Mohwald. Scanning tunneling microscopy of lipid films and embedded biomolecules, *Chem. Phys. Lett.* 145:151-158 (1988).
20. D. Koruga, Neuromolecular computing, *Nanobiology*, 1(1):1-21 (1992).
21. D. Koruga, S. Hameroff, R. Loutfy, M. Sundareshan, "Fullerene C_{60} - From Nanobiology to Nanotechnology," Elsevier, Amsterdam (1993).
22. H. Larsson, M. Wallin, A. Edstrom, Induction of a sheet polymer of tubulin by Zn^{++}, *Exp. Cell Res.*, 100:104-108 (1976).

23. S.M. Lindsay, B. Barris, Imaging deoxyribose nucleic acid molecules on a metal surface under water by scanning tunneling microscopy, *J. Vac. Sci. Technol.* A6:544-547 (1988).
24. S.M. Lindsay, O.F. Sankey, Y. Li, C. Herbst, A. Rupprecht, Pressure and resonance effects in scanning tunneling microscopy of molecular adsorbates, *J. Phys.Chem.*, 94:4655-4659 (1990).
25. E. Mandelkow, E.M. Mandelkow, Image reconstruction of tubulin hoops, *J. Ultrastruct. Res.*, 74:11-33 (1981).
26. E. Mandelkow, J. Thomas, C. Cohen, Microtubule structure at low resolution by x-ray diffraction, *Proc. Natl. Acad. Sci. USA*, 79:3370-3374 (1977).
27. O. Marti, H.O. Ribi, B. Drake, T.R. Albrecht, C.F. Quate, P.K. Hansma, Atomic force microscopy of an organic monolayer, *Science*, 239:50-52 (1988).
28. A.J. Middleton, E.B. Smith, General anaesthetics and bacterial luminescence. II. The effect of diethyl ether on the in vitro light emission of Vibrio fischeri, *Proc. R. Soc. Lond. B.*, 193:173-190 (1976).
29. Y. Simic-Krstic, M. Kelley, C. Schneiker, M. Krasovich, R. McCuskey, D. Koruga, S. Hameroff, Direct observation of microtubules with the scanning tunneling microscope, *FASEB J.*, 3:2184-2189 (1989).
30. M.L. Shelanski, F. Gaskin, C.R. Kantor, Microtubule assembly in the absence of added nucleotides, *Proc. Natl. Acad. Sci. USA*, 70:765-769 (1973).
31. L.K. Tamm, R.H. Crepeau, S.J. Edelstein, Three-dimensional reconstruction of tubulin in zinc-induced sheets, *J. Mol. Biol.*, 130:473-492 (1979).
32. G. Travaglini, H. Rohrer, M. Amrein, H. Gross, Scanning tunneling microscopy on biological matter, *Surface Sci.*, 181:380-390 (1987).
33. L.A. Vernetti, C.L.A. Nowlin, S.R. Hameroff, A.J. Gandolfi, Y.C. Lee, D. Sarid D, STM resolution of surface features on cytokeratin protein is enhanced by prolonged exposure of protein to cold temperatures, *J. Vac. Sci. Technol.* B9:1223-1226 (1991).
34. M.A. Voelker, S.R. Hameroff, J.D. He, E.L. Dereniak, R.S. McCuskey, C.W. Schneiker, T.A. Chvapil, L.S. Bell, L.B. Weiss, STM imaging of molecular collagen and phospholipid membranes, *J. Micros.* 152:557-566 (1988).
35. H. J. Woodland, C.J. Potrikus, S.C. Gupta, M. Kurfurst, J.C. Makemson, Biochemistry and physiology of bioluminescent bacteria, *Adv. Microb. Physiol.*, 26:235-289 (1985).
36. D.C. White, B. Wardley-Smith, G. Adey, The site of action of anaesthetics on bacterial luminescence, *Life Sci.*, 12(II):453-461 (1973).
37. J.A.N. Zasadzinski, J. Schneir, J. Gurley, V.B. Elings, P.K. Hansma, Scanning tunneling microscopy of freeze fracture replicas of biomembranes, *Science*, 239:1013-1015 (1988).
38. M.M. Ziegler, T.O. Baldwin, Biochemistry of bacterial bioluminescence, *Curr. Top. Bioenerget.*, 12:65-113 (1981).

SCALE-AREA ANALYSIS OF SCANNING TUNNELING MICROSCOPY/ATOMIC FORCE MICROSCOPY DATA BY THE PATCHWORK METHOD

Christopher A. Brown and E. Michael Shipulski

Mechanical Engineering Department
Manufacturing Engineering Program
Worcester Polytechnic Institute
Worcester, MA 01609-2280

Abstract: A new method of analyzing large, topographic data sets, such as those generated by scanning probe microscopy (e.g., STM, AFM) is described. The method uses triangular patches to calculate areas of the topographic data set as a function of patch areas, i.e., as a function of scale. This patchwork method is analogous to the coastline method of fractal analysis for profiles, however, because the calculations are based on areas rather than profiles or coastlines, the patchwork method makes use of the information inherent in the proximity of the individual scans. This method develops characterizations for the data sets that have clear applications in engineering design and research. It has the potential for elucidating mechanisms of interactions with surfaces, analyzing surface creation and wear mechanisms, and for designing surface topographies and the manufacturing processes to produce them.

Using the patchwork method, one can identify the smooth-rough crossover in scale which separates the relatively large scales where the surface is smooth and best described by Euclidean geometry, from those smaller scales where the surface is rough and best described by fractal geometry. One also obtains parameters that characterize the intricacies, or complexities, of the topography below the smooth-rough crossover, and the scales over which different levels of geometric complexity exist. These complexity parameters can be related to fractal dimensions; more than one fractal dimension for a topographic data set indicates that the geometry has a multifractal character.

The surface areas, calculated as a function of scale from the topographic data sets, can also be used directly to determine the strengths of interactions with the surface. To determine the strength of interaction (e.g., adhesion) the interaction is modeled as a collection of discrete interactions (e.g., adhesive bonds) of determinable size and strength. The strength of the interaction is then determined from the surface area calculated at the size, or scale, of the interaction.

Atomic Force Microscopy/Scanning Tunneling Microscopy
Edited by S.H. Cohen *et al.*, Plenum Press, New York, 1994

INTRODUCTION

Our objective is to present a new method for characterizing large, topographic data sets, such as those acquired by scanning tunneling or atomic force microscopy (STM or AFM). This new method, called the patchwork method, has been developed to reduce these large topographic data sets to a few parameters, which are indicative of the essence of the topography responsible for its behavior and of the process which created it. The patchwork method has clear physical interpretations for supporting product and process design.

The technologies associated with STM and AFM make it possible to acquire large topographic data sets at scales comparable to or finer than the scales of interaction for topographically dependent phenomena of engineering interest. These instruments are capable of producing images of surface topographies at scales as small as atomic. The images are important aids in developing an understanding of the topographic dependence of interactions with surfaces. Still, it is desirable from the perspective of surface engineering to be able to reduce these large data sets to a small number of parameters for specific engineering applications. Characterization of the geometry of a surface, so that it can be used for rational design of topographies and the processes to create them, is a central problem in surface engineering.

Many current methods of characterizing topographies, even those acquired by STM or AFM, are based on techniques for characterizing profiles acquired by contact stylus profilometry.[1] In a profile the height, z, is a function of position along a line, i.e., $z = z(x)$. These early techniques were developed within the technical expertise of the 1940s, and the limitations that are implied. They were developed with the limitations of analog electronics, without the benefit of high-speed digital computers, and with the technically necessary assumption that the profile is indicative of a complete topographic data set ($z = z(x,y)$). These early techniques were also developed without an understanding of how to characterize the intricacy or complexity of a random or chaotic data set. These limitations may be associated with the difficulties in finding correlations between surface topography and surface behavior. Conventional analysis methods have not been good at correlating profilometry data with surface behavior, except under limited conditions.[2]

The development of fractal geometry has suggested methods of characterizing data sets for their complexity or intricacy.[3-5] Several approaches for applying fractal geometry to profilometry data and finding the fractal dimension have been described.[6-8] Fractal analysis, by capturing the intricacy of the geometry in the characterization parameters, appears to have the potential to yield characterization parameters that will correlate well with surface behavior that relies on intricacy.

The approach in this paper is based on fractal geometry, although it differs from the bulk of the work done to date on characterizing topographies using fractal geometry. The methods reported in the literature tend to concentrate on identifying the fractal dimension of the data set. While the fractal dimension is an accurate representation of the complexity of the data set, it lacks a clear physical interpretation for engineering applications.

The approach reported here is based on finding scale-area relations, i.e., how the surface area of a topographic data set changes with the scale of measurement. The scale-area relations are essential to estimate the strength of interactions with the surface, once the scale of interaction is known. Additionally, the scale-area relations identify the crossover scale, between smooth and rough character of the surface. Interactions with the surface at scales greater than the crossover interact with the surface as if it were smooth; at scales below the crossover the interactions are with a rough surface. The parameters derived from the scale-area relations have clear physical interpretations for engineering applications.

METHOD

The method of determining the scale-area relations is based on determining the surface areas at a number of scales of measurement. It is accomplished by repeatedly tiling over a surface, as represented by a topographic data set ($z = z(x,y)$), with triangles of different areas in a patchwork fashion. For one coverage, or tiling, the areas of the triangles, in three dimensions, are kept strictly the same. The area of the triangle represents the scale of measurement. The corners of the triangles are fixed on topographic data points or on linear interpolations connecting the points. Only complete triangles are fit on the surface, which leaves some topographic data points on the edges of the data set, which are not covered. The measured area of the surface, at a particular scale, is the product of the number of triangles needed to tile over the surface times the area of the triangle. The projected area of the triangles is noted and is divided into the measured area to determine the relative area, which is used in the characterizations.[9]

In such a triangular coverage exercise there are four aspects which might be controlled: the area of the covering triangles, the shape of the triangles, the orientation of the triangles and the percent coverage. All four of these coverage aspects cannot be narrowly controlled simultaneously. In this exercise we have controlled the triangle area precisely, as it represents the scale. The coverage is complete, one hundred percent, except for two edges of the data set where there is insufficient surface for a full triangle. The orientation is controlled by aligning the bases of the triangles along data rows. The shape of the triangles is only controlled loosely, so that they do not become too obtuse. With this strategy for controlling the coverage aspects, this method fulfills the requirement of measuring surface area with respect to the scale of measurement, and it can detect to texture, or special orientations in the surface roughness.

To determine the characterization parameters, the logs of the relative areas are plotted versus the logs of the triangle areas. Two kinds of topographic characterization parameters can be determined from the scale-area plots. One kind of parameter, the slopes, are indicative of the complexity of the surface geometry and can be related to the fractal dimension.[3,4] Different regions of scale may have different slopes, which is to say that the surface can have a multifractal character. The other kind of parameter, the crossover scales, are the scales, i.e., triangle areas, which define the limits of regions of approximately constant slope. These scale-area plots can be used directly to find the relative area as a function of scale of measurement.

An important crossover scale for many engineering applications is the smooth-rough crossover, which defines the limits of the smooth character which most surfaces of engineering interest exhibit at sufficiently large scales. At these large scales the triangles are too large to be sensitive to the topographic details. In this method a threshold in relative areas is set to determine the smooth-rough crossover scale. Starting from the largest triangles and decreasing, the first triangle area which results in a relative area above a predetermined threshold is the smooth-rough crossover scale.

In the present work the relative area threshold for determining the smooth-rough crossover is based on the maximum area of the topographic data set. The maximum area is found by adding the areas of triangles formed by adjacent points on the grid. Triangles formed this way do not have constant areas, so this maximum area is not included on the scale-area plot. The threshold is one plus ten percent of the quantity: maximum area minus the projected area, divided by the projected area. The method of determining the threshold can be adjusted for the specific application.

Slopes of the scale-area plots can be determined in several ways. In this work a best-fit algorithm is used. The slope of the two decade region in triangle sizes with the greatest slope is measured. Alternatively, the size of the region could be adjusted or the region of best fit could be used.

The patchwork method for determining scale-area relations from the topographic data set is performed using a specially developed software (SURFRAX, Burlington Computer Systems, Waterbury Center, VT 05677). It can be applied to any data set that contains heights as a function of position on a regular grid, such as geographic data sets, or sets acquired by scanning profilometry.

The method has been applied to a data set from a diamond coating on a silicon substrate (from the Research Triangle Institute, Charlotte, North Carolina). The topographic data set was acquired by scanning tunneling microscopy on a Nanoscope II at the Precision Engineering Laboratory, University of North Carolina, Charlotte.

The data set consists of altitudes on a 400 x 400 grid (160,000 points), over a surface 10,000 nm x 10,000 nm, resulting in a point spacing of 25 nm. The smallest triangle that is used is the square of the point spacing divided by two, or about 312 nm^2. This patch size represents the smallest areal scale which still contains topographic information for this data set.

RESULTS

Examples of triangular patchwork coverage exercises, with progressively smaller triangles, are shown in Figure 1. The nature of the topography of the diamond coating on the silicon coating can be discerned from these figures.

The area of the triangles, used in the coverage shown in Figure 1 (a), is slightly below the smooth-rough crossover, i.e., just below the point where the triangles are small enough to become sensitive to the topographic details. In Figure 1 (b) the triangle area is near the center of the two-decade region of greatest slope, which corresponds to the region of greatest complexity. In Figure 1 (b) the chaotic nature of the topography at this scale is evident. In Figure 1 (c) the smallest scales in the region of greatest complexity are being approached, and more topographic detail can be discerned here than in than in Figure 1 (b). The triangle area in Figure 1 (d) is smaller than that at the lower end of the two-decade region of greatest complexity. Comparing Figure 1 (d) with Figure 1 (c) reveals a smaller change in topography than comparing Figures 1(b) and 1(c). Overall, the nature of the surface changes from smooth at the large scales, to rough with a relatively high degree of complexity at intermediate scales, to rough with a relatively low degree of complexity at the finer scales.

These changes in the nature, or character, of the surface roughness are evident in the STM images (not shown). At the intermediate scales, a collection of pyramidal structures, typical of discrete nucleation and growth phenomena for diamond coatings, are evident. At the finest scales the complexity of the geometry decreases as there are few topographic features on the facets of the pyramids.

The log-log scale-area plot for the diamond surface is shown in Figure 2. At the large scales, down to approximately 4 μm^2 the character of the surface is essentially smooth. The smooth-rough crossover, as determined by the relative area threshold described above, is 1.97 μm^2. In the two decade region of greatest geometric intricacy, which extends from 1.63 μm^2 to 16,300 nm^2, the slope is 0.0722 μm^{-2}. Below this it is seen that the surface is less intricate, in that the relative area increases less rapidly, with respect to the triangular patch area, than it did at slightly larger scales. At the finest scale at which the data is meaningful, about 312 nm^2, the relative area is the largest observed, 1.57.

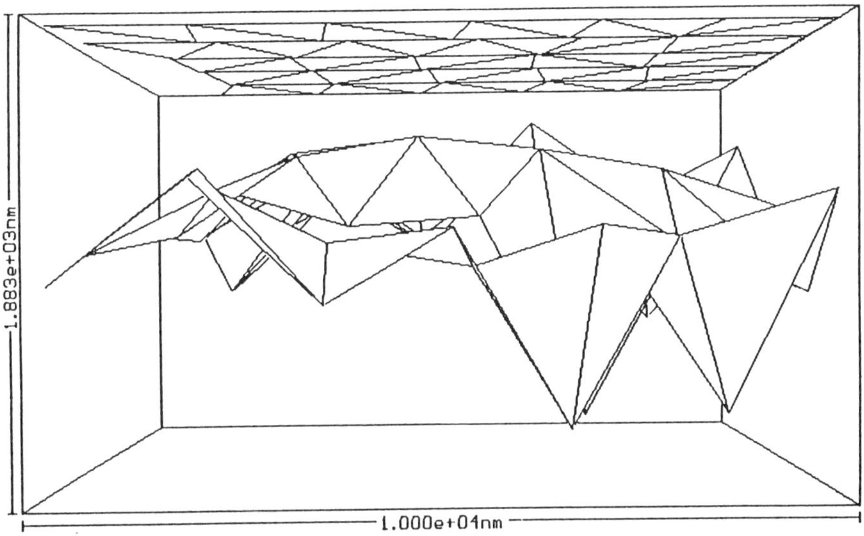

Figure 1. (a)-(d) Representations of approximations of the topography of a diamond coating using triangular patches are shown in the middle of the perspective boxes. The projected area of the surface covered by the triangular patches is shown on the ceiling of the boxes. The fitting, or tiling, exercise used in the approximations begins in the back right corner of the boxes. Note that the magnification in the vertical direction is greater than that in the horizontal directions.

Figure 1. (a) There are 52 triangular patches with areas of 1.63 μm^2 measuring an area 1.056 times larger than the projected area (i.e., the relative area).

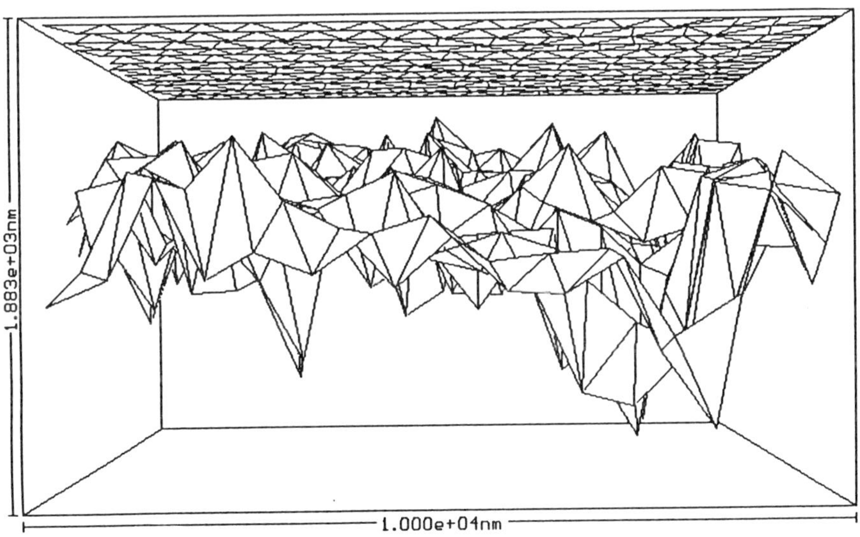

Figure 1. (b) There are 361 triangular patches with areas of 295,000 nm^2 measuring an area 1.174 times larger than the projected area.

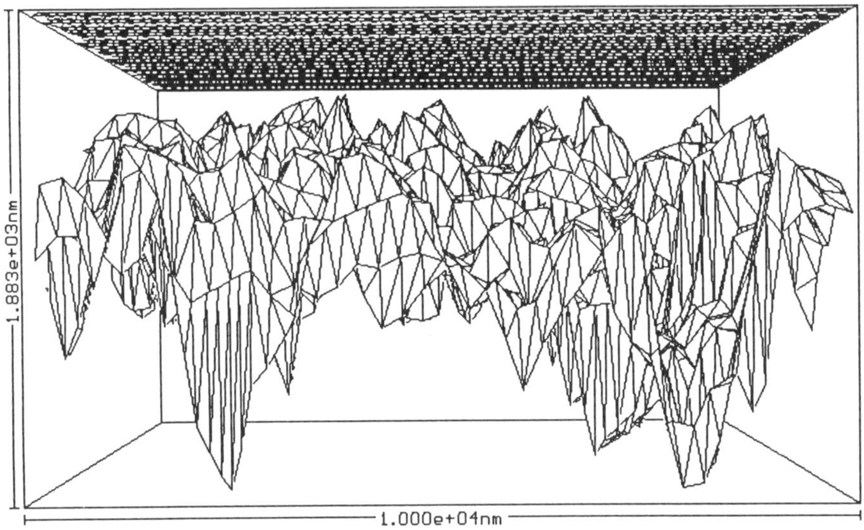

Figure 1. (c) There are 2,408 triangular patches with areas of 53,300 nm^2 measuring an area 1.331 times larger than the projected area.

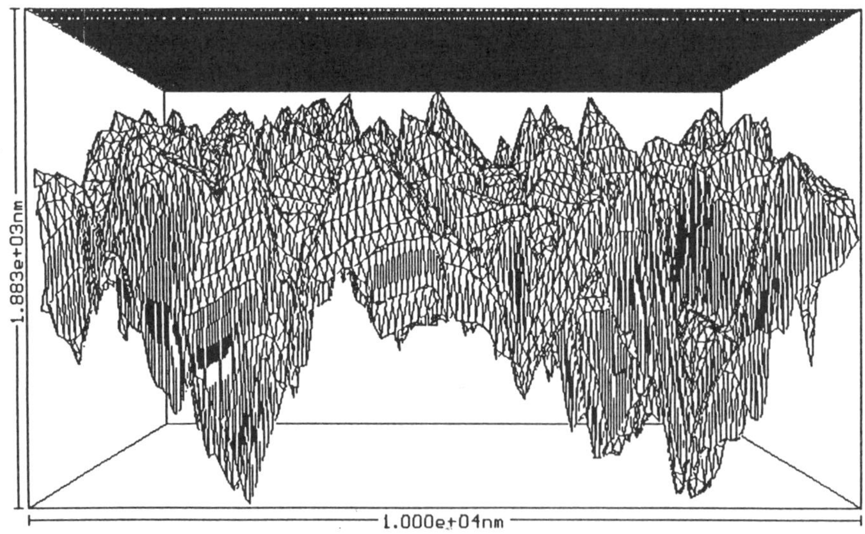

Figure 1. (d) There are 14,988 triangular patches with areas of 9,620 nm^2 measuring an area 1.461 times larger than the projected area.

DISCUSSION

There are a variety of physical interpretations which can be used for different aspects of the scale-area plots, some of them are discussed here. The interpretations are in the

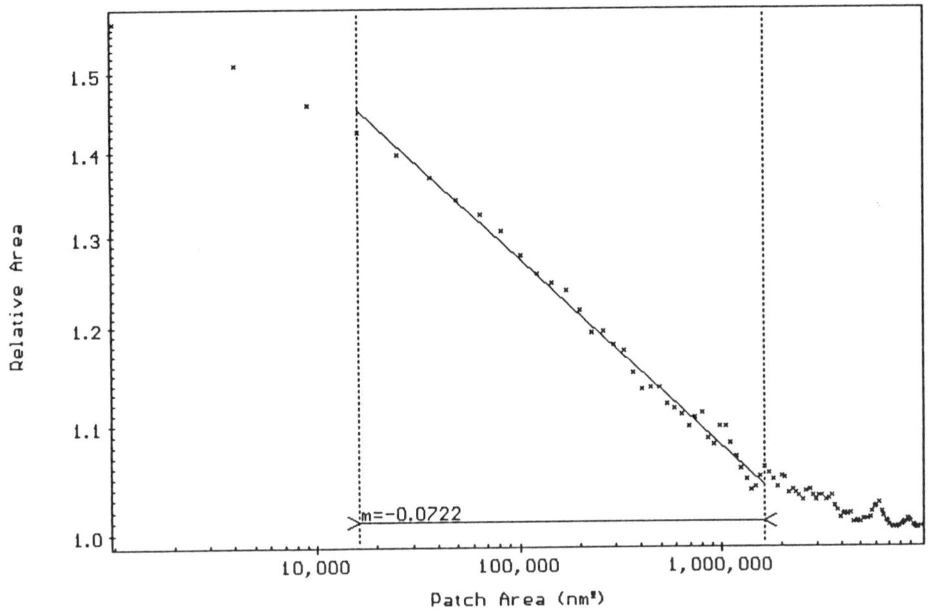

Figure 2. Scale-area plot, with log-log scales, shows the relative areas as a function of scales, or triangular patch areas. The slope of the indicated, two-decade region of greatest geometric complexity, or intricacy, extending from 16,320 nm^2 to 1.632 nm^2 is -0.0722. The smooth-rough crossover scale is 1.97 μm^2.

context of understanding topographically dependent interactions with surfaces in terms of the scale of the interaction. This understanding is applicable insofar as topographically dependent phenomena can be modeled as a collection of discrete interactions, each having a finite strength of interaction and occurring over a finite area.

An essential element in characterizing topographically dependent phenomena is the appropriate scales of interaction. For a diamond coating, one scale of interaction is related to the spacing of the nucleation sites. For a powder coating, a scale of interaction is the powder diameter. For machining, one scale of interaction is the feed. The scale-of-interaction concept can be applied to the processes for the forming, or manufacturing, of the surface, as well as to surface behavior.

The parameter with the most straightforward physical interpretation, in terms of the scale-of-interaction concept, is probably the smooth-rough crossover. Topographically dependent phenomena that interact with the surface at scales above the smooth-rough crossover interact with a surface they see as being smooth, whereas, if the scales of interaction are below the smooth-rough crossover, then the roughness of the surface plays a role in the interaction.

In the case of the diamond coating examined in this paper, the smooth-rough crossover scale is 1.97 μm^2. It is significantly smaller than feeds typically used in machining, so that, from the perspective of the feed per revolution, the roughness of this coating would not interfere with the machining operation. There are other, finer, scales of interaction in machining that might be effected, however, such as those associated with chip formation. This crossover scale is probably related to the spacing of the nucleation sites. If each nucleation site produced one of the pyramidal shapes observed on the surface, then, when the triangular patch area approaches the spacing between the peaks of the pyramids, they will start falling into the troughs between the peaks. The scale at which the triangles start falling into the troughs is the crossover scale.

For some applications a surface must satisfy requirements that require it to be rough and smooth simultaneously. A hard disk for data storage is one such example: it must be smooth for a flying read-write head, to avoid "head crash" and loss of data,[10] and rough for a resting head to avoid start-up stiction.[11] Using the crossover scale, it would be possible to design a surface to physically integrate and functionally separate these two requirements on the same surface, provided the scales of interaction are known, and provided the stiction interaction occurs at a finer scale than the flying interaction. If this is the case, then it suffices to locate the crossover at a scale intermediate to the two scales of interaction.

The other type of characterization parameter, the slope, which can be related to the fractal dimension, can be used by itself or in combination with the crossover type parameters. By itself it is an indication of complexity. Using it with the crossover parameters adds the sophistication of attributing different complexities to different scales. The greatest complexity for the diamond coating is in the two-decade region beginning at 9,120 nm^2. This region corresponds to the scales of the major topographic features, i.e., the pyramidal diamond crystals. The surface is less intricate at finer scales because there are no significant pyramidal crystals below this scale and because topographic features on the facets of the pyramids are relatively few and small.

Relations with physical aspects of the surface geometry for this type of scale-area fractal characterization are clear. This is largely due to combining the notion of crossover scale with regions of complexity. This is not the case in those fractal characterizations that only determine the fractal dimension[8] or are concerned with determining self affinity.[12]

CONCLUSION

Scale-area analysis by the patchwork method has been described and applied to a STM data set acquired from a diamond coating. Relations between parameters derived from this analysis and physical attributes of the surface have been discussed. The analysis method has been shown have strong potential for applications in the engineering design of surfaces' topographies and the processes to create them.

ACKNOWLEDGMENTS

The authors are grateful to Robert Honken and Jim Miller of the Precision Engineering Laboratory, University of North Carolina, Charlotte who provided the topographic data set. We also acknowledge the assistance of the following undergraduate student programmers Pat Charles (Worcester Polytechnic Institute '92), Bill Johnsen (Colorado School of Mines '93), and Bruce Meacham (Worcester Polytechnic Institute '96).

REFERENCES

1. L. Mummery, "Surface Texture Analysis The Handbook," Hommelwerke, Schnurr Druck, West Germany 23-59 (1990).
2. M. Cailler, A. Lahmar, G.H. Lee, Adhesion studies of magnetron-sputtered copper films on chemically etched nickel substrates: effects of the concentration and the temperature of the bath, *Thin Solid Films*, 182:167-184 (1989).
3. B.B. Mandelbrot, Les Objects Fractal, l'Imprimerie Chirat, Saint-Just-la-Pendue (1975).
4. B.B. Mandelbrot, "Fractals Form, Chance, and Dimension," W.H. Freeman and Co., San Francisco (1977).
5. B.B. Mandelbrot, D.E. Passoja, A.J. Paullay, Fractal character of fracture surfaces of metals, *Nature* 308:721-722 (1984).
6. J.J. Gagnepain, C. Roques-Carmes, Fractal approach to two-dimensional and three-dimensional surface roughness, *Wear*, 109:119-126 (1986).
7. E.E. Underwood, K. Banerji, Fractals in fractography, *Mater. Sci. Eng.* 80:1-14 (1986).
8. C. Tricot, J.F. Quiniou, D. Wehbi, C. Roques-Carmes, B. Dubuc, Evaluation de la fractale d'un graphe, *Review Phys. App.* 23:111-124 (1988).
9. C.A. Brown, P.D. Charles, W.A. Johnsen, S. Chesters, Fractal Analysis of topographic data by the patchwork method, *Wear*, 161:61-67 (1993).
10. S. Suzuki, F.E. Kennedy, Friction and temperature at head-disk interface in contact start/stop tests, Tribology and Mechanics of Magnetic Storage Systems. V:30-36 (1988).
11. T.C. Tsai, Y. Mehmandoust, H. Samani, A. Eltoukhy, Tribological phenomena at the head-disk interface of thin-film - rigid disks, *J. Vac. Sci. Technol.* A7:2491 - 2495 (1989).
12. B.B. Mandelbrot, Self-affine fractals and fractal dimension, *Phys. Scr.* 32:257-260 (1985).

IMAGING MATRIX MATERIALS AND FUNDAMENTAL LAMELLAE STRUCTURE OF BIOGENIC ARAGONITE

R.W. Gauldie, G. Raina, S.K. Sharma, and I.F. West

Hawaii Institute of Geophysics
School of Ocean and Earth Science and Technology
University of Hawaii, Honolulu, HI 96822

Abstract: Atomic force microscopy (AFM) has been used to image lamellae at the nanometer level of resolution in broken sections of otoliths of snapper *(Pagrus major)*, hake *(Macruronus novaezelandiae)* and salmon *(Oncorhynchus tshawytscha)* as well as in abiogenic aragonite. In abiogenic aragonite the mean lamellae width is found to be 30 nm. Comparison of the values of lamellae widths obtained from the published literature showed no significant difference between the lamellae widths measured from atomic force and transmission electron microscopy of otoliths of hake *(Macruronus novaezelandiae)*. There are, however, statistically significant differences in lamellae widths between species, that is, the mean lamellae width is in the order of snapper (80 nm) < hake (90 nm) < salmon (130 nm).

INTRODUCTION

Otoliths are calcium carbonate concretions found in each of the pair of endolymphatic sacs of most teleost fishes.[1] There are three otoliths in each sac and they are involved in the physiology of orientation and hearing.[2,3] Once deposited, the calcium carbonate of the otolith is resorbed only in extreme stress.[4] The largest of the three otoliths, the sagitta, has a species-specific shape that is used in taxonomic classification of fishes at the level of species[5] and genus.[6] Otoliths have a complex crystalline structure that results in different patterns of both internal growth fields[7-10] and external growth fields.[11] The patterns of increments within the growth axes of otoliths form the basis of various conventions used in fish age estimation.[12]

The crystals in otoliths commonly occur in the aragonite form of calcium carbonate. Other forms such as calcite[13-15] and vaterite;[13,16] occur as partial or total replacements of aragonite in some otoliths, but such otoliths are uncommon, occurring less frequently than

1 in 100 otoliths. The aragonite crystals in otoliths are twinned[17] as is aragonite in some molluscs.[18]

In this paper we describe the use of atomic force microscopy (AFM) to examine the crystal structure of the otolith as well as the orientation and crystallographic relationships of the mineral and matrix material of the sulcal growth field of the sagittae of three different species of teleosts and to compare otolith crystal structure with inorganic aragonite. The AFM data are then compared with data drawn from transmission electron microscopy (TEM) published in the literature.

MATERIALS AND METHODS

The otoliths used for this study were sagittae taken from snapper (*Pagrus major* (Sparidae) formerly *Chrysophyrs auratus*) collected by RWG in the Bay of Plenty, New Zealand; hake (*Macruronus novaezelandiae* (Merlucciidae)) collected by RWG from the Chatham Rise, New Zealand; juvenile salmon (*Oncorhynchus tshawytscha* (Salmonidae)) taken from a fish-rearing experiment described earlier.[9] In the text sagittae are referred to as otoliths.

The macroscopic crystal structure of the otoliths of snapper has been described at the macroscopic level.[19-21] The otoliths of juvenile salmon have been previously described.[9] The otoliths of hake have also been described in 1993.[22] Although these four species are taxonomically unrelated, their otoliths share a division into two crystalline growth fields, anti-sulcal and sulcal common to many otoliths.[8] This study is restricted to crystals from the sulcal growth field where the opportunity for mistaking true microincrements from crystal lamellae appears to be greatest.[21] Crystalline aragonite was collected from the Minglanilla aragonite facies at Cuenca in Spain in the form of a large (3 cm wide) crystal with exfoliated surface crystals up to 5 mm long. Fragments of the surface crystals were detached and used in this study.

Otoliths and the aragonite crystal were prepared for study in an atomic force microscope (AFM) by shattering the specimens in a plastic bag and then recovering an appropriate piece from the broken, but still loosely in-shape, sample. Samples were glued onto a metal disk which was held down magnetically. The sample was not treated in any way. The loading force applied to the sample at the probe tip was about 10^{-8} newtons. Similar procedures have been used in AFM studies of calcite as well as molluscan and echinoderm material.[22-23] Consequently the AFM method is sensitive to artifacts introduced as regular patterns that can mimic the expected surface pattern.[24-26] Such artifacts are avoided by rotating the target and rescanning from another angle. Target rotation was routinely carried out in all AFM imaging in this study.

RESULTS

AFM images of the surface of the broken aragonite crystal showed lamellae (Figure 1). The aragonite crystal does not show good cleavage planes that facilitate macroscopic orientation of the crystal faces; however, the sample obtained showed some large (~5 mm diameter) cyclic twinned pseudohexagonal crystals. These large pseudohexagons have the characteristic 63°48' twinning angle that allows particular faces to be recognized. The orientation of the sample of crystal with respect to the pseudohexagonal twin face showed that the lamellae lie parallel to the {010} face of the crystal.

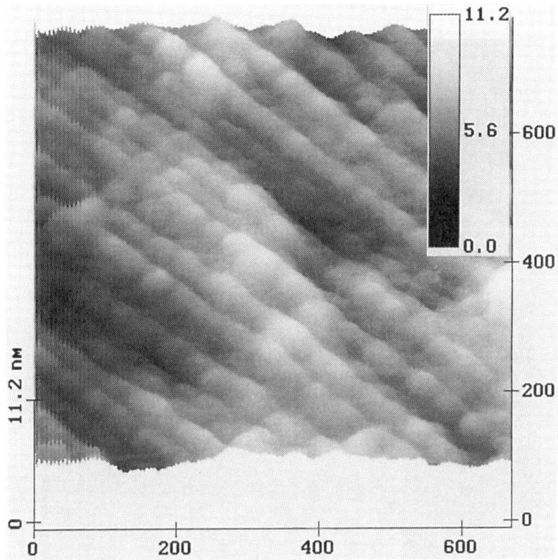

Figure 1. An AFM image of inorganic aragonite shows well-ordered lamellar. The bar is 100 nm.

Otolith Aragonite Crystal

AFM images of the otoliths of hake, snapper and salmon all showed lamellae. The lamellae of the otolith of snapper in AFM sections could be seen in two planes: one vertical to the surface of the image and the other parallel to the surface of the image (Figure 2 (a), (b)). Matrix material lying at an angle to the plane of the lamellae can also be seen in Figures 2 (a), (b). Similar lamellae have been described in TEM sections of the otolith of hake by Gauldie and Nelson.[17] Lamellae widths were measured from AFM images of the otoliths of the three species as well as from inorganic aragonite. Differences and similarities in lamella width can be seen from histograms comparing the lamellae widths observed in TEM and AFM sections of the otolith of hake (Figure 3 (a)); as well as by comparing lamella widths of otoliths between the three species (Figure 3 (b)).

Lamellar Width Frequency Distributions

The most striking feature of the lamellar width frequency distributions in Figures 3 (a) and (b) was the narrow distribution of lamellar widths in the abiogenic mineral aragonite material. To test whether the observed TEM lamellae frequency distributions and the observed AFM lamellae frequency distribution (Figure 3(a)) were consistent with being drawn from a common distribution (i.e., to test that the two methods were measuring the same structures) we used the nonparametric Kolmogorov-Smirnov test (described in Hollander and Wolfe[27] p. 219-227) and the Cramer-Von Mises test (Sprent[28] p. 106). We could find no evidence that the samples came from different distributions. The Kolmogorov-Smirnov asymptotic test statistic was 0.98 and this the lowest significance level at which we could reject the null hypothesis that the TEM and AFM samples came from the same underlying distribution of lamellae widths as 0.29 (i.e., 29%), (see Hollander and Wolfe[27] Table A.24). The conventional level of statistical significance of 0.05 (i.e., 5%) corresponds to an asymptotic Kolmogorov-Smirnov statistics of 1.38. The Cramer-

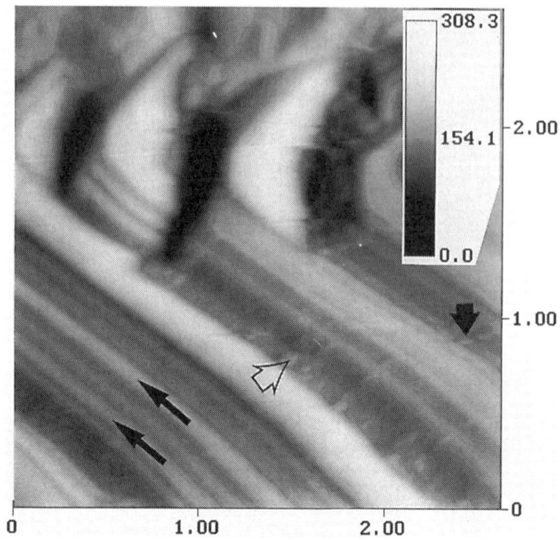

Figure 2. (a) An AFM image of crystals from the sulcal part of the otolith of snapper *(Pagrus major)* shows lamellae (arrows). Marked angles indicate fibrous (arrows) and granular (open arrows) material that is consistently co-planar and lies along the emergent 110 face of the otolith crystal. The bar is 300 nm.

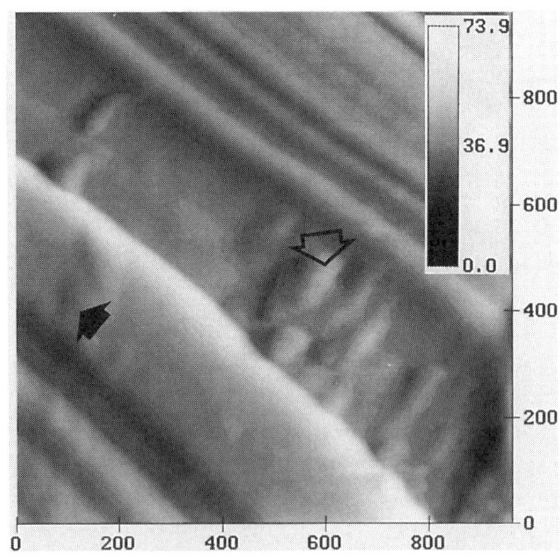

Figure 2. (b) Detail of (a) above shows the positive/negative appearance (open/closed arrows) of the organic material. The bar is 100 nm.

Von Mises test statistic was 0.065 and would have had to extend 0.461 to reject at the 5% level the hypothesis that the TEM and AFM measurements were made on the same underlying population of lamellae. At the 1% level the Cramer-Von Mises statistic would need to exceed 0.743 before rejection of the null hypothesis.

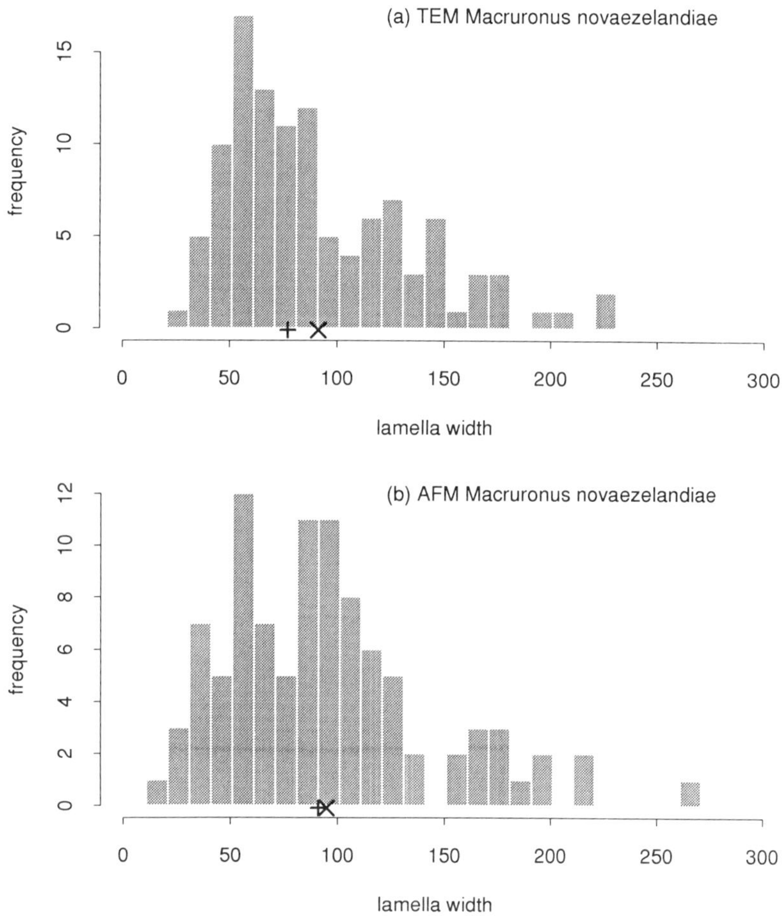

Figure 3. (a) Size frequencies of lamellae for (a) TEM sections and (b) AFM sections of hake *(Macruronus noveazelandiae)*. The + indicates median lamella width and the x indicates mean lamella width.

The mean lamellae widths of Figure 3 (b) suggest some ordering for the sample types such that mean width of 30 nm for inorganic aragonite $\leq$ mean width of 80 nm for snapper (*Pagrus major*) $\leq$ mean width of 90 nm for hake (*Macruronus novaezelandiae*) $\leq$ mean width of 130 nm for salmon (*Oncorhynchus tshawytscha*). We tested the null hypothesis that the means were equal against the alternative hypothesis that they were ordered as above with at least one of the inequalities being strict, by use of the nonparametric

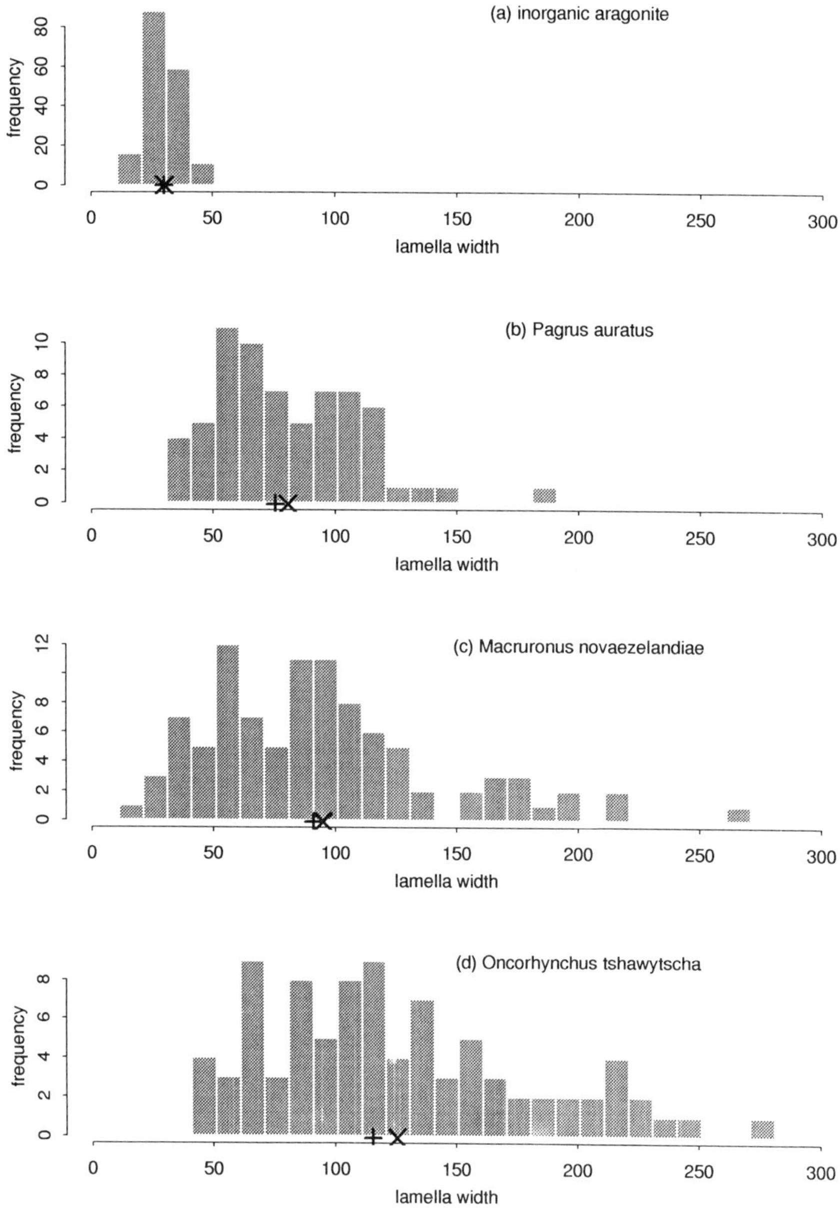

Figure 3. (b) Size frequencies of lamellae are shown as histograms for (a) inorganic aragonite; (b) AFM sections of snapper *(Pagrus major)*; (c) AFM sections of hake *(Macruronus noveazelandiae)*; and (d) AFM sections of *Oncorhynchus tshwaytscha*. The + indicates mean lamella width and the x median lamella width.

Jonckheere's test.[29,30] (See also, Hollander and Wolfe[27] p. 120-123.) Jonckheere's large sample statistic is distributed N(0,1). Jonckheere's large sample statistic was 17.6, so clearly we reject the null hypothesis and accept the hypothesis of ordering. Excluding the inorganic aragonite from the ordering yields a Jonckheere's large sample statistic of 5.54, which is still highly significant.

CONCLUSIONS

AFM and TEM sections of otoliths of hake *(Macruronus novaezelandiae)* give lamella width distributions that are similar, if not identical. We conclude that at the nanometer level of resolution AFM and TEM images give consistent data. Biogenic aragonites retain the lamellar structure of the abiogenic crystal, but some unknown biological effect results in different lamellar widths in the otoliths of different species of fish. The present study shows that AFM is a convenient and powerful tool for investigating biominerals and their submicroscopic variability.

ACKNOWLEDGMENTS

This research was supported by funds from the Hawaii Institute of Geophysics of the University of Hawaii. SOEST Contribution 3277.

REFERENCES

1. T. Dale, The labyrinthine mechanoreceptor organs of the cod *Gadus morhua* L. (Teleostei: Gadidae), *Norw. J. Zool.* 24: 85-125 (1976).
2. E.T. Degens, W.B. Deuser, and R.L. Haedrich, Molecular structure and composition of fish otoliths, *Mar. Biol.* 2:105-113 (1969).
3. R.R. FAY, The goldfish ear codes the axis of acoustic particle motion in three dimensions, *Science* 225:951-953 (1980).
4. Y. Mugiya, and T. Uchimure, Otolith resorption induced by anaerobic stress in the goldfish, *Carassius auratus*, *J. Fish. Biol.* 35: 813-818 (1989).
5. D. Nolf, "Handbook of Paleoichthyology," vol. 10, *Otolithi piscium*, Gustav Fisher Verlag, Stuttgart, New York (1985).
6. P.A.M. Gaemers, Taxonomic position of the cichlidae (Pices, Perciformes) as demonstrated by the morphology of their otoliths, *Neth. J. Zool.* 34: 566-595 (1984).
7. D.G. Dunkelberger, J.M. Dean, and N. Watabe, The ultrastructure of the otolithic membrane and otolith in the juvenile mummichog, *Fundulus heteroclitus*, *J. Morphol.* 163:367-377 (1980).
8. R.W. Gauldie, Phase differences between check ring locations in the orange roughy otolith Hoplostethus atlanticus), *Can. J. Fish. Aquat. Sci.* 47: 760-765 (1990a).
9. R.W. Gauldie, The morphology and periodic structures of the otolith of the chinook salmon *(Oncorhynchus tshawytscha)* and temperature dependant variation in microscopic growth increment width, *Acta. Zool., Stockh.* 72: 159-179 (1991).
10. R. Lecomte-Finiger, The crystalline ultrastructure of otoliths of the eel (*A: anguilla* L. 1758), *J. Fish. Biol.* 40: 181-190 (1992).
11. N.M. Davies, R.W. Gauldie, S.A. Crane, and R.K. Thompson, Otolith ultrastructure of smooth oreo *Psuedocyttus maculatus* and black oreo, *Allocyttus sp.*, species, *Fish. Bull. U.S.* 86: 499-515 (1988).
12. T.B. Bagenal, Aging of fish, *in:* "Proceedings of an International Symposium," Unwin Brothers Ltd., Gresham Press, Old Woking, Surrey, England (1974).
13. D.A. Carlstrom, A crystallographic study of vertebrate otoliths, *Biol. Bull.*, Woods Hole 125:441-463 (1963).
14. M.B. Strong, J.D. Neilson, J.J. Hunt, Aberrant crystallization of pollack (*Pollachius virens*) otoliths, *Can. J. Fish. Aquat. Sci.* 43:1457-1463 (1986).

15. K.P. Mulligan, and R.W. Gauldie, The biological significance of the variation in crystalline morph and habit of otoconia in elasmobranchs, *Copeia* 1989: 856-871 (1989).
16. R.W. GAULDIE, Vaterite otoliths from the opah, *Lampris immaculatus*, and two species of sunfish, *Mola mola* and *M. ramsayi*, *Acta Zool., Stockh.* 71: 193-199 (1990b).
17. R.W. Gauldie, and D.G.A. Nelson, Aragonite twinning and neuroprotein secretion are the cause of daily growth rings in fish otoliths, *Comp. Biochem. Physiol.* 90:510-509 (1988).
18. M.E. Marsh, and R.L. Sass, Aragonite twinning in the molluscan bivalve hinge ligament, Science 208: 1262-1263 (1980).
19. R.W. Gauldie, The fine structure of check rings in the otolith of the New Zealand snapper *Chrysophrys auratus*), *N.Z. J. Mar. Freshw. Res.* 22: 273-278 (1988).
20. R.W. Gauldie, D.G.A. Nelson. Otolith growth in fishes, *Comp. Biochem. Physiol.* 97: 119-135 (1990).
21. R.W. Gauldie, G.C. Coote, K.P. Mulligan and I.F. West, A chemical probe of the microstructure organization of fish otoliths, *Comp. Biochem. Physiol.* 102: 533-545 (1992).
22. R.W. Gauldie, Continuous and discontinuous growth in the otolith of *Macruronus novaezelandiae* (Merlucciidae: Teleostei) *J. Morph.* 216:1-24 (1993).
23. P.E. Hillner, A.J. Gratz, S. Manne, and P.K. Hansma, Atomic-scale imaging of calcite growthd dissolution in real time, *Geology* 20:359-362 (1992).
24. G. Friedbacher, P.K. Hansma, E. Ramli, and G.D. Stucky, Imaging powders with the Atomic force Microscope: from biominerals to commercial materials, *Science* 253:1261-1263 (1991).
25. A.L. Rachlin, G.S. Henderson, and M.C. Goh. An atomic force microscopy (AFM) study of the calcite cleavage plane: image averaging in Fourier space, *Am. Mineral.* 77:904-910 (1992).
26. P.W. Grutter, Zimmerman-Edling and D. Brodbeck, Tip artifacts of microfabricated force sensors for atomic microscopy, *Appl. Phys. Lett.* 60: 2741-2743 (1992).
27. M. Hollandev and D.A. Wolfe, Nonparametric statisical methods, John Wiley and Sons, New York (1973).
28. P. Sprent, "Applied Nonparametric Statistical Methods," Chapman and Hall, London (1989).
29. T.J. Terpstra, The asymptotic normality and consistency of Kendall's test against trend, when ties are present in one ranking, *Indag. Math.* 14:327-333 (1952).
30. A.R. Jonckheere, A distribution-free *k*-sample test against ordered alternatives, *Biometrika* 41:133-145 (1954).

ATOMIC FORCE MICROSCOPY IMAGES OF STARCH POLYMER CRYSTALLINE AND AMORPHOUS STRUCTURES

R.W. Gauldie,[1] G. Raina,[1] S.K. Sharma,[1] and J.-L. Jane[2]

[1]Hawaii Institute of Geophysics
School of Ocean and Earth Science and Technology
University of Hawaii, Honolulu, HI 96822
[2]Department of Food Science and Human Nutrition
Iowa State University of Science and Technology
Ames, Iowa, 50011

Abstract: Atomic force microscopy (AFM) has been used to image the internal structure of broken surfaces of a starch-zein thermoplastic polymer that had been previously pressure-formed into tensile bars. Analysis of the AFM data reveals that the surface of starch-zein polymer consists of a complex mixture of crystalline and amorphous domains. AFM analysis of the crystalline domains of a freshly broken tensile bar surface show highly ordered skein-like arrays. At higher magnifications helical arrays suggestive of a double helix structure are present.

INTRODUCTION

Biodegradable plastic polymers are being developed from a number of different feed-stocks, including starch and other materials derived from corn. Unprocessed corn, particularly the desiccated corn-kernel, is a very durable material that resists the effects of storage for long periods. It is not surprising that replication of such durable properties should be attempted by polymer chemists. One such replication at Iowa State University uses previously purified cornstarch and the corn protein zein with smaller amounts of plasticisers. That research has resulted in the development of a thermoplastic polymer (cornstarch-zein protein polymer: CZP) with potential commercial properties. In this study we describe AFM observations of crystalline and amorphous domain structures in the CZP polymer and crystallized starch fibers.

EXPERIMENTAL METHODS

Atomic Force Microscopy

The AFM used in this study was a Digital Instruments Nanoscope III. All AFM measurements were made in air on sample surfaces from fractured tensile bars that were glued to steel sample plates. The measurements were made after the samples were allowed to equilibrate at room humidity. The high resolution of the AFM has the disadvantage of making regular artifacts difficult to distinguish from genuine regular surface morphologies. High levels of resolution are also subject to varying degrees of noise that often result in lowered signal-to-noise ratios. The techniques for removing such noise, in order to reveal the true signal, require careful judgement to assess the extent to which judicious noise removal can lead to desired, but not spurious, results. To some extent, noise resulting from machine artifacts, such as mechanical vibrations and electronic interference, can be overcome by rotating the specimen, and by changing the magnifications at which the observations are made. These techniques were used in this study; however, some kinds of material by virtue of their instability at the point of scanning always lead to some masking of the signal by noise.

Polymer Composites and Starch Fibers

The basic cornstarch-zein-polymer (CZP) has been described by Jane et al.[1] For this study tensile bars were made in principal component formulations 50/50% cornstarch and zein. Normal cornstarch was a gift of American Maize-Products Company (St. Louis, MO). Zein and formaldehyde were purchased from Sigma Chemical Company (St. Louis, MO). Zein was dissolved in an aqueous alcohol (70% methanol - 1 g zein in 1.3 mL alcohol). Normal cornstarch (starch/zein=4/1, w/w) was added to the zein solution and mixed. The mixture was then treated with formaldehyde solution overnight, washed, and drained by filtration. The dough-like mixture was dried overnight in an air-draft oven at 50° C and ground. The moisture content of the starch-zein mixture was adjusted to 9% and molded by using a compression molding press (Wabash Metal Products, Inc., Wabash, IN) at 2000 psi, 159-165° C for 10 min.

Crystals of starch were produced by adding iodine/potassium iodide solution to high amylose cold-water soluble starch. Long (> 5 μm) fibers of starch were produced in this way. The starch fibers were crystallized onto an atomically flat mica surface so that the microscopic features could be observed without substrate interference.

RESULTS AND DISCUSSION

Although there is as yet little crystallographic theory specific to biodegradable polymers, we can assume that the basic issues of crystalline and amorphous domain will apply to biodegradable plastic as they do in other thermoplastic polymers. The results and discussion that follow describe the organization of internal microcrystalline and amorphous domains in this polymer material as observed by AFM.

50:50 Zein Cornstarch Mixed Polymer

Figure 1 shows an AFM image of the broken surface of a 50:50 CZP polymer. Analysis of a number of AFM images of these broken surfaces of the 50:50 CZP polymer indicate that the material is composed of amorphous and crystalline domains similar to

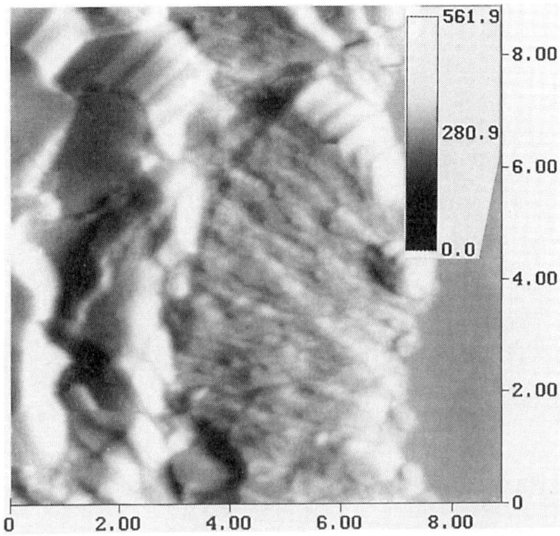

Figure 1. Broken surfaces of the 50:50 CZP polymer in the AFM showing crystalline and amorphous domains. The crystalline domains exhibit a laminated crystal structure, but the amorphous domains show a granular appearance organized along striations. The surface roughness scale is also given.

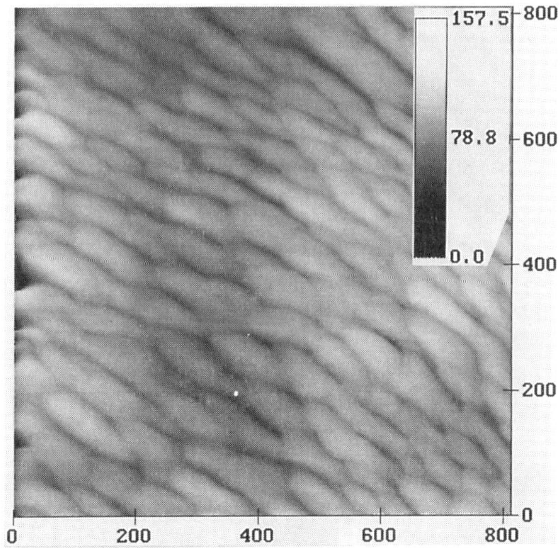

Figure 2. Laminated surfaces of some crystalline domains showing skeins with sequential twists. The surface roughness scale is also given.

those shown in Figure 1. Some parts of the broken surface of the 50:50 CZP tensile bar showed a lamella-like surface that appeared to be formed from folded skeins of material laid in more-or-less parallel rows (Figure 2).

Starch Helices. Higher magnification of the crystalline blocks produced noisy, but highly ordered arrays in the AFM images whose widths were about 20 Å. These highly ordered arrays are similar to those expected from starch double helices. Examination of the frequency spectra of these images by two-dimensional Fourier transform showed an unusual spectrum consisting of two distinct frequency components. Removal of the single high frequency peak in the Fourier spectrum and reconstruction of the original image with the remaining frequency components generated an image with an apparently partly unravelled helical structure (Figure 3).

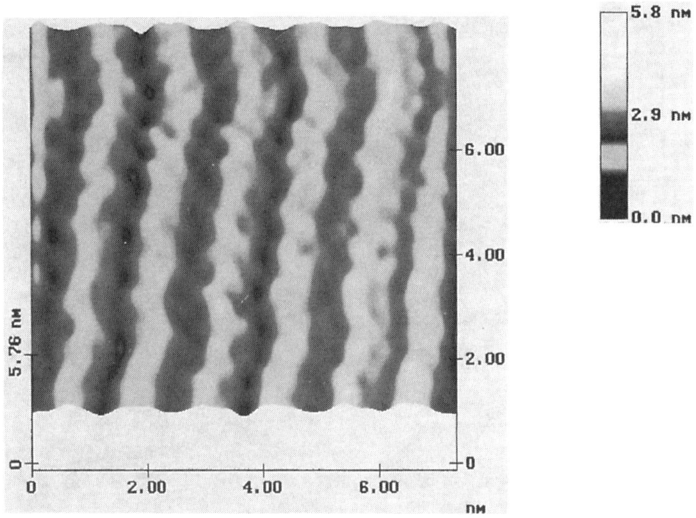

Figure 3. AFM images of the crystalline domains showing a spiral structure similar in appearance to partially unravelled helices.

Starch Fibers. Images of the long (>5 μm) fibers of high amylose cold-water soluble starch produced as described above are shown in Figure 4. The starch fibers crystallized onto an atomically flat mica surface are shown in Figure 5. Although this fiber is about 30 nm wide, we were not able to observe helix-like structures on these precipitated materials.

CONCLUSIONS

The CZP polymer is very typical of the kind of material that is of great interest to the food and packaging industry. It is a mixed polymer with both mixed chemical phases and mixed crystalline and amorphous domain phases. In addition, processing involves a considerable amount of shearing stress followed by heating and rapid cooling resulting in a 'glassy' material composed of a mixture of amorphous and crystal domains. From the AFM microscopist's view, this is a messy target material. Nonetheless, the important

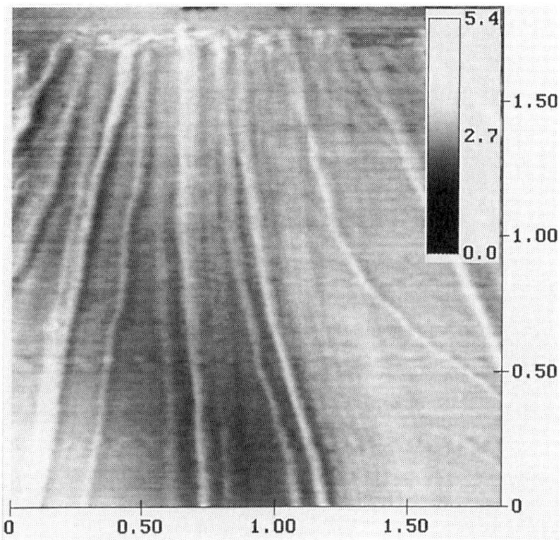

Figure 4. AFM image of starch fibers on a mica surface. Typically these show groups of fibers growing from a discrete line of nucleation points and fanning out as they grow. A surface roughness scale is given.

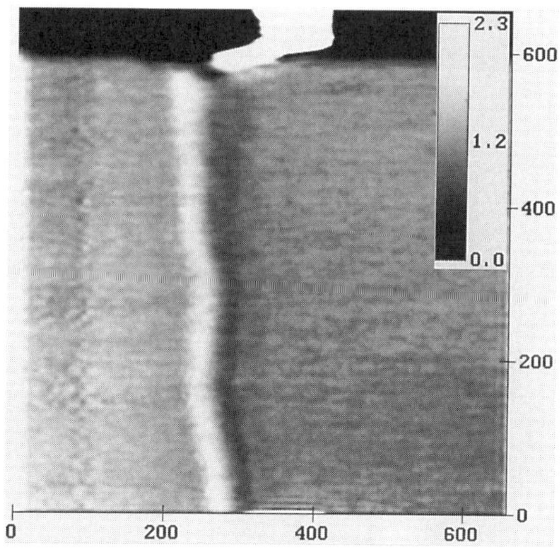

Figure 5. AFM image of a single starch fiber showing a uniform surface structure.

properties of the material will still be determined at the atomic level, probably at the boundaries of crystalline and amorphous domains.

ACKNOWLEDGMENTS

This research was supported by the U.S. Department of Agriculture and the University of Hawaii. SOEST contribution 3278.

REFERENCE

1. J. Jane, R.L. Evangelista, L.Wang, S. Ramattan, J.A. Moore, and R.J.Gelina, Use of modified starches in degradable plastics, "Corn Utilization Conference 3 Proceedings," 4:1-5 (1990).

SCANNING TUNNELING MICROSCOPY STUDIES ON XANTHAN GUM

M.J. Wilkins, M.C. Davies, D.E. Jackson, C.J. Roberts,
S.J.B. Tendler, and P.M. Williams

The VG SPM Laboratory for Biological Applications
Department of Pharmaceutical Sciences, University of Nottingham
University Park, Nottingham NG7 2RD, United Kingdom

Abstract: The application of the scanning tunneling microscope (STM) to imaging biological molecules has generated much interest over recent years. Research has mainly concentrated on obtaining images of DNA, proteins and small organic molecules. However, the work we present concerns polysaccharides, specifically xanthan gum, commonly used in the food and oil industries. We have employed high-resolution metallic shadowing to overcome some of the problems associated with imaging biological molecules by STM, such as tip-sample interaction effects and lack of conduction. The images displayed demonstrate the ability of the STM to image replicas of discrete polysaccharide molecules. We directly correlate and compare these results with electron micrographs of similarly prepared molecules. In addition, the effect of different sample deposition techniques and substrates on the resultant images of xanthan gum is demonstrated. We discuss the implications for sample preparation procedures for future STM studies on biological molecules.

INTRODUCTION

The scanning tunneling microscope (STM) has emerged as a potential biophysical tool for the analysis of biomolecular structure and has already yielded topographical information on a range of macromolecules including DNA[1] and proteins.[2] However, the difficulties of imaging biomolecules by STM are well documented and include the presence of substrate features which resemble biomolecules,[3,4] tip-substrate interactions and tip sweeping of adsorbed biological material,[5,6] and image interpretation of STM images of biomolecules[7]. To overcome the major problem of tip sweeping of adsorbed biomolecules, one of the approaches we have adopted is the use of high resolution rotary shadowing, as employed in the preparation of biological samples for electron microscopy (EM), to immobilize the samples onto a substrate surface.[8] In practice this coating procedure also reduces the

Atomic Force Microscopy/Scanning Tunneling Microscopy
Edited by S.H. Cohen *et al.*, Plenum Press, New York, 1994

number of observed substrate artefacts and more importantly, addresses the problem of image interpretation since STM topographs can be directly correlated with electron micrographs of similarly prepared samples.

In this study, we extend our previous work on platinum/carbon (Pt/C) coated polysaccharide molecules[8] to investigate the effect of sample deposition and the substrate employed on the resultant STM images. We discuss the implications of the need for careful sample preparation for future STM studies on biological materials.

EXPERIMENTAL

Sample Preparation

Throughout these studies, a commercial grade of xanthan gum, Ketrol (Kelco, San Diego, U.S.A.) was used. The Ketrol samples were prepared for STM analysis as described in the following methods.

Method 1 (Droplet deposition and air-drying on highly oriented pyrolytic graphite (HOPG) substrates): The xanthan samples were prepared by dissolving the polysaccharide in deionized water to a concentration of 10 μg/mL. Sample solution droplets of 10 μL were allowed to dry under ambient conditions on freshly cleaved HOPG substrates (Agar Scientific Ltd., Stansted, U.K.).

Method 2 (Droplet deposition and air-drying on mica substrates): The xanthan samples were prepared by dissolving the polysaccharide in deionized water (1 mg/mL) and purifying the solutions by ultracentrifugation at 90000 g for 120 min. The solutions were diluted to 10 μg/mL with deionized water and 0.5 μL aliquots deposited on freshly cleaved mica substrates (Agar Scientific Ltd., Stansted, U.K.). The samples were allowed to dry under ambient conditions.

Method 3 (Spray deposition in aqueous glycerol solutions and vacuum drying on HOPG substrates): Xanthan samples were prepared and purified as in method 2. The solutions were diluted to 10 μg/mL with 60% glycerol and 20 μL aliquots were sprayed onto freshly cleaved HOPG substrates. The samples were dried at 10^{-6} T for 90 min.

Method 4 (Spray deposition in aqueous glycerol solutions and vacuum drying on mica substrates): Xanthan samples were prepared as in method 3, but substituting freshly cleaved mica as the substrate as opposed to HOPG.

All the samples were rotary shadowed with a platinum/carbon (Pt/C (95:5)) layer at an angle of 45° using a Balzers High Vacuum Coating Unit (BA 360M) prior to imaging unless otherwise stated. The BA 360M produced Pt/C coatings of grain size $\leq 5+1$ nm, as measured by the STM. To facilitate imaging by EM,[8] samples prepared by method 4 were rotary shadowed with a Pt/C layer at 5° followed with a carbon layer at 90°. The replica was transferred, via flotation on a water droplet, to a freshly cleaved HOPG substrate prior to scanning with the STM. A similar sample was transferred to a Ni/Cu grid and imaged by transmission EM (TEM).

Sample Analysis

The samples were scanned under ambient conditions using a VG STM 2000 (VG Microtech, Uckfield, U.K.) employing a negatively biased platinum/iridium (80:20) tip. The microscope was operated in constant current mode, with a tunneling current and voltage set at 30 pA and 1.5 V, respectively. This relatively large gap resistance (i.e., large tunneling gap) avoids any problem with tip-surface interactions[5-6] disturbing the Pt/C film.

The sample prepared for analysis by electron microscopy was imaged with a Hitachi 7000 EM. The microscope was in transmission mode at 100 keV accelerating voltage.

RESULTS

The images shown in Figure 1 are representative high and low resolution STM scans taken from different areas within the sample spot of xanthan air-dried on an HOPG substrate (method 1). The topographs show entangled molecules forming a dense network across the substrate surface. The diameter of the strands was found to range from approximately 6±1 nm to 16±1 nm. In addition, clumps of the aggregated polysaccharide may also be observed overlaying some areas of the networks (see Figure 1 (b)).

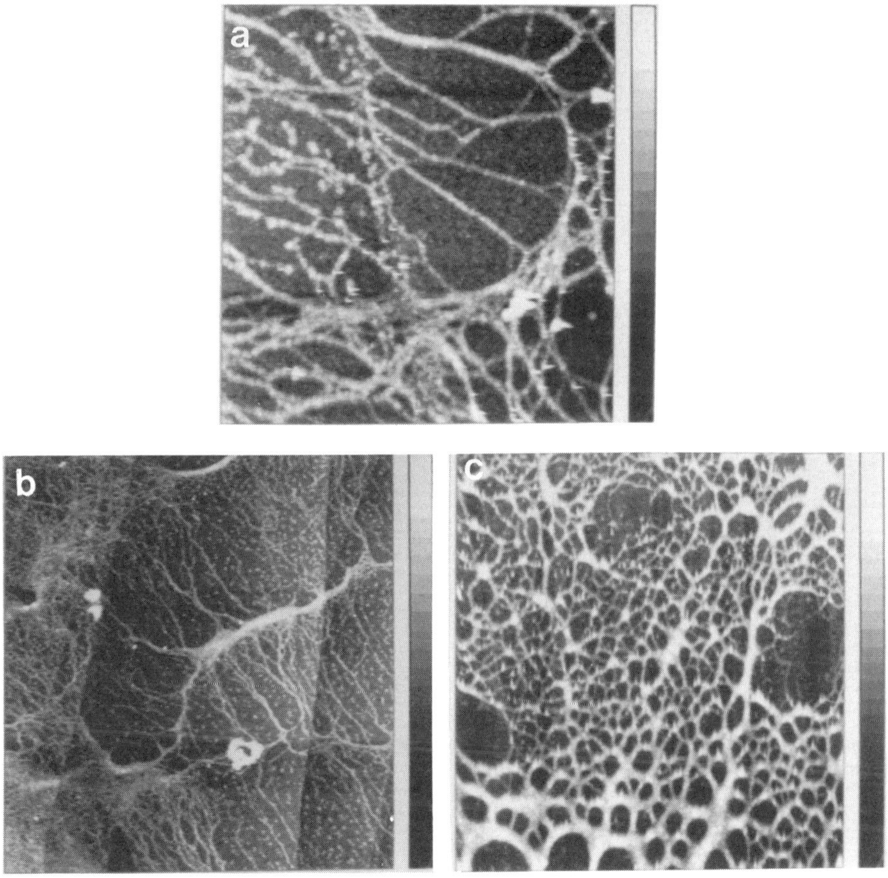

Figure 1. (a) 591 nm x 591 nm STM topograph of xanthan molecules, drop deposited onto an HOPG substrate and dried under ambient conditions. The sample was rotary shadowed with a Pt/C layer of grain size 5±1 nm. The image shows the xanthan molecules forming a network on the substrate surface. Gray scale = 8.42 nm. (b) Low-magnification (1790 nm x 1790 nm) STM image of a xanthan network deposited on HOPG. The image shows alignment of the molecules within the network structure produced during the drying process. Grey scale = 13.07 nm.
(c) 1198 nm x 1198 nm scan of a dense area of the xanthan network on the HOPG surface. Grey scale = 15.87 nm.

Figure 2 shows topographs of purified xanthan, air-dried onto a mica substrate (method 2). Figure 2 (b), which depicts a high resolution scan from the area displayed in Figure 2 (a), shows xanthan molecules forming strands in a close-knit lattice on the substrate surface. The network consists of polysaccharide strands, entangled in random directions across the substrate. The diameters of the strands range from 6±1 nm to 30±1 nm. The scan in Figure. 2 (c) shows another area of the network exhibiting areas of clumped polysaccharide with strands as thin as 6±1 nm emerging from these aggregates.

The images shown in Figure. 3 show individual linear strands of polysaccharide material aligned against the step edges of an HOPG substrate. The sample was prepared

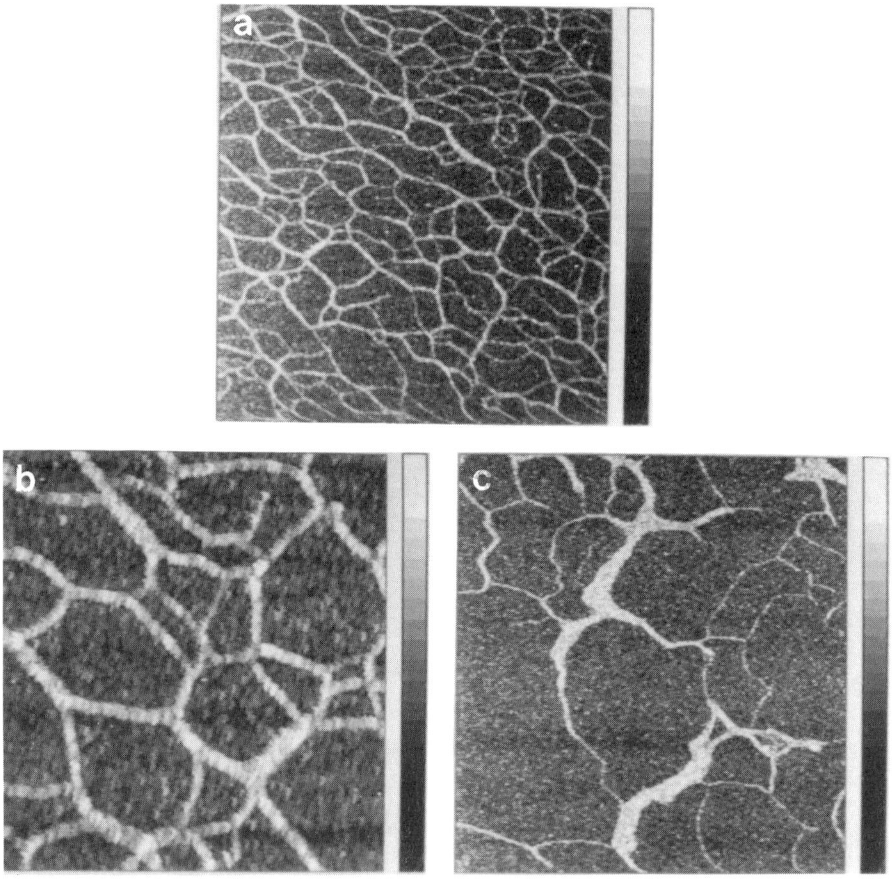

Figure 2. (a) 1587 nm x 1587 nm STM topograph of xanthan molecules, drop deposited and air-dried onto a mica substrate. The sample was rotary shadowed with a Pt/C layer of grain size 5±l nm. This image shows a dense area of the xanthan network with the strands of molecules randomly oriented across the substrate surface. Grey scale = 5.40 nm. (b) A higher-magnification (591 nm x 591 nm) image of the entangled xanthan molecules depicted in Figure 2(a). The strands in the network range in diameter from 6±1 nm to 30±1 nm. Grey scale = 6.07 nm. (c) 1587 nm x 1587 nm STM topograph showing aggregation of xanthan molecules which occurs during the drying process. The aggregates are joined by thin strands (6±1 nm) of polysaccharide material. Grey scale = 3.95 nm.

as described in method 3 resulting in linear macromolecules distributed uniformly across the substrate. Figure 3 (b) shows a high resolution scan of the end of one such strand. The diameter of the strand is 6±1 nm, increasing to 25±1 nm at the terminus.

Figures 4 (a) and 4 (b) contain STM scans of single xanthan molecules prepared by method 4. The polysaccharide molecules were well distributed across the mica substrate and show a variety of linear and convoluted conformations. The molecules also vary in length, indicative of a polydisperse sample population. The xanthan molecules shown in Figure 4 (a) were rotary shadowed with a carbon backing layer in addition to the Pt/C film as described under sample preparation. The grain size of the coarse carbon coating was 18±2 nm corresponding to the observed width of the xanthan molecules. The grain size of the Pt/C coating shown in Figure 4 (b) was 5±1 nm and the diameter of the xanthan molecule, 6±1 nm.

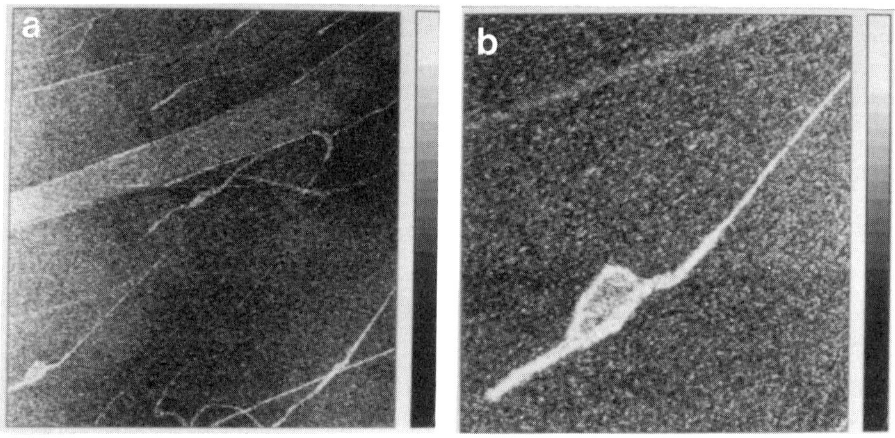

Figure 3 (a) A low-magnification (2599 nm x 2599 nm) STM topograph of xanthan in glycerol solution spray deposited onto HOPG and rotary shadowed with a Pt/C layer. Linear strands of polysaccharide material are aligned against step edges on the substrate surface. Grey scale = 11.98 nm. (b) A higher-magnification (1385 nm x 1385 nm) STM image of a single xanthan molecule taken from the scan shown in Figure 3(a). In the bottom left hand corner of the scan, the xanthan appears to deform or fold back upon itself, contrasting with the molecules shown in Figures 4(a) and (b). Grey scale - 5.11 nm.

The electron micrograph in Figure 5 shows a single xanthan molecule prepared by method 4. The sample was rotary shadowed with a Pt/C layer at 5° followed by a carbon layer at 90°, as described under sample preparation. Individual polysaccharide molecules were found distributed across the substrate surface. The diameter of the xanthan molecule shown is 9±1 nm.

DISCUSSION

Droplet Deposition and Effect of Substrate on Sample Conformation

A comparison of the STM topographs presented leads to the conclusion that the conformation adopted by the polysaccharide molecules is dependent on the choice of

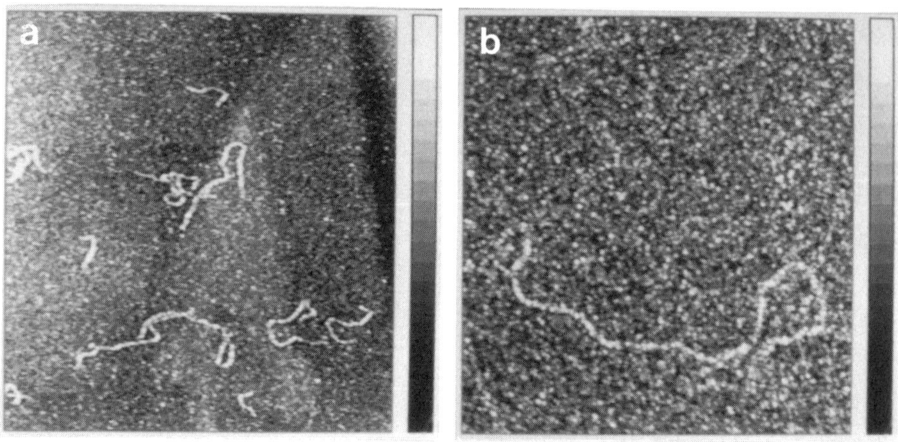

Figure 4. (a) An STM image (1385 nm x 1385 nm) showing xanthan molecules uniformly distributed on a mica substrate. The sample was rotary shadowed with a Pt/C layer and a coarse carbon backing layer with grain size 16±2 nm, corresponding to the observed width of the molecules.This image exhibits the polydisperse nature of the xanthan solution. Grey scale = 6.84 nm. (b) A high-magnification (389 nm x 389 nm) STM topograph of a single xanthan molecule sprayed on mica and coated with a Pt/C layer. The grain size of the coating was 5±1 nm, and the width of the molecule 6±1 nm. The image exhibits the convoluted conformation of the xanthan. Grey scale = 3.03 nm.

substrate and preparation technique employed. From the images displayed in Figures 1 and 2, the xanthan molecules are seen to become entangled when sample solutions are pipette deposited and air-dried onto a substrate, consistent with previous EM data.[9] As water evaporates from the substrate, the proximity of the molecules in solution increases causing entanglement and the formation of networks. The density of the networks on both substrates varied due to the nonuniform distribution of material within a given sample spot. This variation indicates the absence of a strong substrate-molecular interaction on both the hydrophobic HOPG and the hydrophilic mica, which would promote uniform distribution of the xanthan. Rather, the intermolecular associations within the sample solution dominate, causing local aggregation of polysaccharide material.

Apart from xanthan we have observed network formation with a range of biological (e.g., DNA) and organic polymers (e.g., poly(-benzylglutamate), poly(-ethylene oxide)[10] and poly(-ethylene glycol)) when air-dried onto STM substrates. Together with similar images of polystyrene deposited on silicon surfaces,[11] these results strongly suggest that network images of polymers are an artefact of the drying process after drop deposition.

Spray Deposition and Effect of Substrate on Sample Conformation

Spray deposition of xanthan in glycerol solutions has already been successfully employed in previous electron microscopy and STM studies to image individual molecules.[8,12,13] The high viscosity of the glycerol solution resists deformation of the polysaccharide during the vacuum drying process by restricting the movement of the xanthan molecules in solution, preventing network formation during solvent evaporation.[13]

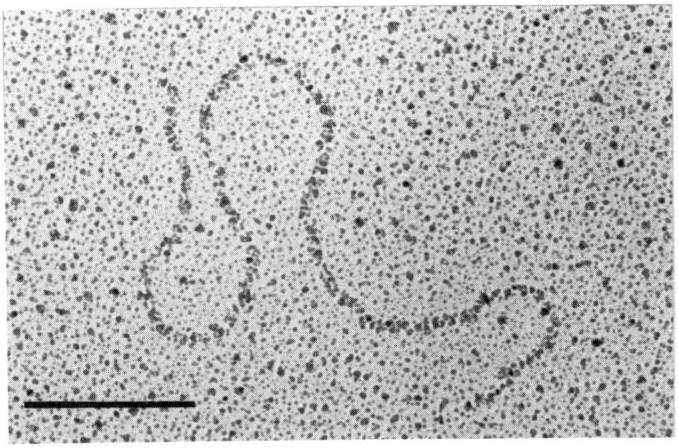

Figure 5. An electron micrograph of xanthan gum spray deposited on a mica substrate. The sample was rotary shadowed with a Pt/C layer at a 5° angle followed by a carbon backing layer. This image shows a single xanthan molecule, correlating with the STM topographs in Figures 4(a) and 4(b). Magnification 280,000x Scale Bar = 100 nm.

Here, the choice of substrate is shown to influence the structure of the individual molecules observed after spray deposition, due to its surface affinity for the xanthan gum.

As shown in Figure 3 (a), the molecules have sufficient mobility on the HOPG surface to align themselves along step edges during the final stages of the drying process. This alignment forces the molecules to adopt a linear conformation. Further deformation of the polysaccharide molecule on HOPG is observed in Figure 3 (b). The diameter of the xanthan terminus is disproportionally large compared to the rest of the molecule, suggesting that the molecule is folding back upon itself or that the 3D structure has been distorted. This contrasts with the images shown in Figures 4 and 5, where the molecules, deposited on mica, exhibit a variety of convoluted conformations in agreement with previous studies.[8,12,13] This resistance to the types of deformations observed on HOPG is most likely due to the comparative rarity of step edges on mica surfaces together with the greater attraction of hydrophilic xanthan molecules for mica as compared to HOPG. Measurements of the molecular diameter, taken at various points from the xanthan molecules in Figures. 4(a) and 4(b), yielded a uniform end-to-end width. This observation implies that the conformation of the molecules is retained when sprayed onto mica.

The apparent width of the xanthan molecules is effected by the grain size of the conductive coating.[8] A comparison of the images in Figures 4 (a) and 4 (b) reveals that high resolution Pt/C layers produce dimensions comparable to TEM measurements, whereas the addition of a relatively coarse grain carbon layer increases the apparent size of the molecules. The electron micrograph in Figure 5 of a single xanthan molecule is comparable to the STM image in Figure 4 (b). The discrepancy in molecular dimensions can be accounted for by the difference in shadowing angles used when preparing the conductive coatings.[14] Further developments in high resolution shadowing techniques[14,15] producing finer grain coatings should reveal greater molecular detail.

The presence of individual molecules distributed across both substrates is indicative of the uniform coverage produced by spray deposition. In practice this reduces the time for location of samples, which allows immediate identification of polysaccharide material upon scanning new areas of the sample.

CONCLUSIONS

We have explored the effect of substrate and sample deposition on image formation in STM studies on polysaccharides, specifically xanthan gum. Pipette deposition and air drying of xanthan solutions has been shown to produce network artefacts. This preparation technique has promoted aggregation of material and is therefore not ideal for producing uniform substrate coverage. The alternative approach of spray deposition produces images of individual molecules. In addition, uniform sample distribution as a result of spraying allows for rapid sample location when using the STM.

The choice of substrate has been found to influence the size and conformation of the xanthan gum depending upon its surface affinity for the molecules. HOPG has been shown to be an unsuitable substrate for STM studies on polysaccharides. The poor binding affinity for the hydrophilic molecules accounts for specimen movement and deformation during the drying process. This distortion of the molecular structure will therefore produce misleading conformational information, emphasizing the need for data validation.

Spray deposition of the polysaccharide onto mica substrates in conjunction with high resolution Pt/C shadowing revealed the most useful and accurate information concerning molecular size, shape and contour length distribution. This latter combination of sample preparation also has the advantage that the replica can be imaged by EM as well as STM, thus allowing direct comparison of images. This study emphasizes the need for the systematic examination of deposition method and substrate employed in the search for the most optimum approach for STM analysis.

ACKNOWLEDGMENTS

We wish to thank the MAFF/DTI hydration (HYDRA) Link programme for their continued support throughout these studies.

REFERENCES

1. R.J. Driscoll, M.G. Youngquist, and J.D. Baldeschwieler, Atomic-scale imaging of DNA using scanning tunneling microscopy, *Nature* 346:294-296 (1990).
2. C.J. Roberts, M. Sekowski, M.C. Davies, D.E. Jackson, M.R. Price and S.J.B. Tendler, Topographical investigations of human ovarian carcinoma polymorphic epithelial Mucin, *Biochem. J.* 283, 181 - 185 (1992).
3. C.R. Clemmer and T.P. Beebe, Jr., Graphite: a mimic for DNA and other biomolecules in scanning tunneling microscope studies, *Science* 251: 640 - 642 (1991).
4. M.C. Davies, D.E. Jackson, K.M. Kreusel, C.J. Roberts, S.J.B. Tendler, M.J. Wilkins and P.M. Williams, Polymer surfaces III scanning probe microscopy: current issues in the analysis of polymeric biomolecules, *Polymer Surfaces* II:227-247, J. Feast, ed., Wiley.
5. C.J. Roberts, M.J. Wilkins, M.C. Davies, D.E. Jackson and S.J.B. Tendler, Surface damage of sputtered gold films at the high and low gap resistance settings of a scanning tunneling microscope, *Surf. Sci. Lett.* 261: L29-L34 (1992).
6. T.E. Wilson, M.N. Murray, D.F. Ogletree, M.D. Bednarski, C.R. Cantor, and M.B. Salmeron, Scanning tunneling microscopy at high gap resistances and on chemically modified silicon surfaces, *J. Vac. Sci. Technol.* B9, 1171-1176 (1991).
7. S.M. Lindsay, T. Thundat, and L. Nagahara, Adsorbate deformation as a contrast mechanism in STM images of biopolymers in an aqueous environment: images of the unstained, hydrated DNA double helix, *J. Microsc.* 152:213-220 (1988).
8. M.J. Wilkins, M.C. Davies, D.E. Jackson, J.R. Mitchell, C.J. Roberts, B.T. Stokke, and S.J.B. Tendler, Comparison of scanning tunneling microscopy and transmission electron microscopy image data of a microbial polysaccharide, *Ultramicroscopy* 48:197-201 (1993).
9. T. Harada, Y. Kanzawa, K. Kanenaga, A. Koreeda, and A. Harada, Electron microscopic studies on

the ultrastructure of curdlan and other polysaccharides in gels used in foods, *Food Structure* 10, 1-18 (1991).
10. G.J. Leggett, M.J. Wilkins, M.C. Davies, D.E. Jackson, C.J. Roberts and S.J.B. Tendler, A scanning tunneling microscopy and transmission electron microscopy study of poly(ethylene oxide) films, *Langmuir* in press.
11. T.G. Stange, R. Matthew, D.F. Evans and W.A. Hendrickson, Scanning tunneling microscopy and atomic force microscopy characterization of polystyrene spin-coated onto silicon surfaces, *Langmuir* 8:920-926 (1992).
12. B.T. Stokke, A. Elsæter, G. Skjåk-Bræk and O. Smidsrød, The molecular size and shape of xanthan, xylinan, bronchial mucin, alginate, and amylose as revealed by electron microscopy, *Carbohydrate Res.*, 160:13-28 (1987).
13. B.T. Stokke and A. Elgsæter, Electron microscopy of carbohydrate polymers, "Advances in Carbohydrate Analysis," Vol.1, 195-247 JAI Press, Ltd. (1991).
14. G.C. Ruben, Ultrathin (1 nm) vertically shadowed platinum-carbon replicas for imaging individual molecules in freeze-etched biological DNA and material science metal and plastic specimens. *J. Elect. Microsc.Tech.* 13:335-354 (1989).
15. R. Wepf, M. Amrein, U. Burkli and H. Gross, Platinum/iridium/carbon: a high-resolution shadowing material for TEM, STM and SEM of biological macromolecular structures, *J. Microsc.* 163, 51-64 (1991).

ATOMIC FORCE MICROSCOPIC IMAGING OF BIOMINERAL POWDER SAMPLES FORMED BY DEPOSITS FROM ETHANOLIC SUSPENSIONS

Lorraine M. Siperko[1] and William J. Landis[2]

[1]IBM Microelectronics Division
D675/14-3
P.O. Box 8003, Endicott, NY 13760-5533
[2]Harvard Medical School and Children's Hospital
Enders Bldg. Rm. 284
300 Longwood Avenue, Boston, MA 02115

Abstract. A sample preparation method was developed that facilitates imaging of powders by force microscopy. The feasibility of the method was tested by imaging a NaCl sample prepared in a like manner. The characteristic face-centered cubic (fcc) structure common to the sodium halide salts was apparent in NaCl atomic scale images. A periodicity of 0.60 nm, which is within 6% of the value of the NaCl lattice constant, was measured.

By suspending biomineral powders in ethanol and depositing an aliquot onto a suitable substrate, the images of hydroxyapatite ($[Ca_{10}(PO_4)_6(OH)_2]$) and brushite ($[CaHPO_4 \cdot 2H_2O]$) were obtained. On glass substrates, brushite formed flat platelets. Atomic spacings were found to be 0.45 nm and 0.60 nm, which agree well with published values for its <110> crystal plane. In contrast, hydroxyapatite primarily formed well-isolated clusters with atomic spacings of 0.43 nm and 0.68 nm, in agreement with published values for its <110> and <001> crystal planes, respectively.

Preliminary results indicate that substrates affect the structure of the deposits. Structural differences of the two mineral deposits on glass and mica were observed.

INTRODUCTION

The major inorganic component found normally in the bones, teeth and other hard tissues of vertebrates is the calcium phosphate salt, hydroxyapatite $[Ca_{10}(PO_4)_6(OH)_2]$, HA. Because of its critical role in the skeleton and elsewhere, HA has been extensively studied by many analytical methods including X-ray diffraction and transmission electron

Atomic Force Microscopy/Scanning Tunneling Microscopy
Edited by S.H. Cohen *et al.*, Plenum Press, New York, 1994

microscopy.[1-5] The surface structure of the mineral is, however, virtually unknown. Another calcium phosphate salt, brushite [$CaHPO_4.2H_2O$], DCPD, resembles HA structurally and is also biologically important. Bulk DCPD has also been examined,[6-8] but not in as much detail as HA.

One of the specific structural similarities between hydroxyapatite and brushite is that the positions of Ca atoms in two of the brushite crystal planes are nearly coincident with those of Ca in hydroxyapatite.[9] On this basis, brushite has been suggested as a template to guide the subsequent formation of hydroxyapatite,[9] which appears as the mature form of mineral in all vertebrate tissues. The consideration that brushite and hydroxyapatite may help define events in the sequence of biological tissue mineralization implies a significance for determination of the surface structure of these two materials. Knowledge that many of the processes occurring in living systems take place at surfaces or interfaces reinforces the need for surface studies of HA and DCPD.

METHOD

Both HA and DCPD may be conveniently prepared in a powder form but powder samples, in general, may be difficult to image by atomic force microscopy (AFM). The technology relies on the deflection of a cantilever of known force constant to generate a topographical map over each point on the surface of a sample. The sample is rastered beneath the cantilever probe (tip) by means of a piezoelectric drive. The cantilever deflection, which is indicative of surface topography, or structure, is recorded for each x,y coordinate. For powder samples to be examined in this manner, they must clearly be constrained from moving under the cantilever tip during rastering.

A method for sample preparation that facilitates AFM imaging of HA and DCPD has been recently reported.[10] By suspending small amounts (1 mg/mL) of the powders in a liquid that evaporates rapidly, in this instance ethanol, and depositing an aliquot onto a suitable substrate, the powder is sufficiently anchored to the substrate to allow imaging. The method was tested by preparing a NaCl sample on a glass slide. The sample was dried in air and imaged with a Digital Instruments (Santa Barbara, CA) Nanoscope II Atomic Force Microscope. Silicon nitride tips were used for the measurements, which were acquired in air.

No special sample treatments (for example, ultra-high vacuum, oil) were needed to obtain the images. A typical NaCl atomic scale image showed a periodicity consistent with the fcc structure of NaCl (0.60 nm).[10] This value is within 6% of that of the NaCl lattice constant (0.564 nm). Any three closely spaced Cl surface atoms (the larger of the two present on the crystal surface) form an angle of 62^0, a value close to that calculated on the basis of the NaCl crystal symmetry (60^0). The image thus reported compared well with images of sodium chloride crystals obtained under oil droplets or in ultra-high vacuum.[11,12]

RESULTS

An atomic-scale image of HA on glass obtained by this method had atomic spacings of 0.68 nm and 0.43 nm, values that correspond to the HA <001> and <110> planes, respectively.[10] On glass, a DCPD atomic-scale image consisted of atomic spacings of 0.45 nm and 0.60 nm, which appeared to correspond to atomic spacings of the DCPD <110> plane.[10] When these two images of HA and DCPD were superposed graphically, it was shown that their respective crystal planes are geometrically similar at the atomic perspective.[13] Since the atoms at the vertices of the crystal planes observed on the surface of HA and DCPD are positioned to within 0.09 nm of each other, the surface structure of these materials is consistent with the concept that DCPD may serve as a template for the deposition of HA.[13]

The microscopic features of these minerals were found to differ when mica was substituted for glass as a substrate. Like glass, mica has a very smooth surface. However, in contrast to a characteristic disordered glass surface, mica exhibits an atomic periodicity of 0.57 nm. On glass, HA deposits tend to form well-isolated crystals, while DCPD deposits tend to form large flat platelets when viewed over a 0.5 µm x 0.5 µm area.

An atomic-scale AFM image of HA deposited on freshly cleaved mica is shown in Figure 1. A square structure with a periodicity of 0.78 nm, representing the closest atomic positions, is seen. Although the atoms at the top of the image are not clearly resolved, they still retain the 0.78 nm periodicity. The distance measured between the more widely spaced atoms (across the diagonal of the square) is 1.03 nm. These spacings are

Figure 1. Atomic image of HA on mica. x,y = 3.83 nm.

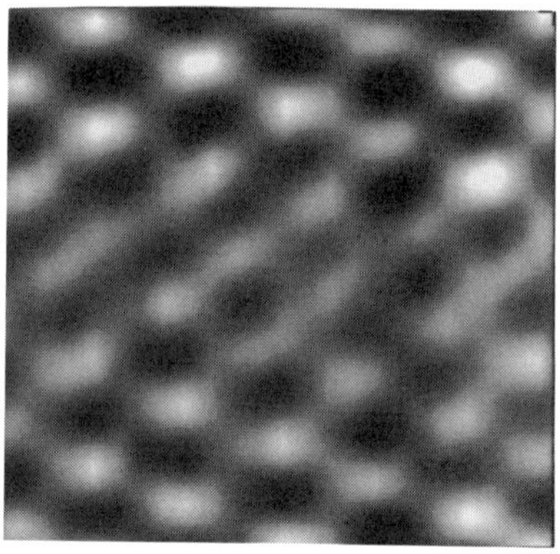

Figure 2. Atomic image of DCPD on mica. x,y = 3,83 nm.

significantly larger than those measured for HA on glass (0.68 nm and 0.43 nm) and indicate that HA deposits on mica consist of a crystal orientation different from that formed on glass.

Figure 2 is the atomic-scale AFM image of DCPD on mica. An hexagonal pattern is clearly seen at the top of the image. The closest atomic spacings were determined to be 0.37 nm. The more distant atomic spacings measured 0.43 nm. Only the larger spacings compare with the spacings most frequently measured for DCPD on glass (0.45 nm). Figure 2 also shows that, at the bottom of the image (proceeding from right to left), the atomic pattern changes, a result indicating another crystal plane. The measurements in this region demonstrate that the 0.37 nm spacing is retained, while a still more closely spaced atomic pattern emerges (0.23 nm). These spacings are, in this instance, smaller than those observed for DCPD on glass (0.68 nm and 0.43 nm),[10] but the 0.37 nm distance compares well with the reported shortest Ca-Ca distance in a calcium phosphate corrugated sheet structure.[14] These data suggest that the mica substrate has an effect on the crystallographic orientation of DCPD as it apparently has with HA as well.

Figure. 3 is the image of a 10 µm x 10 µm scan of HA deposited on freshly cleaved mica. Some isolated spherical structures ranging from 500 nm - 1000 nm can be seen.

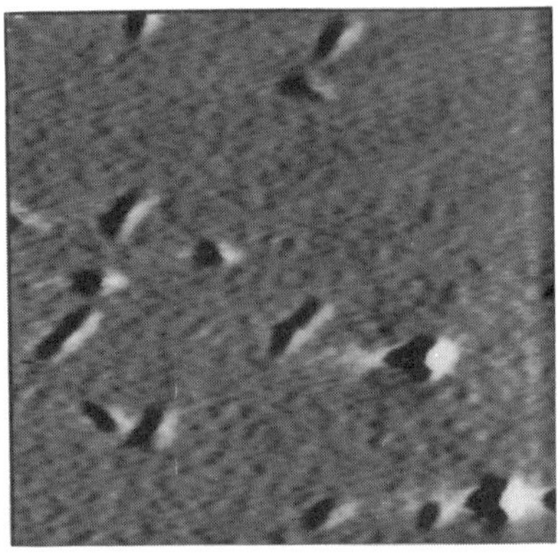

Figure 3. HA on mica. *x,y* = 10 μm x 10 μm

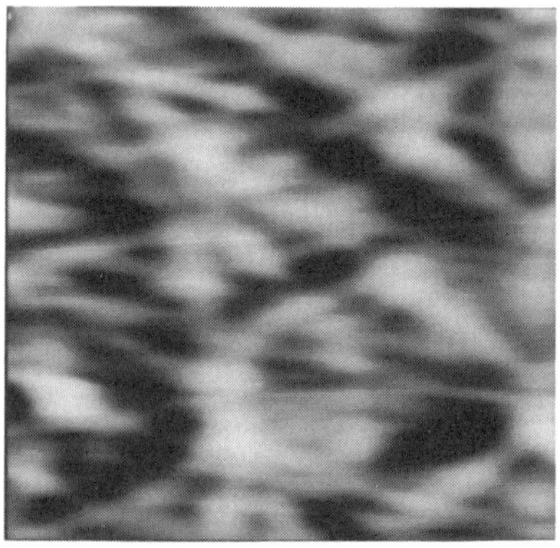

Figure 4. HA on mica. *x,y* = 0.5 μm x 0.5 μm

There are also elongated structures measuring 700 nm - 1500 nm in length. A 0.5 μm x 0.5 μm scan range image taken in the same area is shown in Figure 4. Deposits with a fan-like structure emanating from small crystals are observed at this magnification. In Figure 5 is shown the image of 10 μm x 10 μm scan of DCPD on mica. Isolated clusters approximately 500 nm - 1200 nm are found. Figure 6 shows that the background of DCPD on mica over a 0.5 μm x 0.5 μm scan consists of web-like features without small crystals present.

These various AFM images suggest a substrate effect on the microscopic structure of HA and DCPD deposits. The exact nature of the effect or effects has not yet been determined. Obviously, surface energies and wetting characteristics of the different substrates would contribute to the chemicophysical interaction of the HA or DCPD suspension, thus affecting the nature of the deposit.

Further studies are in progress to address both the chemical and physical interaction between various substrates and HA and DCPD as well as collagen, a protein important

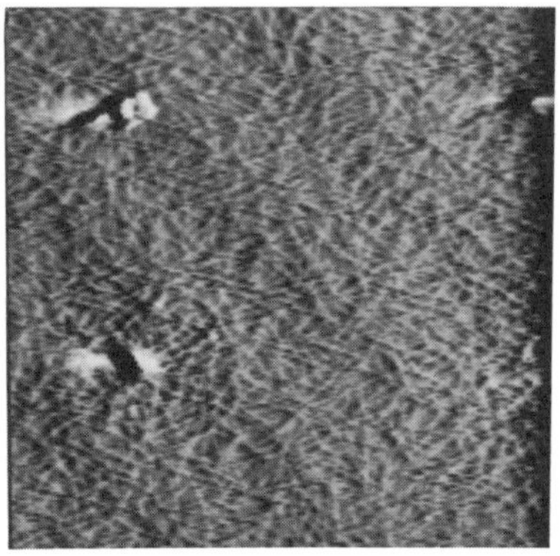

Figure 5. DCPD on mica. *x,y* = 10 μm x 10 μm.

Figure 6. DCPD on mica. *x,y* = 0.5 μm x 0.5 μm.

in biological mineralization. Preliminary results indicate the substrate effects on collagen are not as obvious as those observed with HA and DCPD. These observations indicate the need for closely controlled experiments if biologically significant information concerning vertebrate mineral deposition is to be gained.

REFERENCES

1. H. Ji, P.M. Marquis, Modification of hydroxyapatite during transmission electron microscopy, *J. Mat. Sci. Lett.* 10:132-134 (1991).
2. W.J. Landis, J. Moradian-Oldak, S. Weiner, Topographic imaging of mineral and collagen in the calcifying turkey tendon, *Connect. Tiss. Res.* 25:181-196 (1991).
3. W.J. Landis, M.J. Glimcher, Electron diffraction and electron probe microanalysis of the mineral phase of bone tissue prepared by anhydrous techniques, *J. Ultrastruct. Res.* 63:188-223 (1978).
4. M.I. Kay, R.A. Young, A.S. Posner, Crystal structure of hydroxyapatite, *Nature* 204:1050-1052 (1964).
5. H.C.W. Skinner, H.T. Hunt, J. Griswold, Automatic scanning and analysis of multiple-sample Guinier X-ray powder diffraction films, *J. Phys. E: Sci. Instrum.* 13:74-79 (1980).
6. N.A. Curry, D.W. Jones, The application of an empirical correction for absorption and secondary extinction to neutron data for crystals of brushite, $CaHPO_4.2H_2O$. Z. Kristallogr. 181:205-214 (1987).
7. R.H. Plovnick, Crystallization of brushite from EDTA-chelated calcium in agar gels, *J. Cryst. Growth* 114:22-26 (1991).
8. M. Ohta, M. Tsutsumi, The relationship between the morphology of brushite crystals grown rapidly in silica gel and its structure, *J. Cryst. Growth* 56:652-658 (1981).
9. M.D. Francis, N.C. Webb, Hydroxyapatite formation from a hydrated calcium monohydrogen phosphate precursor, *Calc. Tiss. Res.* 6:335-342 (1971).
10. L.M. Siperko, W.J. Landis, Atomic scale imaging of hydroxyapatite and brushite in air by force microscopy, *Appl. Phys. Lett.* 61:2610-2612 (1992).
11. O. Marti, B. Drake, P.K. Hansma, Atomic force microscopy of liquid-covered surfaces: Atomic resolution images, *Appl. Phys. Lett.* 51:484-486 (1987).
12. G. Meyer, N.M. Amer, Optical-beam-deflection atomic force microscopy: The NaCl (001) surface. *Appl. Phys. Lett.* 56:2100-2101 (1990).
13. L.M. Siperko, W.J. Landis, Atomic force microscopic imaging of biologically important materials. Materials Research Society Fall Meeting, Boston, MA, November 1992 (proceedings at press).
14. C.A. Beevers, The crystal structure of dicalcium phosphate dihydrate, $CaHPO_4.2H_2O$, *Acta Cryst.* 11:273-277 (1958).

NANOSTRUCTURE OF MATERIALS

SCANNING TUNNELING MICROSCOPY STUDIES OF FULLERENE C_{60}

Djuro Koruga,[1,2,3] Jovana Simic-Krstic,[3] Mirko Trifunovic[3]
Svetlana Jankovic,[3] Stuart Hameroff[2] James C. Withers,[1]
and Raoulf O. Loutfy[1]

[1]MER Corporation, 7960 S. Kolb Road, Tucson AZ 85706, USA
[2]Advanced Biotechnology Laboratory, Department of Anesthesiology,
University of Arizona Health Sciences Center, Tucson, AZ 85724
[3]Molecular Machines Research Center, University of Belgrade
27 Marta 80, 1100 Belgrade, Yugoslavia

Abstract: Fullerene C_{60} was purified and imaged utilizing scanning tunneling microscopy (STM) in a constant current mode. By fixing the fullerenes on the substrate ("frozen state" - no movement or rotation), direct imaging of C_{60} with atomic resolution was possible, showing pentagon and hexagon carbon rings of C_{60}. A high voltage STM pulse of C_{60} appears to have created a defect resulting in C_{58} which can dimerize to C_{116}. C_{116} has potential utility as a molecular switching device.

INTRODUCTION

Science and technology related to fullerene C_{60} ("Buckyballs") and similar compounds have attracted the attention and research interest of hundreds, perhaps thousands, of scientists and engineers over the past several years. This attraction stems from the fullerenes' beautiful symmetry, dazzling chemical and physical properties, and potential utilization in nanotechnology.

This paper describes a method for stabilization of C_{60} on a substrate permitting STM imaging with atomic resolution, as well as STM induction of a defect creating C_{58} which can form potentially useful C_{116}.[7]

Mixed molecular weight fullerenes were synthesized using the Huffman/Kratschmer procedure[8] and C_{60} was purified using a chromatographic procedure (MER Corporation, Tucson, Arizona, USA). A three-layered substrate was designed and prepared. The upper layer, a gold substrate Au(110), was sputter cleaned, examined by XPS/UPS and imaged by a scanning tunnelling microscope (STM). In control studies of the cleaned gold layer by STM in air, no structures or irregularities were identified. Before being applied to the gold layer, the dry purified fullerene C_{60} samples were redissolved in toluene. A

Atomic Force Microscopy/Scanning Tunneling Microscopy
Edited by S.H. Cohen *et al.*, Plenum Press, New York, 1994

NanoScope II STM (Digital Instruments Inc., Santa Barbara, California, USA) was used. The STM tip was mechanically cut from Pt/Ir wire at an angle of 15°. The experiment was done on three samples at room temperature. The NanoScope II parameters for tunnelling current and bias voltages were 0.1-5 nA and 0.02-1 V (Z = 20.0 A/V, XY = 18.2 A/V) with effective "gap resistance" of about 10^8 ohm. The acquisition time for the image was 20 s (400/scan). Under certain conditions (three layers, electromagnetic and mechanical properties of C_{60}, bias voltage of STM), we fixed C_{60} on the substrate (no movement, no rotation) and observed its internal structure. Optimal parameters for imaging fullerene C_{60} with internal structure resolution was 1.0 nA (tunnelling current) and 20.1 mV (bias voltage). Previous reports of STM imaging of fullerenes (C_{60} and C_{70})[9,10,14-16] fail to show atomic resolution.

Fullerenes in solution rotate (rotational diffusion constant D = 1.8 x 10^{10} per second)[6] and move around and bounce on the substrate (speed near 3 x 10^5 cm/s)[3] as shown in Figures 1(a), (b), (c), and (d). All pentagonal rings rotate in the same direction,[13] while

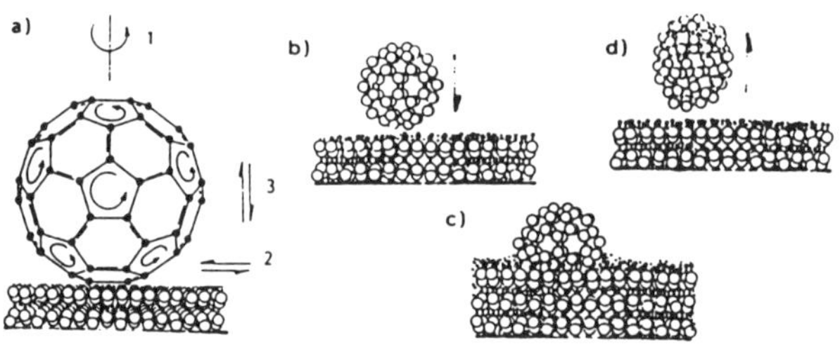

Figure 1. (a)-(d) In solution fullerene has three main movements: rotation, sliding, and bouncing (Figure 1(a) - adapted from Ref. 12). Based on molecular dynamics simulation, interaction between the fullerene and the substrate is shown as "bouncing" (Figure 1(b),(c), and (d) adapted from Ref. 3).

the "ring current" (delocalized π electron) during fullerene rotation generates a magnetic field. When we used a two-layer substrate (gold-silicon), the fullerene movement and bouncing on the substrate was muffled (through hybridization of the highest occupied molecular orbital (HOMO) and lowest unoccupied molecular orbital (LUMO) of the fullerene and HOMO of the golden substrate) and we observed fullerene structure as in Figure 2. Under these conditions, it is not possible to observe the internal structure of the fullerenes (the fullerene rotates and bounces slightly). Based on magnetic susceptibility data[5] of C_{60} and the STM current, we predicted the effects of magnetic induction of the substrate on C_{60}. With a three-layer substrate, C_{60} movement was eliminated (through high-level hybridization of molecular orbitals of fullerene and the substrate, resulting in electromagnetic effects of the alloy layer on fullerene and its mechanical deformation) such that C_{60} was "stuck on" the substrate (Figures 3 and 4).

Using a three-layer substrate (silicon-alloy-gold) fixed the fullerene on the substrate (Figure 1 (c)) and reveals the internal structure of an individual fullerene as rings which may be construed as pentagons and hexagons (Figure 5). Viewing Figure 6 (a), pentagons appear white and hexagons black. The STM constant current mode views primarily the high-lying π electrons on the molecular energy surface of C_{60}. This suggests that the molecular energy of C_{60} is concentrated in the (white) pentagons. Surface energy images (Figure 6 (a)) and structural images (Figure 6 (b)) differ slightly. Because two carbon atoms are missing (7 and 11, Figure 6(b)), the energy surface is broken (Y, Figure 6(a)).

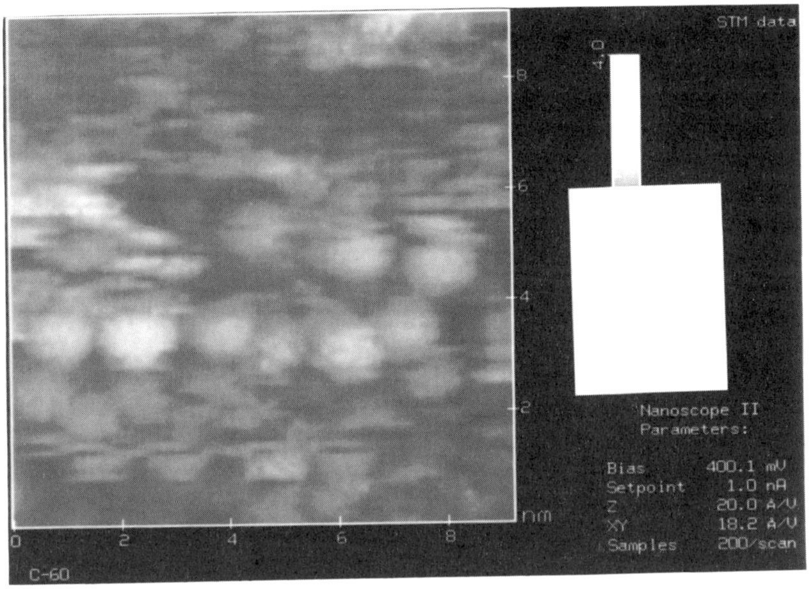

Figure 2. With a low level of molecular orbital hybridization between fullerene and the substrate, the fullerene STM image shows no internal structure.

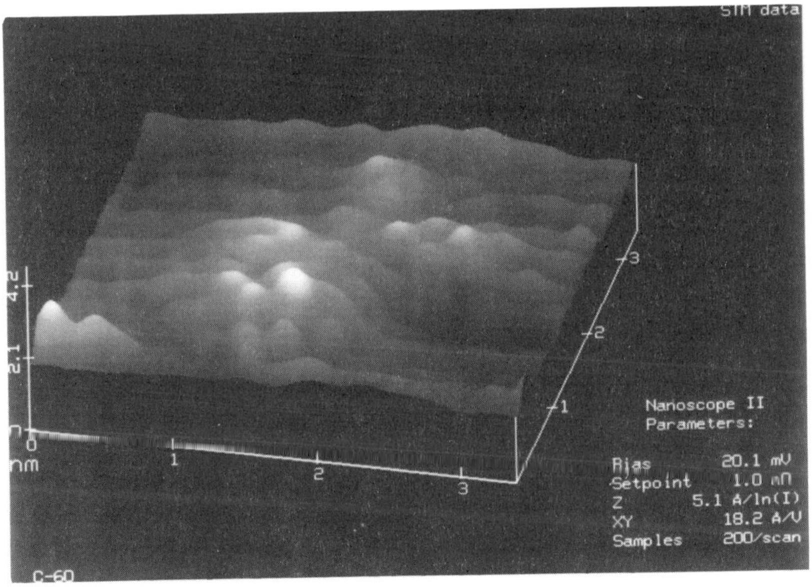

Figure 3. Several C_{60} molecules adhered to surface revealing inner structure.

This suggests that fullerene C_{60} presented in Figure 6 (a) has a defect and is missing two carbon atoms, C_7 and C_{11} connected by a double bond, resulting in C_{58} (Figure 6 (c)). The minimum energy necessary for the dissociation process ($C_{60} \rightarrow C_{58} + C_2$) is 4.6 eV.[10] The center pentagon (C_1-C_5) in Figure 6 (a) is distorted compared to Figure 6 (b). This could be due to a surface defect Y and an energy dislocation. The energy value for

C_{60} is 0.4383 eV/atom,[2] while for C_{58} (C_{60} with defect) it is much higher: 0.7168 eV/atom. This energy value is based on quantum molecular dynamics calculations[1,12] as approximate energies per atom (relative to single-plane graphite) of fullerene molecules, and the average π-angle of the molecule assuming a perfect sphere. Because the fullerenes were fixed on the substrate ("frozen state"), C_{58} has the basic structure of C_{60} (C_{58} is C_{60} with a hole).

We believe that the STM tip may have severed the bonds during imaging, because if C_{58} occurred during synthesis, it seems more likely to occur "closed," rather than as C_{60} with a hole. The form of the defect (Y) is presented in Figure 6 (c). We suggest that it is possible for two C_{58} molecules to join together to give a new dimer form: fullerene C_{116} (Figure 6 (d)). The energy value of C_{116} is 0.4473 eV/atom, which is a little higher than C_{60}. If the C_{116} can be "doped" with an ion that can "switch" between the two monomers, C_{116} promises to have an exciting future as a molecular switch in computational technology.

Figure 5 (a) shows an image of C_{60} with diameter about 1 nm which has been low-pass filtered (to reduce high-frequency noise). The STM image shows what appear to be pentagon and hexagon carbon rings of C_{60}. The other parts of the C_{60} image are fuzzy. This may be due to interaction of the C_{60} with the three-layer substrate ("frozen state"), the position of C_{60} on the substrate and/or its relationship with the STM tip during imaging.

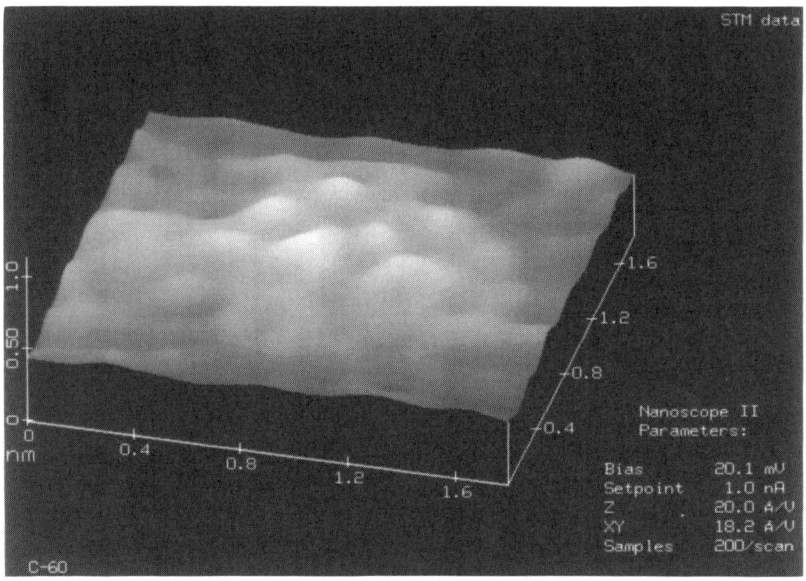

Figure 4. Single C_{60} molecule adhered to surface revealing inner structure.

Figures 5 (b) and 5 (c) show a hexagonal "energy hole" (under C_4-C_5) through which electrons can tunnel into the C_{60} molecule. This could explain the view of two carbon atoms from the opposite side of C_{60}. The hexagonal "energy hole" may be used as a "door" to introduce small atoms like hydrogen or ions for doping of C_{60} (or C_{116}) by electrochemical methods or STM techniques. Electron affinity of C_{60}, based on the photoelectron spectrum, is given by the threshold energy for the appearance of photoelectrons.[4]

Based on these data, a rather large difference between highest occupied molecular orbital ("HOMO") and lowest occupied molecular orbital ("LUMO") occurs between 1.7 - 1.9 eV. This result indicates that bias voltages for STM imaging of the isolated C_{60} should be about 1 V. However, our experimental result showed that the optimal bias voltage for STM imaging of C_{60} on the three-layer substrate with internal structure resolution was 0.0201 V. This discrepancy may be explained by the interaction between C_{60} and the substrate. The energy separation between the HOMO and LUMO is calculated for an *isolated* fullerene[4], while the results of a much lower STM bias voltage are for *non-isolated* C_{60}. Thus, hybridization of the HOMO and LUMO of the C_{60} and the HOMO of the substrate can account for electrons tunnelling into and out of the C_{60} with a lower STM bias voltage and account for atomic resolution observation of C_{60} molecules. These experimental results demonstrate that internal and atomic resolution of C_{60} is possible at room temperature under certain conditions. Based on the present research, we foresee STM-based fullerene information nanotechnology, which, combined with electrochemical methods, may lead to self-assembling molecular computers.

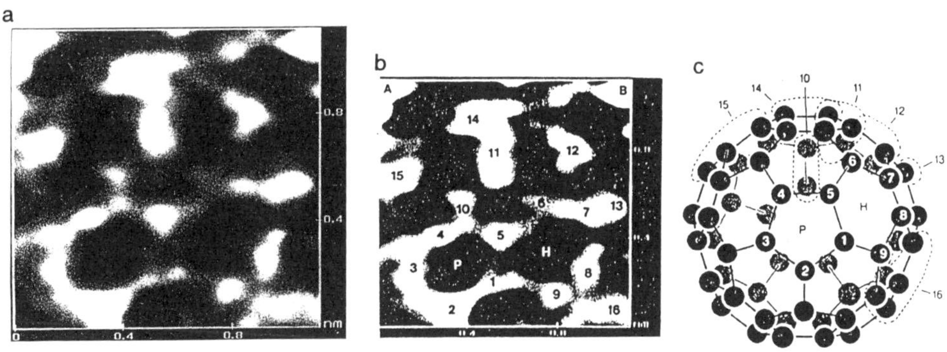

Figure 5. (a) STM image of C_{60} showing ring structure. (b) Regions A and B are from neighboring fullerenes. (c) comparative "stick" model.

ACKNOWLEDGMENTS

D. Koruga thanks L. Lamb from Huffman's Lab, University of Arizona, for providing the starting material for this research, and O. Sankey from Arizona State University for stimulating discussions. The authors are also indebted to Richard Hofstadt and Daphne Anthony for technical support.

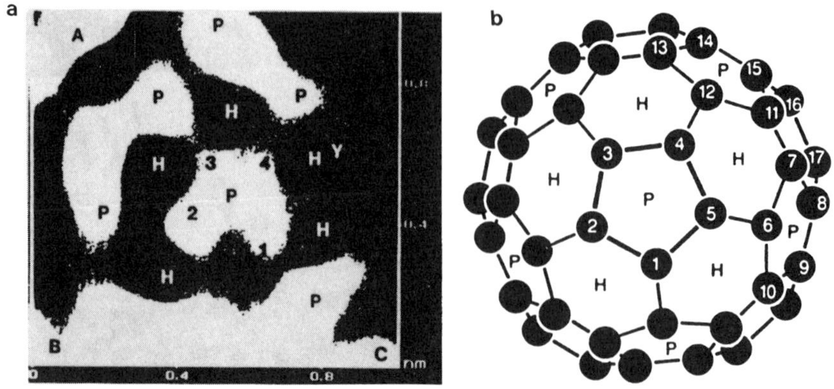

Figure 6. (a)-(b) Fullerene C_{58} as a defect of C_{60}. With middle-level hybridization of the HOMO and LUMO of the fullerene and HOMO of the three-layer substrate, atomic resolution of the Fullerene is not possible, but internal structure (pentagons and hexagons) are visible. (a) Under the influence of surface energy at the fullerene position (as in Figure 1(c)) and the defect (Y), the pentagons (P) and hexagons (H) are not ideal as in Figure 6 (b) (X as energy dislocation). Areas A, B, and C are from neighboring fullerenes. (b) Stick model of C_{60}. Two atoms, C_7 and C_{11}, are missing (C_{58}), producing defect Y.

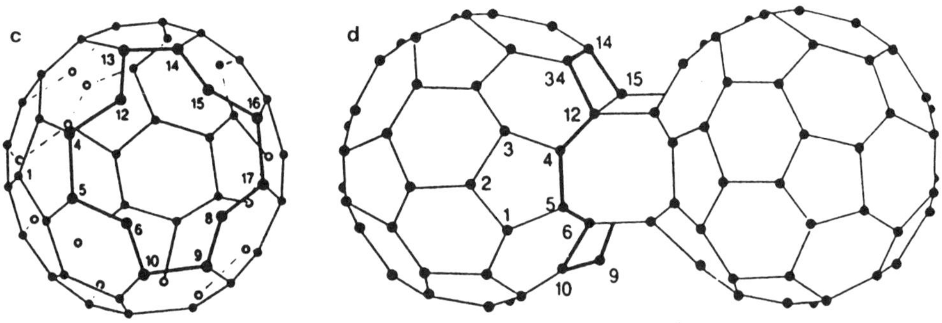

Figure 6. (c)-(d) (c) Front view of the shape of defect Y (Figure 3 (a)) when two atoms (C_7 and C_{11}) are removed (Figure 3 (b)). (d) Two fullerenes C_{60} with defect Y, as C_{58}, may join together and make C_{116} as a fullerene dimer.

REFERENCES

1. G.B. Adams, J.B. Page, O.F. Sankey, K. Sinha, J. Meneder, D.R. Huffman, First principles quantum-molecular dynamics study of the vibrations of icosahedral C_{60}, *Phys. Rev.* B., 44:4052-4055 (1991).
2. G.B. Adams, O.F. Sankey, J.B. Page, M. O'Keeffe, D.A. Drabold, Energetics of large fullerenes: balls, tubes and capsules, *Science*, 256:1792-1795 (1992).
3. R.D. Beck, P. St. John, M.M. Alvarez, F. Diederich, R.L. Whetten, Resilience of all-carbon molecules C_{60}, C_{70}, and C_{84}: A surface-scattering time-of-flight investigation, *J. Phys. Chem.*, 95:8402-8409 (1991).
4. R.F. Curl, R.E. Smalley, Probing C_{60}, *Science*, 242:1017-1022 (1988).
5. R.C. Haddon, V. Elser, Icosahedral C_{60} revisited: an aromatic molecule with a vanishingly small ring current magnet susceptibility, *Chem. Phys. Letters*, 169:362 (1990).

6. R.D. Johnson, C.S. Yannoni, H.C. Dorn, J.R. Salem, D.S. Bethune, C_{60} rotation in the solid state: dynamics of a faceted spherical top, *Science*, 255:1235-1238 (1992).
7. D. Koruga, J. Simic-Krstic, M. Trifunovic, S. Jankovic, S. Hameroff, J.C. Withers, R. Loutfy, Imaging fullerene C_{60} with atomic resolution using a scanning tunnelling microscope, *Fullerene Sci. Tech.* 1:93-100 (1993).
8. W. Krätschmer, L. Lamb, K. Fostiropoulos, D. Huffman, Solid C_{60}: a new form of carbon, *Nature*, 347:354-358 (1990).
9. D.L. Lamb, D.R. Huffman, R.K. Workman, S. Howells, T. Chen, D. Sarid, R.F. Ziolo, Extraction and STM imaging of spherical giant fullerenes, *Science*, 255:413-416 (1992).
10. Y.Z. Li, M. Chander, J.C. Patrin, J.H. Weaver, L.D.F. Chibante, R.E. Smalley, Order and disorder in C_{60} and K_x C_{60} multilayers: direct imaging with scanning tunneling microscopy, *Science* 253:429-433 (1991).
11. P.P.Radi, M-T Hsu, T.L. Bunn, P.R. Kemper, M.T. Bowers, On the structure, reactivity and relative stability of the large carbon cluster ions C_{62}+, C_{60}+ and C_{58}+, *Chem. Phys. Letters*, 74:223-227 (1990).
12. O.F. Sankey, D.J. Niklewski, Ab initio multicenter tight-binding model for molecular-dynamics simulations and other applications in covalent systems, *Phys. Rev. B.*, 40:3979-3995 (1989).
13. R.E. Stanton, M.D. Newton, Normal vibrational modes of buckminster fullerene, *J. Phys. Chem.*, 92:2141-2145 (1988).
14. R.J. Wilson, G. Meijer, D.S. Bethune, R.D. Johanson, D.D. Chambliss, M.S. DeVries, H.E. Hunziker, H.R. Wendt, Imaging C_{60} clusters on a surface using a scanning tunneling microscope, *Nature* 348:621-622 (1990).
15. J.L. Wragg, J.E. Chamberlain, H.W. White, W. Krätschmer, D.R. Huffman, Scanning tunnelling microscopy of solid C_{60}/C_{70}, *Nature* 348:623-624 (1990).
16. Y. Zhang, X. Gao, J.M. Weaver, Scanning tunneling microscopy of C_{60} and C_{70} on ordered Au(111) and Au(110): Molecular structure and electron transmission, *J. Phys. Chem.*, 96:510-513 (1992).

SCANNING TUNNELING MICROSCOPY STUDIES OF ALCOHOL/ALKANE MIXTURES ADSORBED ON GRAPHITE SURFACES

Bhawani Venkataraman, John J. Breen, and George W. Flynn

Department of Chemistry and Columbia Radiation Laboratory
Columbia University
New York, NY 10027

Abstract: Mixtures of triacontanol/triacontane adsorbed on graphite surfaces have been investigated using STM in order to understand the relative importance of functional groups in determining the adsorbate/surface sticking probability. The images observed for these mixtures indicate that the alcohol and alkanes form separate regions on the surface and that the sticking probability is dependent on the liquid phase environment above the graphite surface. The mixtures exhibit interesting reorganization dynamics during which the alcohols and alkanes separate out on the surface.

INTRODUCTION

Scanning tunneling microscopy studies of molecular adsorbates reveal details of the packing order for the adsorbate molecules on a surface. From this information the nature of molecule-molecule and molecule-surface interactions that govern the packing arrangement and the forces responsible for surface adsorption can be understood. Data of this type provide an understanding of processes like molecular epitaxy, chromatography, thin film growth and the molecular details of friction and lubrication. An improved picture of the relationship between the molecular details of the packing arrangements that determine the formation of thin films could help in the development of thin polymer films suitable for microelectronic packaging. Since the STM can be used to image monolayers, the packing order of a molecular layer in direct contact with a surface, the layer most directly involved in friction and lubrication, can be determined.

Long chain hydrocarbons adsorbed on graphite have been studied to model adsorption processes and to help determine the nature of the forces governing adsorption.[1-3] Thermodynamic measurements suggest that the long-chain hydrocarbons have a high affinity for graphite and are adsorbed with their molecular axis parallel to the surface.[3] These results have been confirmed by STM studies, which reveal a highly ordered arrangement of the molecules on the graphite surface. Experiments performed with n-alkanes and terminally substituted alcohols and acids reveal the importance of functional groups in determining the short and long range packing order of these molecules on graphite.[4-9] Differences in the packing order can be used to differentiate molecular types. Recent experiments indicate that changes in the contrast ratio within a molecule, as a function of the bias voltage applied between the tip and the sample, can be related to the presence of different functional groups in the molecule.[10]

The interplay between the entropy and the energy of adsorption in determining the packing order for long molecular chains has been investigated by molecular dynamics simulations.[11] The dynamics of the reorganization of domains for didodecylbenzene on graphite surfaces have been studied by the use of fast scanning and imaging techniques.[12-13] These studies, along with molecular dynamics simulations of these processes, help to provide an understanding of the nature of the driving forces behind these cooperative molecular motions. Images of alkane bilayers have also been obtained showing how the second layer packs relative to the first layer in direct contact with the surface.[8-9]

The experiments presented here investigate further the importance of molecule-molecule and molecule-surface interactions in adsorption and the role played by the strength of these interactions in determining the probability of sticking for a hydrocarbon molecule on a graphite surface. By studying two-component mixtures, where the components are the same except for one functional group, the relative importance of functional groups on the adsorbate in determining the sticking probability can be investigated. The role of intermolecular interactions in determining the packing arrangement can also be probed in these studies. For example, the relative tendency of the solution components to separate out on the surface and form "pure" regions on the surface can be observed. Finally, since these studies are performed under a liquid layer, the competition between sticking to the surface and the dynamics of solvation can be investigated.

Solutions of triacontanol ($C_{30}H_{61}OH$) and triacontane ($C_{30}H_{62}$), in phenyloctane were used to test the resolution of our system. Mixtures of 1:1 triacontanol/triacontane and 1:10 triacontanol/triacontane were used in the studies aimed at investigating the relative importance of the alcohol functional group in determining the sticking probability of the two molecules on the graphite surface.

EXPERIMENTAL

Experiments were performed with a Digital Instruments Nanoscope III STM under ambient conditions. Triacontane and triacontanol were purchased from Aldrich and used without further purification. For single-component solutions, concentrations of approximately 1 mg/mL were prepared in phenyloctane (Aldrich). For two-component alcohol/alkane solutions, concentrations of either 1:1 or 1:10, with the alcohol at lower concentration, were prepared. A drop of the solution was deposited on a piece of freshly cleaved highly ordered pyrolytic graphite (HOPG), purchased from Praxair Surface Technologies. Images were obtained by immersing a 0.01" diameter Pt/Rh (87/13) tip, snipped with wire cutters, in the solution. Typical tunneling conditions were 1.2 - 1.5 V (sample negative) and 60 - 80 pA with the STM operating in the constant current mode. At higher tunneling currents, the tip tends to move the molecules resulting in poor or no images. This effect has been noted by others.[14] The scan rates used varied depending on the size of the area being imaged. Images were obtained with different tips and samples to check for reproducibility and to ensure that the images were free from artifacts which can arise from the tip or sample.

RESULTS

Single-Component Solutions

STM images of triacontanol and triacontane reveal that these molecules form highly ordered structures on the graphite surface with the molecular axis lying parallel to the surface. However, they exhibit different short-range and long-range packing order. The alcohols lie at 60° (or 120°) with respect to the troughs between two rows of molecules. The rows form "zig-zag" patterns where a row changes its direction by 60° or 120°. The triacontane molecules on the other hand lie at 90° with respect to the troughs between two rows and form long straight rows over areas on the order of 300 nm x 300 nm, which is the largest area that can be scanned and still resolve the rows of molecules. All of the above results agree with previously reported STM images of these molecules.[4-9]

Two-Component Mixtures

1:1 Alcohol/Alkane Mixtures. The images obtained with the 1:1 triacontane/triacontanol mixture seem to indicate that there is no interaction between the two components since images of either alcohol regions or alkane regions are observed. These "pure" regions have dimensions at least on the order of 300 nm x 300 nm, the largest area scanned while still resolving the rows of molecules. Even though the solutions contained an equal number of alcohol and alkane molecules, the alcohol appeared about 70% of the time in the STM images. On occasion, when an alkane region was scanned, the image would blur, and after settling, the images showed that the alkanes had been replaced by alcohols. Since the alcohols seemed to cover a larger area of the surface, it was hard to observe the interface between the alcohol and alkane, but on a few occasions the interface was observed. Figure 1 shows an image of the alcohol/alkane interface with the two regions differentiated by their molecular packing arrangement; the lower right area of the figure is the alkane region while the upper left is the alcohol region.

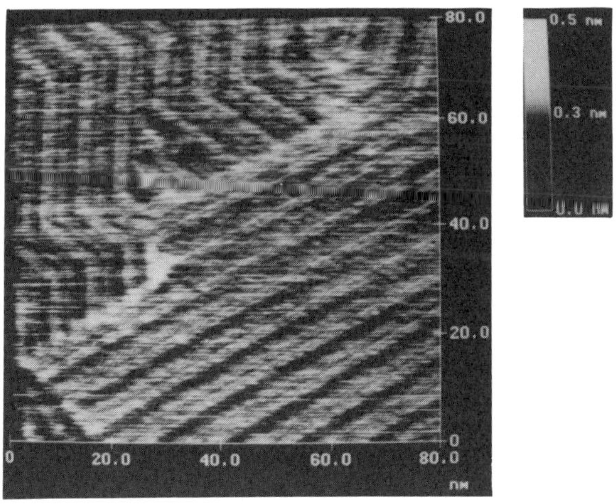

Figure 1. STM image of 1:1 triacontanol/triacontane mixture showing the alcohol-alkane interface. Tunneling conditions were -1200 mV (sample negative) and 60 pA.

The alcohol region shows a "zig-zag" pattern while the alkanes align in straight rows. At the interface, the orientation of the alcohol molecules are at 60° or 120° with respect to the alkane molecules. When attempts were made to obtain a higher resolution image of the packing arrangement of the molecules along this interface region, the images tended to blur and the alkanes were replaced by alcohols. The interface is probably not a stable region and it is possible that tip-surface interactions cause further instability in this area. This is particularly noticeable when the area being scanned is small.

1:10 Alcohol/Alkane Mixtures. Since the alcohol regions were imaged more often than the alkane regions in the 1:1 mixtures, 1:10 alcohol/alkane solutions were also studied. This was done to increase the probability of imaging the alcohol/alkane interface. Once again regions of "pure" alcohol or "pure" alkanes were observed, with the alkane now being imaged more often than with the 1:1 solutions. However, as with the 1:1 solutions, the alcohols were observed more often than 1 out of 10 times, once again suggesting a preference for the alcohol molecules to be on the graphite surface compared with the alkanes. On a few occasions, areas with domains on the order of 30 nm x 30 nm were observed (Figure 2), with the orientation of molecules in a given domain differing from that of the neighboring domains. These regions were not observed with the 1:1 mixtures. The

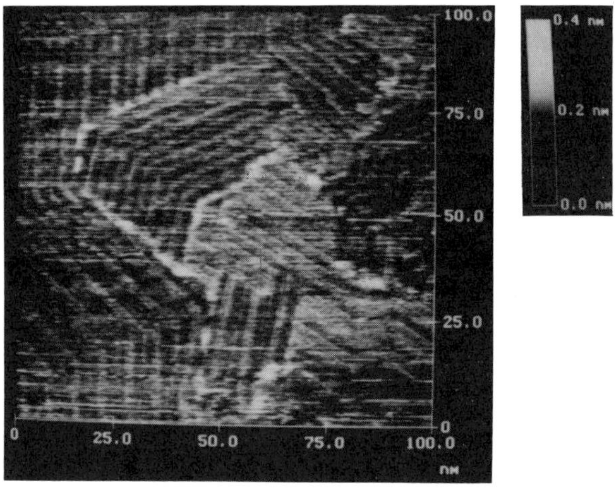

Figure 2. STM image of 1:10 triacontanol/triacontane mixture. Note the presence of domains which are absent in the 1:1 mixtures. Tunneling conditions were -1400 mV and 60 pA.

domains are comprised of either alcohol or alkane molecules, as determined by the arrangement of the molecules in a domain. Some domains show definite boundaries, while others have relatively undefined boundaries. The molecules in neighboring domains once again seem to be oriented at 60° or 120° with respect to each other, as can be seen in Figure 3.

In general, the two-component mixtures are harder to image compared to the single-component solutions. Often the images observed right after tunneling starts seem more disordered than do the images of the single-component solutions. Occasionally, both alcohol and alkane molecules can be observed in the images. After some time, which varies from a few minutes to tens of minutes, the images settle and the ordered arrangement of molecules is observed. Since the initial images change so rapidly and the STM tube piezo requires some time to stabilize once tunneling starts, these images could not be captured for storage.

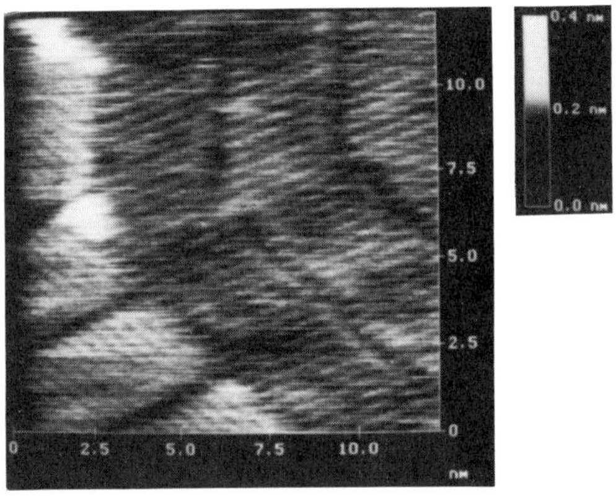

Figure 3. STM image of 1:10 triacontanol/triacontane mixtures showing the orientation of the domain boundaries. Tunneling conditions were -1400 mV (sample negative) and 60 pA.

DISCUSSION

The stabilization of long chain hydrocarbons on graphite can be attributed to the almost perfect match of the distance between alternate methylene groups along the hydrocarbon chain and between carbon atom gaps in the graphite lattice.[3] Thermodynamic measurements indicate that the longer the hydrocarbon chain, the higher the heat of adsorption on graphite.[1] For the terminally substituted alcohols, additional stabilization is achieved through hydrogen bonding between molecules in adjacent rows. The "zig-zag" pattern observed in the packing arrangement of the alcohol molecules is due to a combination of the symmetry of the underlying graphite surface and the fact that the alcohol molecules lie at 60° (or 120°) with respect to the troughs between two rows (Figure 4 (a)). This permits the rows to change direction by 60° or 120° yet maintain the dense packing of alcohol molecules, the a lattice match between the hydrocarbon chains and the gaps in the graphite lattice and the network of hydrogen bonds between the molecules. Thus, the molecules can remain parallel to one another which provides for stabilization through lateral interactions, while changing direction. The alkanes and acids cannot accommodate any change in orientation of rows since the molecules lie at roughly 90° with respect to the troughs (Figure 4 (b)). Hence they are not able to maintain dense packing if the rows change direction.

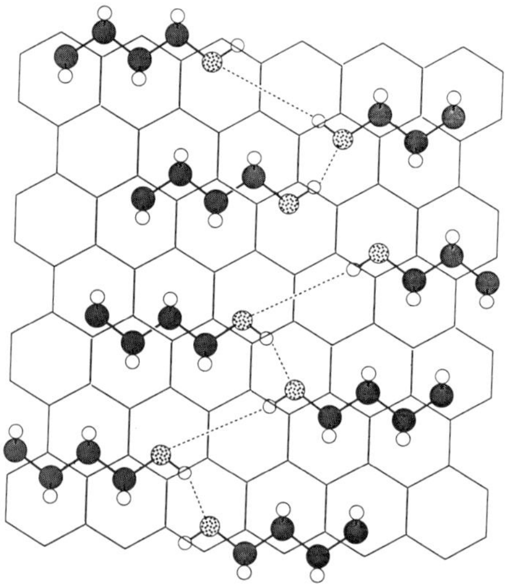

Figure 4. (a) Orientation of the alcohol molecules with respect to each other and the graphite substrate (the dotted circle indicates the oxygen atom). Note the "zig-zag" patterns and the hydrogen bonding between the molecules (denoted by the dotted lines).

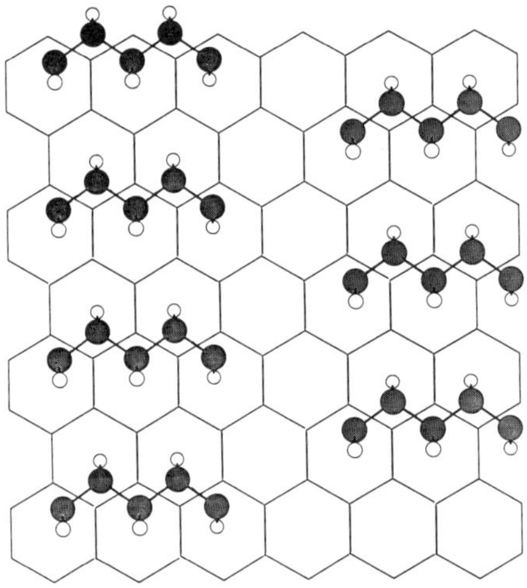

Figure 4. (b) Orientation of the alkane molecules relative to each other and the graphite substrate.

The origin of the 90° angle between the troughs and the long chains for the alkanes probably arises because the main stabilizing mechanism between the surface adsorbed molecules is due to lateral attractive interactions. In the case of the alcohols, the preference for lateral interactions is somewhat mitigated by the stabilizing influence of hydrogen bonding by the end groups, leading to the 60° or 120° angle between the long chains and the troughs.

Measurements of the surface excess mass of hydrocarbons on graphite suggest that for short carbon chains ($n < 18$), the terminally substituted alcohols have a larger surface excess mass (and hence heat of adsorption) than an n-alkane of the same chain length, due to the network of hydrogen bonding formed between the alcohols.[1] If the methylene groups lie in the gaps of the graphite lattice, then the resulting distance between the OH groups is such that the formation of strong hydrogen bonds is favored.

As the carbon chain gets longer, the contribution from the methylene groups should dominate over any stabilization gained from hydrogen bonding in the alcohols; hence, the heat of adsorption for alcohols and alkanes should tend to the same value as the chain length gets longer. This has been shown to be the case for $n>18$.[1] If this is true, then in the STM experiments reported here, one would expect the probability of observing the alkane should be the same as that for observing the alcohol on the graphite surface for a 1:1 triacontanol/triacontane mixture. As mentioned above, the alcohols were imaged about twice as often as the alkanes. Even with 1:10 solutions, the alcohols were imaged more than 10% of the time. This could imply that the triacontanol molecules have a higher heat of adsorption than the alkanes, and that even at $n = 30$ the presence of the OH groups provides an important contribution to the stabilization of the alcohol molecules on the surface. While this may be the case, it is possible that the liquid phase environment above the surface is affecting the dynamic equilibrium between surface adsorbed molecules and those in solution.

The STM images reported here were obtained using phenyloctane as the solvent, which is a low dielectric medium. Such a solvent provides an unfavorable medium to support hydrogen bonding, and hence the alcohols are likely to remain unassociated in solution. On the other hand, by being on the surface, the alcohol molecules can form a network of hydrogen bonds leading to an associated molecular form rather than an unassociated form in solution. This suggests that the separation of the alcohols and alkanes observed in the STM images must be happening on the surface. Since the alcohols are unassociated in solution, they must find another alcohol molecule on the surface with which to form a hydrogen bond, rather than depositing on the surface together as a hydrogen bonded moiety. This would explain the initial disorder observed in the STM images and the time required for settling down before ordered images are observed.

We are planning low temperature STM studies to slow down the dynamics of surface adsorption, which may provide information about the re-ordering processes that lead to separate alcohol and alkane regions on the graphite surface. Solvents with different dielectric constants and surfaces with different types of lattices and lattice constants can also be used to probe the effects of the solvent and substrate on the packing and organization of these mixtures. In the 1:10 alcohol/alkane mixtures, the presence of the observed domains is probably due to the early-time dynamics of re-organization into separate alcohol and alkane regions. Figure 5 shows regions of what appear to be alkanes encroaching into an alcohol region. The two regions can be differentiated from the orientation of the rows; the alkanes forming straight rows while the alcohols form the "zig-zag" pattern. Since for the 1:10 mixtures there is a higher probability for an alkane molecule to be on the surface compared with the 1:1 mixtures, the alcohol molecules encounter alkane molecules more often and cannot migrate as easily towards an alcohol region. This may be the origin of the domains observed in Figures 2 and 5 where there are both alcohol and alkane domains.

CONCLUSIONS

The experiments presented here demonstrate that the local environment is important in determining the sticking probability of a molecule on a surface. Alcohol molecules

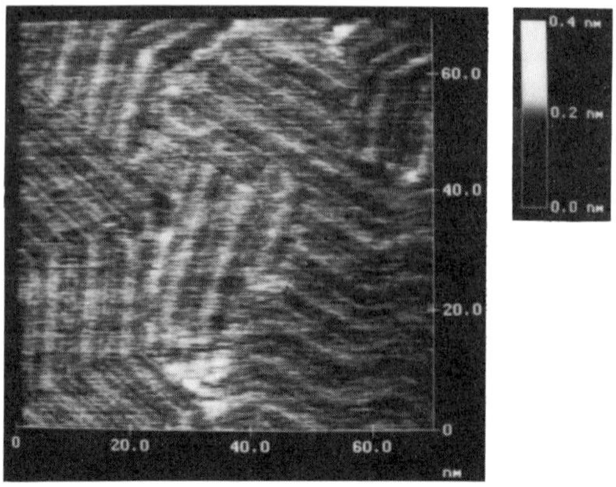

Figure 5. STM image of 1:10 triacontanol/triacontane solution showing an alkane region and an alcohol region. Tunneling conditions were -1400 mV (sample negative) and 60 pA.

prefer to hydrogen bond to each other on the surface, rather than remain dissolved in a low dielectric medium that does not support hydrogen bonding. For triacontanol/triacontane mixtures on the graphite surface, the STM images indicate that the molecules undergo reorganization until the alcohol and alkane molecules separate to form "pure" regions.

ACKNOWLEDGMENTS

This work is supported by The Joint Services Electronics Program (U.S. Army and U.S. Air Force, DAAL03-91-C-0016) and the Donors of the Petroleum Research Fund administered by the American Chemical Society. Equipment support provided by the National Science Foundation (CHE-91-18782).

REFERENCES

1. G.H. Findenegg, Ordered layers of aliphatic alcohols and carboxylic acids at the pure liquid/graphite surface, *J. Chem. Soc. Faraday Trans.* 69:1069-1078 (1973).
2. G.H. Findenegg, Order-disorder transitions at the liquid/solid interface, *J. Chem. Soc. Faraday Trans.* 68:1799-1806 (1972).
3. A.J. Groszek, Selective adsorption at graphite/hydrocarbon interface, *Proc. Roy. Soc. Lond.* A 314: 473-498 (1970).
4. G.C. McGonigal, R.H. Bernhardt, D.J. Thomson, Imaging alkane layers at the liquid/graphite interface with the scanning tunneling microscope, *Appl. Phys. Lett.* 57:28-30 (1990).
5. G.C. McGonigal, R.H. Bernhardt, Y.H.Yeo, D.J. Thomson, Observation of highly ordered, two-dimensional n-alkane and n-alkanol structures on graphite, *J. Vac. Sci. Technol.* B9: 1107-1109 (1991).
6. J.P. Rabe, S. Buchholz, Commensurability and mobility in two dimensional molecular patterns on graphite, *Science*, 253:424-426 (1991).
7. S. Buchholz, J.P. Rabe, Molecular imaging of alkanol monolayers on graphite, *Angew. Chem. Int. Ed. Engl.* 31:189-191 (1992).
8. G. Watel, F. Thibaudau, J. Cousty, Direct observation of long chain alkane bilayer films on graphite by scanning tunneling microscopy, *Surf. Sci. Lett.* 281:L297-L302 (1993).
9. F. Thibaudau, G. Watel, J. Cousty, Scanning tunneling microscopy imaging of alkane bilayers adsorbed on graphite: mechanism of contrast, *Surf. Sci. Lett.* 281:L303-L307 (1993).

10. Y.H. Yeo, K. Yachoboski, G.C. McGonigal, D.J. Thomson, Intramolecular imaging of physisorbed molecules with the scanning tunneling microscope at the liquid/graphite interface, *J. Vac. Sci. Technol.* A10:600-602 (1992).
11. R. Hentschke, B.L. Schurmann, J.P. Rabe, Molecular dynamics simulations of ordered alkane chains physisorbed on graphite, *J. Chem. Phys.* 96:6213-6221 (1992).
12. S. Buchholz, J.P. Rabe, Conformation, packing, defects and molecular dynamics in monolayers of dialkyl-substituted benzenes, *J. Vac. Sci. Technol.* B9:1126-1128 (1991).
13. J.P. Rabe, S. Buchholz. Direct observation of molecular structure and dynamics at the interface between a solid wall and an organic solution by scanning tunneling microscope, *Phys. Rev. Lett.* 66:2076-2099 (1991).
14. D.P.E. Smith, W.M. Heckl, H.A. Klagges, Ordering of alkylcyanobiphenyl molecules at N_oS_2 and graphite surfaces studied by tunneling microscopy, *Surf. Sci.* 278:166-174 (1992).

FRACTURE SURFACE TOPOGRAPHY OF TNT USING ATOMIC FORCE MICROSCOPY

M. Yvonne D. Lanzerotti,[1] A.G. Rinzler,[1] James J. Pinto,[1,2]
D.J. Thomson,[3] and A. Wolfe[4]

[1]U. S. Army Armament Research, Development and Engineering Center
Picatinny Arsenal, NJ 07806 5000
[2]National Research Council Postdoctoral Associate
[3]AT&T Bell Laboratories, Murray Hill, NJ 07974
[4]New York City Technical College, Brooklyn, NY 11201

Abstract: Height profiles spaced 0.008 μm apart across the fracture surface of TNT were obtained with an atomic force microscope (AFM). Spatial power spectra (wavelengths of 0.016 μm to 4.2 μm) were calculated using a prolate spheroidal data window in the horizontal space domain prior to using a fast Fourier transform algorithm.[1] Preliminary findings based on the first 13 profiles across the surface are as follows. The power spectral density of the individual fracture surface profiles is found to decrease with increasing spatial frequency over the region examined, $\approx 1.0\ \mu m^{-1}$ to $\approx 10.0\ \mu m^{-1}$. Power spectral slopes ≤ -3 indicate deterministic fracture. Harmonic distribution of peaks in the power spectra defines the cluster size across the full scan. The variation of the power amplitudes at given frequencies across the profile set defines the cluster size across the 13 profiles. A typical TNT cluster size is found to be approximately 10^6 TNT molecules.

INTRODUCTION

Energetic materials are of significant interest for scientific and practical reasons in the extraction (mining) industry, space propulsion, and ordnance. The nature of the fracture process of such materials under high acceleration is of particular interest, especially in ordnance. We have underway a program for studying the fracture behavior of TNT, Composition B, and Octol using an ultracentrifuge[1-3] and appropriate diagnostic techniques. This includes study of the characteristics of the fracture surfaces of the materials. In this paper the information obtained from AFM studies of the fracture surface topography along several parallel scans across the fracture surface of TNT are described.

Atomic Force Microscopy/Scanning Tunneling Microscopy
Edited by S.H. Cohen *et al.*, Plenum Press, New York, 1994

TECHNIQUE

The fracture surfaces of the energetic materials under study are obtained by accelerating prepared samples in a Beckman preparative model L8-80 ultracentrifuge. The sample under study can be rotated up to 60,000 rpm. When the tensile or shear strength of the object material is exceeded, a fracture surface is obtained. Previously, the fracture surface topographies of TNT, Composition B, and Octol have been studied with a diamond stylus profilometer.[2,3] Fractal analysis of the spatial power spectra (wavelengths of 1.0 μm to 1.0 cm) indicates the regions of deterministic, intergranular failure of the materials as well as the regions where nondeterministic transgranular failure for TNT, RDX, or HMX grains occurred.[2,3]

In this present study, a Nanoscope III atomic force microscope has been used to measure the topography of the TNT fracture surface at higher resolution than obtainable with the profilometer. Each scan profile produces 512 x 512 data points on a total scan size of 4.2 μm. Therefore, the individual data points and the profiles are spaced ≈ 0.008 μm apart across the fracture surface of TNT. The silicon nitride cantilevers have tips with a radius of between 20 and 50 nm.[4] The finite dimension of the stylus imposes a cutoff wavelength approximately equal to the tip radius.[5-7]

We report here preliminary results based upon analyses on studies of the first 13 topographical profiles. Spatial wavelengths of 0.016 μm to 4.2 μm have been characterized. Spatial power spectra have been calculated from the data of the individual traces using a prolate spheroidal data window[8,9] applied in the horizontal space domain prior to using a fast Fourier transform algorithm[10]. The prolate spheroidal window is used because it is superior to more commonly used windows in analyses of short series[9].

RESULTS

The sample of melt-cast TNT under study was prepared in a 9-mm i.d. steel tube. It was found to fracture at an acceleration of 35,000 g at 25°C. The TNT fracture surface topography is shown in Figure 1. As noted above, the scan size for this image is 4.2 μm with the total number of samples being 512 x 512. The first data point in the file corresponds to the foreground corner of the image; the subsequent data points fill out the scan line before starting on the next scan line (left to right, up one line, left to right, etc.)

A typical surface height profile of one trace (profile 5) across the fracture surface of the TNT is shown in Figure 2. The vertical topographic displacement is plotted as a function of sample number. The Z,X coordinates of this profile have been measured with a horizontal spacing (mean sampling interval) of 0.008 μm. Standard silicon nitride probes with spring constant k of 0.58 N/m have been used.

The spatial power spectra of the fracture surface profiles across traces 6 and 12 are compared in Figure 3 (a). The power spectra for both traces decrease with increasing frequency. The solid and starred lines superimposed on the spectra traces are linear fits to the sections of the spectra. Figure 3 (a) illustrates the change in the amplitude of the power spectral density of the peak at 6.0 μm^{-1}.

The spatial power spectra of the fracture surface profiles across traces 6 and 11 are compared in Figure 3 (b). The power spectra for trace 11 also decreases with increasing frequency, as do the spectra for all of the traces. Figure 3 (b) illustrates the change in the amplitude of the power spectral density of the peak at 12.0 μm^{-1}.

The variation of the spectral slopes across the surface is of considerable interest. In the frequency interval 10^0 μm^{-1} to 10^1 μm^{-1} the spectral slope of trace 6 is -3.14; in the frequency interval 10^1 μm^{-1} to 60 μm^{-1} the spectral slope of this trace is -3.92; the slope of

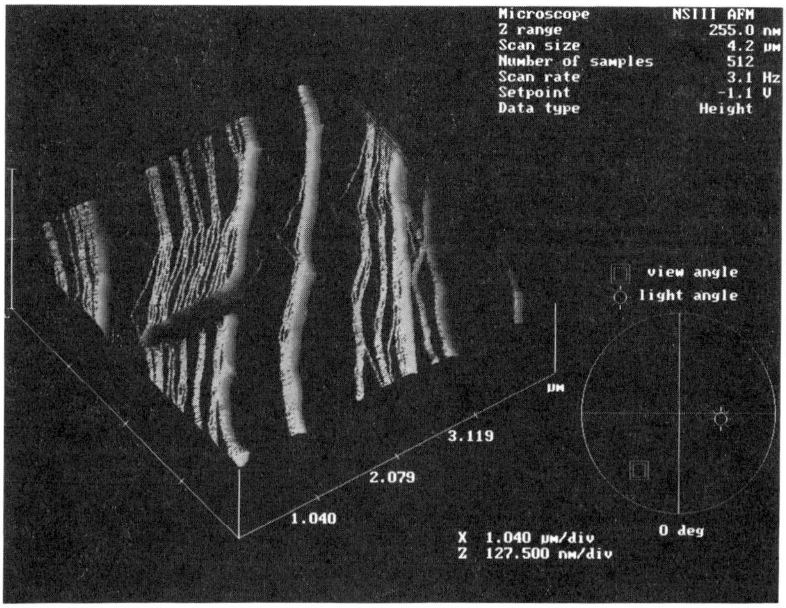

Figure 1. TNT fracture surface

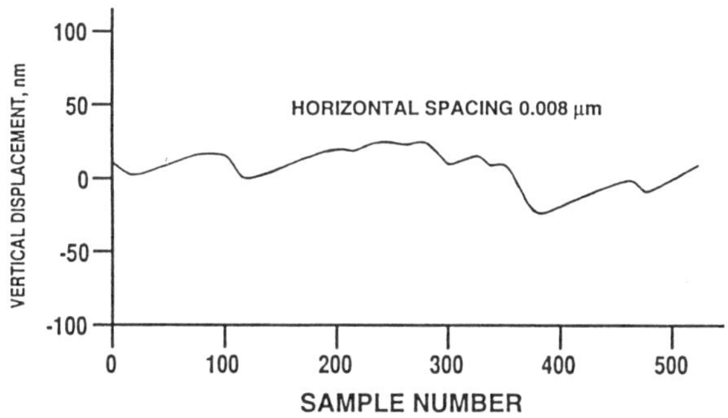

Figure 2. Vertical displacement as a function of sample number.

the overall spectrum is -3.83. Spectral slopes ≤ -3 are obtained for both frequency intervals as well as for the over-all slopes for each of the sample traces 1 through 13, as shown in Figure 4. Typical errors in the slope are shown.

For spectral slopes ≤ -3, the fractal dimension is 1.[11,12] Such a slope suggests that the failure process is deterministic. As we have concluded from our lower resolution studies[2,3], this interpretation is consistent with the occurrence in the spectra of quasiharmonic peaks.

For these higher resolution spectra, peaks are seen to occur at spatial frequencies ≈ 3.0 μm^{-1}, ≈ 6.0 μm^{-1}, and ≈ 12.0 μm^{-1}, are highlighted in Figure 2. These peaks essentially define the size of the fractured material across the full scan. For these spectra, the fractured material can be thought of as large clusters of TNT that are released by the fracture.

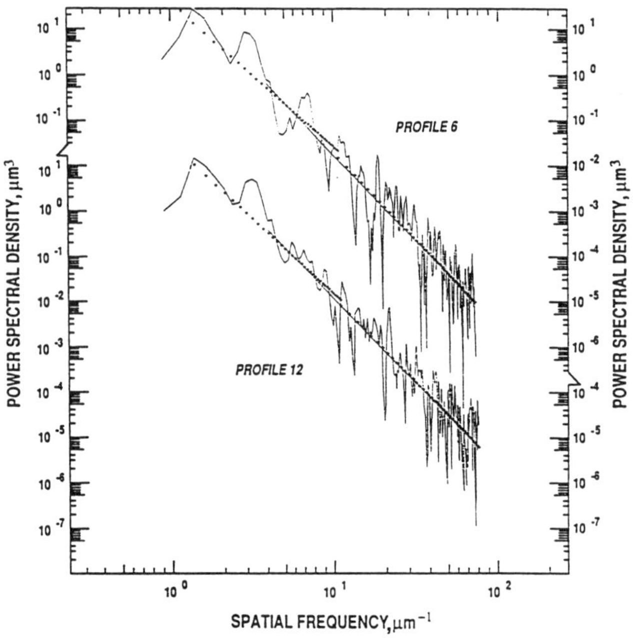

Figure 3. (a) Spatial power spectra of the fracture surface profiles for sample traces 6 and 12 of TNT. The horizontal spacing is 0.008 μm.

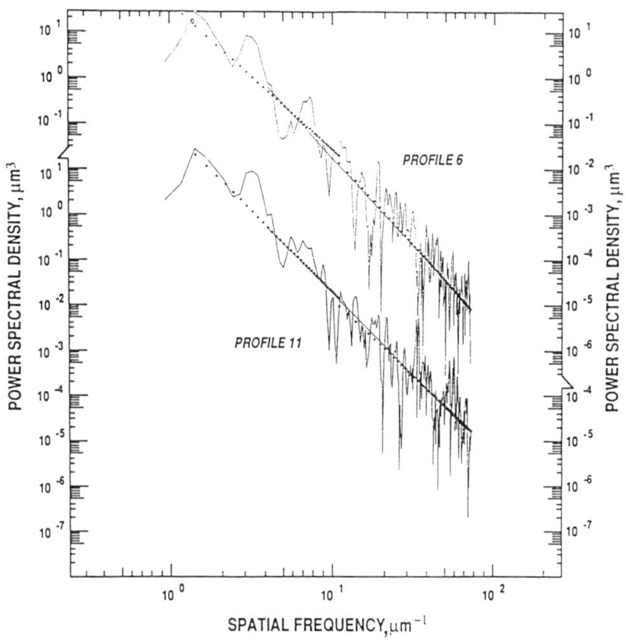

Figure 3. (b) Spatial power spectra of the fracture surface profiles for sample traces 6 and 11 of TNT. The horizontal spacing is 0.008 μm.

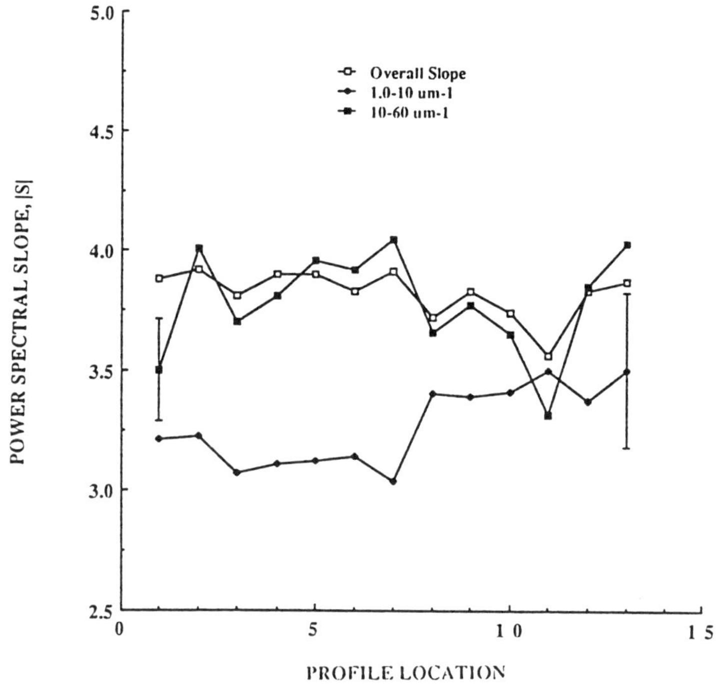

Figure 4. Spectral slopes versus profile location 1 through 13 of TNT.

Also of importance is the nature of the spectral variations across the fracture surface. This has been investigated by determining the amplitudes of the spectral peaks in each of the spectra computed from profile data taken for the sample traces 1 through 13. The power amplitudes for each of the dominant three peaks ($\approx 3.0\ \mu m^{-1}$, $\approx 6.0\ \mu m^{-1}$, $\approx 12.0\ \mu m^{-1}$) are plotted in Figures 5-7 as a function of the profile location. The power amplitudes define the cluster size across the 13 profiles. Cluster sizes of ≈ 800 Å, ≈900 Å, and ≈ 500 Å are suggested by the results of Figures 5-7, respectively.

DISCUSSION

The spatial frequencies and amplitudes of the peaks are consistent and define TNT clusters of the order of 1000 Å in size. Since the TNT molecules have a spacing of about 5 Å[13,14] a typical cluster would then consist of about 10^6 TNT molecules. In summary atomic force microscopy measurements are beginning to provide data and insights into TNT processes on the molecular scale. The relationship between the microscopic clusters and macroscopic fracture remains to be explored, for TNT and other energetic materials.

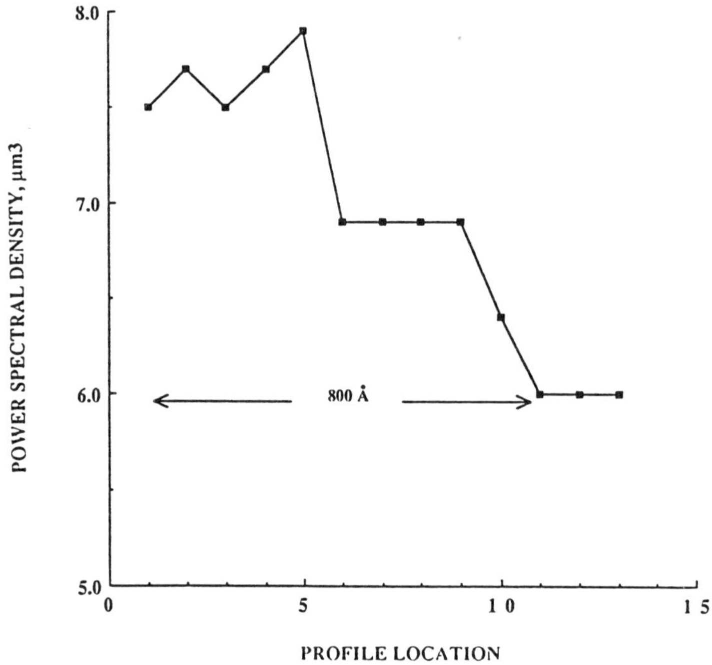

Figure 5. Power amplitudes for spatial frequency 3.0 μm^{-1} as a function of profile location

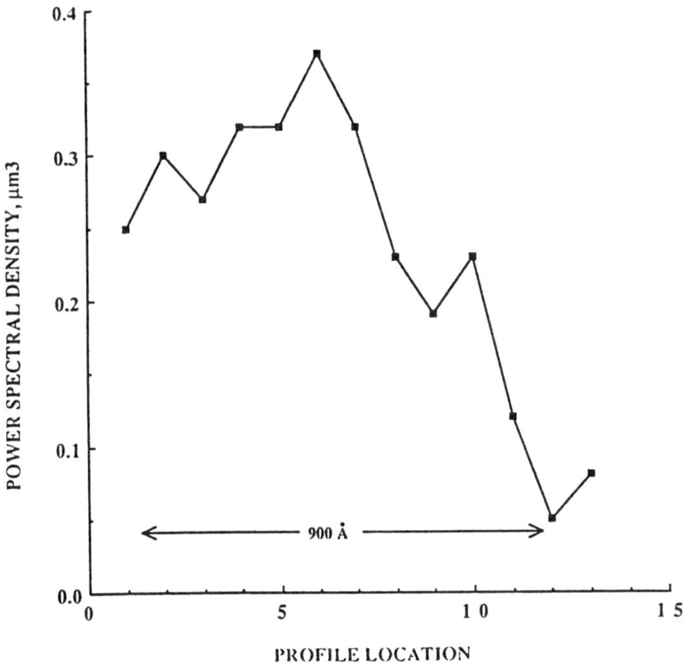

Figure 6. Power amplitudes for spatial frequency 6.0 μm^{-1} as a function of profile location.

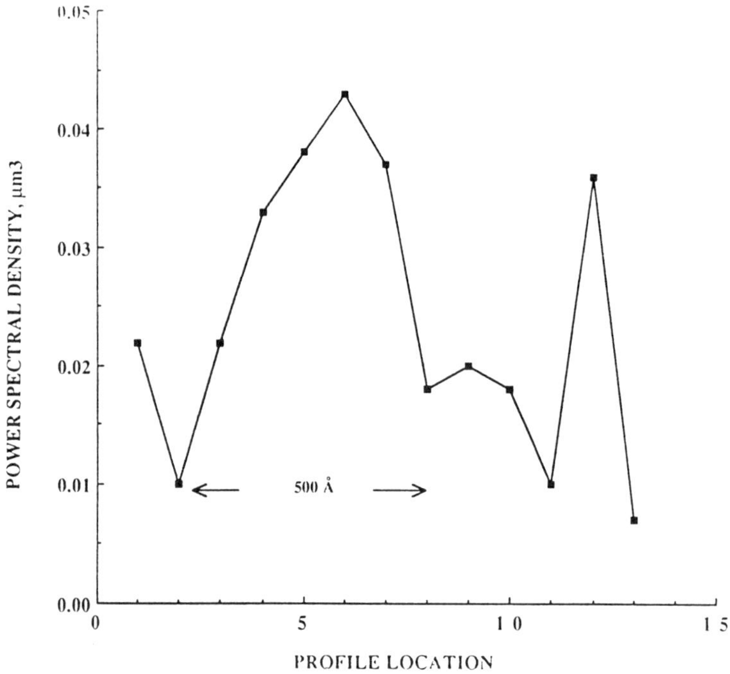

Figure 7. Power amplitudes for spatial frequency 12.0 μm^{-1} as a function of profile location.

ACKNOWLEDGMENT

We would like to thank C. G. Maclennan and J. S. Kraus, AT&T Bell Laboratories, Murray Hill, NJ for helpful discussions regarding the data analysis.

REFERENCES

1. M.Y.D. Lanzerotti, J. Sharma, Brittle behavior of explosives during high acceleration, *App. Phys. Lett.* 39:455-457 (1981).
2. M.Y.D. Lanzerotti, J. Pinto, A. Wolfe, Broad bandwidth study of the topography of the fracture surfaces of explosives, "Ninth Symposium (International) on Detonation," Vol., 1:355-361(1989).
3. M.Y.D. Lanzerotti, J. Pinto, A. Wolfe, Fracture surface topography of TNT, Composition B, and Octol,"Tenth Symposium (International) on Detonation," in press.
4. Digital Instruments Inc., Private communication (1993).
5. A.G.M. Hunter, E.A. Smith, Measurement of surface roughness, *Wear*, 59:383-386 (1980).
6. E.L. Church, P.X. Takacs, Effects of the nonvanishing tip size in mechanical profile measurements, Proc SPIE (SPIE--International Society for Optical Engineering, Bellingham, WA), 1332:504-515 (1990).
7. J.E. Griffith, D.A. Grigg, Dimensional metrology with scanning probe microscopes, *Phys. Rev.* B, in press.
8. D.J. Thomson, Spectrum estimation and harmonic analysis, Proc. IEEE. 70:1055-1096 (1982).
9. D.J. Thomson, Quadratic-inverse spectrum estimates: Applications to paleoclimatology, *Philos.Trans. R. Soc. London*, A 332:539-597 (1990).
10. D.J. Thomson, M.F. Robbins, C.G. Maclennan, L.J. Lanzerotti, Spectral and windowing techniques

in power spectral analyses of geomagnetic data, *Physics of the Earth and Planetary Interiors* 12:217-231 (1976).
11. S.R. Brown and C.H. Scholz, Broad bandwidth study of the topography of natural rock surfaces, *J. Geophys. Rev.* 90:12,575-12,582 (1985).
12. S.E. Hough, On the use of spectral methods for the determination of fractal dimension, *Geophys. Res. Lett.* 16:673-676 (1989).
13. C. Choi, Private communication (1993).
14. H.G. Gallagher and J.N. Sherwood, The growth and perfection of single crystals of trinitrotoluene (TNT), *in*: "Structure and Properties of Energetic Materials," R.W. Armstrong, J.J. Gilman, and D.H. Liebenberg. eds., Materials Research Society Volume 296, in press.

SCANNING TUNNELING MICROSCOPY AND SPECTROSCOPY OF CARBON NANOTUBES

W. Rivera,[1,2] J.M. Perez,[1] R.S. Ruoff,[3] D.C. Lorents,[3] R. Malhotra,[3] S. Lim,[4] Y.G. Rho,[4] E.G. Jacobs,[4] and R.F. Pinizzotto[4]

[1]Department of Physics, University of North Texas
Denton, TX 76203
[2]also: Universidad del Cauca, Popayán, Colombia
[3]SRI International, Menlo Park, CA, 94025
[4]Center for Materials Characterization, University of North Texas
Denton, TX, 76203

Abstract: Scanning tunneling microscopy (STM) and spectroscopy have been used to obtain images and current-voltage (I-V) curves of carbon nanotubes. The nanotubes are 5-7 nm in diameter and up to 1 μm in length. The I-V curves indicate that carbon nanotubes with diameters of approximately 5.0 nm are metallic. This observation is consistent with recent theoretical predictions concerning the electronic structure of carbon nanotubes in relation to their diameter and helicity. Using STM, we also observe graphene sheets in the carbon soot that is produced during the arc discharge. The STM images of nanotubes are in good agreement with transmission electron microscopy images. Potential applications of carbon nanotubes include novel nanometer-size electronic devices and fiber-reinforced materials.

INTRODUCTION

Carbon nanotubes consisting of sheets of carbon atoms rolled into a tube only a few nanometers in diameter were recently discovered in the soot produced from the arc discharge of carbon electrodes.[1] Methods of producing these nanotubes in macroscopic quantities have also recently been reported.[2,3] These developments have stimulated much theoretical and experimental research concerning the structural and electronic properties of these structures.[4] Potential applications of carbon nanotubes and other fullerenes include nucleation sites for diamond film growth.[5,6] One of the most interesting theoretical predictions concerning the electronic structure of carbon nanotubes is that nanotubes are metallic or semiconducting depending on the diameter and helicity of the tube.[7,8] It has been predicted that, for nanotubes with diameters on the order of C_{60} fullerenes, approximately one third are metallic while the other two thirds are semiconducting.[8] In this paper, we report results of scanning tunneling microscopy (STM) and STM current-voltage (I-V) spectroscopy which show, for the first time, nanotubes with diameters of approximately 5.0 nm and greater which are metallic.

METHOD

The STM images and I-V spectroscopy curves were obtained using a commercial scanning tunneling microscope.[9] The carbon nanotubes were grown using arc discharge of carbon electrodes at 500 Torr of He. Growth on the carbon electrodes was removed, crushed, and then placed in methanol under ultrasonic agitation. A few drops were placed on a highly-oriented-pyrolytic graphite (HOPG) substrate and allowed to dry. The STM imaging and I-V spectroscopy were performed in air using a mechanically cut Pt-Ir tip. Typical sample voltages and tunneling currents used for imaging were 75 mV and 1.0 nA, respectively. The I-V spectroscopy data were obtained by disengaging the tip-sample distance feedback loop and measuring the tunneling current as a function of tip-sample voltage at a fixed tip-sample distance. The tunneling current was measured using a 12 bit analog to digital converter and stored in a computer. The use of HOPG, which is metallic, as the substrate instead of a semiconductor such as Si[10] allowed us to measure the I-V curves of nanotubes without interference from the electronic structure of the substrate.

Transmission electron microscopy (TEM) images of the carbon nanotubes were obtained using a JEOL 100CX and a Hitachi H9000 high resolution 300 keV transmission electron microscopes. The carbon nanotube samples were prepared for TEM imaging as described above for STM imaging. However, the substrate used for TEM imaging was a thin amorphous carbon film.

RESULTS

Figure 1 (a) shows a TEM image of carbon nanotubes with diameters of approximately 5.0 nm and greater. Figure 1 (b) shows a TEM image of carbon nanotubes protruding from the carbon soot material which is produced during the arc discharge.

In Figure 2 (a) is a top view STM image of a 1 x 1 μm area of the carbon soot from the arc discharge deposited on HOPG as described above. The STM image shows a strong contrast between the carbon soot and the atomically flat surface of the HOPG. The carbon soot appears in clumps, as observed in the TEM image of Figure 1 (b). Figure 2 (a) shows, in the region indicated by the arrow, aligned nanotubes approximately 110-140 nm in length. The nanotubes protrude from the carbon soot in a manner similar to that observed in the TEM image of Figure 1 (b). Figure 2 (b) has a close-up three-dimensional topographic STM image of the region indicated by the arrow in Figure 2 (a). From Figure 2 (b), the nanotubes are measured to have diameters between 5-7 nm.

Figure 3 (a) shows a top view STM image of a sample with several aligned nanotubes having different diameters. Figure 3 (b) shows a close-up STM image of the nanotubes in Figure 3 (a). The diameter of the smallest nanotube observed is approximately 5.0 nm, in good agreement with the diameter of nanotubes reported using high-resolution TEM.[1]

The fact that the above nanotubes were observed using tunneling currents of 1.0 nA at tip-sample voltages of only 75 mV suggests that these nanotubes are metallic. To test this hypothesis, STM I-V spectroscopy was performed at the four locations indicated in Figure 3 (b). From I-V spectroscopy, one can determine the local electronic density of

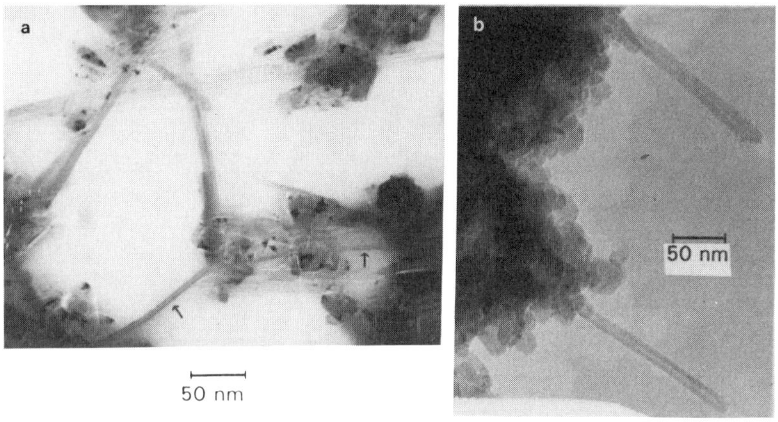

Figure 1. (a) Transmission electron microscopy (TEM) image of carbon nanotubes. (b) TEM image of carbon nanotubes protruding from the carbon soot.

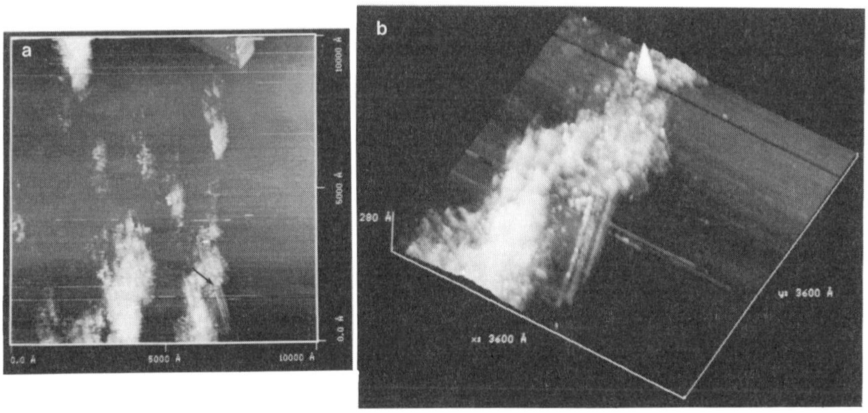

Figure 2. (a) STM image of carbon soot from the arc discharge deposited on an HOPG substrate. The arrow indicates a region where carbon nanotubes protrude from the carbon soot. (b) Close-up three-dimensional STM image of the region indicated by the arrow in Figure 2(a). The observed nanotubes have diameters between 5-7 nm.

states of the surface of a material.[11] In particular, one can determine if the surface is metallic or semiconducting. A metallic surface has a linear I-V curve near 0 V, whereas a semiconducting surface has no measurable tunneling current near 0 V due to the band gap. In Figure 3 (b), location 4 corresponds to the HOPG substrate, and locations 1, 2, and 3 to carbon nanotubes having diameters of 5.0, 6.5, and 7.0 nm, respectively. The I-V curves measured from -80 to 80 mV at all the above locations were linear near 0 V, indicating a metallic density of states. Figure 4(a) shows the I-V curve for the HOPG graphite substrate, and Figure 4(b) shows the I-V curve for the carbon nanotube with diameter of 5.0 nm.

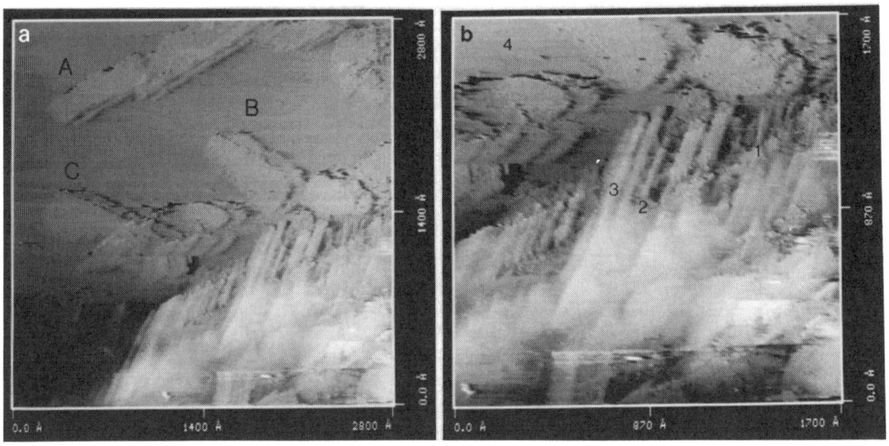

Figure 3. (a) STM image of aligned nanotubes. Regions labelled A,B, and C contain graphene sheets as discussed in the text. (b) Expanded top view STM image of the nanotubes in Figure 3(a). Numbers 1-4 denote locations where STM I-V spectroscopy data were obtained.

Indicated in Figure 3(a) by the labels A, B, and C, are large flat rectangular structures not normally observed on clean HOPG. We identify these structures as graphene sheets produced during the arc discharge. These structures are on the order of 50 nm in width and several hundred nanometers in length. The heights of all observed sheets were approximately 13.5 Å, in good agreement with the thickness of four graphite layers (the interlayer distance in graphite being 3.4 Å). Graphene sheets have been proposed to be produced with nanotubes during the arc discharge,[1] but have not yet been reported. Such atomically thin sheets on the substrate would be difficult to observe using TEM because of the substrate. However, such sheets would be easily observed using STM which images the surface topography.

In conclusion, we have used STM to study the structural and electronic properties of individual carbon nanotubes. Our results confirm the existence of metallic nanotubes with diameters on the order of 5.0 nm and greater, consistent with recent theoretical predictions. We also observe graphene sheets which are four atomic graphite layers thick.

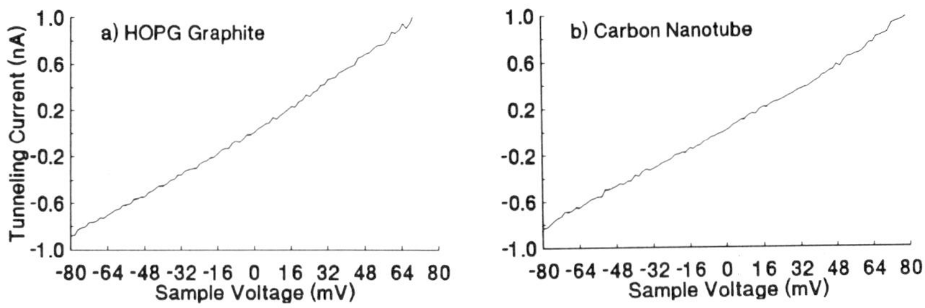

Figure 4. (a) Tunneling current versus sample voltage curve for the HOPG substrate (location 4 in Figure 3(b)), showing linear behavior near 0 V indicating metallic behavior. (b) Tunneling current versus sample voltage for the carbon nanotube with a diameter of 5.0 nm (location 2 in Figure 3(b)), showing linear behavior near 0 V indicating metallic behavior.

ACKNOWLEDGMENTS

We gratefully acknowledge Natalia Murillo and Veronica Criado for their assistance in this work. This work was supported in part by the National Aeronautics and Space Administration under Award No. NAG-1-1468.

REFERENCES

1. S. Iijima, Helical microtubules of graphitic carbon, *Nature* 354:56-58 (1991).
2. T. Ebbesen and P. Ajayan, Large-scale synthesis of carbon nanotubes, *Nature* 358:220-222 (1992).
3. M. Jose-Yacaman, M. Miki-Yoshida, and L. Rendon, Catalytic growth of carbon microtubules with fullerene structure, *Appl. Phys. Lett.* 62: 657-659 (1993).
4. See, for example, Materials Research Society Meeting, Spring 1993, Symposium on Fullerenes and Related Materials.
5. M. Regueiro, P. Monceau, and J. Hodeau, Crushing C_{60} to diamond at room temperature, *Nature* 355: 237-239 (1992).
6. Private communication, Ulvick Industries, Inc., Houston, TX.
7. J. Mintmire, B. Dunlap, and C. White, Are fullerene tubules metallic? *Phys. Rev. Lett.* 68:631-634 (1992).
8. R. Saito, M. Fujita, G. Dresselhaus, and M. Dresselhaus, Electronic structure of chiral tubules, *Appl. Phys. Lett.* 60:2204-2206 (1992).
9. Burleigh Instruments, Inc., Rochester, NY.
10. M. Gallagher, D. Chen, B. Jacobsen, D. Sarid, L. Lamb, F. Tinker, J. Jiao, D. Huffman, S. Seraphin, and D. Zhou, Characterization of carbon nanotubes by scanning probe microscopy, *Surf. Sci. Lett.* 281: L335-L340 (1993).
11. R. Feenstra, Scanning tunneling microscopy: semiconductor surfaces, adsorption, and epitaxy, *in:* "Scanning Tunneling Microscopy and Related Methods," R.J. Behm, N. Garcia, and H. Rohrer, eds., Boston: Kluwer, 211-240 (1990).

SCANNING TUNNELING MICROSCOPY AND ATOMIC FORCE MICROSCOPY INVESTIGATIONS ON ORGANIC MATERIAL THIN FILMS AND ADSORBATE PARTICLES IN AIR

Michael Hietschold,[1] Olaf Pester,[1] Waltraud Vollmann,[1] Andreas Heilmann,[1] Peter Stabeler,[1] Hartmut Sbosny,[1] Xina Grahlert,[1] Hans-Ulrich Sonntag,[1] Anna Bruska,[2] Bettina Winzer,[3] Thomas Schimmel,[4] and Ludger Koenders[5]

[1]Department of Physics, Technical University of Chemnitz-Zwickau, Germany
[2]Institute of Physics, N. Copernicus University, Torun, Poland
[3]Max Planck Gesellschaft, Heterogeneous Solids Group, Dresden, Germany
[4]Institute of Experimental Physics VI, University of Bayreuth, Germany
[5]Physikalisch-Technische Bundesanstalt Braunschweig, Germany

Abstract: We present investigations on polycrystalline thin films of TCNQ and of metal-phthalocyanines as well as of metal-polymer composite material films by scanning tunneling and atomic force microscopy at ambient conditions. The influences of preparation conditions (usual thermal evaporation in high vacuum) as well as external influences such as applied external pressure on the crystallite morphology, surface topography and imaging conditions are discussed. Using a liquid cell in an atomic force microscope, we have been able to resolve the molecular structure of copper- and of lead-phthalocyanine crystals. For rough surfaces as discussed above the influence of the mesoscopic tip shape has to be taken into account. We report on theoretical modelling as well as on experimental efforts related to STM imaging using proper tips. In the case of ultrathin organic films, there is often the problem of fixing the molecules to the substrate. We show that in this case after some rather harsh sweeping by the tip, one can observe fairly ordered molecular structures near to the lower part of surface steps on layered material substrates.

INTRODUCTION

Scanning tunneling microscopy (STM) and atomic force microscopy (AFM) have been shown to be very useful in high-resolution imaging of organic materials (for a review see Reference 1). The reason for this is the completely different nature of the characteristic binding forces in the surface region of such substances. There are dominant van der Waals forces and the corresponding binding energies are of the order of 0.1 eV or even much smaller, which is in contrast to 1 ... 10 eV encountered in covalently bonded organic

Atomic Force Microscopy/Scanning Tunneling Microscopy
Edited by S.H. Cohen *et al.*, Plenum Press, New York, 1994

solids. This circumstance allows imaging of many organic materials at ambient conditions. The force exerted by the probing tip to the sample surface is then just sufficient to remove surface contaminations and weak enough not to destroy the bulk cohesion. Embedding of the sample in a proper liquid may support the specific dissolution of the surface contamination layer without weakening the sample stability. We have used this phenomenon for imaging different organic samples with STM and AFM in ambient conditions. In the second section we report on investigations of polycrystalline organic thin films on a mesoscopic scale. The surface topography of TCNQ, lead-phthalocyanine, and metal-polymer composite films is related to the deposition conditions and to other external influences. On a special composite structure even spectroscopy in atmosphere is useful. Next we discuss high-resolution attempts to single crystal surfaces of two metal-phthalocyanines. The attempts revealed a direct image of the molecular stacking at the crystallographic (001) surfaces. We then represent a model that explains some theoretical aspects of the topographic imaging process that are especially important in the case of mesoscopical rough surfaces. The last section is devoted to molecular agglomerations near surface defects of layered materials as, e.g., lead-phthalocyanine molecules on HOPG.

TOPOGRAPHY OF POLYCRYSTALLINE ORGANIC THIN FILMS

In this section we report on topographic imaging on polycrystalline thin films, which have been prepared by thermal evaporation in high vacuum.

TCNQ

TCNQ (7,7,8,8-tetracyanoquinodimethane) films of 1.6 μm thickness have been fabricated by thermal evaporation of TCNQ powder on polished steel substrates. Thin films of this type are used in a special force sensor.[2] Two of them are put face to face whereas their substrate backsides are directly exposed to an external force. This device shows a characteristic dependence of the electrical conductivity across the films from the external force applied. Figure 1 shows the AFM top view of a TCNQ film before use in a force sensor.

Figure 2 demonstrates what happens to the film after using it as part of a force sensor. In the large-scale AFM top view of Figure 2 (a) there are two different regions: those that are definitely disturbed due to the external compression (a result of some hydrodynamic flow of the crystallites, which seem to be baked together) and others that exhibit crystallites clearly separated from each other and that are at the first glance not seriously disturbed. More information can be obtained using a sharply etched tip in a STM to zoom onto the crystallite slopes. Figure 2 (b) is a three-dimensional representation of the region between two adjacent crystallites for an uncompressed sample (as shown in Figure 1). No characteristic features are found either on the flanks of the crystallites or in the intercrystallite region. Figure 2 (c) shows a crystallite flank side (after tilt removal) from the seemingly undisturbed region of Figure 2 (a). One clearly observes some wrinkles resulting from the mechanical stress that has been applied and also some holes. (The very nature of these features is not yet completely clear. It might be that these are dislocations entering the surfaces, which stabilize the crystallite morphology after the first compressions and allow reproducible results with the sensor.)

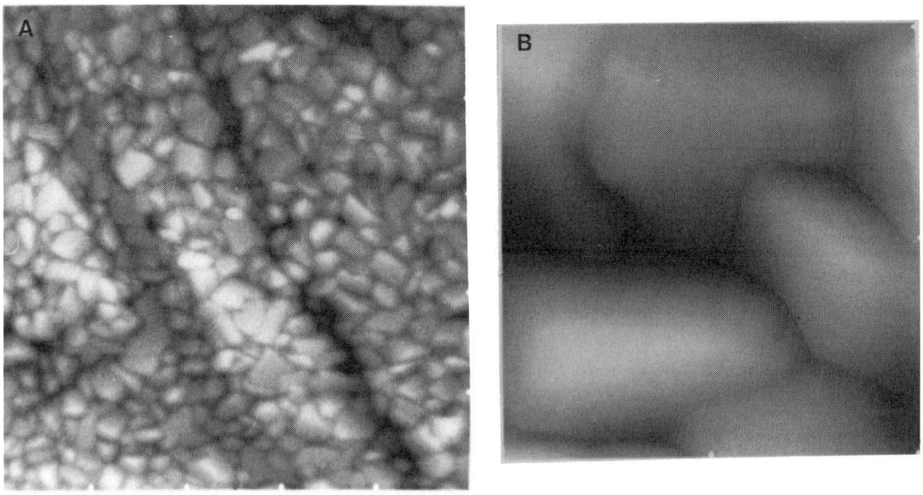

Figure 1. The AFM top view of a TCNQ film before using in a force sensor: (a) A large-scale image showing the polycrystalline organic film deposited on the substrate just above some scratch of the latter; (b) A zoomed part of Figure I (a), which visualizes the complete coverage of the substrate by irregularly shaped crystals. (This crystal shape encountered here is different from that of isolated TCNQ crystals or of films deposited on a protective layer, which is probably due to metal atom diffusion from the substrate followed by the formation of metal-TCNQ complexes.[3])

Lead-Phthalocyanine Films

Polycrystalline thin films of lead-phthalocyanine evaporated in high vacuum are known to exhibit various crystalline modifications depending on the deposition conditions.[4] Scanning probe microscopy (SPM) methods allow one to judge the crystalline modification encountered in a very direct manner. Figures 3 (a) and (b) show AFM top views of thin films, which are, according to diffraction results, mainly amorphous (a) and purely triclinic (b). In the latter case a large crystallite is exposed. The film has been evaporated on silicon substrates at temperatures of 20°C and 220°C, respectively.

It has to be stressed that such a flat and relatively structureless surface topography as shown in Figure 3 (a) exhibits no features at all in a SEM - a shining example for the superior vertical resolution of the SPM methods. Furthermore, it should be mentioned that it is seemingly impossible to image these phthalocyanine thin films by STM or AFM at ambient atmosphere without at least slightly but continuously eroding the sample surface: We have always observed "windows" at the place of the former scanning area. This phenomenon is not very serious in imaging on a mesoscopic scale (because in spite of the erosion the surface topography is reflected very well at least by the first scans) but it is prohibitive for high resolution imaging.

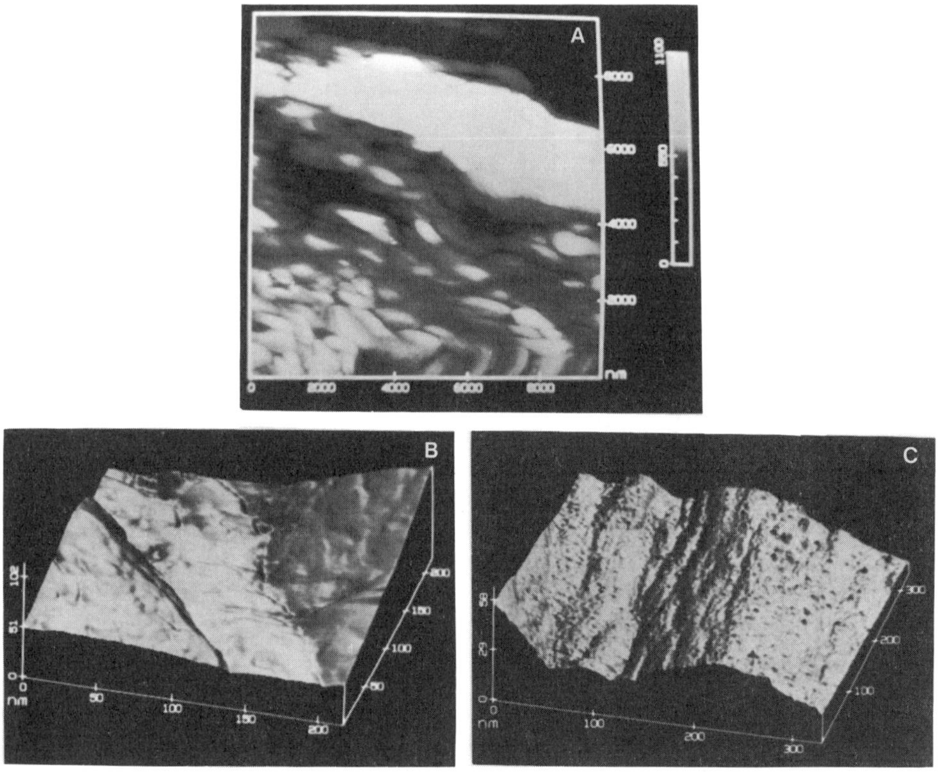

Figure 2. (a) Film after use as part of a force sensor: two different regions; (b) two crystallite flanks and the region between of an uncompressed sample; (c) Crystallite flank side from seemingly undisturbed region of Figure 2 (a).

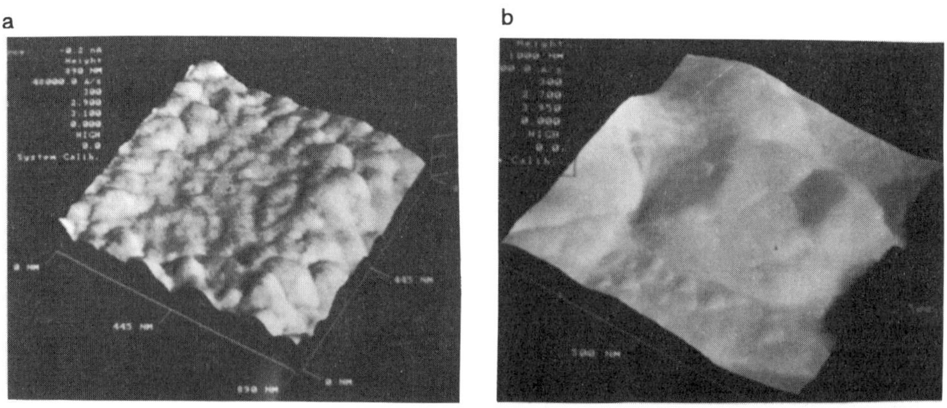

Figure 3. (a) Top view of thin film, mainly amorphous; (b) Top view of thin film, purely triclinic.

Metal-Polymer Composite Films

Very interesting samples are thin films consisting of small metal particles embedded in a polymer matrix. The latter is fabricated by dc plasma polymerization using hexamethyldisilazane as a monomer. The metal is evaporated simultaneously to the plasma polymerization but in the final period there is only polymer deposited.[5] The metal-containing layer exhibits a percolation transition from the insulating to a metallic state. Films prepared along this line can be investigated by STM in air due to some surface conductivity which results from contamination. But there is no imaging of pure topography since metal clusters embedded close to the surface create peaks in the local conductivity. Figure 4 is a combined STM/STS picture. On the left hand side, there is shown the surface topography obtained in the constant current mode at biases U_T = 60 mV and U_T = -30 mV in the upper and the lower quadrant, respectively. On the right hand side, the differential local surface conductivity dU_T/dI_T at the same U_T is imaged. One can distinguish between features that are present in all of these images - they belong to the real surface topography. On the other hand, there are spots which appear bright on one and dark on the other image - these features are closely related to metal particles that are approaching the sample surface.

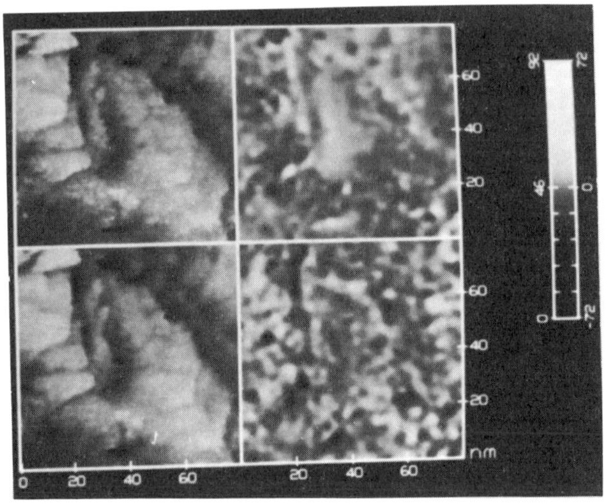

Figure 4. Metal polymer composite film: combined STM/STS view.

MOLECULAR STACKING AT METAL-PHTHALOCYANINE CRYSTALS

We have been able to image the molecular stacking in Cu- and Pb-phthalocyanine single crystals.[6] Due to the sample surface erosion observed when scanned in ambient air, mentioned above, we tried to get AFM images in a liquid cell. Ethanol has been used as an imaging liquid which has two effects: cleaning the sample and tip surfaces and removing the adhesion layer always present in ambient atmosphere due to natural humidity. This adhesion layer strongly disturbs gentle imaging with well defined small repulsive forces.

The samples used are Cu-phthalocyanine single crystals grown from the gas phase. They are needle-shaped paralellepipeds belonging to the monoclinic phase. The largest crystallographic face exhibited is (001).

Figure 5 (a) is a 10 x 10 nm^2 image of such a (001) surface. There is a pronounced periodic arrangement of rows with 11 Å periodicity. A closer look reveals an additional 6 Å period along the rows. Figure 5 (b) is a two-dimensional Fourier transform of Figure 5 (a). One clearly observes both of the periodicities mentioned from the real space image. The periodicities agree fairly well to the distances between adjacent molecules on the (001) surface. The surface unit cell is larger by a factor of two in each direction which is due to different orientation of the adjacent molecules. The molecular features observed here are mainly due to the electron density relatively far away from the molecule which is going to become essentially more and more spherically shaped with increasing distance from the molecule's center. (We have recently established this by first results from Hartree-Fock molecular calculations.[7]) We have also been able to resolve the molecular stacking on a (001) Pb-phthalocyanine surface of the monoclinic crystal phase.[6]

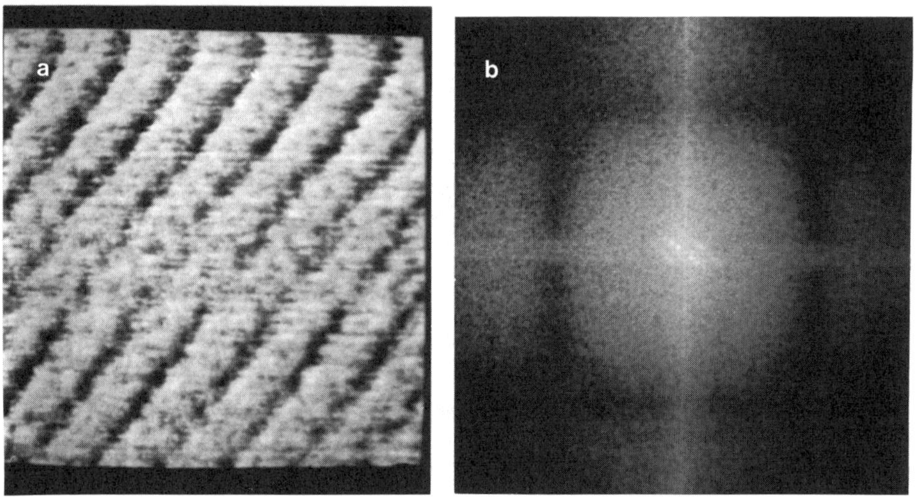

Figure 5. (a) 10 x 10 nm^2 image of (001) surface; (b) two-dimensional Fourier transform of Figure 5 (a).

THEORETICAL ASPECTS OF THE TOPOGRAPHIC IMAGING PROCESS

Especially in the case of imaging surfaces that are rough on a mesocopic scale (as e.g., in the second section above), the influence of the tip shape on the image has to be considered. We have developed a very simple but extremely versatile model[8,9] to estimate, at least qualitatively, these effects. The model for describing tip and sample electrodes is essentially a Sommerfeld model consisting of independent electrons enclosed in a square -well potential. The shape of the electrodes can be arbitrary. We have solved Schroedinger's equation for the electron state considered numerically either in two or three dimensions. Using the standard model of STM theory,[10] we have been able to calculate constant current as well as constant height image profiles for almost arbitrary shape of both tip and sample electrode. Furthermore, this model includes tip-sample interaction, since both electrodes can be treated at once (that means that the electron state common to the whole system is calculated).

Figure 6 demonstrates the tip-sample interaction on a two-dimensional example between a spherical tip and a rectangular sample electrode. There is represented the density of one special electron state belonging to the whole system which is responsible for the STM imaging with certain tunneling parameters.

Figure 7 visualizes a constant current profile taken over a sample notch. From this image one can get an impression of the phenomenon that images of rough surface profiles are very often strongly falsificated due to the finite geometrical extensions and the particular shape of the actual tip used within the instrument.

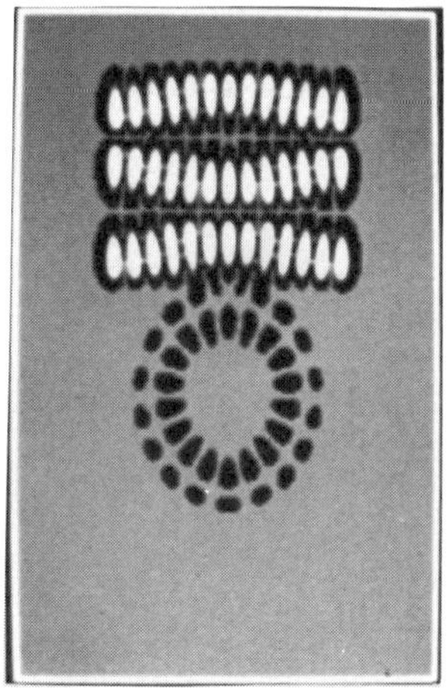

Figure 6. Tip sample interaction between spherical tip and rectangular sample, 2-D.

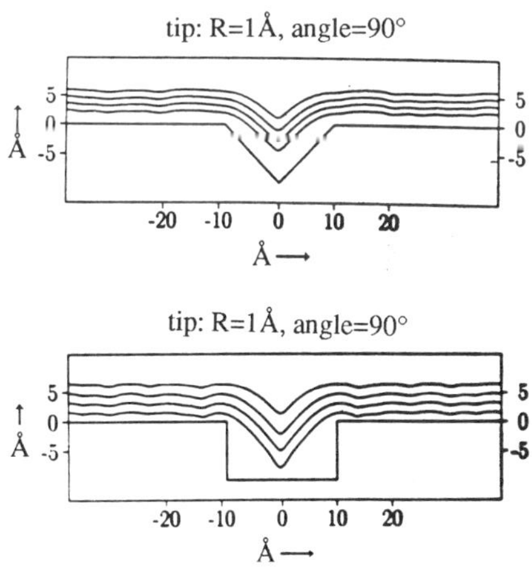

Figure 7. Drawing of constant current profiles taken over different sample notches.

MOLECULAR AGGLOMERATIONS OF ORGANIC MOLECULES ON SURFACE DEFECTS OF LAYERED MATERIALS

As mentioned already, some organic thin films can not be imaged without surface erosion. We have consequently removed a Pb-phthalocyanine thin film from a HOPG substrate by continuous scanning so that finally the graphite atomic structure could be observed. Near the step edges of the substrate sometimes additional structures could be seen.[11]

Figure 8 shows as an example the STM image of a step system on a otherwise atomically flat HOPG surface (the scanning area is 40 x 40 nm^2). Included is a linear arrangement of single Pb-phthalocyanine molecules. We could even watch hopping of single molecules out of this chain and back in subsequent scans from the same area.

The possibility to modify such structures is indicated on Figure 9. A bundle of adjacent molecular chains (roughly 30 nm broad) is imaged after a voltage pulse has been applied to the z-piezo of the STM pushing it onto the surface. Whereas the very elastic graphite layers are not disturbed by this procedure, the bundle is cut on a molecular level.

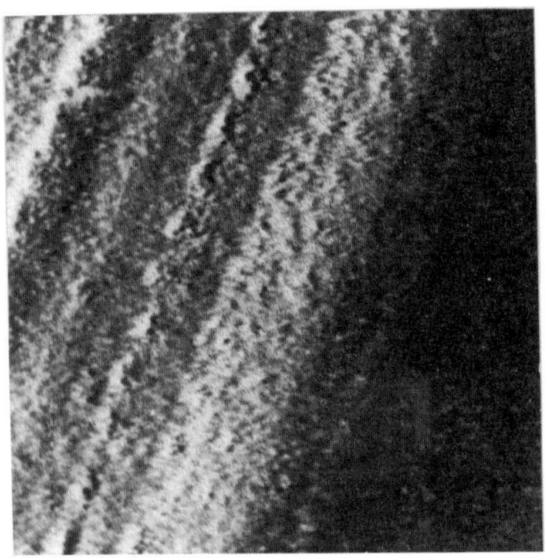

Figure 8. STM image of step system on flat HOPG surface, scanning area 40 x 40 nm^2. The chain of dots is made of single Pb-phthalocyanine molecules.

CONCLUSIONS

We have reported on a variety of cases of organic materials where STM and AFM imaging at ambient conditions can give valuable results. The systems investigated are of different chemical composition (metal-phthalocyanines, TCNQ, and a metal-polymer composite system) as well as of various physical states (thin films, crystals and single

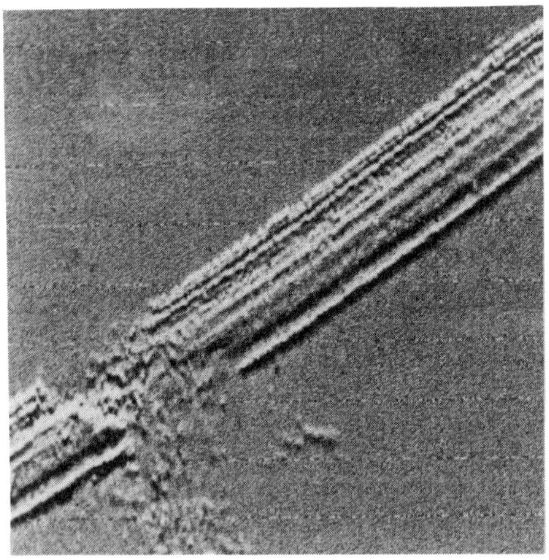

Figure 9. Image of adjacent molecular chains after a voltage pulse.

physisorbed molecules). This relates especially to those investigations concerned with mesoscopic dimensions. High resolution molecular imaging is often prevented by surface contamination and adhesion effects in ambient atmosphere. In such cases AFM in a liquid cell may be successful. For a correct interpretation of images as well on the molecular as on the mesocopic level, a theoretical modeling turns out to be very helpful.

ACKNOWLEDGMENTS

The authors from Chemnitz are grateful to TopoMetrix for lending a AFM TMX 2000. Part of the theoretical work concerning the imaging process has been carried out in terms of an agreement with the Physikalisch-Technische Bundesanstalt Braunschweig. One of us (M.H.) acknowledges the kind hospitality at the University of California at Santa Barbara where part of this work was carried out.

REFERENCES

1. G. Bar, S.N. Magonov, H.-J. Cantow, J. Gmelner, and M. Schwoerer, Atomic-scale imaging of anisotropic organic conductors by scanning probe techniques (STM/AFM), *Ultramicrosc.* 42:644 (1992).
2. W. Vollmann and G. Adam, Druckabhangigkeit der elektrischen Leitfahigkeit organischer TCNQ-Komplexschichten als Basis fur Druck-und Kraftsensoren, *Wiss. Zeitschr.* TU Karl-Marx-Stadt 32:85 (1990).
3. M. Hietschold, W. Vollmann, A. Mrwa, A. Heilmann, and P.K. Hansma, STM and AFM investigations on organic solids, *phys. stat. sol.* (a) 131:59 (1991).
4. A. Mrwa, H. Giegengack, H.-J. Hohne, M. Starke and C. Hamann, Zur Struktur von Bleiphthalocyanin-Dunnschichten, *Wiss. Zeitschr.* TU Karl-Marx-Stadt 31:229 (1989).
5. A. Heilmann and C. Hamann, Deposition, structure, and properties of plasma polymer metal composite films, *Progr. Colloid Polym. Sci.* 85:102 (1991).
6. O. Pester, M. Hietschold, and D. Porezag, Molecular resolution images of copper- and lead-phthalocyanine single crystals by atomic force microscopy, submitted to *Ultramicroscopy*.
7. M. Hietschold, G. Wolff, and D. Porezag, to be published.

8. M. Hietschold, and H. Sbosny, Simple model for the topographic imaging process in a scanning tunneling microscope. *Ultramicrosc*. 42-44:200 (1992).
9. H. Sbosny, M. Hietschold, and L. Koenders, to be published.
10. J. Tersoff and D.R. Hamann, Theory of the scanning tunneling microscope, *Phys. Rev.* B 31:805 (1985).
11. B. Winzer, Th. Schimmel, M. Hietschold, J. Kuppers, and A. Mrwa, to be published.

IMAGE CONTRAST MECHANISMS AND TOPOLOGY OF POLYETHYLENE SINGLE CRYSTALS: LOW-VOLTAGE, HIGH-RESOLUTION SCANNING ELECTRON MICROSCOPY AND ATOMIC FORCE MICROSCOPY

Hao Jiang,[1] S.S. Patnaik,[2] P. Haaland,[1] D.L. Vezie,[3] T. Bunning,[4] J. Williams,[5] and W. Wade Adams[4]

[1]Lawrence Associates Incorporated
Dayton OH 45431
[2]University of Virginia
Charlottesville, VA 22904
[3]Massachusetts Institute of Technology
Cambridge, MA 02139
[4]Wright Laboratory/MLPJ, Bldg 651
Wright Patterson Air Force Base, OH 45433-7702
[5]Universal Energy Systems, Inc.
Dayton OH 45432

Abstract: Polyethylene (PE) single crystals were chosen to study the interaction of low energy (0.8 - 5 keV) electron beams with polymeric materials using a low-voltage, high-resolution scanning electron microscope (LVHRSEM). The charge contrast mechanism was investigated as a function of the conductivity of the substrates, the accelerating voltage, and the sample thickness (from monolayer to multilayers). Comparison of a series of substrates from insulators (such as mica) to conductors (such as aluminum) showed a wide range of image contrast. The best contrast case was PE on carbon film on copper grids (normally used for transmission electron microscopy (TEM)) at an accelerating voltage of 1 kV. The effect of the incident voltage on sample contamination and radiation damage of PE samples was also studied. In addition, topological features of PE single crystals as observed by atomic force microscope (AFM) were compared with the results from LVHRSEM.

INTRODUCTION

In recent years, the evolution of low-voltage high-resolution scanning electron microscopy (LVHRSEM) has provided a new technique for the study of the morphology of organic macromolecules. Although conventional transmission electron microscopy (TEM) and scanning electron microscopy (SEM) have been extensively used for observing morphological structure, many organic materials are easily radiation damaged by the electron beam. Polyethylene (PE) single crystals, are damaged to varying extents when conventional accelerating voltages of 30 to 120 keV are used.[1]

The advantages of LVHRSEM as a polymer imaging technique are the much lower operating voltages and high brightness of the field emission gun, which can reduce and even eliminate charging in many uncoated insulating polymeric organic materials to yield topographical images with lateral resolution on the order of 5.0 nm. In the present work, PE single crystals, whose morphology has been extensively investigated[1-5] were chosen to study the interaction of the electron beam with organic materials by LVHRSEM. Included in this study are investigations of radiation damage, contamination and contrast mechanisms at low accelerating voltages (from 800 eV to 5.0 keV).

With the advent of atomic force microscopy (AFM), another novel technique has become available which can be used to visualize the surface topography of materials. It has been reported that AFM can profile surfaces at resolutions on the order of the molecular and even the atomic levels. The spatial resolution achieved in AFM images is beyond the limits of conventional electron microscopy (EM) techniques.[6,7] Because charging effects are not problems, AFM offers a unique perspective for the study of solid morphology of nonconductive materials such as polymers. The surface morphology of PE single crystals was imaged by AFM and the topological features were compared with those observed by LVHRSEM.

SAMPLE PREPARATION

PE single crystals were grown by the self-seeding technique.[2-5] Linear PE (Marlex 6009 with molecular weight around 200,000) was dissolved in xylene to form a 0.1 (wt)% solution. The solution was boiled under nitrogen atmosphere for an hour. Crystal suspensions were prepared by quenching the solution to room temperature. The suspension solution was then heated slowly (10° C/h) to 100° C (clearing temperature) and kept at 100° C for 1 h. Crystals were quickly grown by rapidly quenching the solution to 80° C (crystallization temperature) and holding at this temperature for 24 h to allow crystal growth. The crystals (in solution) were then deposited on various substrates at room temperature. The substrates used were mica, carbon coated mica, TEM copper and gold grids, carbon film coated TEM copper grids, single crystal graphite and aluminum foils. The crystals were examined with a Hitachi S-900 HRSEM, and Park Instruments AFM under ambient conditions.

RESULTS AND DISCUSSION

It is well-known that PE single crystals are sensitive to electron damage as they change their shapes and structure under high voltage irradiation.[1] It has been reported that with conventional TEM at 100 keV, PE single crystals deposited on a soft collodion film (which allowed motion) could expand by over 50% in area (equally in both a and b axis direction),[8] and reduce their thickness (c axis direction) by about 40% with radiation damage.[1] Figure 1 shows that at a 1.0 keV accelerating voltage in LVHRSEM and with different exposure times (up to 50 min.), the lateral sizes of PE single crystals (deposited on collodion-film coated TEM copper grids) remained unchanged. We also found that at low accelerating voltages (from 800 eV to 2.0 keV), the uncoated PE crystals did not change their apparent shapes. Higher accelerating voltages (above 2.0 keV) could not be used due to excessive sample charging. At a very low accelerating voltage (800 eV), contamination on the surface of the PE single crystals could be seen for high magnifications (>30,000) (Figure 2). This means that even with LVHRSEM, electron irradiation gives rise to scission and crosslinking of hydrocarbon molecules with formation of carbon-rich contamination features on the surface of the insulating organic materials.

An insulating specimen observed by SEM can cause charging either positively or negatively, depending on the total electron yield. The so-called crossover voltage (E_2) is the accelerating voltage at which the number of primary electrons incident to the specimen equals the total number of electrons exiting the specimen (electron yield = 1). Theoretically, no charging should occur at the crossover voltage. If more electrons strike

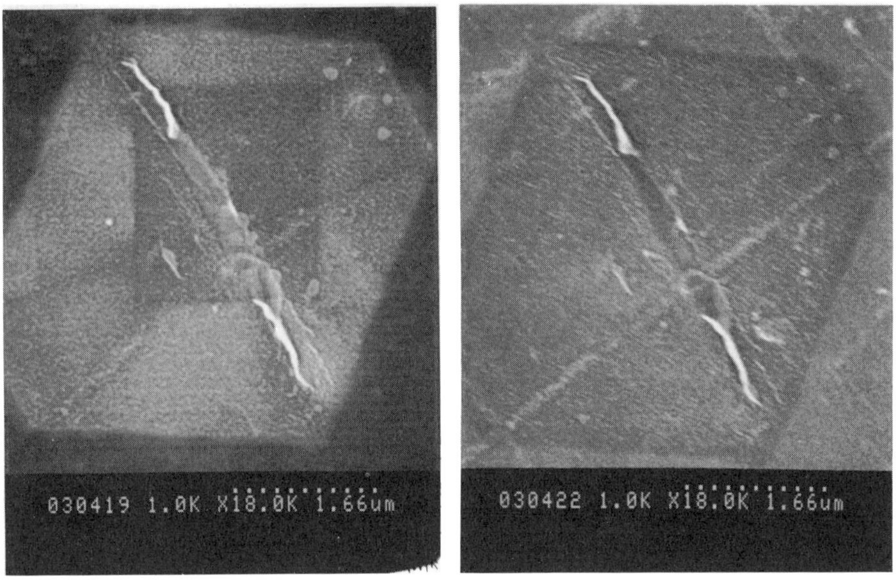

Figure 1. With 1.0 keV accelerating voltage and different exposure times (up to 50 min), the LVHRSEM images of PE single crystals (deposited on collodion film coated TEM copper grids) are similar to each other.

the specimen than can leave (electron yield < 1), the sample becomes negatively charged (image is bright); similarly, if more electrons are emitted than hit the sample (electron yield > 1), the excess emission leads to positive charging of the specimen surface (image is dark). Charging results in poor resolution and produces image artifacts. Generally, charging can be avoided by coating the insulating samples with a conducting layer (metal or carbon). Unfortunately, in the case of high magnification and high resolution, the coating layer can obscure underlying morphological details. One of the advantages of LVHRSEM is the ability to examine samples at low voltage without significant loss of resolution. The small probe size (approximately 4.0 nm at 1.0 keV for the S900) and the small interaction volume at low beam energy combine to provide good resolution with excellent topological image contrast. This can often decrease and even eliminate the usual

Figure 2. Contamination on the surface of a PE single crystal (on carbon-coated TEM grid) caused by electron irradiation at 1.0 keV.

charging problem. However, in some cases, charging can be used to enhance the image contrast. For example, PE single crystals, like many organic and polymeric materials, are inherently difficult to image with conventional SEM techniques because of their small morphological features (heights typically 10 nm) and low atomic number, which results in poor electron scattering. When looking for these thin platelet crystals at low magnifications, negative sample charging is helpful! Therefore, the control of the amount and type of charging is desirable. In order to understand the contrast mechanism of organic materials observed by LVHRSEM, the image contrast of PE single crystals at low voltage was investigated as a function of accelerating voltage and different substrates.

Figure 3 shows the influence of the accelerating voltage (from 0.8 to 5.0 keV) on image contrast in PE single crystals on carbon-coated TEM copper grids. The SEM image brightness changes and becomes gradually darker with increasing voltage, but even at 5.0 keV the image is still bright.

The substrate can also affect the amount of charging in specimens that are thinner than the electron range. Figure 4 shows that on insulating mica (coated with an approximately 10 Å carbon layer), the LVRHSEM image taken at 1.0 keV accelerating voltage is bright (corresponding to the negatively charged sample), but on aluminum foil, the image is dark (corresponding to the positively charged crystal). Table 1 lists the different contrasts of the PE single crystals observed with different substrates. All the metal substrates were found to make the deposited monolayer crystals positively charged.

In addition to the accelerating volting and the sample substrates, the exposure time of the electron irradiation also affects the image contrast. For example, at 3 to 5 keV accelerating voltage, after exposure for several minutes, the initial bright crystal image changed to a darkened contrast. Several factors may affect this phenomenon, including the formation of a contamination layer. Therefore, many factors, including the accelerating voltage, substrates and the exposure time, affect the image contrast of insulating organic macromolecules.[9] Our preliminary results indicate that with a Hitachi S-900 HRSEM, the best choice for clear images at an accelerating voltage of 1.0 keV is to deposit the PE single crystals on carbon-coated TEM copper grids.

Table 1. Different contrasts of PE crystals on different substrates (examined at accelerating voltage of 1.0 keV)

Substrate	Contrast caused by charging
Mica	bright, low contrast, difficult to observe
Carbon coated mica	bright, fair contrast
carbon coated copper TEM grids	bright, good contrast, easy to observe
Single crystal graphite	bright, poor contrast, difficult to observe
Gold TEM grids	dark, difficult to observe
Copper TEM grids	dark
Aluminum foil	deep dark, very difficult to observe

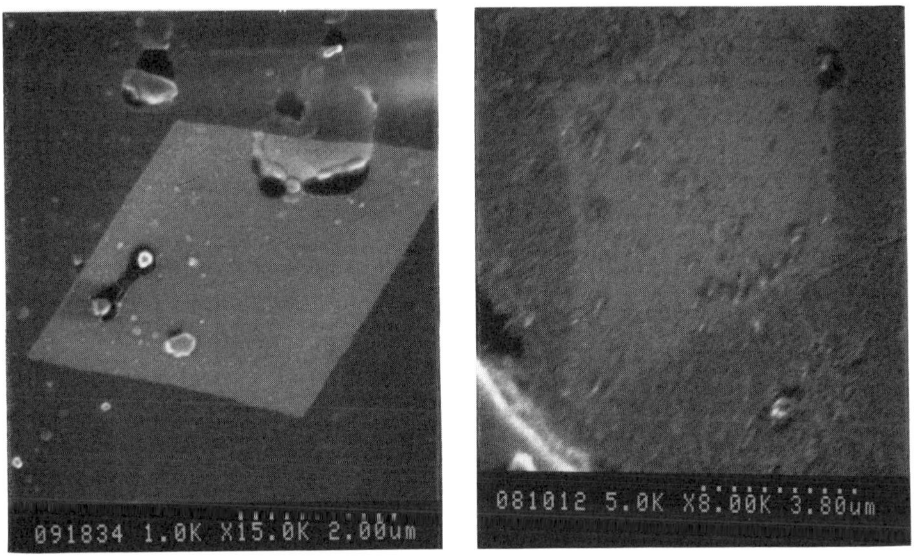

Figure 3. The influence of the accelerating voltage: with increasing voltage from 0.8 to 5.0 keV, the LVHRSEM image brightness of PE single crystals on carbon-coated TEM copper grids decreased.

The surface morphology of the PE single crystals was also investigated by AFM and compared with the LVRHSEM images. Both the LVHRSEM and AFM (Figure 5) images show that most crystals have bumps and undulations on their surface. Such surface feature has been observed elsewhere in the literature.[10] Some areas have pronounced bumps, whereas other surfaces are comparatively smooth. It is interesting to note that the AFM disclosed a periodic structural feature on the crystal surface (Figure 6) with periodicity of

Figure 4. Different charge contrast: on insulating mica (coated about 10 Å carbon) the LVHRSEM image, taken at 1.0 keV accelerating voltage, is bright (corresponding to the negatively charged crystals), but on aluminum foil, it is dark (corresponding to the positively charged crystals).

about 100 Å. The source of the periodicity is unclear and is the subject of work in progress.

Figure 7 shows the regular rhomboid shape expected for the PE crystal as well as multilayer spiral terraces of the PE folded chain lamellae growing on basal folded chain lamellae. The thickness of most crystal lamellae was found to be approximately 80 to 150 Å (Figure 8). However, in one case, a 15 - 20 Å growth spiral layer was also observed (Figure 9). This indicates that the original crystal growth may be different from the growth in later stages. The mechanism for basal lamellae growth (monolayers) may differ from that for the successive over-growth layers. Some single crystals exhibit clear boundaries between different sectors on both SEM and AFM micrographs (Figure 10), which could be caused by the collapse of the hollow pyramidal lamellae. Some appear as flattened platelets.[2] Although the fold domain boundaries are visible in some cases, no difference in height between the adjacent crystal sectors was measurable. This implies that when deposited on the substrate, many single crystals collapsed and the crystal sectors slipped along the substrate, flattening the crystals.[2] Some AFM images (Figure 11) show thicker regions at the edge of the crystals. The PE crystals were cooled from 80 °C to room temperature when they were deposited on the substrate. Low-molecular-weight chains would crystallize at the edges of the crystals, but one would expect that the edge thickness would be thinner than the central regions. This phenomenon has also been noted by other workers,[11] and further experiments are planned to resolve this question. However, it is important to note that when working with AFM, use of improper scanning rate and feedback parameters can produce artifacts, such as time constant anomalies at edges. The convolution of the tip dimensions with the sample dimension may also confuse or mislead the image interpretation.[10] In addition, the electrostatic force induced between the insulating sample and the AFM tip could lead to a deviation of the image from the real topology.[12] Therefore great care must be taken in making sure that the images are the real topological features and not artifacts.

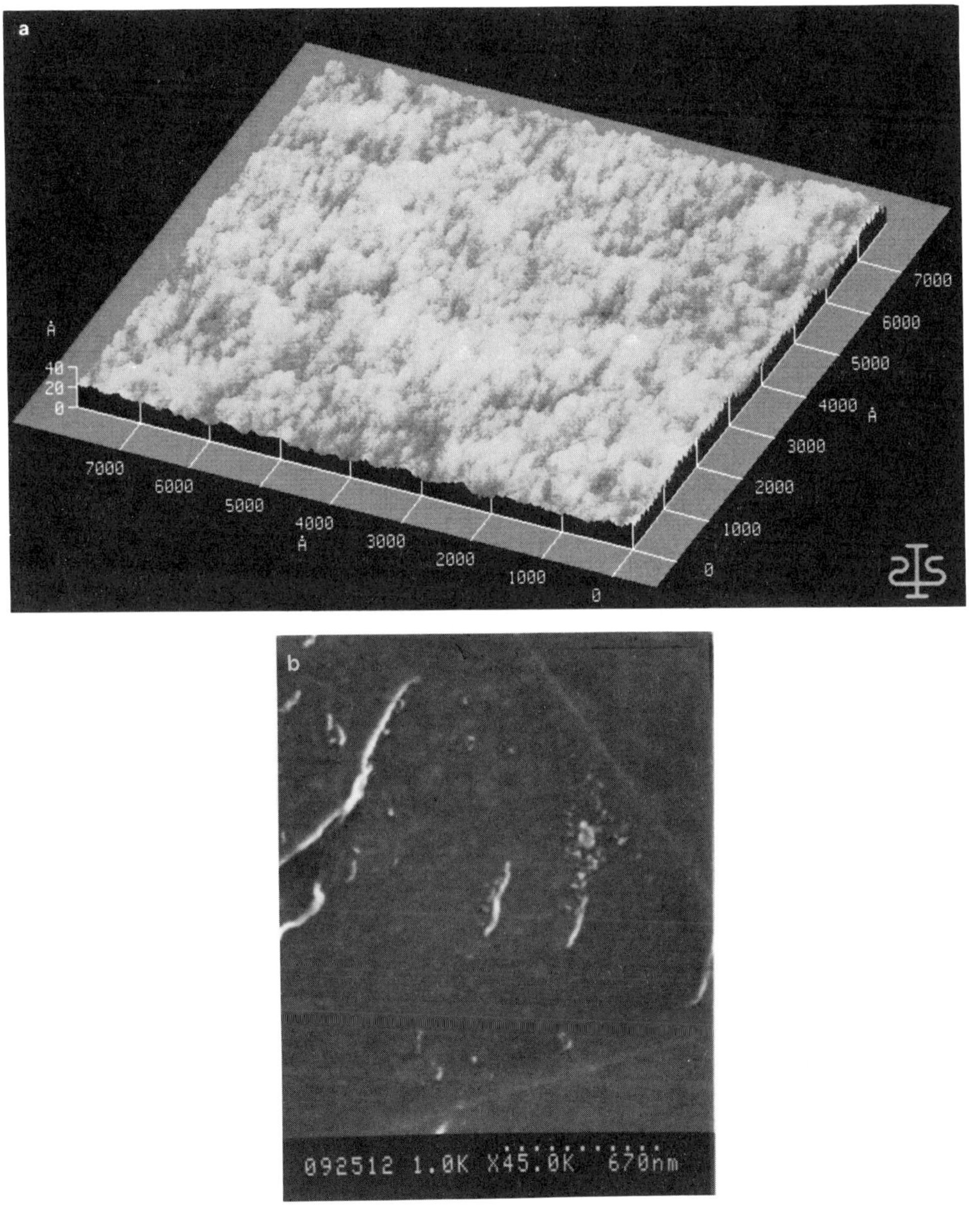

Figure 5. Both (a) AFM and (b) LVHRSEM images show that most crystals have bumps and undulations on their surface.

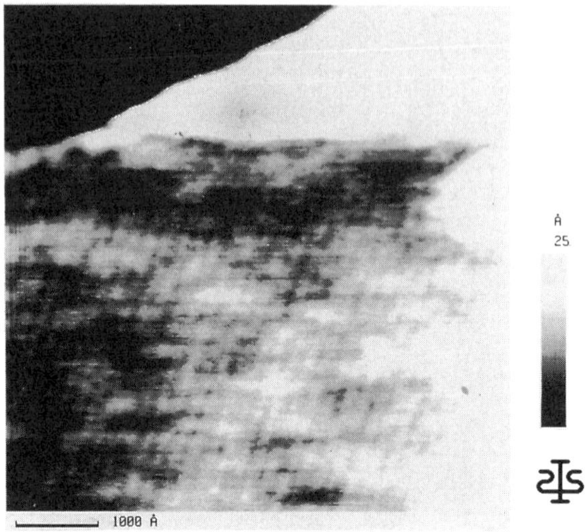

Figure 6. A periodic structure on the surface of PE single crystal AFM image.

Figure 7. The spiral growth of PE single crystals.

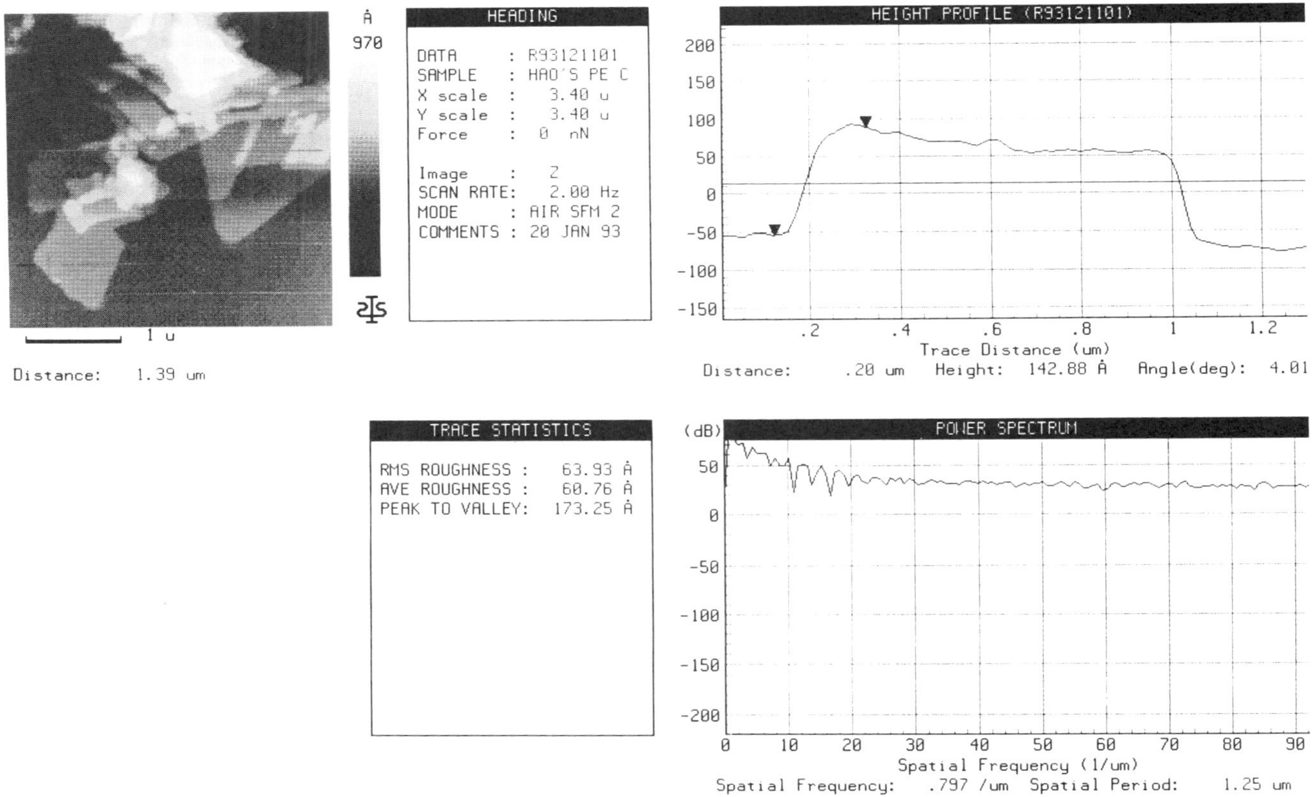

Figure 8. Thickness of most crystal lamellae around 80 to 150 Å.

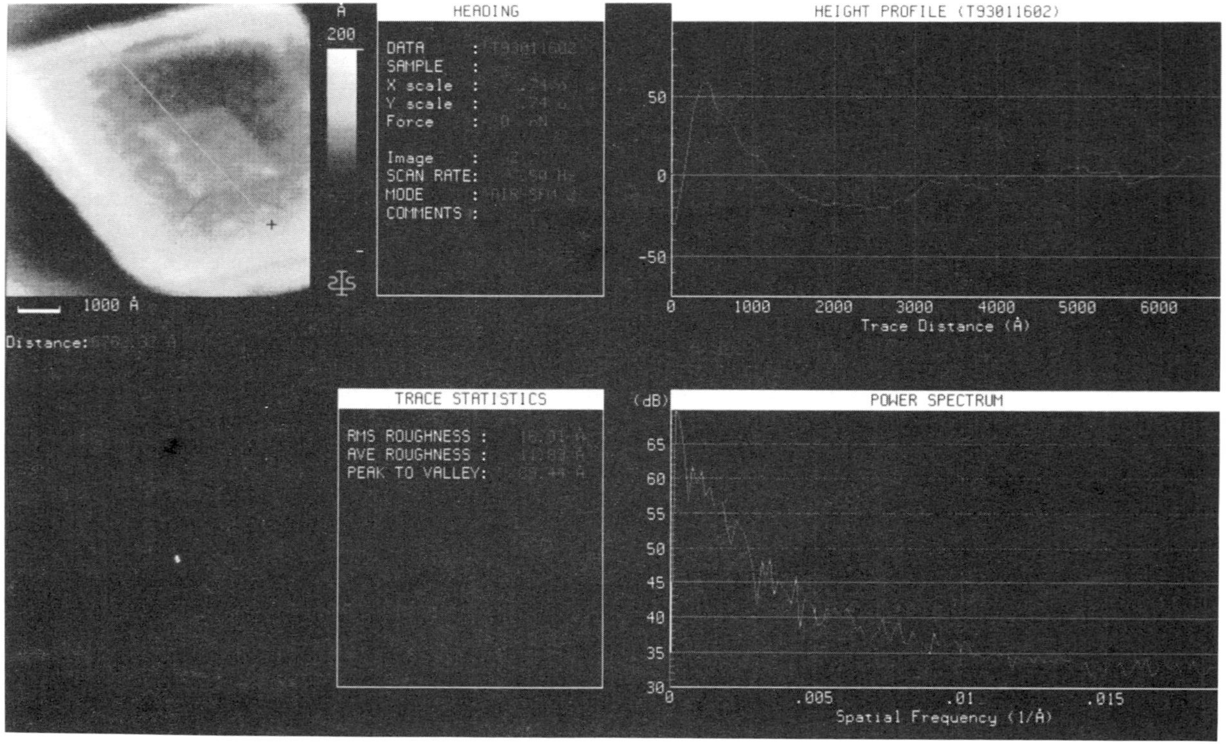

Figure 9. One growth spiral layer is one about 15 to 20 Å.

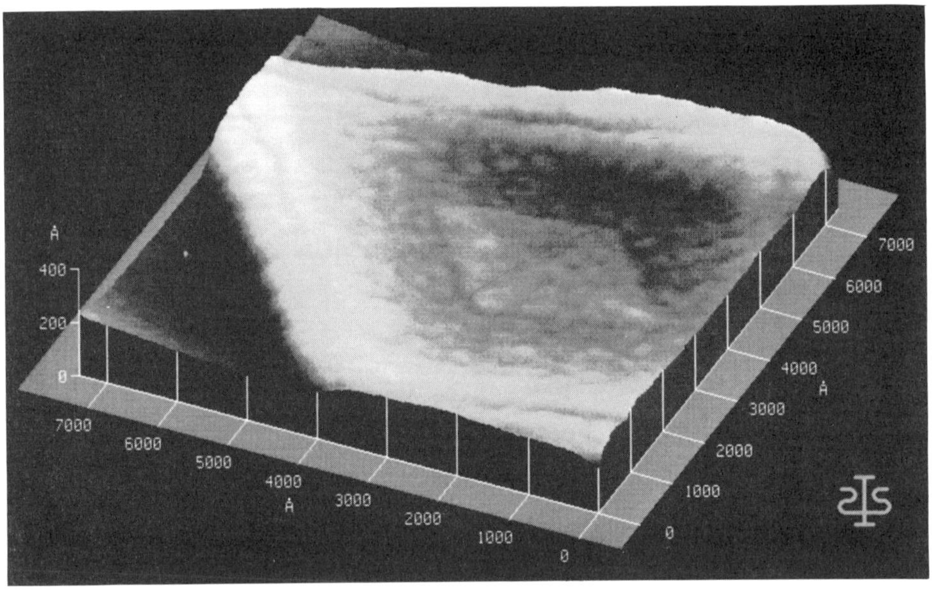

Figure 10. Almost all the single crystal AFM images showing thicker edges and thinner flat center part.

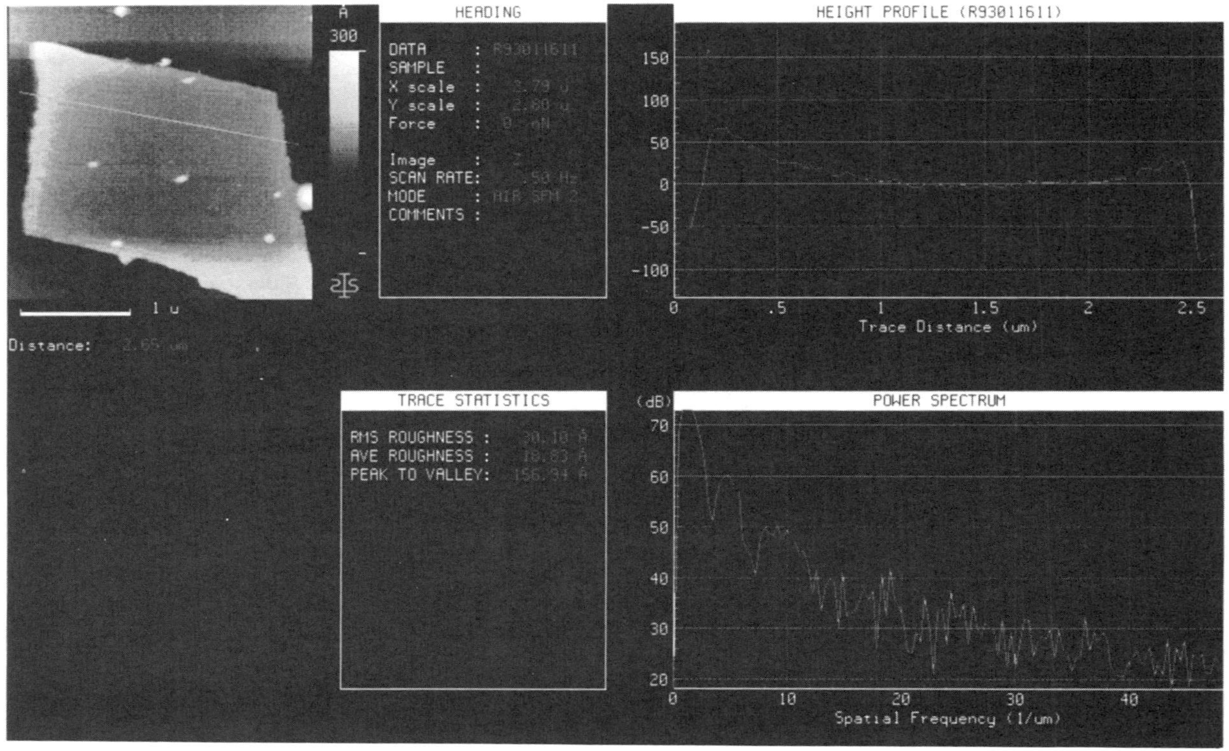

Figure 11. Almost all the single crystals AFM images showing thicker edges and thinner flat center part.

SUMMARY

Contrary to conventional EM, the electron irradiation using LVHRSEM does not change the apparent shape of the PE single crystals. However, contamination still occurs at high magnification (>30,000). The contrast mechanisms in these images are comparatively complicated and the subjects of further study. The crossover voltage (E_2) depends on several factors including the accelerating voltage, substrate, morphology of the specimen and the exposure time, all of which affect the image contrast. The clearest images from the Hitachi S-900 HRSEM at 1.0 keV are obtained from the samples deposited on carbon film coated TEM copper grids.

The morphological features of the PE single crystals as observed by LVHRSEM resemble those from earlier studies using TEM.[2-5] AFM images exhibit new characteristics which were not found with conventional EM techniques. The study of these features such as the periodic structure on the crystal surface, and the thicker edge of the crystals are in progress. It is clear that AFM is a valuable complementary technique to EM and X-ray diffraction in the study of morphology of organic and polymeric materials. Although AFM opens a new area of investigation of three-dimensional topological features, a thorough understanding of this technique is necessary to interpret properly its images since many factors could contribute to artifacts.

REFERENCES

1. D.T. Grubb, Review: Radiation damage and electron microscopy of organic polymers, *J. Mat.Sci.*, 9:1715 - 1736 (1974).
2. B. Wunderlich, "Macromolecular Physics", vol. 1, Academic Press, N. Y, (1973).
3. A. Keller, Some new habit features in crystals of long chain compounds, part I. paraffins, *Phil. Mag.* 6:329-343 (1961).

EXAMINATION OF PLAIN CARBON STEELS USING AN ATOMIC FORCE MICROSCOPE

Thomas L. Altshuler

Advanced Materials Laboratory, Inc.
Concord, Massachusetts 01742-2112

Abstract: Atomic force microscope (AFM) examination was performed on plain carbon steels. A 1018 steel that was austenized and furnace cooled revealed raised grain boundaries when etched with nital. It was concluded that these grain boundaries consist of walls of cementite formed during cooling below the eutectoid temperature. Examination using the AFM was done on 1045 and 1095 steels that had been water quenched after austenizing. Lath martensite and also martensite platelets were etched less than bainite using nital. These structures were seen more clearly than with optical microscopy.

INTRODUCTION

Microscopic examination of steels began with the work of Dr. Sorby in 1886. Albert Sauveur[1,2] expanded the field of metallography of steels in 1891 in the United States and published his famous book in 1912. Thus, the field of metallography of steel began. With the advent of electron microscopy and scanning electron microscopy, greater detail of the microstructure of steel was revealed.[3,4] These microscopes extended the observation of the human eye in two cartesian dimensions. With the advent of the scanning tunneling microscope by Binnig and Rohrer[5] and the atomic force microscope by Binnig, Quate, and Gerber[6], the surface of materials could be examined in the three cartesian dimensions. This has led to observation of the topography of material surfaces including a low carbon steel that had been prepared metallographically[7]. In this study, the surface topography of several plain carbon steels has been examined with the atomic force microscope.

EXPERIMENTAL

A number of plain carbon steel metallographic test specimens were obtained from Buehler Ltd., which were subsequently polished and lightly etched with 2% nital. The specimens were then indented with a Knoop diamond indentor at the center of each specimen with a 25 pound load. These specimens were:

1. Catalogue # 81-1810-001/1018 Steel (0.185%C, 0.79% Mn, 0.02%P, 0.04%S)
Austenized (heated to γFe phase) for 75 minutes at 1650°F and furnace cooled.
Specimen was hexagonal 0.50 inch across the flats.
The ferrite grain Knoop Hardness Number = 142, equivalent to Rockwell B = 77.3[8]

2. Catalogue #81-1810-008/1045 Steel (0.44%C, 0.87% Mn, 0.35%Si, 0.03%P, 0.04%S)
Austenized for 75 minutes at 1650°F and water quenched.
Specimen was a square 0.50 inch on its side.
Knoop Hardness Number = 550 in the martensite, equivalent to Rockwell C = 52.4[8]

3. Catalogue #81-1810-013/1095 Steel (0.97%C, 0.40% Mn, 0.18% Si, 0.14% P 0.05%S)
Austenized for 75 minutes at 1650°F and water quenched.
Specimen was round with 0.50 inch diameter.
Knoop Hardness Number = 855 in the martensite, equivalent to Rockwell C = 65.7[8]

The specimens were subsequently micrographed using bright field illumination, cross-polarized light, and Differential Interference Contrast in the region of the Knoop indentation. The same region of each specimen was examined with a NanoScope III Stand Alone Atomic Force Microscope J scanner[9] (AFM), Digital Instruments, Inc., Santa Barbara, CA.

RESULTS

1018 Steel Furnace Cooled

Figure 1 is an optical micrograph using Differential Interference Contrast of this steel showing pearlite and ferrite grains with grain boundaries. Also, one can see the Knoop indentation. Upon examination of this specimen under polarized light as suggested by Martin Wells,[10] the grain boundaries appeared to rotate polarized light the same as the cementite in the pearlitic region of the specimen.

Subsequently, the specimen was examined using the AFM. Figure 2 shows the same region of the specimen in which the Knoop indentation appears dark indicating that it is lower than the average surface viewed. The color bar to the right of the micrograph can be used to identify the relative height of topographic features by the value (degree of lightness) of the color. The cementite (Fe_3C) platelets appear almost white because the nital did not etch this material nearly as rapidly as the ferrite grains. It is interesting that the grain boundaries appear nearly white. Various ferrite grains have different values, which show that the etchant attacks each grain at a different rate depending upon the crystallographic orientation of the grain.

In Figure 3, one can see the pearlite, shown by arrow (1) in Figure 2, where the cementite platelets are the white strips. Figure 4 presents an AFM surface view of region (2) in Figure 2. Here, the Knoop indentation is clearly visible as a depression. The pearlite is at the right of the indentation while the grain boundaries are at the left of the indentation.

Figure 1. 1018 steel, nital etch, magnified 1000x.

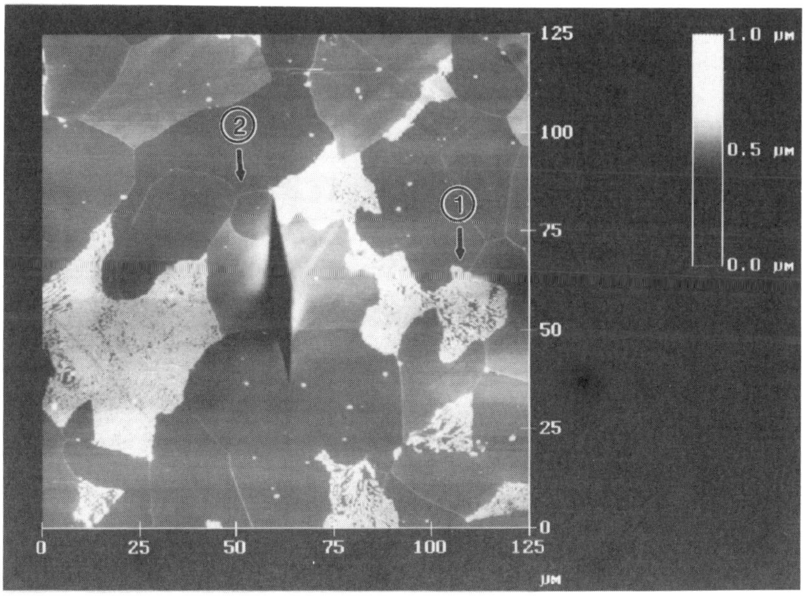

Figure 2. AFM top view of 1018 steel, same region as in Figure 1.

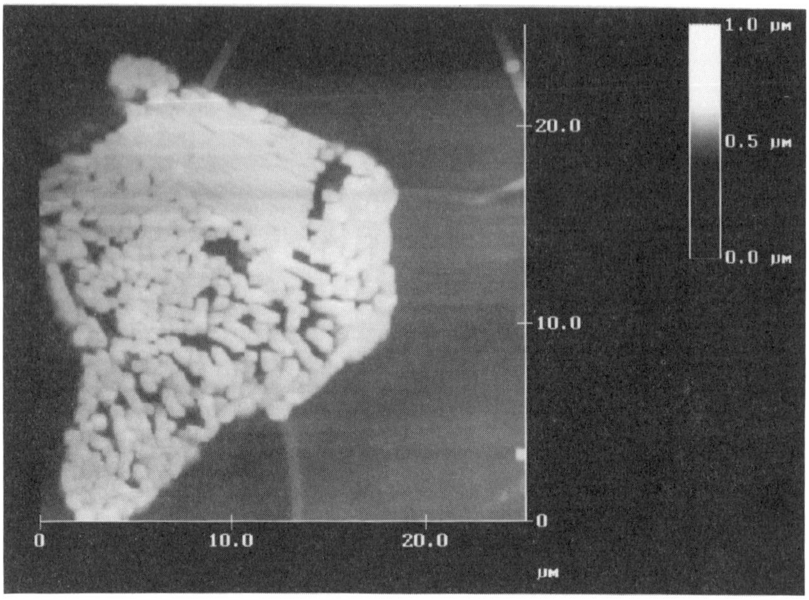

Figure 3. AFM top view of 1018 steel, pearlite.

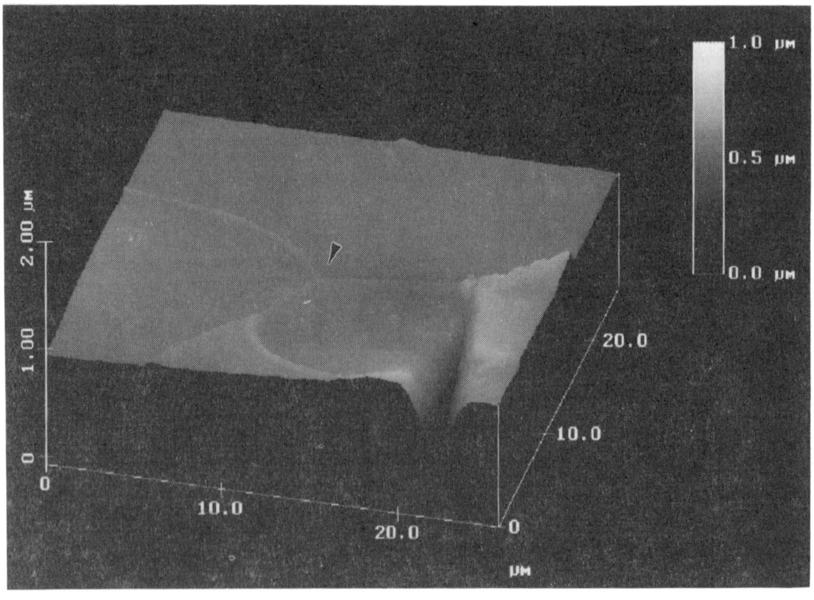

Figure 4. AFM surface view of 1018 steel, pearlite.

At higher magnification, Figure 5, one can see the three intersecting grain boundaries shown at the arrow of Figure 4. This AFM surface view, Figure 5, shows that the grain boundaries are higher than the surrounding grains. Also, one can see that each grain consists of small patches surrounded by small depressed grooves. These might be subgrains in which the grooves may be low-angle grain boundaries. The three grain boundaries probably consist of walls of cementite that do not etch as rapidly as the surrounding ferrite grains. On the other hand, if the patches are surrounded by low angle grain boundaries, they should be attacked more readily by the etchant than the surrounding ferrite.

Figure 6 is a cross-sectional view of one of the three grain boundaries. The micrograph at the lower left part of the Figure is a top view of the region examined in which the white line is the cross-section line. In the graph at the upper left part of the Figure, one can see the cross-section of the grain boundary, where the abscissa represents the cross-section line in the **xy** plane while the **z** height is the ordinate. Here, the width of the grain boundary is 0.313 micrometers (μm) while the height of the grain boundary is 34.659 nanometers (nm). Figure 7 is a magnified view of a grain boundary from another specimen of 1018 steel in which the arrows, labelled (3), lie along the top of the grain boundary. A crack runs across this grain boundary and into the neighboring ferrite grain; see the arrow labelled (4).

1045 Steel Water Quenched

An optical micrograph of the 1045 steel is shown in Figure 8 under bright field illumination. Here one can see the Knoop microhardness indentation in the martensite, the white region. The dark patches are bainite. Atomic force microscopy of the same region can be seen in Figure 9 where the Knoop indentation appears black with the light-colored

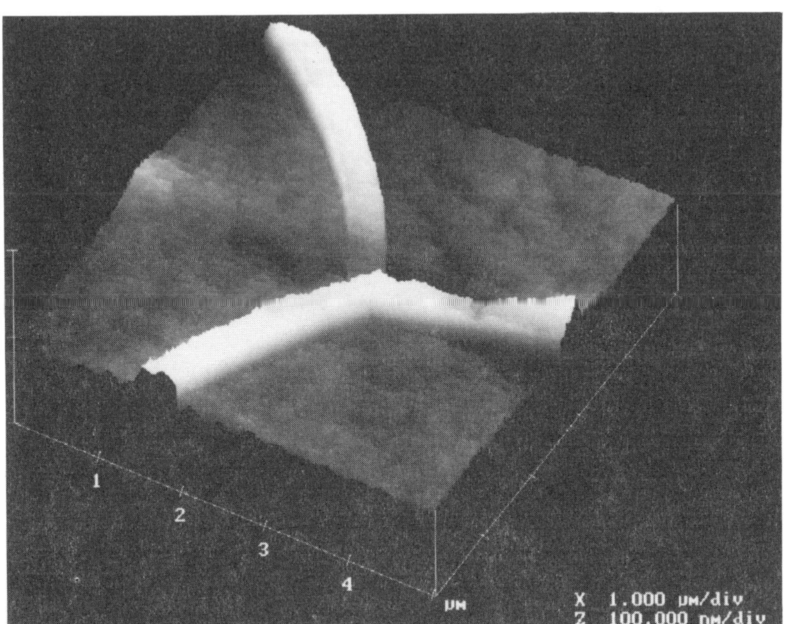

Figure 5. AFM surface view of 1018 steel, three intersecting ferrite grains.

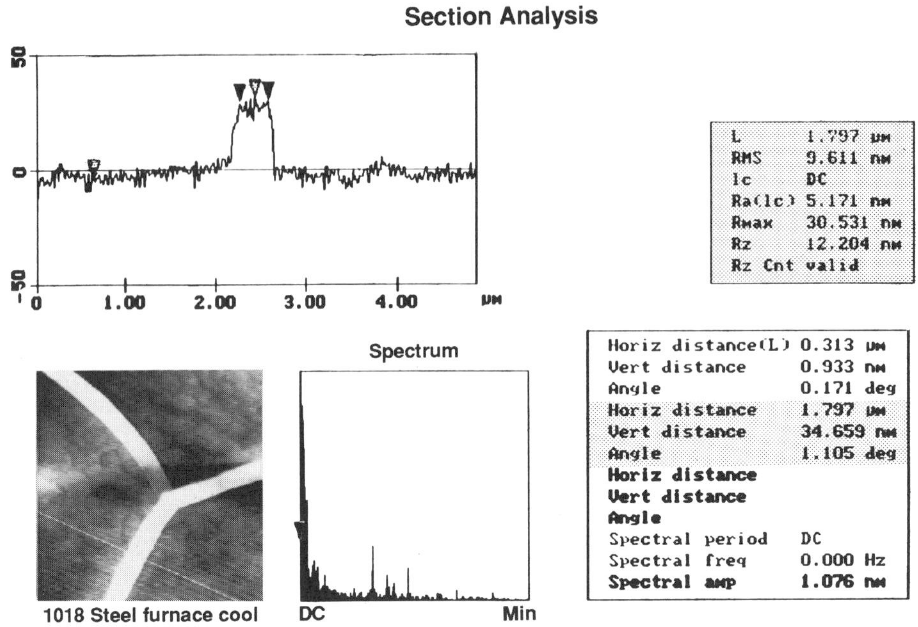

Figure 6. AFM cross-section of 1018 steel, same region as in Figure 7.

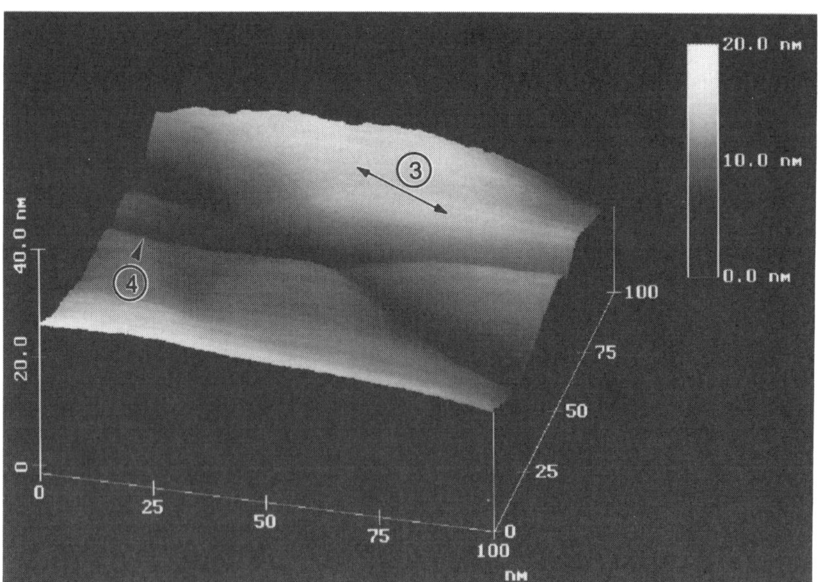

Figure 7. AFM surface view of 1018 steel showing a crack across a grain boundary.

Figure 8. 1045 steel, nital etch, magnified 1000x

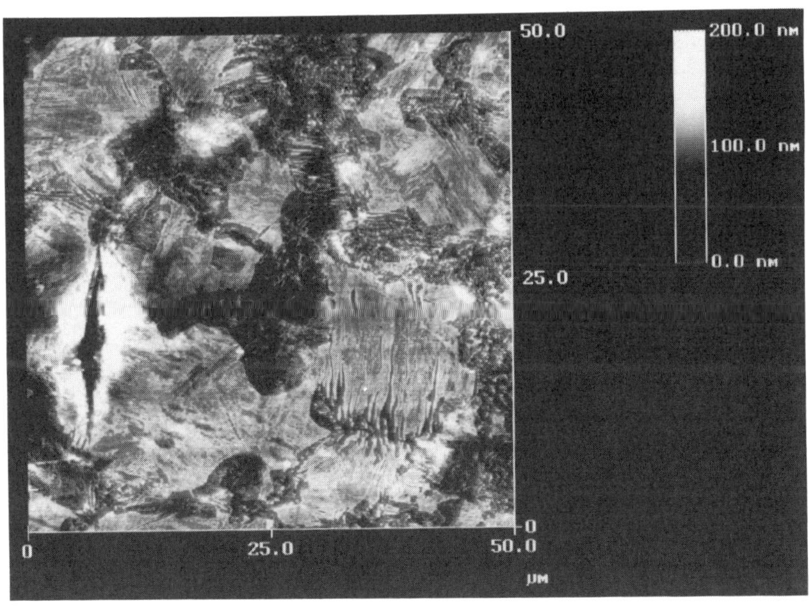

Figure 9. AFM top view of 1045 steel at center, same view as Figure 8.

raised metal on either side. Also, the martensite is shown as the lighter striated material that was etched less than the darker area, which is bainite. The long thin wavy white vertical areas at x = 30 μm and y = 15 μm, probably lath martensite, can be seen magnified in Figure 10. The captured image of the bainite area (see the arrow in Figure 10) was magnified in Figure 11. Here the bainite appears to consist of small parallel light-colored plates of cementite about 0.1 μm long within a ferrite matrix.[11] Figure 12 shows an AFM scan of a region of the specimen near the outside edge where the cooling rate was much greater than the center. Much less bainite is present in this case, shown as dark patches in the martensite matrix.

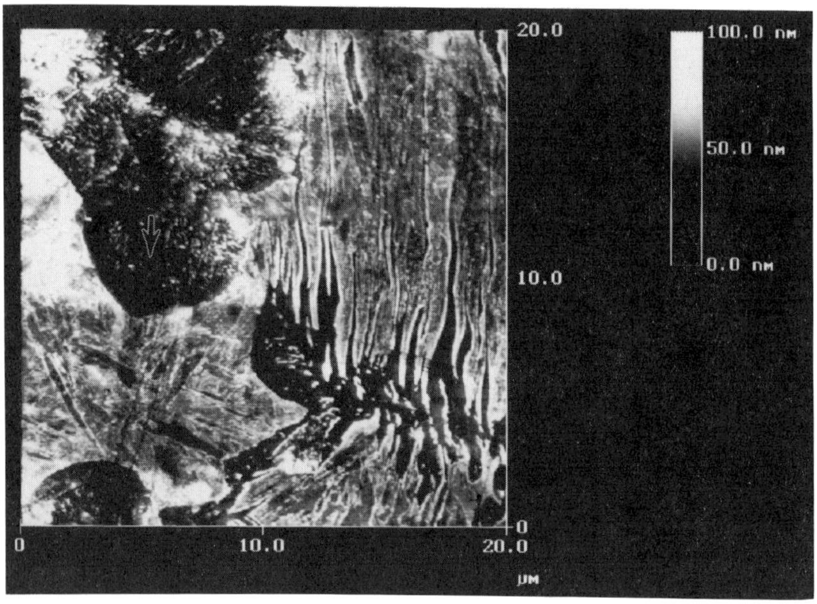

Figure 10. AFM top view of 1045 steel, magnified central part of Figure 9.

1095 Steel, Austenized and Water Quenched

This steel was etched in nital and then had a Knoop indentation at its central part. It was then examined with an optical microscope under bright field illumination (see Figure 13). The structure of the surface consisted of martensite, which is the white needle-like structure with darker small patches of bainite. Upon examination with an AFM, Figure 14 resulted. The white patch at x = 25 μm and y = 10 μm was further magnified as shown in Figure 15. It is not known what the tiny white lines are in this region. Perhaps they are needles of martensite. The steel was then polished with 0.05 μm gamma alumina removing

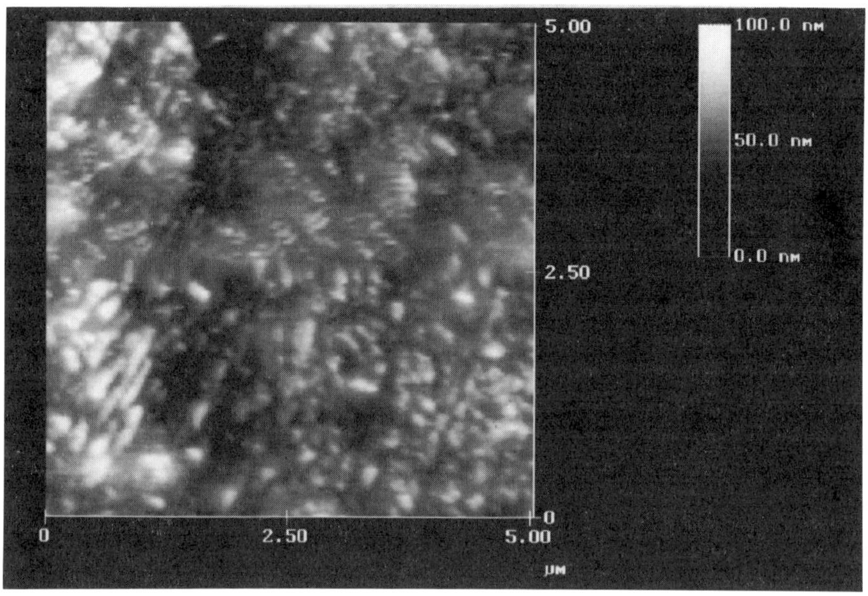

Figure 11. AFM top view of 1045 steel, bainite. Magnified region at arrow, Figure 10.

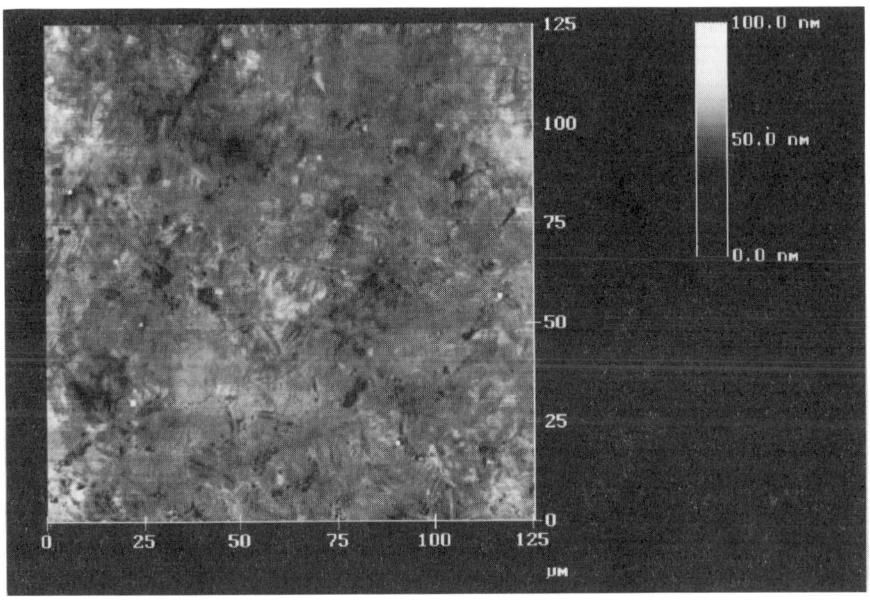

Figure 12. AFM top view of 1045 steel near outside edge.

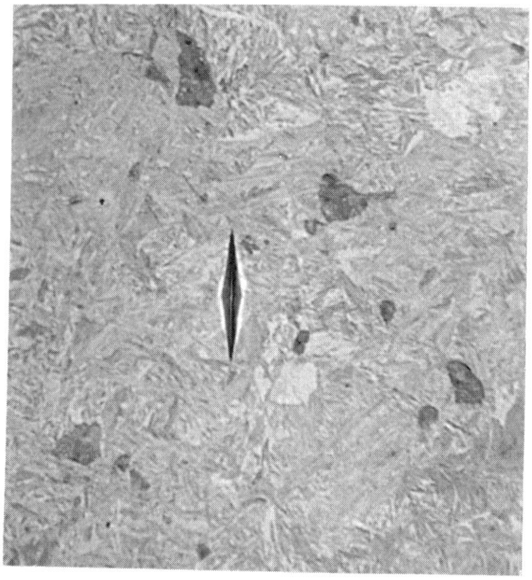

Figure 13. 1095 steel, nital etch magnified 1000 x.

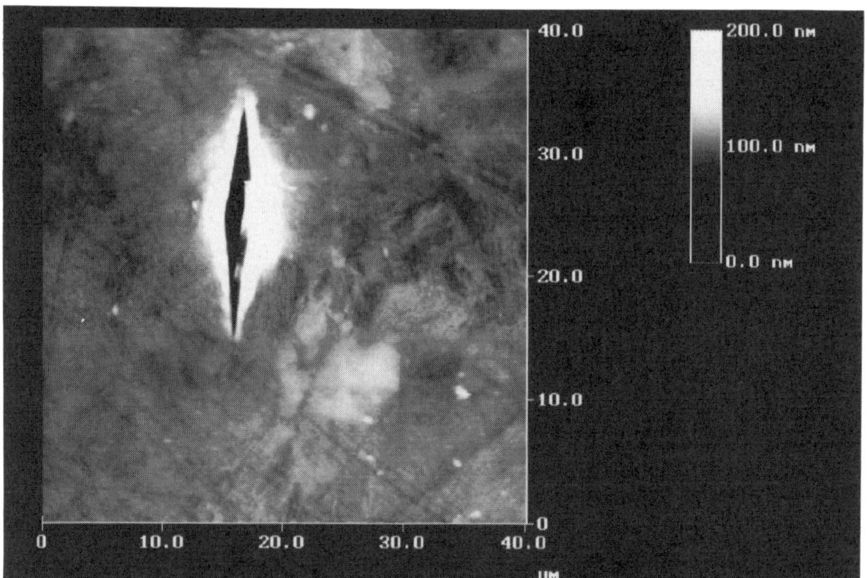

Figure 14. AFM top view of 1095 steel, nital etch. Same view as in Figure 13.

about 0.3 μm of the surface. It was then etched with picral for 30 seconds. An optical micrograph using differential interference contrast is shown in Figure 16. Figure 17 shows the same region when examined with the AFM. The light patch at x = 25 μm and y = 10 μm is the same one shown in Figure 14 at those coordinates. A magnification of that patch is shown in Figure 18. Here the material looks quite bumpy and not as clearly defined as with the nital etch.

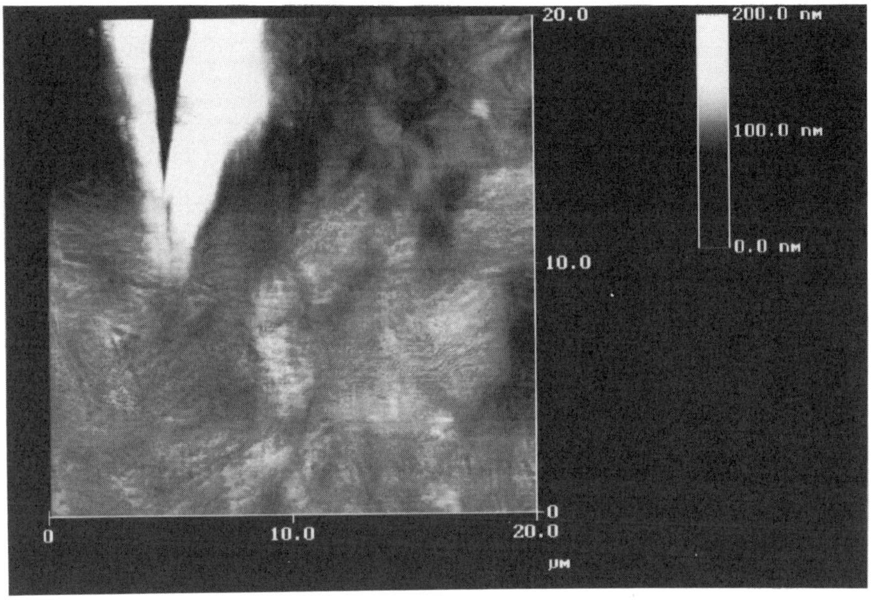

Figure 15. AFM top view of 1095 steel, magnified view of Figure 14.

Figure 16. 1095 steel, picral etch, magnified 1000x

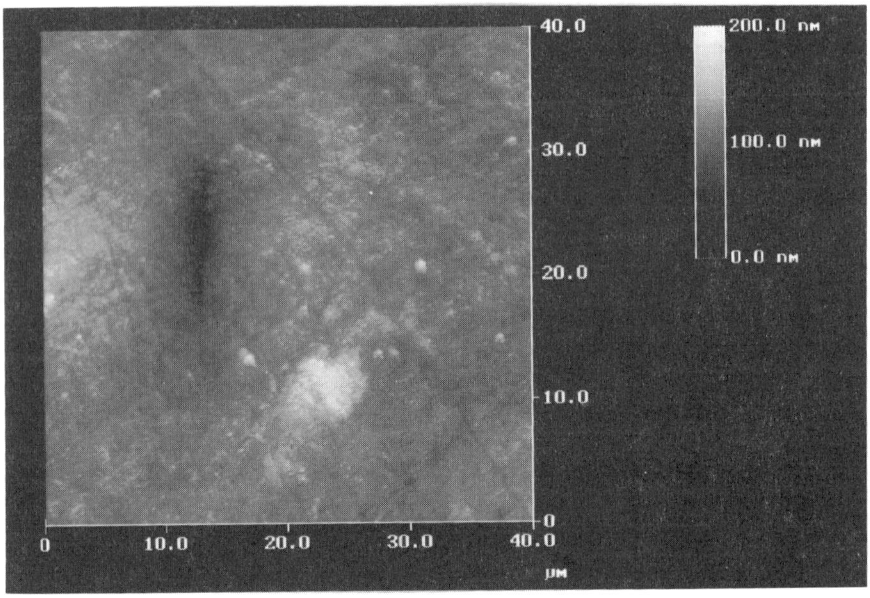

Figure 17. AFM top view of 1095 steel water quenched, picral etch.

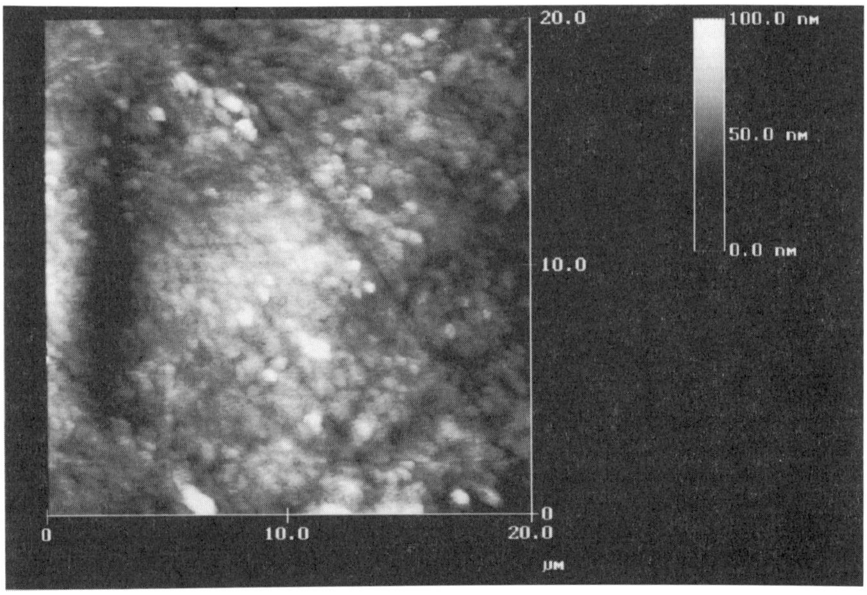

Figure 18. AFM top view of 1095 steel, magnified view shown in Figure 17.

CONCLUSIONS

1018 Steel Austenized and Furnace Cooled

This steel revealed pearlite, ferrite grains, and grain boundaries between the ferrite grains with the atomic force microscope. The pearlite consisted of platelets of cementite that etched more slowly than the surrounding matrix of ferrite. The grain boundaries were not etched at the same rate as the ferrite grains and appear to consist of walls of cementite. This conclusion is supported by three factors. In the first place, the etch rate of the nital was about the same as that of the cementite in the pearlite. In the second, the grain boundaries rotated polarized light in an optical microscope to the same degree that the cementite did. In the third place, a microcrack was discovered intersecting a grain boundary. McMahon and Cohen[12] found cracks within carbide in the grain boundaries of 0.005% C and 0.035% C iron.

It is believed that when the steel was cooled below the eutectoid temperature, carbon diffused to the grain boundaries during further cooling as well as to the cementite in the pearlite. Microcracks across the grain boundaries served as crack arresters that might run along a grain boundary, thereby preventing steel from failing in a brittle manner with intergranular cracking at room temperature. Also, the carbide in the grain boundary would serve as a barrier to dislocation glide, making a large yield stress dependence upon the grain size according to the Petch equation,[13] even though iron has many slip systems.

1045 Steel Austenized and Water Quenched

This steel had both martensite and bainite present. The martensite did not etch as rapidly as the bainite and thus there was considerable contrast between martensite and bainite. A wavy structure, probably lath martensite, showed up more clearly with the AFM than optical microscopy.

1095 Steel Austenized and Water Quenched

After nital etching this steel, one could see plate martensite under the optical microscope as well as light areas. Upon examination of these light areas with the AFM, fine wavy lines approximately 50 nm wide and 1 to 2 μm long were revealed. These might be needles of martensite, but their identity has not been established. Etching with picral resulted in a bumpy structure, but resolution did not appear to be enhanced.

REFERENCES

1. A. Sauveur, "The Metallography and Heat Treatment of Iron and Steel," 1st Edition, New York: McGraw Hill Book Co., Inc. (1912).
2. A. Sauveur, "The Metallography and Heat Treatment of Iron and Steel," 4th Edition, New York: McGraw Hill Book Co., Inc., 60-65 (1935).
3. K. Mills, *in* "Metals Handbook," 9th Edition, Vol. 9, T. D. Cooper, ed., Metals Park, OH: American Society for Metals, 88-112 (1985).
4. J. W. Christian, "The Theory of Transformations in Metals and Alloys," Oxford: Pergamon Press, 906-909 (1965).
5. G. Binnig, H. Rohrer, C. Gerber, E. Weibele, Surface study by scanning tunneling microscopy, *Phys.Rev. Lett.*, 49:57 (1982).

6. G. Binnig, C. F. Quate, C. Gerber, Atomic force microscope, *Phys. Rev. Lett.*, 56: 930-933 (1986).
7. T. L. Altshuler, Atomic-scale materials characterization *in* "Advanced Materials and Processes," 130:3, 18-23 (1991).
8. ASTM Standard E 140, "Annual Book of ASTM Standards," 03.01: 338-339 (1992).
9. Digital Instuments, Inc., 520 E. Montecito St., Santa Barbara, CA 93103.
10. Martin Wells, personal communication, February 1993.
11. American Society for Metals, "Metals Handbook," 9th Edition, Metals Park, OH: American Society for Metals, 9th Edition, 662-672 (1985).
12. C. J. McMahon, Jr., M. Cohen, The fracture of polycrystalline iron, *in* "Proceedings of the First International Conference on Fracture," 2:779-812 (1965).
13. N. J. Petch, The cleavage strength of polycrystals, *J. Iron and Steel Inst.*, 174:25-28 (1953).

SCANNING TUNNELING MICROSCOPY OF POROUS SILICON-BASED SURFACES

Gerardo B. Amisola,[1] Ralf Behrensmeier,[1] James M. Galligan,[1] Fred A. Otter,[1] Fereydoon Namavar,[2] and Nadar M. Kalkhoran,[2]

[1]Institute of Materials Science
University of Connecticut, Storrs, CT 06269-9012
[2]Spire Corporation
Bedford Massachusetts 01730

Abstract: The ability of porous silicon to emit visible light of different wavelengths has been the subject of intense study. While many groups have reported visible light emission from this material, no consistent theory has emerged to explain the effect. Many agree though that understanding the surface structure of this material is an important step in determining the physical processes responsible for luminescence, particularly for the quantum confinement models. Therefore, the scanning tunneling microscope (STM) and the atomic force microscope (AFM) are used to characterize the surface morphology of several porous silicon-based surfaces. A 7 nm gold coating was initially used to get an accurate estimate of vertical dimensions in the porous silicon layer. While a second coating of a 150 nm indium tin oxide on porous silicon was too thick to gain any structural data, changes in the tunnel current (I) and bias voltage (V) curves with respect to ambient light intensity showed that a photovoltaic effect was present in this heterojunction device. Finally, STM and AFM scans (in an inert He atmosphere) of bare photoluminescent porous silicon formed on p- and p+ Si wafers showed that these surfaces are composed of a random network of nanometer-scale features. However, no features of the size and geometry consistent with the quantum wire model for luminescence were observed in all the scans taken of these surfaces.

INTRODUCTION

Porous silicon is a material formed by the anodic etching of a silicon single crystal surface under a controlled current in a HF-based solution. While a relatively simple material to prepare, porous silicon has been the subject of recent, intense study. This is because the discovery of visible light emission from anodically etched porous silicon[1] has brought closer the promise of silicon-based optoelectronic devices. Such silicon-based devices would be cheaper and easier to fabricate than the conventional compound semiconductor-based optoelectronic devices used today.

However before any of the device applications of this material are realized, the physical processes that allow porous silicon to luminesce must be understood. Two major theories have emerged to explain this effect. The first is based on the principle of quantum confinement.[1] This model explains visible light emission from porous silicon through the formation of silicon nanocrystallites in the porous layer while the crystalline surface is being etched. If the nanocrystallite size is comparable to the free exciton radius in Si (about 1-5 nm), then confinement effects can increase porous silicon's bandgap energy to 1.6-1.8 eV (visible red-orange light) in contrast to crystalline silicon's band gap of 1.1 eV (infrared light). The actual size and geometry of such structures are still not known but the quantum wire or quantum dot configurations are the most prominent in the literature.[2]

Another model is based on a change in surface chemistry of this material. This model attributes light emission from silicon to the presence of some optically active Si-based compound formed on its surface during anodic etching. The identity of this compound is not yet known but some of the likely possibilities are siloxene[3] or some form of a silicon hydride species.[4]

It is evident from the above models that understanding the light emission properties of porous silicon requires a complete characterization of both its microstructural and chemical properties. Standard analysis techniques such as XPS[5] and FTIR[6] have been used to determine the chemical nature of the porous surface. On the other hand, the results of standard microscopy techniques in the determining the microstructure (or nanostructure) of this material have not been conclusive. The resolution limit of scanning electron microscopy (SEM)[7] prevents this technique from observing the proposed nanostructure of the quantum confinement model. Transmission electron microscopy (TEM) certainly has the required resolution but controversy still exists as to how TEM sample preparation techniques affect the fragile porous layer.[8] We have therefore used scanning probe microscopy (SPM), specifically, scanning tunneling microscopy (STM) and atomic force microscopy (AFM) to characterize the structure of the porous silicon surface in this study. These microscopes have the unique ability to image a surface with nanometer to atomic scale resolution without extensive sample preparation. STM has the added capability of being able to measure the sample density of states (DOS), which is an important parameter in determining the energy band structure of porous silicon.

EXPERIMENTAL PROCEDURE

Porous silicon (PS) was formed by anodically etching a (100) wafer of 13.5 Ωcm p^- Si until visible red-orange photoluminescence was observed. This wafer was etched in an HF:Ethanol (1:1) solution at 2.52 mA/cm^2 for 2 h. Three different etching zones were identified on the its surface: An outer unetched ring (Region I) of crystalline Si, an inner weakly etched ring (Region II) of porous silicon and a heavily etched center region (Region III). Photoluminescence was observed only in the highly etched zone (Region III). This region was then imaged with the STM in an inert He atmosphere. Prior to imaging, this surface was first dipped into 48% HF and a buffered HF etch for 20 seconds each to remove the native oxide. The wafer was then mounted in an Angstrom Technology TAK 3.0 STM where the STM chamber was continually purged with inert, ultrahigh purity He gas. All STM imaging and spectroscopy were then done in this He atmosphere.

A halogen light source was used to vary the intensity of light incident on the PS surface through a light pipe incorporated into an opaque STM cover. All imaging was done in the constant current mode (bias V = -3V; set point I = 1 nA) using chemically etched Pt-Ir tips.

Despite the steps taken to prevent sample oxidation and contamination, the STM was still not able to image the surface reproducibly. This may have been due to a partial delamination of the porous layer itself from its single crystal substrate or a removal of carriers from the porous surface. These factors would represent fundamental limits on the STM since the instrument requires a conducting surface to operate. Initial efforts to image the porous surface in spite of these limitations included sputter depositing different surface coatings to improve both surface conductivity and integrity. These included a 7 nm gold film and a 150 nm indium tin oxide (ITO) layer. Although these coatings did indeed provide more reproducible STM images, they also masked the actual dimensions of the surface so that an accurate determination of the feature size and geometry was not possible.

These limitations were finally overcome in two ways. The most obvious solution was to use an AFM to image the highly etched Region III of the p^- wafer. While successful in reproducibly imaging Region III, this technique did not allow the use of tunneling spectroscopy to examine the surface electronic states. A more useful solution was to use a heavily doped (100) 0.002 Ωcm p^+ Si as a substrate. Photoluminescent porous silicon was formed by anodically etching this wafer in the previous HF solution for 10 min at a current density of 40 mA/cm^2. After etching, this surface was observed to photoluminesce uniformly through the entire etched zone. The surface was cleaned with the above procedure then rinsed in methanol and double distilled water and loaded immediately into the He-purged STM for imaging.

RESULTS AND DISCUSSION

The porous silicon surface provides unique challenges to structural characterization by scanning tunneling microscopy. One of the most important considerations in studying this material is how to keep the surface clean of its native surface oxide particularly because of the large surface area in the heavily etched zone. Also, the surface density of states has to be significantly reduced so that the measured IV curves are actually representative of the bulk states.[9] Several researchers have determined that a dilute HF treatment would be sufficient to passivate the surface of silicon by saturating all the remaining dangling bonds with hydrogen.[10] This would also have the effect of keeping the surface free of oxides at least for short periods of time.[9] An ambient He atmosphere was therefore used to prevent further oxidation while imaging.

Although the surface may be free from oxides, reproducible imaging may still not be possible if the surface conductivity is very small. This is certainly another consideration for imaging porous silicon because recent work on characterizing its electronic properties has shown that the porous surface is depleted of holes during etching.[11] A 7 nm gold film was then deposited on this highly etched porous surface to provide a conducting layer. A representative image is shown in Figure 1. This surface shows several large protrusions measuring 20 to 40 nm high with lateral extensions of 50-200 nm.

While the observed vertical displacements are within the range expected for the porous silicon nanostructures (shown by some TEM investigations), the lateral extensions are approximately two orders of magnitude larger than the expected dimensions even if the gold coating is taken into account. It is very probable then that the large lateral dimensions are actually artifacts of the sputter deposition process. The 7 nm gold film masks the real lateral dimensions of the underlying individual structures so that they become less distinct and would therefore appear as wide surface protrusions.

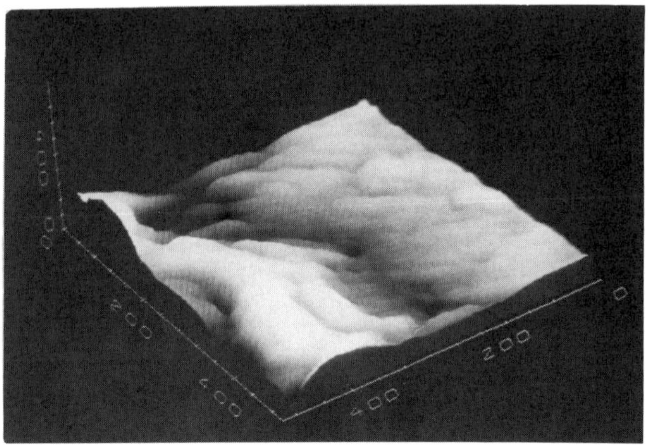

Figure 1. STM scan of porous silicon surface covered by 7 nm gold layer.

The porous surface was also sputter deposited with 150 nm ITO film to provide a source of minority carriers that can be injected into the porous silicon layer under an applied bias to produce an electroluminescent device.[12] Because of the thickness of this layer, the resulting surface morphology is probably more characteristic of the sputtered ITO film rather than the underlying porous layer. However, an interesting effect on the tunnel current was observed when this surface was imaged. There was a clear increase in the tunnel current when the sample was illuminated with the halogen lamp compared with the sample at ambient light conditions (Figure 2).

Another similar phenomenon was observed when the tip bias was set to zero after tunneling position had been reached in the STM. By increasing light intensity, it was possible to increase the tunnel current from 0.0 nA to 100 nA as measured by the STM. It is important to note that these IV changes were not observed in a counterpart ITO-unetched silicon device. These measurements indicate that the presence of light causes the ITO-porous silicon device to supply the additional bias voltage necessary to increase the tunnel current independent of the STM's tip bias. In effect, it acts as a photocell. In addition to its role in making an efficient silicon-based light emitting diode, porous silicon may also be a novel material used in photovoltaic technology.

The final samples imaged were the bare, photoluminescent porous silicon surfaces themselves. For the lightly doped p^- wafer, the STM was used to image both Regions I and II. Note that these weakly etched regions were not observed to photoluminesce. In Region I, the surface is essentially flat with the several large pores (20-50 nm in diameter) starting to appear (Figure 3).

In Region II, we find that the smallest scale features have lateral dimensions of 20-30 nm while their vertical dimensions are approximately 0.5 to 1 nm (Figure 4). In the photoluminescent, highly etched Region III, the combination of hole depletion and layer delamination required the use of an AFM to image the surface. AFM scans show that the smallest scale features on this surface have lateral dimensions of 20-30 nm while the vertical dimensions are 5-10 nm (Figure 5).

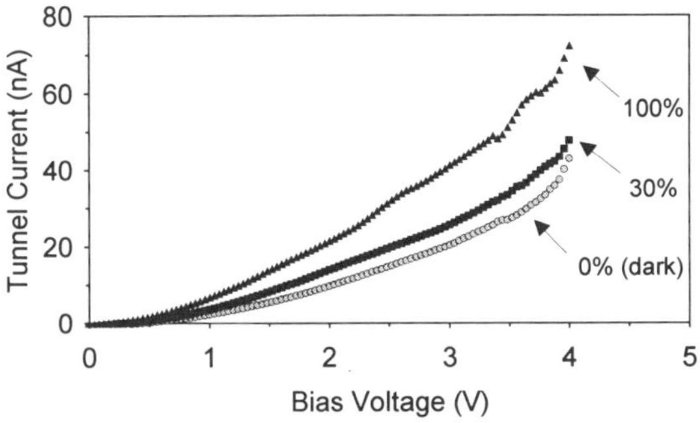

Figure 2. Tunnel current and tip bias voltage curve of ITO-porous silicon surface. A clear increase in the tunnel current is observed when the surface is illuminated.

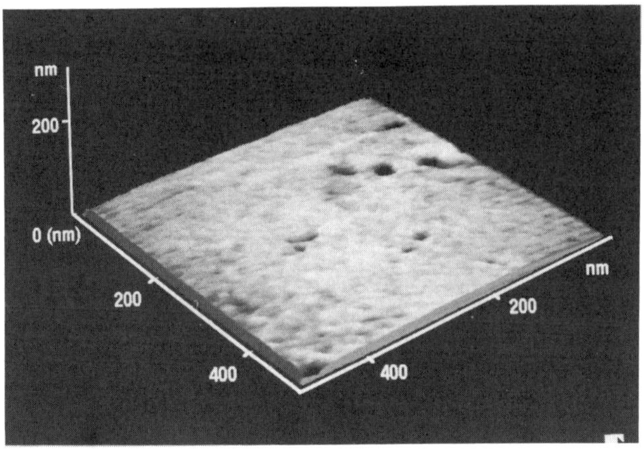

Figure 3. STM scan of Region I. The surface shows atomic scale corrugations with several macropores appearing.

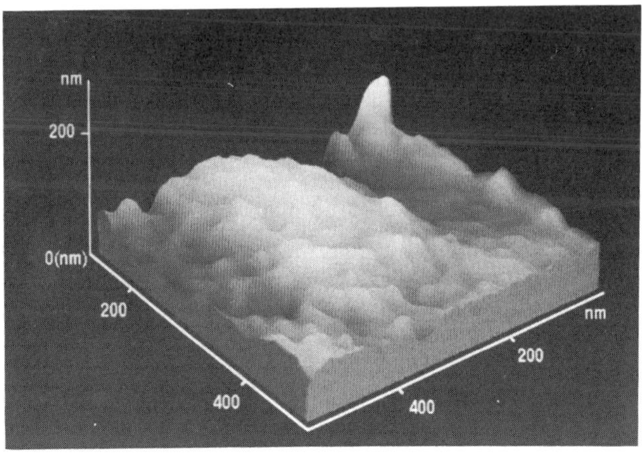

Figure 4. STM scan of weakly etched Region II.

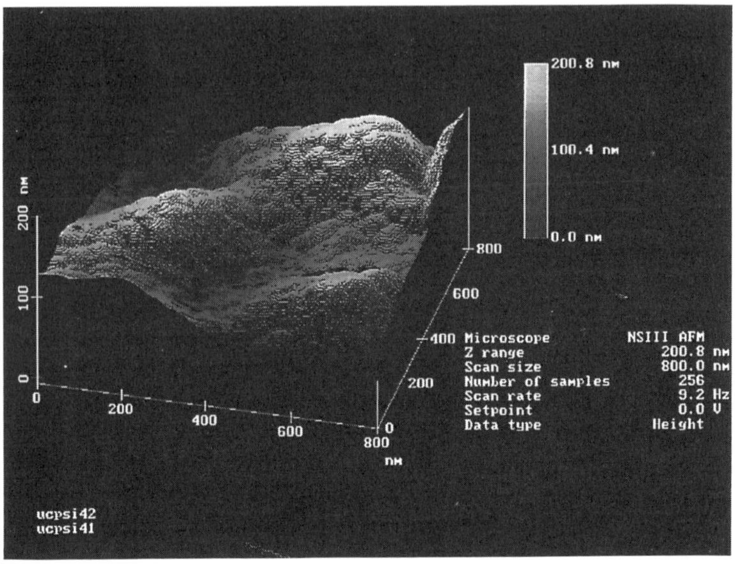

Figure 5. AFM scan of highly etched photoluminescent Region III from p$^-$ Si. Compared to Figure 4, the measured vertical dimensions of surface features have increased by a factor of 10 while the lateral dimensions have remained almost constant.

It is of interest to compare the smallest scale features in Region III with the smallest scale features in Region II. These measurements reveal that while the vertical dimensions in going from Region II to Region III have increased by a factor of 10, their lateral dimensions have remained constant. These observations agree with the etching model suggested by Lehmann, Cerva and Gosele[13] where the sides of the quantum wires become resistant to further etching because of the increase in bandgap energy.

For the heavily doped p$^+$ wafer, the combination of shorter etching time and greater hole concentration leave the surface with enough conductivity and enough contact with its crystalline substrate to allow STM imaging even after etching. These STM scans show that the surface is composed of wire-like structures and large pores. These "wires" measure 10-30 nm wide and 5-10 nm high while the pores are 20-100 nm wide and about 10-20 nm deep. While this surface morphology is more reminiscent of a "quantum wire" array than the observed morphology of porous silicon formed from the p$^-$ wafer, the measured dimensions are still an order of magnitude larger than the dimensions required for quantum confinement. Tunneling spectroscopy performed on porous silicon formed on both wafers shows that porous silicon formed from p$^-$ Si acts like an n-type semiconductor while porous silicon formed from p$^+$ Si retains its p-type character[14] . These observations are again consistent with the surface being depleted of holes as outlined in the etching model of Lehmann and Gosele[12].

In this study, the use of the STM and AFM to image the surface of porous silicon has not revealed the presence of the nanometer-scale features that have been suggested as responsible for luminescence in this material. At present, however, it is still not possible to completely rule out their existence. Although free standing, individualized quantum wires do not seem to be the operational geometry required for luminescence in porous silicon, other geometries, such as quantum dots may still be present. If this is the case, the quantum dots themselves may be hidden inside the observed surface features or may simply exist in a subsurface layer that is inaccessible to the tip probes of either STM and AFM.

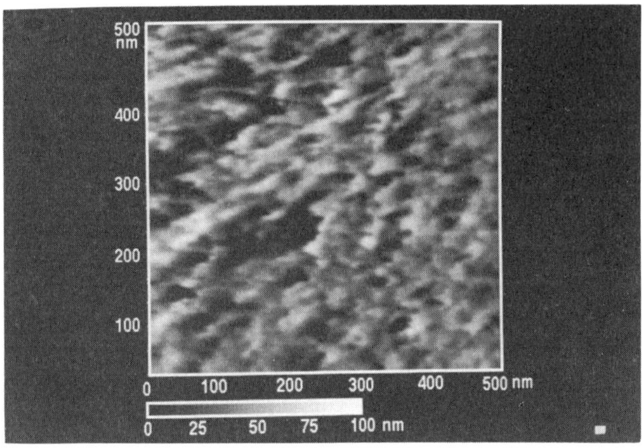

Figure 6. STM scan of photoluminescent porous silicon from p^+ Si.

CONCLUSION

The STM and AFM have been used to characterize several porous silicon-based surfaces. A 7 nm gold film sputter deposited on a porous surface provided an accurate estimate of vertical displacements of features in the porous layer but not of their lateral dimensions. Measurements of changes in the tunnel current with respect to ambient light conditions in an electroluminescent ITO-porous silicon heterojunction diode indicate that such a device can also be used as a photocell. Finally, STM and AFM studies of porous silicon surface formed from p^+ and p^- wafers show the presence of random nanometer-scale features whose dimensions measured at different etching stages agree with known etching models. However, structures of the size and geometry proposed by the quantum wire model were not observed on these surfaces.

REFERENCES

1. L.T. Canham, Silicon quantum wire array fabrication by electrochemical and chemical dissolution of wafers, *Appl. Phys. Lett.* 57:1046-1048 (1990).
2. J. C. Vial, A. Bsiesy, F. Gaspard, R. Herino, M. Ligeon, F. Muller R. Romenstain, and R.M. Macfarlane, Mechanisms of visible light from electro-oxidized porous silicon, *Phys. Rev. B.* 14:171-176 (1992).
3. P. Deak, M. Rosenbauer, M. Stutzmann, J. Weber, and M.S. Brandt, Siloxene:Chemical quantum confinement due to oxygen in a silicon matrix, *Phys. Rev. Lett.* 69:2531-2534 (1992).
4. S.M. Prokes, W.E. Carlos, and V.M. Bermudez, Luminescence cycling and defect density measurements in porous silicon: Evidence for hydride based model, *Appl. Phys. Lett.* 60: 1447-1449 (1992).
5. R.P. Vasquez, R.W. Fathauer, T. George, A. Ksendov, and T.L. Lin, Electronic structure of light emitting porous silicon, *Appl. Phys. Lett.* 60:1004-1006 (1992).
6. A. Nakajima, T. Itakura, S. Watanabe and N. Nakayama, Photoluminescence of porous silicon, oxidized then deoxidized chemically, *Appl. Phys. Lett.* 61:46-48 (1992).
7. J. M. Macaulay, F.M. Ross, P.C. Searson, S.K Sputz, R. People, and L.E. Friedersdorf, Microstructural characterization of photoluminescent porous silicon Light, *in* "Emission from Silicon," S.S. Iyer, L. T. Canham and R.P. Collins eds., Mat. Res. Soc., Pittsburgh PA, 256: 47-51(1992).

8. Y.H. Xie, W.L Wilson, F.M. Ross, J.A. Mucha, E.A. Fitzgerald, J.M. Macaulay, and T.D. Harris, Luminescence and structural study of porous silicon films, *J. Appl. Phys.* 71:2403-2407 (1992).
9. L.D. Bell, W. J. Kaiser, M.H. Hecht, and F.J. Grunthaner, Direct control and charaterization of a Schottky barrier by scanning tunneling microscopy, *Appl. Phys. Lett.* 52:278-280 (1988).
10. E. Yablonovitch, D.L. Allara, C.C. Chang, T. Gmitter, and T.B. Bright, Unusually low surface-recombination velocity on silicon and germanium surfaces, *Phys. Rev. B.* 57:249-252 (1986).
11. R.C. Anderson, R.S. Muller, and C.W. Tobias, Investigations of the electrical, properties of porous silicon, *J. Electrochem. Soc.* 138:3406-3411 (1991).
12. F. Namavar, H.P. Maruska, and N.M. Kalkhoran, Visible electroluminescence from porous silicon np heterojunction diodes, *Appl. Phys. Lett.* 60: 2514-2516 (1992).
13. V. Lehmann, H. Cerva, and U. Gosele, Pore formation and propagation mechanism in porous silicon, *in* "Light Emission from Silicon," *in* S.S. Iyer, L.T. Canham and R.P. Collins, eds. Mat. Res. Soc., Pittsburgh PA, 256: 3-6 (1992).
14. G. B. Amisola, R. Behrensmeier, J. M. Galligan, F. A. Otter, F. Namavar, and N.M. Kalkhoran, Scanning probe microscopy and scanning tunneling spectroscopy of porous silicon, *Appl. Phys Lett.* 61: 2595-2597 (1992).

ATOMIC FORCE MICROSCOPE STUDY OF FERROELASTIC DOMAINS

Amar S. Bhalla,[1] Gargi Raina,[2] and Shiv K. Sharma[2]

[1]Materials Research Laboratory, The Pennsylvania State University
University Park, PA 16802
[2]Hawaii Institute of Geophysics, School of Ocean and Earth Science and Technology, University of Hawaii
Honolulu, HI 96822

Abstract: In the present study, the authors have employed, for the first time, atomic force microscope (AFM) for studying domains in ferroelastic materials. Studies of ferroelectric/ferroelastic domains is important for understanding the acousto-optic, opto-electronic properties and the aging behavior of the ferroelectric/ferroelastic devices. In this paper we report our results on ferroelastic lead vanadate, $Pb_8V_2O_{13}$ crystals. Both stripe and needle shaped domains are observed. The typical width of the tip of the needle domains is ~100 Å. Sharp and round tips of needles have been detected very clearly with developed surface features.

INTRODUCTION

Ferroelastic material has ferroelastic states that are reorientable under the external stress. In a ferroelastic material the ferroelastic states occur when a material is cooled from the paraelastic state through the transition temperature, T_c. As a result, domain wall occurs between the ferroelastic states below the T_c and these states are reorientable from one state to the other under the influence of externally applied stresses. These domains are typically designated as 90° domains and the walls as 90° domain walls. In several cases where the material is also a ferroelectric, in addition to the 90° walls, 180° walls and domains occur in the materials. Depending upon the crystal structure and the order parameter, ferroelectricity and ferroelasticity in a crystal could be fully or partially coupled or completely uncoupled.

In various ferroelectric/electrooptic/ferroelastic devices and basic phenomena, 90° and 180° domains play an important role, e.g., dielectric loss mechanism in capacitors, electrooptic and switching devices, ferroic substrates for high T_c superconductor films. It is shown in various studies that the presence and the movement of 90° domain walls in a crystal influence the device performance to a great extent.

Atomic Force Microscopy/Scanning Tunneling Microscopy
Edited by S.H. Cohen *et al.*, Plenum Press, New York, 1994

A considerable number of different techniques have been applied in the past for the study of 180° domain structure in ferroelectric crystals, e.g., optical microscopy[1], (ferroelastic:ferroelectric domain), electron scanning and imaging techniques,[2-4] chemical etching,[5] x-ray topography,[6] decoration techniques,[7-13] pyroelectric[14] electroluminescent technique,[15] optical second harmonic generation,[16] force microscopy[17] (detection of static surface charge), etc. In most techniques, the surface polish or a special sample preparation is required and in the process the real surface features and domain walls get disturbed. Also at present there is a lack of full understanding of the role of 90° domain walls, their thickness and movement in the crystal and hence the associated effects on the device performance. Due to the associated elastic shear stresses with the domain states, there are lattice distortions, which may possibly be revealed in the surface morphology of the pristine surfaces of a ferroelastically twinned crystal. The present paper reports the first such observation of the ferroelastically twinned Pb vanadate crystal surface morphology.

Lead vanadate, $Pb_8V_2O_{13}$ (PV) is a ferroelastic and has a point group symmetry 2/m below the transition temperature, 160° C. Above the T_c it probably is orthorhombic. At room temperature it has lattice constants a = 10.647 Å, b = 10.616 Å, c = 28.47 Å, and ß = 90.36.

EXPERIMENTAL METHOD

High quality surfaces of PV were examined with a contact mode atomic force microscope (Nanoscope III from Digital Instruments, Inc.). A silicon nitride tip mounted on a 100 μm microcantilever was used to scan the surface with an estimated tracing force of no greater than 10^{-7} N. The microscope was run in the constant force mode, i.e. keeping the tip stationary by restoring the sample in x, y, and z (viz., the variation in the z-movement of the piezo element required to keep the deflection signal constant, is measured).

RESULTS AND DISCUSSIONS

Figure 1 shows a typical ferroelastic domain pattern on (001) surface of $Pb_8V_2O_{13}$ single crystal observed under the crossed polarized light. Two ferroelastic states, separated by the 90° domain wall, are represented by the white and dark contrast in the photograph.

Figure 1. (a) Ferroelastic domains in single crystal of lead vanadate (1 mm x 0.75 mm area).

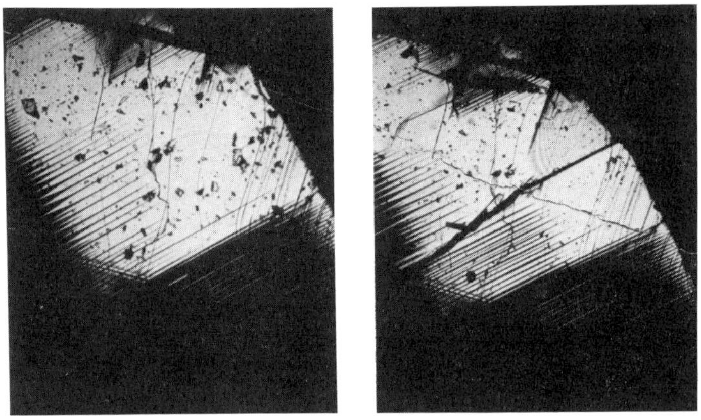

Figure 1. (b) Change in domain state under the externally applied stress of different magnitudes in the case of two pictures.

Figure 2 shows the typical domain features (top view and surface plots) revealed by the AFM over a 3 μm scan on the surface. The typical needle shaped domains of fine structure are revealed. It is found that these domains are not of uniform depth as indicated by the section analysis of the domain features on the surface.

Invariably the relative differences in the surface levels of the two domain areas on the order of 20 Å were measured. These differences tapered off along the length of the domain towards the tip of the needle shape domains.

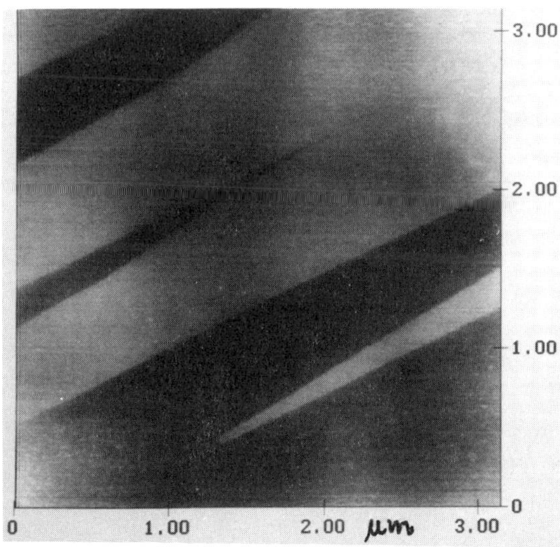

Figure 2. (a) Ferroelastic domains as revealed by AFM over a 3 μm x 3 μm area.

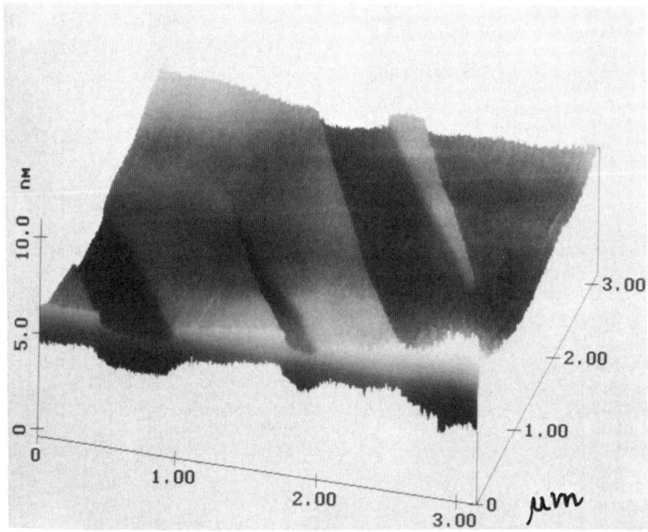

Figure 2. (b) Edge view illustrating the differences in the surface levels of the two domain regions that resulted from the ferroelastic distortions. Dark and white contrasts represent two ferroelastic states.

Figure 3 illustrates such section analysis study done on one set of needle domains. It suggests that the lattice displacement is less than a unit cell height (c~28.47 Å) in an area where the straight domain wall dominates but the relative difference or the domain wall height in the surface levels gradually vanishes approaching the tip of the needle-shaped domain. Such a large magnitude of relative depth or height differences in the domain areas is difficult to explain. Also these features can be most undesirable for various device applications. It also raises the question how the domain wall movement takes place under externally applied stresses when such large lattice distortions are present. (As shown in Figure 1, the ferroelastic domain states do get affected by the external stresses.) To answer these questions more detailed studies on the atomic scale are in progress. Under high magnification the typical domain tip width on the order of ~50-100 Å was observed.

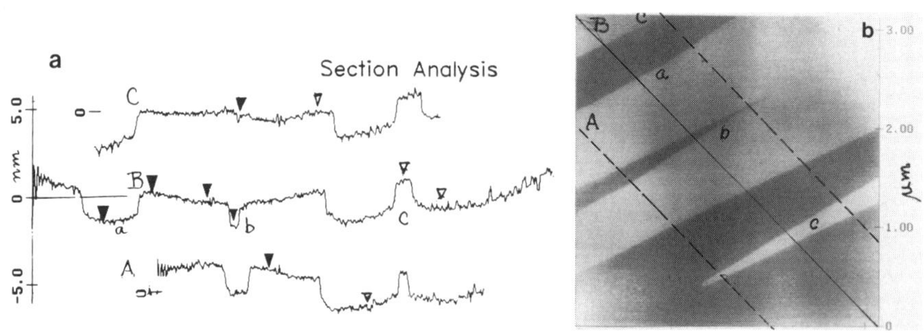

Figure 3. Section study at various parts of the domains. It is clear that needle shape domain is shallower towards the tip.

This observation suggests that the ferroelastic domain wall thickness is ≈ 25 Å (~2-3 unit cells), as seen in Figure 4. Such order of magnitudes has also been suggested in the case of $BaTiO_3$ (25 Å ≈ 5 unit cells) studied under TEM.

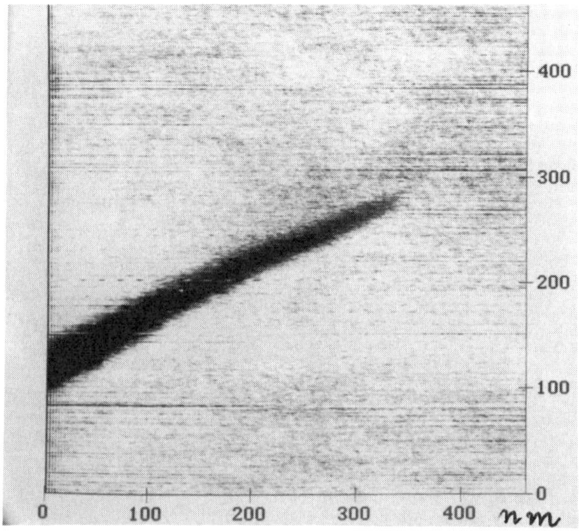

Figure 4. Magnified image of the domain tip, typically of 50 Å width suggests the domain walls on the order of 25 Å.

CONCLUSIONS

We have demonstrated that AFM can be used for imaging the ferroelastic domains on the (001) pristine surfaces of the single crystal of lead vanadate. The domain wall height and thickness can be precisely studied by this technique. The surface morphological differences on unit cell scale in the two domain regions, which finally resulted as a contrast in the AFM scans of the surface, may provide a unique insight in understanding the ferroelastic distortions of the lattice in crystals. Further study on the unit cell level distortions is in progress and the results will be reported in the forthcoming papers.

ACKNOWLEDGMENT

The financial assistance for the purchase of AFM was provided by the Office of Naval Research, the UH Research Council and the School of Ocean and Earth Science and Technology.

REFERENCES

1. T. Mitsui, and J. Furuichi, Domain structure of Rochelle salt and KH_2PO_4, *Phys. Rev.* 90:193-202 (1953).

2. M. Tanaka and G. Honjo, Electron optical studies of $BaTiO_3$ single crystal films, *J. Phys. Soc. Japan* 19:954-970 (1964).
3. G.V. Spivak, E. Igras, and I.S. Zheludev, Electron optical investigation and electron microscopy of the domain structure of ferroelectric single crystals, Intern. Kongr. Electronenmikroskopie, 4., Berlin, Verhandl. 1:502-506 (1958).
4. H. Blank and S. Amelinckx, Direct observation of ferroelectric domains in barrium titanate by means of the electron microscope, *Appl. Phys. Lett.* 2:140-142 (1963).
5. J.A. Hooton and W.J. Merz, Etch patterns and ferroelectric domains in $BaTiO_3$ single crystals, *Phys. Rev.* 98:409-413 (1955).
6. N. Niizeki and M. Hasegawa, Direct observation of antiparallel 180° domains in $BaTiO_3$ by X-ray anomalous dispersion method, *J. Phys. Soc. Jap.* 19:550-554 (1964).
7. G.L. Pearson and W.L. Feldman, Powder-pattern techniques for delineating ferroelastic domains, *J. Phys. Chem. Solids* 9:28-30 (1959).
8. Y. Furuhata and K. Toriyama, New liquid-crystal method for revealing ferroelectric domains, *Appl. Phys. Lett.* 23:361- 362 (1973).
9. J. Fousek, M. Safrankova, and J. Kaczer, A new method for revealing ferroelectric domains, *J. Appl. Phys. Lett.* 8:192-194 (1966).
10. M. Takagi, S. Suzuki, and K. Tanaka, Evaporated metal films on ferroelectric domains, *J. Phys. Soc.* Jap. 23:134 (1967).
11. G.I. Distler, Y.M. Konstantionava, Y.M. Gerasionov, and G.A.Tohnacheva, Interaction of defect and domain structures of triglycine sulphate crystals in ferroelectric and paraelectric states, *Nature* 218:762-765 (1968).
12. A.S. Bhalla, and L.E. Cross, A simple technique for decorating ferroelectric domains in gadolinium molybdate, *J. Mater. Sci.*, 12:2346-2347 (1977).
13. G.I. Distler, and L.A. Shenyavskaya, Anisotropy of the real structure of triglycine sulfate domains, *Ferroelectrics* 9:237-239 (1975).
14. J.C. Burfoot and R.V. Latham, A new method for studying movements of electric domain walls, *Brit. J. Appl. Phys.* 14:933-934 (1963).
15. V.A.Yurin, A.S. Baberkin, and I.S. Zheludiev, Effect of γ-radiation on the ferroelectric properties of guanidine $Al_2(SO_4)_3$ crystals, *Kristallografiya* 7:147-150 (1962).
16. J. Lajzerowicz, private communication.
17. F. Saurenbach and B.D. Terris, Imaging of ferroelectric domain walls by force microscopy, *Appl. Phys. Lett.* 56:1703-1705 (1990).

ATOMIC SCALE IMAGING OF MINERALS WITH THE ATOMIC FORCE MICROSCOPE

G. Raina, R.W. Gauldie, S.K. Sharma, and C.E. Helsley

Hawaii Institute of Geophysics, School of Ocean Earth Science and Technology
University of Hawaii at Manoa
Honolulu, HI 96822

Abstract: We present in this paper the surface structure of minerals like mica, calcite and aragonite observed with the atomic force microscope (AFM). Surface images at atomic resolution on mica and calcite were achieved but it was not possible for us to achieve atomic resolution on aragonite crystals. We believe this is due to the poor cleavage in aragonite. Results of an experiment performed on mica to find the variability in the measurement of atomic distances are also presented. In the case of calcite (Iceland spar), we provide an analysis of the surface atomic structure on the $(10\bar{1}1)$ perfect cleavage as observed with the AFM and compare it with X-ray diffraction data.

INTRODUCTION

Atomic force microscopy (AFM) offers the possibility of examining surface structure of a variety of materials, conducting and nonconducting, with very high resolution. Both scanning tunneling microscopy (STM) and AFM have been used successfully for analyzing mineral surfaces.[1-8] Minerals with imperfect cleavage and biogenic minerals present a challenge in imaging of their surfaces with atomic or molecular resolution, with the AFM. In few AFM studies, the atomic lattice of mineral surfaces has been resolved.[4,7,9,10] In our study we have chosen calcite and aragonite with a view of forming a basis for a related study of biogenic minerals.

EXPERIMENTAL METHODS

A Nanoscope III AFM (Digital Instruments, Inc.) with a silicon nitride integral tip mounted on a microcantilever of length 100 μm was used to probe a freshly cleaved surface of mica (muscovite) in air. An experiment was performed to gauge the variability in the atomic measurements of mica images. A set of mica images were captured by:

1. Keeping the scan size, scan rate, and all other parameters constant with time and imaging with the same tip;
2. Keeping scan rate and all other parameters constant with changing scan size and imaging with the same tip;
3. Imaging with the same tip moving to different region but keeping the scan rate and other parameters constant; and
4. Imaging with a different tip on a different mica surface.

Mica was used as a standard material since it has been well studied and has well known properties. Variability caused by experimental noise was examined by making repeated measurements of the same surface under the conditions stated above. The results of 30 successive scans of the same sample surface were then converted into Fourier space and the amplitudes of the peaks at fixed Fourier periods were measured. These yielded spacings of 3.66, 4.13, 4.49, 4.68 and 4.98 Å, which are in good agreement with the spacings expected for mica within three percent. An atomically resolved surface of mica is shown in Figure 1 and demonstrates the quality of image that our equipment acquires.

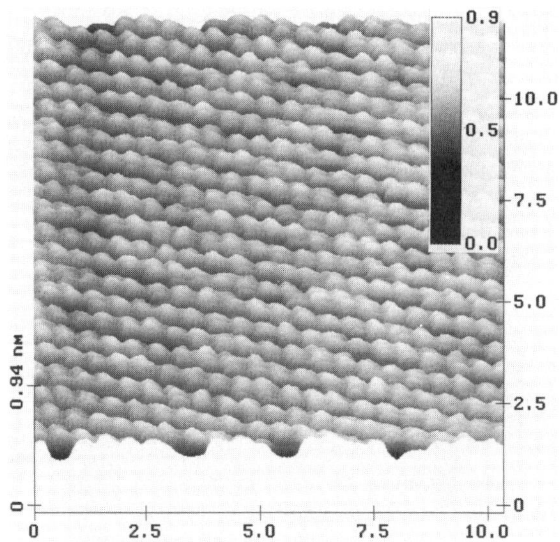

Figure 1. An AFM image showing atomic resolution on mica (muscovite); scale is in nm.

CALCITE RESULTS AND DISCUSSION

A freshly cleaved (10$\bar{1}$1) surface of optical quality (Iceland spar) calcite was imaged in a fluid cell for a period of three hours using a constant tracking force of 8.76×10^{-8} N. The set point voltage was continuously adjusted from 0 to -1 volts to maintain image stability and to minimize scanning artifacts. Figure 2 shows the resultant AFM image of the $(10\bar{1}1)$ cleavage plane of this calcite. The image of the calcite surface shows features at atomic scale, is periodic, and clearly shows multidirectional ordered structures. The presence of the "plowed field" character is unexpected and its origin is not fully understood.

There is some inconsistency in the nomenclature of the cleavage plane of calcite imaged with the AFM. For the AFM work, the choice of the morphological (cleavage rhomb) pseudo cell is essential, as we deal only with the morphological aspects of the sample being examined. A projection of the surface structure at the $(10\bar{1}1)$ cleavage plane of calcite is shown in Figure 3. This projection was constructed using the morphological indexing of calcite with cell parameter a = 6.41 Å and alpha = 101.92° (reference 11, p. 127). The corners and the center of the unit cell are occupied by Ca atoms, and the mid-points of the sides of the cell are occupied by carbonate molecules. We calculate from the projection in Figure 3, the distance between the nearest O sites to be 4.98 Å and the distance between the next-nearest O sites to be 8.08 Å. The angle between the intersection of the line containing the nearest-neighbor CO_3 group and the line containing the next-nearest-neighbor CO_3 group would be 90°. However, the AFM yields the image of the projecting O atoms of the CO_3 groups. The observed angle theta between the lines through the projecting nearest-neighbor and the next-nearest-neighbor O atoms is not the same as that of the CO_3 group as a whole and is 69°.

Direct measurements from the AFM image (Figure 2) give distance between nearest O atoms of 4.99 Å (±0.18 Å) compared with a value of 4.98 Å from the projection of the cleavage plane (Figure 3), and between next-nearest O atoms of 8.92 Å (±1.06 Å) compared with the expected value of 8.08 Å from Figure 3. Furthermore, the AFM image gives an average angle theta of 74° compared with an expected angle of 69° from Figure 3. The CO_3 groups are triangular and, in the $(10\bar{1}1)$ cleavage plane, there are two distinct

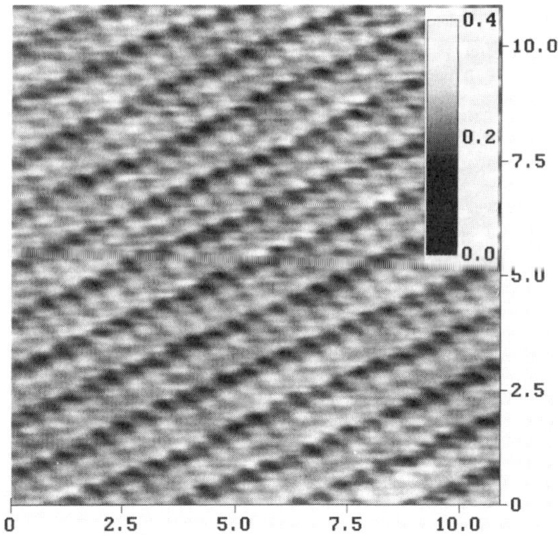

Figure 2. AFM image of the surface structure of calcite in the $(10\bar{1}1)$ plane showing surficial rows of the oxygen atoms (bright spots) of the carbonate group with parallel dark rows (Xs) formed by the subsurface presence of calcium atoms. The vertical inset in this and subsequent figures provides a scale of surface roughness in nm; the scale is in nm.

types of CO_3 oriented ions: these two CO_3 groups are rotated 60° relative to each other. This difference in orientation of the nearest-neighbor CO_3 groups results in the observed offset in the line joining the successive projecting nearest O atoms, as depicted by angle D of 19.5° in Figure 3; the measured offset D from our AFM images varied from 8.00° to 19.5° with an average value of 11.7° (±3.7°).

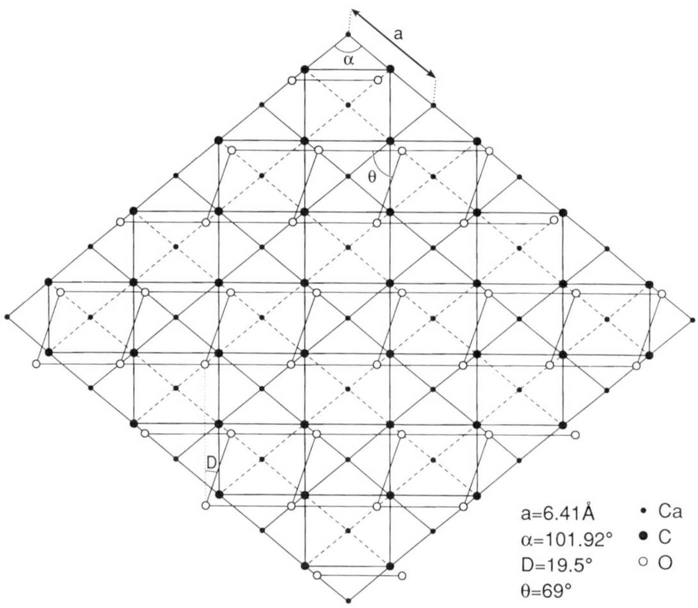

Figure 3. Projection of the $(10\bar{1}1)$ cleavage plane of calcite drawn such that alpha = 101.92°, D = 19.5°, and a = 6.41Å.

Another atomically resolved AFM image of $(10\bar{1}1)$ cleavage plane of calcite was obtained in a different run of a similar experiment. Measurements of 4.33 Å (±0.50) for the atomic spacing between nearest O atoms and 9.33 Å (±0.46) for next-nearest O atoms were obtained for this run. These spacings match closely to those obtained by Stipp and Hochella (1991) using LEED, but differ from the spacings obtained from the first run. Our explanation for the differences observed in our experiments, and perhaps some of the variability in AFM studies of calcite literature, is some sort of surface effect. Surface effects caused by H_2O accumulation have been discussed by Stipp and Hochella.[12] An alternative surface effect is the possibility of a surface reorganization to lower the high free energy levels caused by exposed charges at the cleaved surface.

In Figures 4 and 5 we show a computer simulation of the geometry of an internal surface slightly oblique to the idealized Bragg surface based on X-ray data geometry (Figure 4), and the subsequent reconfiguration that was present on the free surface as a result of free energy reduction at the surface introduced by the model calculation embodied in the HYPER CHEM computer program (Figure 5).

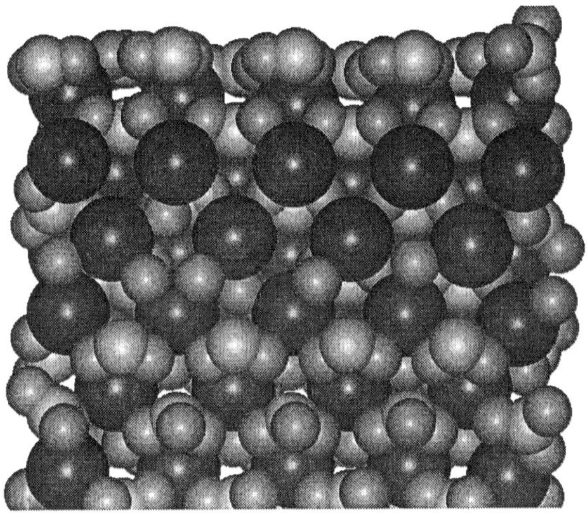

Figure 4. Computer model of calcite showing a cut slightly oblique to the $(10\bar{1}1)$ plane. Note the parallel alignment of the carbonate groups which is characteristic of the Bragg model for calcite. Ca is dark and O is the smallest of the light atoms.

CONCLUSIONS

Our results show that it is possible to obtain good atomic resolution images of minerals with perfect cleavage like mica and calcite. The observation of different set of surface atomic spacings for calcite may be an inherent property of calcite. Atomic scale AFM imaging of minerals that have less than perfect cleavage remains a challenging problem.

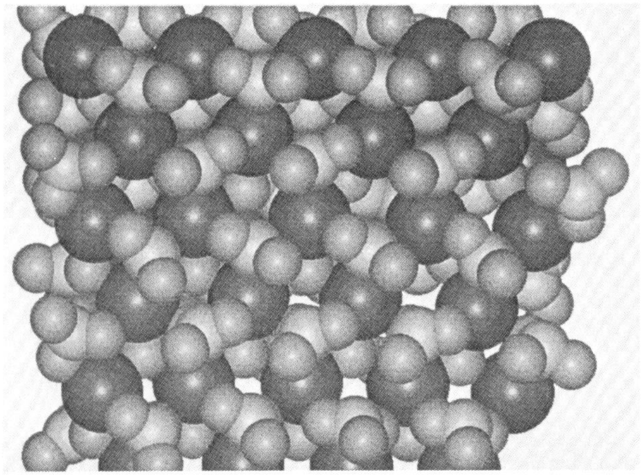

Figure 5. Computer model of the arrangement of atoms on the free cleavage surface of a calcite crystal. The model and the calculation are identical to that depicted in Figure 4 but the display is now of the surface instead of the interior of the model crystal. Note that adjacent rows of carbonate groups in the center of the picture are no longer parallel. Edge effects near the boundary of the model crystal should be ignored.

ACKNOWLEDGMENTS

This research was partially supported by the Biodegradable Packaging Program of the U.S. Department of Agriculture. Financial assistance was also provided by the Office of Naval Research, the University of Hawaii Research Council, and the School of Ocean and Earth Science and Technology for the purchase of AFM equipment. The authors thank Dr. A. Jayaraman for his comments. The SOEST contribution number is 3229.

REFERENCES

1. M. F. Hochella, C. M. Eggleston, V.B. Ehlings, and M.S. Thompson, Atomic structure and morphology of the albite {010} surface: An atomic force microcopy and electron diffraction study. *American Mineralogist*, 75:723-730 (1990).
2. B. Drake, and R. Hellman, Atomic force microscopy imaging of the albite (010) surface, *American Mineralogist*, 76:1773-1776 (1991).
3. C.M. Eggleston and M.F. Hochella, Jr. Scanning tunneling microscopy and spectroscopy of sulfide surfaces, *Geochimica et Cosmochimica Acta*, 54:1511-1517 (1990).
4. P.E Hillner, A.J. Gratz, S. Manne, and P.K. Hansma, Atomic scale imaging of calcite growth and dissolution in real time, *Geology*,20: 359-362 (1992).
5. R. Wiesendanger, I.V. Shvets, D. Bürgler, G. Tach, H.J. Güntherodt, J.M.D. Coey, and S. Gräser, Topographic and Magnetic Sensitive Scanning Tunneling Microcopy Study of Magnetite, *Science*, 255:583-586 (1992)
6. P.A. Johnsson, C.M. Eggleston, M.F. Hochella, Jr. Imaging molecular scale structure and microtopography of hematite with the atomic force microscope. *American Mineralogist*, 76:1442-1445 (1991).
7. F.J. Wicks, K. Kjoller, G.S. Henderson, Imaging the hydroxyl surface of Lizardite at atomic resolution with the atomic force microscope, *Canadian Mineralogist*, 30:83-91 (1992).
8. H. Lindgreen, J. Garnaes, P.L. Hansen, F. Besenbacher, E. Laegsgaard, I. Stensgaard, S.A.C. Gould, and P.K Hansma, Ultrafine particles of North Sea illite/smectite clay minerals investigated by STM and AFM, *American Mineralogist*, 76, 1218-1222 (1991).
9. G. Friedbacher, P.K Hansma, E. Ramli, and G.D. Stucky, Imaging Powders with the atomic force microscope: From biominerals to commercial materials, *Science*, 20:1261-1263 (1991).

10. A.L. Rachlin, G.S. Henderson, and M.C. Goh, An atomic force microscope (AFM) study of the calcite cleavage place: Image averaging in Fourier space. *American Mineralogist*, 77:904-910 (1992).
11. W.L. Bragg, G.F. Claringbull, and W.H. Taylor, "The Crystalline State, vol. III Crystalline Structure of Minerals," Cornell University Press, Ithaca, New York (1965).
12. S.L. Stipp and M.F. Hochella, Jr. Structure and bonding environments at the calcite surface as observed with x-ray photoelectron spectroscopy (XPS) and low energy electron diffraction (LEED), *Geochimica et Cosmochimica Acta*, 55: 1723-1736 (1991).

SCANNING TUNNELING MICROSCOPY OF THE STRUCTURAL AND ELECTRONIC PROPERTIES OF CHEMICAL-VAPOR DEPOSITED DIAMOND FILMS

J.M. Perez,[1] W. Rivera,[1,2] C. Lin,[1] R.C. Hyer,[3] M. Green,[3]
S.C. Sharma,[3] D.R. Chopra,[4] and A.R. Chourasia[4]

[1]Department of Physics
University of North Texas, Denton, Texas 76203.
[2]also: Universidad del Cauca, Popayan, Colombia
[3]Department of Physics
The University of Texas at Arlington, Arlington, Texas 76019
[4]Department of Physics
East Texas State University, Commerce
Texas 75429

Abstract: Scanning tunneling microscopy (STM) and spectroscopy have been used to characterize the structural and electronic properties of diamond films grown using hot tungsten filament and microwave plasma chemical vapor deposition. The hot-filament-grown films contained microcrystallites measuring 50 nm, while the microwave-plasma-grown films contained larger crystallites measuring 500 nm. STM tunneling current versus voltage (I-V) curves for the hot-filament-grown films exhibit a zero-current region about the Fermi level corresponding to a surface band gap of 4.1 eV, to be compared with the bulk band gap of diamond of 5.45 eV. The surface electronic density of states computed from these I-V curves is in good agreement with x-ray photoelectron and appearance potential spectroscopies. The I-V curves for the microwave plasma grown films exhibit rectifying behavior in good agreement with a Schottky model for surface band bending.

INTRODUCTION

Recently, it has been shown that diamond films can be grown using low-pressure chemical vapor deposition (CVD) at temperatures of less than 800 °C.[1] This discovery has triggered much research in CVD diamond growth and its commercial applications. Potential applications of diamond films include protective coatings, heat sinks, optical windows, and transistors.[2,3] For applications involving transistors, knowledge of the

Atomic Force Microscopy/Scanning Tunneling Microscopy
Edited by S.H. Cohen *et al.*, Plenum Press, New York, 1994

electronic structure of the surface of the film is crucial. Recently, the surface morphology of diamond films has been studied on a nanometer scale using scanning tunneling microscopy (STM) and atomic force microscopy (AFM).[4-6] It is also possible to study the electronic density of states of surfaces using STM by a technique referred to as scanning tunneling spectroscopy (STS).[7,8] In this technique, the tunneling current is measured at a particular sample location as a function of tip-sample voltage. In this paper, we present STM images and tunneling current versus voltage (I-V) spectroscopy curves of diamond films. The films were grown using hot-tungsten filament and microwave plasma CVD. The hot-filament grown films are found to exhibit a symmetric I-V curve with a zero-current region corresponding to a band gap of 4.1 eV. The surface electronic density of states computed from the I-V curves is in good agreement with x-ray photoelectron spectroscopy (XPS) and appearance potential spectroscopy (APS) data. The microwave plasma grown films exhibit a rectifying I-V curve in good agreement with a simple model for surface band bending. We also describe the ultrahigh vacuum (UHV) STM system at the University of North Texas currently being used to study diamond films with atomic resolution.

METHOD

The hot-filament grown films were grown on Si(111) substrates using a hydrogen gas flow of 350 sccm, methane gas flow of 1.75 sccm, growth chamber pressure of 25 torr, sample temperature of 800 °C, filament temperature of 2000 °C, and growth time of two hours. The microwave plasma grown film was grown on a graphite substrate. The films were approximately 1 μm thick. Raman spectra of the hot-filament grown films have been discussed in detail elsewhere.[9] Both types of films exhibit predominantly sp^3 type bonding with an sp^3/sp^2 bonding ratio of approximately 9.0. The microwave plasma grown film has a narrower diamond Raman peak at 1332 R cm^{-1}, indicating larger diamond crystallites. The full-width at half-maximum of the diamond peaks are 4.15 ± 0.02 cm^{-1} and 5.62 ± 0.08 cm^{-1} for the microwave plasma and hot-filament grown films, respectively.

STM images and I-V spectroscopy curves are obtained using a commercial STM.[10] The experiments are performed in air using mechanically cut Pt-Ir tips, and electrochemically etched tungsten and gold tips. Typical tip-sample voltages and tunneling currents used for imaging are 3-4 V and 0.5-1.0 nA, respectively. I-V spectroscopy curves are measured by disengaging the tip-sample distance feedback loop and measuring the tunneling current as a function of tip-sample voltage at a fixed tip-sample distance. The tunneling current is measured using a 12 bit analog to digital converter and stored in a computer. Several I-V curves are measured at the same sample location and averaged to obtain a high signal-to-noise ratio. The surface electronic density of states is computed from the I-V curves by numerical differentiation using the expression for the normalized conductivity.[8]

In addition to I-V spectroscopy, we also performed tunneling current versus tip-sample separation (I-S) measurements. In this technique, the tunneling current is measured at a fixed tip-sample voltage as a function of tip-sample separation. These measurements showed a strong dependence of the tunneling current on separation with a barrier height of approximately 5.0 eV, consistent with tunneling across a vacuum gap.

RESULTS

For the initial tip approach, we used tip-sample voltages of 3-4 V and tunneling currents of 0.5-1.0 nA. It is possible to gradually reduce the tip-sample voltage during tunneling to 0.1 V while maintaining a tunneling current of 2-4 nA. Atomic resolution imaging of the diamond surface is possible under these conditions and will be the subject of a future publication. Although natural diamond is an insulator, we believe that diamond films contain impurities introduced during the growth process and amorphous carbon that make the films conducting.

Figures 1(a) and (b) show a top view and corresponding three-dimensional topographic STM image of a 2 X 2 μm area of a hot-filament-grown diamond film. Large crystallites on the order of 0.5 μm are observed at this resolution. Both triangular (111) and square (100) diamond faces are observed. The film is relatively smooth with a root-mean-squared roughness of 0.29 μm. At higher resolution, microcrystallites measuring approximately 50 nm in width and 50 nm in height are observed, as shown in Figures 1(c) and (d). These findings are in agreement with scanning electron microscopy (SEM) and high resolution transmission electron microscopy images of the same film.[11] Figures 1(e) and (f) show a top view STM image and corresponding three-dimensional topographic image of a 500 X 500 nm area of the microwave plasma grown film. SEM images show that the crystallites in this film are wider and taller than those in the hot-filament grown film, which makes large-area STM imaging of the microwave plasma grown film more difficult. The smallest crystallites observed in this type of film measured approximately 500 nm in width and 500 nm in height, as shown in Fig. 1(f). The crystallite observed in Fig. 1(f) has triangular (111) faces.

I-V spectroscopy curves were measured at the sample locations indicated by X's in Figures 1(c) and (e), as well as other locations. The I-V curves measured at different locations were similar. Figure 2(a) shows a typical I-V curve for the hot-filament grown diamond film. The tunneling current is symmetric about the Fermi level at zero volts and has a well defined zero-current region. This type of I-V curve having a zero-current region has also been observed for Si and GaAs.[12,13] The zero-current region is due to the band gap which is absent of electronic states which are necessary for tunneling. The surface band gap of the film measured from the width of the zero-current region is 4.1 eV, to be compared with the bulk band gap of diamond of 5.45 eV. Figure 2(b) shows a typical I-V curve for the microwave-plasma-grown diamond film. The I-V curves in this case are rectifying, showing a tunneling current only for negative sample voltages. The same rectifying behavior was observed using platinum, tungsten and gold tips, making tip-induced band bending unlikely. A similar phenomenon has been reported for I-V curves of ZnO[14], hydrogen-terminated Si(111)[15], and sulfur-terminated GaAs.[16] The rectification in these cases is attributed to surface band bending and the forward current is accurately modelled by the expression[14-16]

$$J = J_0\{\exp[q(V_{BO} - V_{Bi})/kT - 1]\} \tag{1}$$

for the forward current of a Schottky barrier where V_{BO} is the barrier height and $V_{Bi} = V_{BO}\{[(c+1)^2 - V/V_{BO}]^{1/2} - c\}^2$ with c a constant is the band bending as a function of sample voltage. Figure 2(c) shows a fit to the forward current shown in Figure 2 (b) using the above expression, showing good agreement with the model. Band bending in ZnO is attributed to water on the surface, and in Si and GaAs to hydrogen and sulfur termination, respectively. Rectifying point contacts on diamond films have been reported to produce larger rectification in diamond films with larger diamond crystallites.[17] Therefore, the

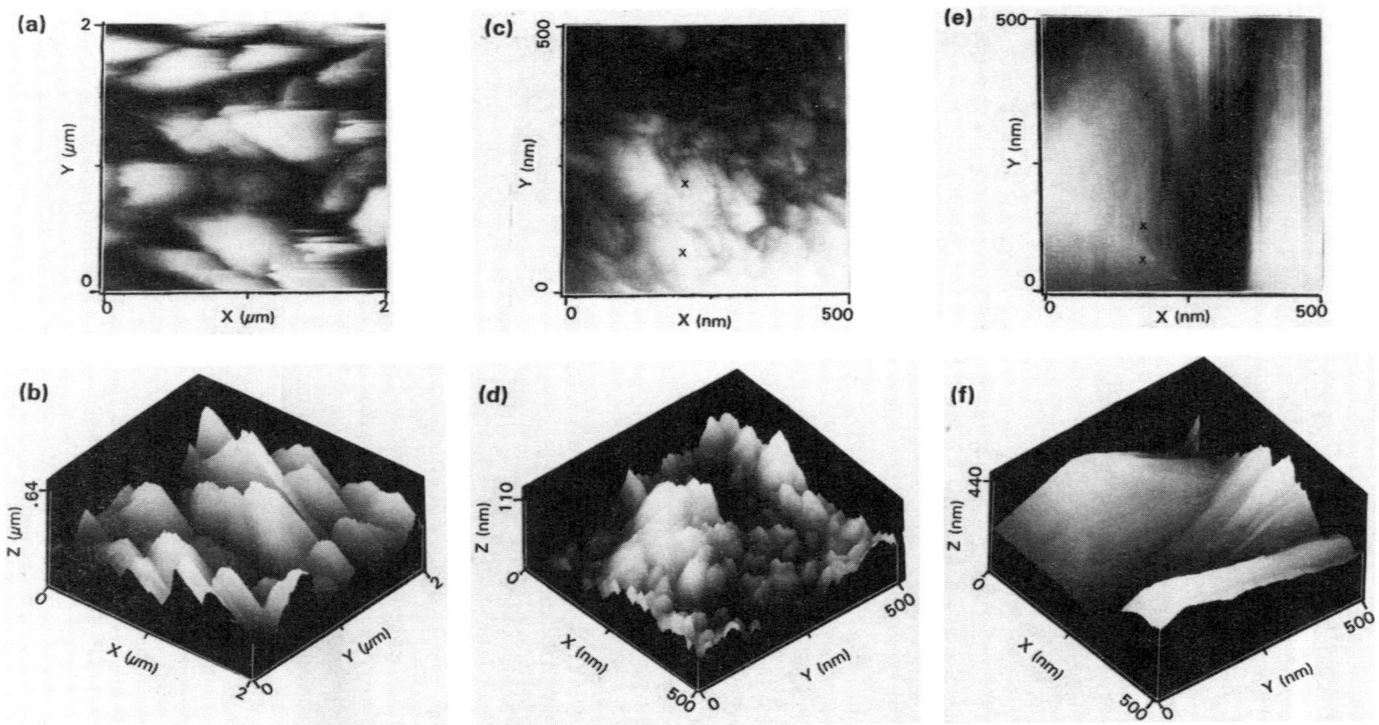

Figure 1. (a) STM image of the hot-filament grown diamond film. (b) Three-dimensional image of (a). (c) High resolution STM image of the same hot-filament grown diamond film as in (a). X's denote locations where current vs. voltage curves were measured. (d) Three-dimensional image of (c). (e) STM image of the microwave plasma grown diamond film. X's denote locations where current vs. voltage curves were measured. (f) Three-dimensional image of (e). The triangular faces of the crystallite are in the (111) direction.

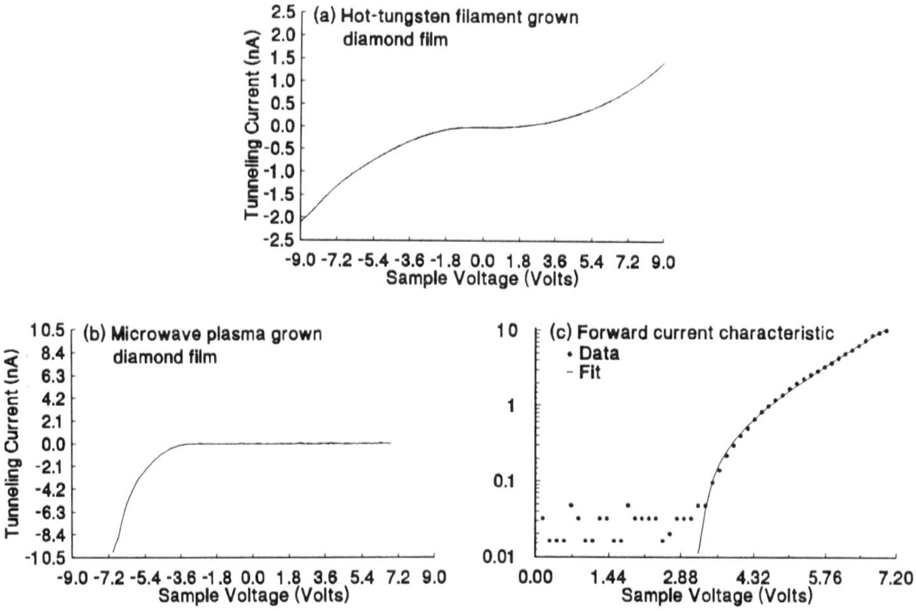

Figure 2. (a) Tunneling current vs. sample voltage curve for the hot-filament grown diamond film. From the width of the zero-current region, a surface band gap of approximately 4.1 eV is measured. (b) Tunneling current vs. sample voltage for the microwave-plasma-grown diamond film. (c) Filled circles show a semilog plot of the measured I-V curve for the microwave plasma grown. Solid line shows a fit to the forward current part of the I-V curve using the Schottky junction equation in the text with a barrier height of 4 volts.

rectification in the microwave plasma grown film is consistent with the film's larger crystallites.

We obtained the surface electronic density of states, ρ, of the hot-filament grown diamond film using the expression[8]

$$\rho \propto \frac{dI/dV}{I/V} \tag{2}$$

where the right-hand-side is referred to as the normalized conductivity. This expression is valid for non-zero tunneling current, I, but is divergent in the band gap region where I is zero. The singularity is simply removed by setting the entire expression equal to zero in the band gap region, or using an average non-zero value for I/V in the band gap region.[8,18] Similar results were obtained using these normalization procedures. Figure 3 shows ρ as a function of V computed by numerically differentiating the I-V curve shown in Figure 2(a) and setting ρ equal to zero in the band gap region. The peaks at -2.8 eV and 3.7 eV result from the numerical differentiation. These peaks are not observed in XPS or APS data, as discussed below. This type of overshoot occurs when computing the normalized conductivity from I-V curves.[10] The peaks in the valence band observed at -6.1 eV and -8.1 eV are in good agreement with XPS data for the same film.[19] A peak in the conduction band is observed at 7.0 eV. The conduction band surface density of states of the sample was measured using APS.[20] The APS data showed a peak at 6.5 eV above the Fermi level in agreement with the 7.0 eV peak in the STM surface density of states.

We are currently studying CVD-grown diamond films using STM in UHV. In particular, we are studying the nucleation and initial stages of growth of the films with atomic resolution. As mentioned previously, it is possible to lower the tip-sample voltage to approximately 0.1 V and obtain atomic resolution imaging. A schematic of the hot-tungsten filament CVD diamond growth system attached to the UHV STM system at the University of North Texas is shown in Figure 4. The diamond growth chamber consists

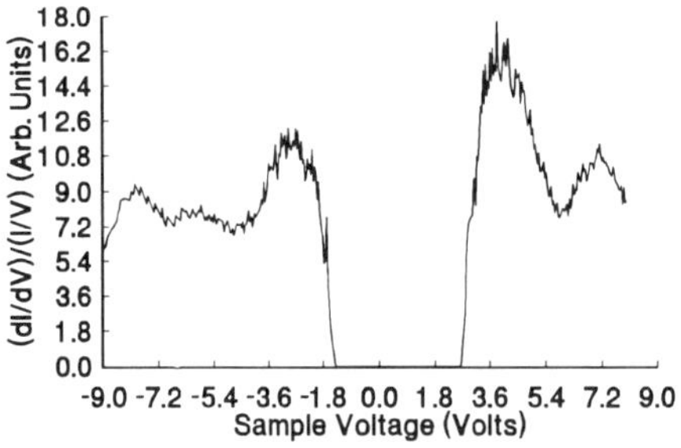

Figure 3. Surface electronic density of states of the hot-filament grown diamond film. The Fermi level is at zero volts. Peaks in the valence band corresponding to filled states at -6.1 eV and -8.1 eV are in agreement with x-ray photoelectron spectroscopy data. The peak at 7.0 eV in the conduction band is in agreement with appearance potential spectroscopy.

of a 5-way water-cooled cross. Two linear translators are used to transfer samples from the growth chamber to the UHV STM chamber without exposing the samples to air. *In situ* Raman spectroscopy is performed in the growth chamber or the UHV STM chamber.

In summary, we have studied the structural and electronic properties of CVD-grown diamond films using STM. Hot-filament grown films showed a well-defined surface band gap of approximately 4.1 eV and a surface electronic density of states in good agreement with XPS and APS data. The microwave plasma grown film contained larger diamond crystallites, as determined from STM, SEM, and Raman spectroscopy. These films exhibited rectifying behavior which was well modelled by a simple model for surface band bending. Atomic resolution imaging of diamond films is possible at low tip-sample voltages and related experiments are currently under way.

ACKNOWLEDGMENTS

We thank Dr. Earl W. Stromberg for the plasma grown diamond film. Work at the University of North Texas was supported in part by the National Aeronautics and Space Administration under Award No. NAG-1-1468. Work at East Texas State University was supported in part by Robert A. Welch Foundation Grant No. T-643.

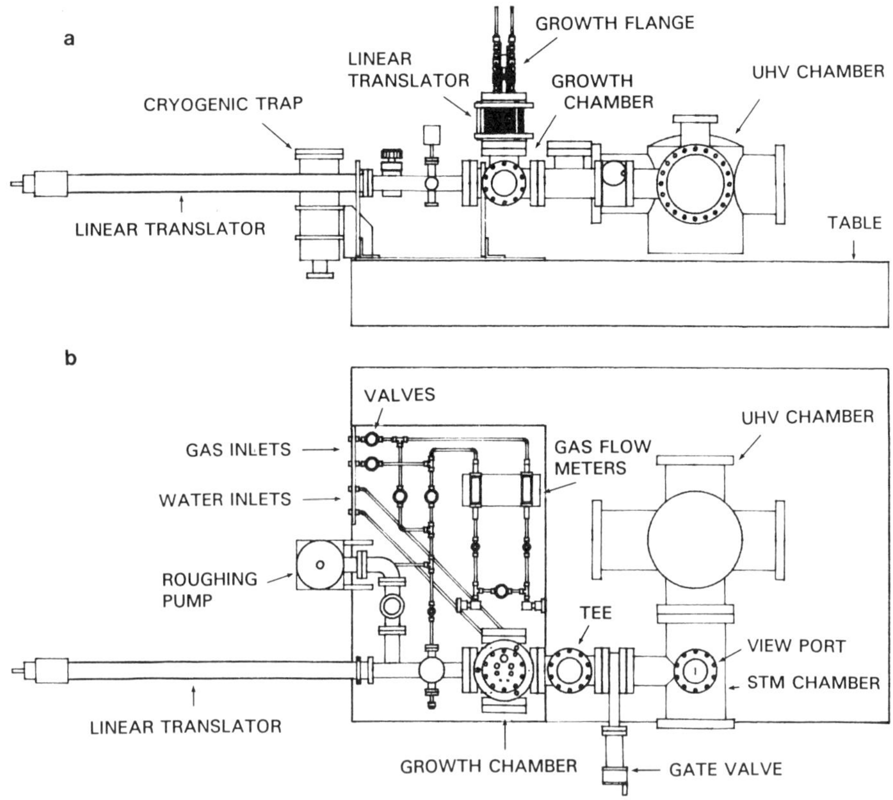

Figure 4. Hot-tungsten filament chemical vapor deposition diamond growth system attached to the ultrahigh vacuum scanning tunneling microscopy system. (a) Side view and (b) top view.

REFERENCES

1. B. Deryaguin, D. Fedoseev, N. Polyanskaya, and E. Statenkova, Epitaxial diamond-graphite graphite films, *Krystallografiya* 21:433-434 (1976).
2. W. Yarbrough and R. Messier, Current issues and problems in the chemical vapor deposition of diamond, *Science* 242:688-695 (1990).
3. J. Mort, M. Machonkin, and K. Okumura, Density of states distribution in diamond thin films, *Appl Phys Lett* 59:455-457 (1992).
4. H. Busman, H. Sprang, I. Herlel , W. Zimmermann-Edling, H. Güntherodt, Scanning tunneling microscopy on chemical vapor deposited diamond films, *Appl. Phys. Lett.* 59:295-297 (1991).
5. V. Baranauskas, M. Fukui, C. Rodrigues, N. Parizotto, V. Trava-Airoldi, Direct observation of chemical vapor deposited diamond films by atomic force microscopy, *Appl. Phys. Lett.* 60:1567-1569 (1992).
6. L. Sutcu, M. Thompson, C. Chu, R. Haugue, J. Margrave, and M. D'Evelyn, Nanometer-scale morphology of homoepitaxial diamond films by atomic force microscopy, *Appl. Phys. Lett.* 60: 1685-1687 (1992).
7. F. Salvan, H. Fuchs, A. Baratoff, G. Binnig, Characterization by tunneling microscopy and spectroscopy, *Surf. Sci.* 162:634-639 (1985).
8. R. Feenstra, Scanning tunneling microscopy: semiconductor surfaces, adsorption, and epitaxy. *Scanning Tunneling Microscopy and Related Methods*, R.J. Behm, N. Garcia, and R. Rohrer, eds. (Kluwer, Boston) 211-240 (1990).

9. S. Sharma, M. Green, R. Hyer, C. Dark, T. Black, A. Chourasia, D. Chopra, and K. Mishra, Growth of diamond films and characterization by Raman, scanning electron microscopy, and x-ray photoelectron spectroscopy, *J. Mater. Res.* 5:2424-2432 (1990).
10. Burleigh Instruments, Inc., Fishers, NY 14453.
11. R. Pinizzotto, private communication.
12. J. Stroscio, R. Feenstra, and A. Fein, Atom-selective imaging of the GaAs (110) surface, *Phys. Rev. Lett.* 58:1192-1195 (1987).
13. R. Feenstra, J. Stroscio, Tunneling spectroscopy of the GaAs (110) surface, *J.Vac. Sci.Technol.* B. 5:923-929 (1987).
14. G. Rohrer, D. Bonnell, Probing the surface chemistry of polycrystalline ZnO with scanning tunneling microscopy and tunneling spectroscopy, *J. Vac. Sci. Technol.* B. 9:783-788 (1991).
15. W. Kaiser, L. Bell, M. Hecht, F. Grunthaner. Scanning tunneling microscopy characterization of the geometric and electronic structure of hydrogen-terminated silicon surfaces, *J. Vac. Sci. Technol.* A. 6:519-523 (1988).
16. V. Berkovits, L. Ivantsov, I. Makarenko, T. L'vova , R. Khasieva, V. Safarov, Scanning tunnel microscope investigation of the surface of gallium arsenide passivated in an aqueous solution of Na_2S, *Sov. Phys. Semicond.* 25: 31-234 (1991).
17. K. Miyata, Y. Matsui, K. Kumagai, S. Miauchi, K. Kobashi, A. Nakaue, Characteristics of metal point contacts on diamond thin films, *New Diamond Science and Technology*, edited by R. Messier J. Glass, J. Butler, R. Roy (Materials Research Society, Pittsburgh), 981-986 (1991).
18. P. Martensson, R. Feenstra, Geometric and electronic structure of antimony on the GaAs (110) surface studied by scanning tunneling microscopy, *Phys. Rev.* B. 39:7744-7753 (1989).
19. A. Chourasia, D. Chopra, S. Sharma, M. Green, C. Dark, R. Hyer, Characterization of low pressure deposited diamond films by X-ray photoelectron spectroscopy, *Thin Solid Films* 193/194: 1079-1086 (1990).
20. D. Chopra, A. Chourasia, Appearance potential spectroscopy of solid surfaces, *Scanning Microscopy* 2: 677-702 (1988).

COMBINED SCANNING TUNNELING MICROSCOPE AND QUARTZ MICROBALANCE STUDY OF MOLECULARLY THIN WATER LAYERS

J. Krim, A. Dayo, and C. Daly

Physics Department, Northeastern University
Boston, MA 02115

Abstract: A combined quartz crystal microbalance and scanning tunneling microscope has been used to monitor frictional drag forces on the microbalance, which are due to the presence of the microscope's scan tip. The experiments have been carried out with varying thicknesses of water adsorbed on the surface of the microbalance. Increased friction between the tip and the substrate is observed at moderate film thicknesses.

INTRODUCTION

Adhesion, the binding of different materials at an interface, is of general importance to a variety of areas ranging from tribology to microelectronics and manufacturing.[1] In principle, an exact knowledge of the bonding forces at an interface allows one to predict the actual conditions that will cause the interface to fail or to separate. The relation of such forces to actual failure conditions is however a matter of great complexity.[2] Research on adhesion therefore tends to focus on either the fundamental nature of interfacial bonding forces[3] or, alternatively, on more practical aspects of interfacial failure.[4] Experimental measurements of fundamental contact forces frequently involve the lowering of a tip, such as that of a field emission or atomic force microscope[5,6], into contact with a surface while measuring the force to bring it in and out of contact. In-plane mapping of either the contact[7] or shear[8,9] force is also possible through suitable configuration of an atomic force microscope (AFM).

Despite much recent progress, measurements of fundamental adhesive forces are subject to a variety of experimental complications. For example, when a tip is lowered into surface contact, deformation of the tip or substrate may occur, and the two materials comprising the interface may mix.[10] Tip asperities and/or nonuniform loading of the tip may further complicate the notion of an interface with a well-defined contact area. When

Atomic Force Microscopy/Scanning Tunneling Microscopy
Edited by S.H. Cohen *et al.*, Plenum Press, New York, 1994

separation occurs, the interface is typically subjected to shear as well as to a normal force, and there is no guarantee that the actual separation will occur at the original interface. Finally, there may be adsorbed monolayers or submonolayers present at the interface that may significantly influence the experimental results.[5] Knowledge of whether or not such layers are present and information concerning the adhesive properties of such layers are essential to the interpretation of experimental results.

We report here a quartz microbalance investigation of the effect of molecularly thin water layers on the shear force recorded in a tip-substrate contact point geometry. The tip of a scanning tunneling microscope is first lowered into near contact with the surface electrode of the oscillator. Monolayers of water are then adsorbed onto the surface, and the shear force before and after adsorption is recorded. Since adsorbed films are known to greatly influence out-of-plane atomic force measurements,[5] our experiment sheds light on the manner in which such layers influence all experimental geometries involving tip-substrate contact points.

EXPERIMENTAL METHOD

A quartz microbalance consists of a single crystal of quartz which oscillates in transverse shear motion with a quality factor Q near 10^5. The driving force (supplied by a Pierce circuit) has constant magnitude and is periodic with frequency f = 5 MHz, the series resonant frequency of the oscillator. Each major face is plated with a metal electrode, which serves as the substrate for adsorption. Water adsorption measurements are carried out at room temperature, after transferring the microbalance crystal to the stage of an STM, which is enclosed in a vacuum chamber. The STM has been adapted so as to operate in conjunction with the microbalance. A detailed description of the apparatus and the technique are presented elsewhere.[11] Adsorption occurs under *equilibrium* conditions as water vapor is very slowly admitted to the chamber. This produces a change in the resonant frequency of the microbalance, which is proportional to the mass of the condensed film coupled to the motion of the oscillator. (Sensitivities on the order of tenths of monolayers are readily achieved.) Since the condensation occurs under equilibrium conditions, the film thickness is dependent only on the vapor pressure of adsorbate gas surrounding the sample. An additional consequence of the equilibrium adsorption conditions is the prevention of three-dimensional droplet, or crystal condensation at pressures less than the three-dimensional saturation pressure. (As will be discussed below, capillary condensation effects between the tip and the surface may, however, still be present.) Room temperature adsorption of water is in the form of a uniform film (assumed to be liquid) whose thickness increases monotonically with pressure up to at least 45-100 Å, before the possible condensation of droplets at the bulk saturation point.[12,13] Film adsorption onto the microbalance produces shifts in both the frequency and the amplitude of vibration, which are simultaneously recorded as a function of pressure. Amplitude shifts are due to frictional shear forces, which are exerted on the surface electrode by the adsorbed film (or alternatively by the STM tip or a three dimensional vapor).The microbalance in its present arrangement detects shear forces in excess of 2.5×10^{-7} N.

Interfacial (i.e., film-substrate) friction coefficients η_2 are deduced for the various film coverages according to the relations[12]

$$\delta(Q^{-1}) = 4\pi\tau\delta f \qquad \eta_2 = \rho_2/\tau \tag{1}$$

where ρ_2 is the mass per unit area of the adsorbed film deduced from the shift in frequency of the oscillator, and τ is a characteristic film slip time. (Amplitude shifts are converted to quality factor shifts $\delta(Q^{-1})$ through calibration with a gas that does not condense at 77.4 K). The coefficient η_2 corresponds to the force per unit area required to slide the film at a speed of 1 cm/s. It is a quantity that in theory can be calculated from first principles, assuming a knowledge of the strength and corrugation of the interlayer adhesive forces.

RESULTS

Figure 1 shows the frequency (1 (a)) and amplitude data (1 (b)) for water adsorption as vapor pressure is varied from 0 - 15 torr, for the tip both engaged (tunneling current held constant at 2 nA) and disengaged. At low pressure, the frequency shift data for the two cases are very similar. In contrast, as soon as the tip is engaged, the inverse amplitude shift jumps by 0.75 volts^{-1}, corresponding to an acoustic resistance of 0.58 g cm^{-2}s^{-2}. For the pressure range 6-13 torr, the frequency shift for the tip-engaged case becomes

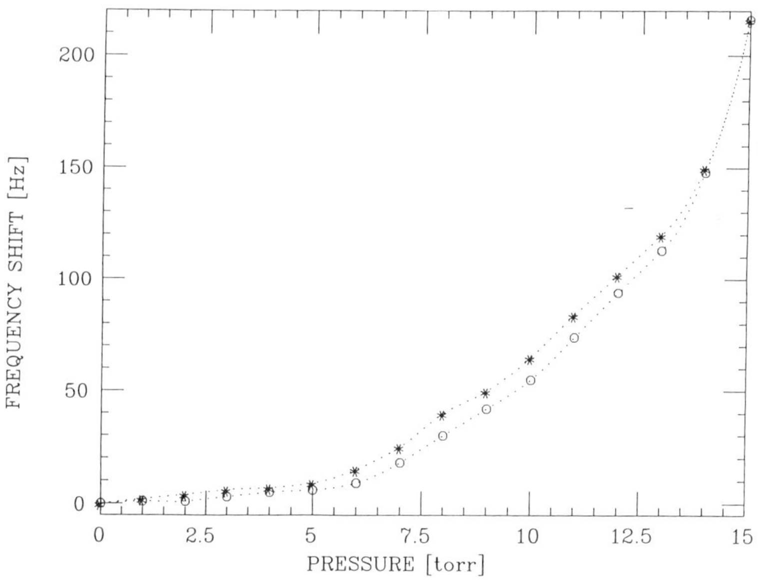

Figure 1. (a) Frequency shift vs. vapor pressure for tip engaged (*) and tip not engaged (o). The estimated mass per unit area of the monolayer is 31 ng/cm^2 (3.5 Hz or 0.10 molecules/Å^2).

noticeably higher than that of the tip-disengaged case. The difference is on the order of 10 Hz, corresponding to an additional film thickness of 9.1 Å. This (hypothetical) thickness is estimated by assuming the film to have the same density as the bulk liquid, and to coat both sides of the microbalance crystal uniformly. A bump is typically observed in both frequency and amplitude shifts at 8 Torr. It is more pronounced for the inverse amplitude curve. As vapor pressure approaches 15 torr, the data curves again start to approach those for the corresponding tip-disengaged cases. Beyond 15 torr, very high humidity can cause the tip to oscillate or crash.

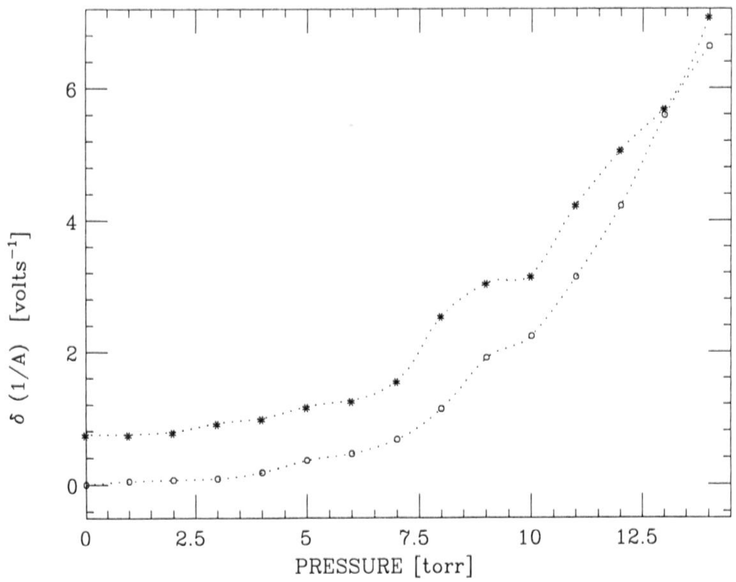

Figure 1. (b) Change in inverse amplitude vs. vapor pressure for tip engaged (*) and tip not engaged (o). A shift in inverse amplitude of 1 volt^{-1} corresponds to 0.77 g cm^{-2} s^{-1} of acoustic resistance.

After the recording of adsorption measurements, the oscillator is turned off and the surface imaged in order to investigate tip and/or surface damage. None is observed.

DISCUSSION

As the relative humidity inside the chamber ranges from 30-70%, a capillary bridge is believed to form between the tip and the substrate.[14] The mass of this bridge could in principle contribute to the additional increase in frequency shifts observed. However, calculations of the mass based on estimated dimensions of the tip and average tip-substrate separations show that the mass of any such bridge would be too small to account for the 10 Hz observed difference in frequency shifts between the tip-engaged and tip-disengaged cases. We believe that for small film thicknesses, a certain *stiffness* or solid-like quality develops in the film held between the tip and the substrate. The additional "mass" detected by the microbalance is likely to be that of the scan tip itself, partially coupled to the motion of the oscillator. The bumps observed in the frequency and amplitude curves both occur at water vapor pressures around 8 torr. We believe that this is a result of condensation of water in the wedge formed between the scan tip and the substrate. As relative humidity increases, water fills up the space around the sides of the tip near the relatively flat substrate, increasing the size of the capillary bridge.

Theoretical models show that the tunneling current is sensitive to the size of the bridge,[14] though its exact dependence is yet to be studied in detail. Because the STM is operated in the constant-current mode, the increase in the size of the bridge and the corresponding increase in the tunneling current will induce the tip to gradually lift. As this happens, the stiffness in the water film disappears. This is consistent with the observed approach of both frequency and amplitude curves, at high vapor pressures, to those of the corresponding cases where the tip is absent.

CONCLUSIONS

The level of shear force for this experiment is of the same order of magnitude as that for recent AFM experiments by McClelland, which mapped the in-plane shear forces of a graphite surface.[8] In AFM experiments, it is frequently unclear whether an adsorbed film is present on the tip or the substrate. As noted earlier, such a film can significantly alter the experimentally measured friction levels. We observe that although such films may indeed be able to support loads exerted by tips, it is unlikely that such films will have a major impact at low relative humidity levels.

ACKNOWLEDGMENTS

This work has been supported by the ONR, Grant No. N00014-89-J-1853, and NSF Grant DMR-86-57211.

REFERENCES

1. D. M. Mattox, et al., eds. Adhesion in solids, "Mat. Res. Soc. Symp. Proc.", vol. 119 (1988).
2. D. H. Kaeble, "Physical Chemistry of Adhesion," Wiley Interscience, New York (1971).
3. J. Ferrante, et al., "Mat. Res. Soc. Symp. Proc." vol. 119 (1988).
4. J. J. Bikerman, "The Science of Adhesive Joints," Academic Press, New York (1968).
5. N. A. Burnham, D. D. Dominguez, R. L. Mowery, R. Colton, Probing the surface forces of monolayer films with an atomic force microscope, *Phys. Rev. Lett.* 64: 1931-1934 (1990).
6. N. A. Burnham, R. J. Colton, Measuring the nanomechanical properties and surface forces of materials using an atomic force microscope, *J. Vac. Sci. Tech.* A 7: 2906-2913 (1989).
7. P. K. Hansma, V. B. Elings, O. Marti, C. E. Bracker, Scanning Tunneling Microsscopy and Atomic Force Microscopy: Application to Biology and Technology, *Science* 242: 209 (1988).
8. C. M. Mate, G. M. McClelland, R. Erlandsson., S. Chang, Atomic-scale friction of a tungsten tip o a graphite surface, *Phys. Rev. Lett.* 59: 1942-1945 (1987).
9. R. Erlandsson., G. Hadzioannou, C. M. Mate, G. M. McClelland, S. Chiang, Atomic-scale friction between the muscovite mica cleavage plane and a tungsten tip, *J. Chem. Phys.* 89: 5190-5193 (1988).
10. U. Landman, W. D. Luedtke, N. A. Burnham, R. J. Colton, Atomistic mechanisms and dynamics of adhesion, nanoindentation, and fracture, *Science* 248: 454-461 (1990).
11. C. Daly, J. Krim, Applications of atomic scale friction measurements with a combined quartz microbalance and scanning tunneling microscope, this publication.
12. J. Krim, E. T. Watts., J. Digel, Slippage of simple liquid films adsorbed on silver and gold substrates *J. Vac. Sci. Tech.* A 8: 3417-3420 (1990).
13. R. Chiarello, "Roughness Characterization of Gold and Silver Surfaces and the Wetting Behavior Of Physisorbed Films Studied by a Combined Quartz Crystal Microbalance and X-Ray Reflectivity Technique," Ph.D. thesis, Northeastern University, 1990.
14. J. Y. Yuan, Z. Shao, C. Gao, Alternative method of imaging surface topologies of nonconducting bulk specimens by scanning tunneling microscopy, *Phys. Rev. Lett.* 67: 863-866 (1991).

ATOMIC FORCE MICROSCOPY OF POLYMER DROPLETS

David C. Martin,[1] Jaime R. Ojeda,[2] J. Philip Anderson,[3] and Gopal Pingali [4]

[1] Materials Science and Engineering
[1,2,3] Macromolecular Science and Engineering
[4]Electrical and Computer Engineering
The University of Michigan
Ann Arbor, MI 48109-2136

Abstract: The near-surface structure of polymers is important for understanding such macroscopic properties as adhesion, friction, and fracture initiation. Studies of polymers near surfaces by conventional transmission electron microscopy (CTEM), selected area electron diffraction (SAED), and high resolution electron microscopy (HREM) have provided local information about the reorganization of macromolecules near constraining surfaces. Atomic Force Microscopy (AFM) is now emerging as a powerful technique for characterizing local surface structure but AFM is only reliable when used in conjunction with these other more traditional microscopic techniques. It is important to establish that features observed in AFM can be corroborated by other experiments. It is also necessary to compare experimental data with theoretical descriptions which take account of the geometry of the probe-sample interaction. Computer simulations indicate that artifacts can be observed in the images including high frequency "ringing" and "beating."

Our work has focused on the characterization of the poly(imide) PMDA-ODA in thin films and solution grown single crystals. We have also investigated the structural evolution of genetically engineered silk-like poly(peptides) as a function of processing condition. The shape of the probe can be discerned from images of model structures such as poly(styrene) spheres.

INTRODUCTION

Scanning probe microscopy (SPM) has revolutionized the high resolution microstructural analysis of surfaces. In SPM a sharp probe is rastered across the surface of a sample. Information about microstructure is obtained by monitoring the interaction of this probe with the object of interest. In scanning tunneling microscopy (STM) there is a current due to an applied potential difference between the sample and the probe. This tunneling

Atomic Force Microscopy/Scanning Tunneling Microscopy
Edited by S.H. Cohen *et al.*, Plenum Press, New York, 1994

current is an exponential function of the sample-tip separation, making it possible for STM to measure extremely small height fluctuations (down to ~0.01 Å) in the z-direction. However, STM is of limited use for nonconductive materials such as organic polymers and many biological systems. Atomic force microscopy (AFM) measures the mechanical interaction of a probe with the surface and therefore does not require an electrically conductive sample.

Because of the potential for imaging microstructure at the molecular level, there is considerable interest in the use of AFM to characterize nonconducting materials surfaces.[1] AFM has several significant advantages over other microscopic characterization techniques. Sample preparation is relatively easy, and imaging can be performed in vacuum, gaseous, or liquid environments. The capital investment for an AFM is significantly less than either a scanning electron microscope (SEM) or transmission electron microscope (TEM), and the maintenance costs are lower as well. However, AFM should not be considered as a complete replacement for SEM, TEM or other techniques of surface analysis. Rather, it is necessary to determine the limitations of the AFM technique and employ it as an extension of current capabilities. Combining the AFM, SEM, and TEM imaging techniques is a powerful means to explore both surface and bulk morphology of polymers.

AFM is already well established as an important and convenient method for the analysis of surface microstructure. However, without complimentary data from other techniques, it can be difficult to establish the reliability of conclusions drawn from AFM information alone. At present, the limits of the quantitative interpretability of AFM images are still being determined.

We have been using AFM to compliment our efforts to characterize the microstructure of polymer thin films and interfaces. In order to investigate the influence of surfaces on structural reorganization in macromolecules, we have been studying transitions in microstructure that occur near the periphery of polymer droplets deposited from solution onto solid substrates. This sample geometry makes it possible to address the influence of various processing parameters on the local organization of polymer systems. Previous experiments have indicated that the molecular structure near the edge of a polymer droplet can be substantially different from that found in the bulk.[2] The droplet edge provides a useful reference point and makes it possible to compare the surface topography of the sample with that of the substrate. Here, we discuss the morphology of two systems of current interest to our research group: (1) poly(imides) and (2) genetically engineered poly(peptides).

In the course of this work we have become particularly concerned with the relationship between frequency information in the sample and frequency infornnation in the experimental image. In other words, we hope to determine the transfer function of the AFM (i.e., the relationship between the theoretical structure and experimental image) as a function of probe geometry. To this end, we have been AFM imaging well-defined standard objects (i.e., uniform spheres of poly(styrene) prepared by latex polymerization) deposited on a flat substrate. Comparison of experiment with theoretically predicted images of these spheres provides an opportunity to quickly and unambiguously determine the probe tip geometry. With this information in hand, it is then possible to reconstruct the actual structure of the sample in a manner that minimizes artifacts.[3,4]

In the critical evaluation of AFM, it is particularly important to identify potential artifacts which can arise due to idiosyncrasies of the imaging process. These artifacts include effects due to the finite size and shape of the probes used in the AFM experiment. Recently, it has been shown that frequency artifacts can arise from the inability of the probe tip to correctly follow the shape of the sample surface.[5,6] It was shown, for example, that for a sample composed of two primary sinusoidal frequencies neither frequency was present in the final image.[6] Here, we also address the origin of these frequency artifacts, and in particular consider the question of when it is possible to reliably interpret frequency information seen in an experimental AFM image.

BACKGROUND

The number of studies which have employed AFM to investigate surface structure is growing rapidly.[7] Several studies have compared the results of AFM studies with those obtained by more well-established microscopic techniques. Schwark et al. contrasted results from AFM of block copolymer images with those obtained from low voltage scanning electron microscopy (LVSEM).[8] Annis et al. compared AFM results from surfaces of extended-chain poly(ethylene) with replicas imaged in TEM.[9] The AFM images were described as more "rounded",[9] suggesting that the AFM resolution was poorer (i.e., high frequency information was lost).

Now that the technique is becoming widely available, the reliability of the information from the AFM is of particular interest. This is particularly true when studying defects, crystal edges, and other nonperiodic features near molecular resolution. The perturbing influence of the probe size and shape on AFM images is now recognized but not clearly understood. For example, Annis et al. noted images of the probe tip created from protrusions sticking up on the sample surface.[9]

Studies have also confirmed that the AFM tip can introduce modifications of the sample surface. Certain groups have begun to use this phenomena for examining near surface mechanical and rheological properties. For example, Leung and Goh have observed roughening of the surface of amorphous poly(styrene) during scanning.[10] Meyers et al. measured the size and shape of this roughening as a function of time and sample molecular weight.[11] The roughness increases with time in a manner which depends on the poly(styrene) molecular weight. A transition in the morphology of the surface structure was seen at a molecular weight corresponding to the onset of physical entanglements.[11]

PROCEDURE

In our laboratory we are interested in the microstructure of polymer solids and thin films. We have been focusing attention on the transitions in organization which occur near the periphery of polymer droplets deposited onto substrates. The nature of the microstructure near the droplet periphery gives new information about the arrangement of the material in the presence of constraining surfaces. Our polymer samples were prepared by atomizing solutions onto solid substrates including amorphous carbon, single crystal silicon, and poly(styrene). This method of sample preparation makes it possible to control the concentration of polymer, environment, and substrate. The result of this processing method is isolated droplets of polymer on a solid surface. Our interest has focused on transitions in microstructure which occur near the droplet periphery.

The poly(imide) we are studying is PMDA-ODA, prepared as a poly(amic acid) from pyromellitic dianhydride (PMDA) and oxydianiline (ODA). Droplets were prepared from a dilute solution of the poly(amic acid) in N-methyl pyrrolidinone (NMP). Samples of PMDAODA poly(amic acid) in solution were originally obtained from Dr. Larry Berger, DuPont Central Research, Wilmingon, DE. The genetically engineered protein we are currently working with is SLPF, a silk-like protein with fibronectin functionality. This material is prepared by the genetic engineering of *E. coli* bacteria and was supplied to us by Dr. Joseph Cappello of Protein Polymer Technologies, Inc., San Diego, CA. Droplets of SLPF were prepared from dilute solutions in formic acid.

Our AFM imaging has been performed with a Digital Instruments Nanoscope III located in the University of Michigan Electron Microbeam Analysis Laboratory (EMAL). The images were obtained with the D and J heads with a typical scanning frequency of 2 Hz. We have used Si_3N_4 pyramidal tips as well as etched single-crystal silicon tips. We have also grown sharp tips with well-defined geometries using our environmental SEM.

We have used a variety of techniques to compliment our AFM data including optical microscopy (Olympus BH-2), wide angle X-ray scattering (Rigaku Geigerflex theta-theta diffractometer), scanning electron microscopy (Hitachi S-800 with FEG), transmission electron microscopy (Philips 420 120 kV, JEOL 2000 EX 200 kV), and high resolution electron microscopy (JEOL 4000 EX 400 kV).

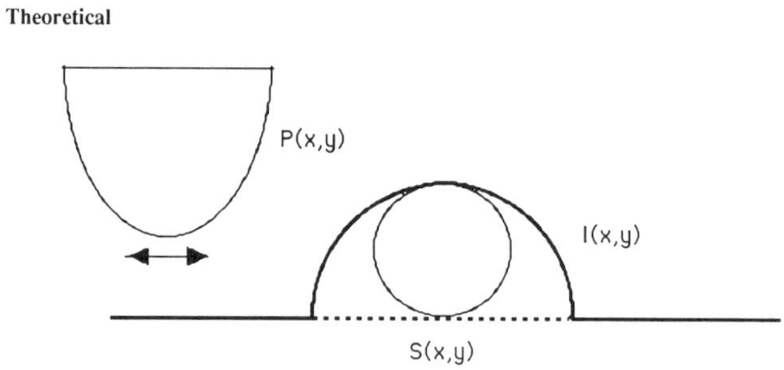

Figure 1. Schematic diagram of an image I(x,y) formed when scanning a surface S(x,y) with a probe of geometry P(x,y).

THEORETICAL

Here we discuss the theoretical relationship between the structure of a sample and an AFM image. This approach has been described in some detail previously.[3,4] The interaction between the sample and probe tip is shown in Figure 1. From geometrical considerations it is possible to predict the theoretical image expected when a probe of known shape P(x,y) scans over a known sample surface S(x,y). The analysis assumes only that P and S are single valued functions over finite domains of the lateral coordinates x and y. Our description of the imaging process is given in the convenient notation of mathematical morphology.[12] The image I(x,y) is given in contact mode imaging by

$$I(x,y) = [S \oplus P^{\wedge}](x,y) \tag{1}$$

where $\oplus$ corresponds to the morphological operation of gray scale Minkowski addition or gray scale dilation. For two functions f and g this operation is defined as:

$$f \oplus g = \max_{x,y \in D_f} (f(x,y) + g_{x,y}). \tag{2}$$

Here we use the notation $g_{a,b}$ to denote the translation of the function g by (a,b). The operation ^ corresponds to reflection about the origin, i.e.,

$$f^{\wedge}(x,y) = -f(-x,-y) \tag{3}$$

Equation (1) describes the contact mode image as the morphological dilation of the surface shape with the reflection of the pro'oe shape about the origin. In noncontact mode the probe is assumed to stay a constant distance d away from the sample. In this case,

$$IN(x,y) = [S \oplus g^d \oplus P^\wedge](x,y) \tag{4}$$

where g^d denotes a positive hemisphere of radius d centered at the origin. Thus, in the non-contact imaging mode the imaging process can be described in two stages. The first step involves the dilation of the original surface with a positive hemisphere of radius d centered at the origin. The second step involves the dilation with the reflection of the probe shape about the origin as for the contact mode of imaging. These results make it possible for us to readily simulate a theoretical AFM image I(x,y) given a function to describe our sample surface S(x,y) and information about the probe shape P(x,y).

RESULTS

In Figure 2 we show a theoretical sample with a single, sinusoidally varying periodicity.

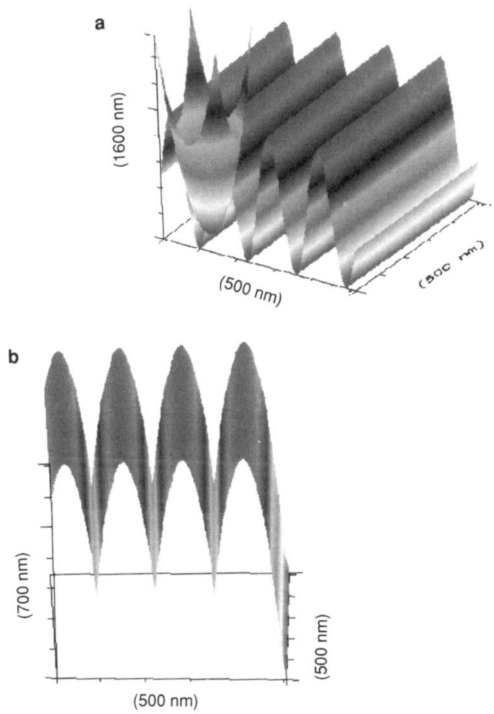

Figure 2. (a) Sinusoidal surface and parabloidal probe. (b) Theoretical image obtained from the surface and probe shown in Figure 2 (a). The probe is unable to penetrate into the valleys, leading to cusps in the image. These cusps correspond to high frequency artifacts.

The Fourier transform of this sample shows a single peak at the frequency corresponding to the spacing between peaks in the sample. Also shown in Figure 2 is the theoretical probe, which we here assume to be a paraboloid with a characteristic radius. The image shows the resulting theoretical image expected when the sample is scanned with a probe having a radius of curvature similar to that of the sinusoidal waveform. Note that the probe is unable to penetrate into the valleys of the wave form, leading to "cusps" in the image.[5,6]

This phenomenon leads to high-frequency artifacts, which are shown as a function of the ratio of probe radius of curvature to the frequency of the sine wave in Figure 3. For sharp tips, the surface is reproduced exactly. For broad tips, the image appears featureless. At intermediate amounts of curvature there are frequency artifacts which occur at harmonics of the primary frequency. These artifacts are most pronounced when the radius of the probe is on the same order of size as the periodicities in the image.

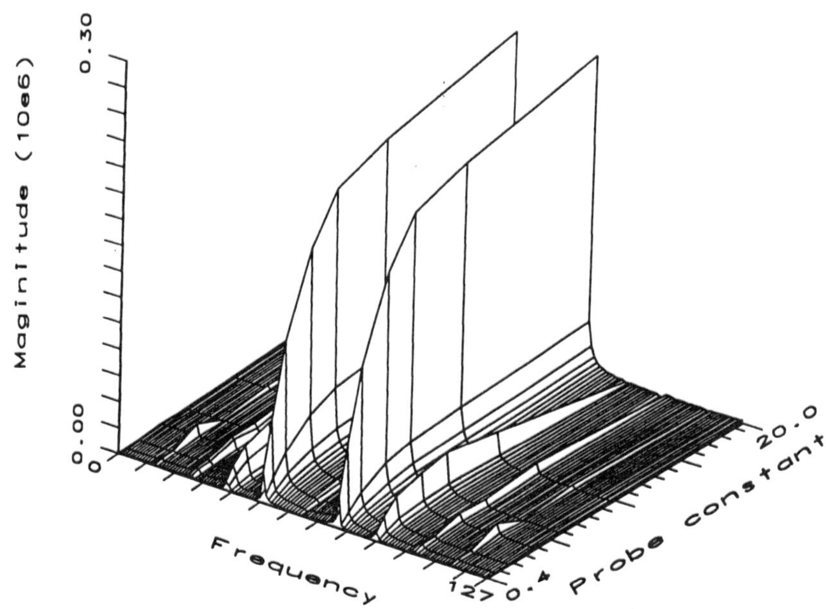

Figure 3. Fourier spectra of theoretical images as a function of the probe constant: the ratio between the frequency of the surface and the radius of curvature of the probe. A high probe constant corresponds to an infinitely sharp probe (in which case the surface is reproduced exactly), whereas a low probe constant corresponds to an infinitely blunt probe (in which case no information is extracted). The frequency harmonics arise at intermediate values of curvature. The maximum distortion occurs when the curvature of the probe is of the same order of magnitude as the periodic spacing in the sample.

A more insidious artifact can be observed when the theoretical sample surface is constructed from the superposition of two primary frequencies. The Fourier spectra of the sample now consists of two independent peaks. In this case, situations can be found where the peaks corresponding to the original sample structure are no longer observed in the image.[6] The precise manner in which the various frequencies in the sample relate to those in the image is clearly an area which needs considerably more theoretical work, but it is possible to induce frequency artifacts in AFM images simply from considerations of the sample structure and the shape of the probe. This result has important implications for the proper interpretation of AFM images, particularly when considering images of defects such as grain boundaries, rippling phenomena, and the proper assignment of crystallographic models of structure. It also indicates the need to simulate a theoretical AFM image given an estimate of the probe shape and a model for the sample surface.

Clearly, successful high resolution AFM imaging will require more detailed information about the shape of the probe tip. This can be obtained by imaging an object of known structure. For this purpose, we have been examining images of uniform spheres deposited onto solid surfaces. Figure 4 (a) shows the theoretical image of an isolated sphere expected when imaging with a pyramidal probe tip. Figure 4 (b) shows an experimental image of a 482 nm diameter poly(styrene) sphere deposited from water suspension onto a silicon surface. The sphere tends to aggregate into domains. At the edge of one of these domains, the experimental image shows a distortion similar to that predicted by the theoretical image. Clearly, the imaging of individual spheres provides a powerful means to evaluate the geometry of a given probe shape.

Figure 5 (a) shows an AFM image of a PMDA-ODA droplet deposited onto a silicon substrate. It is possible to see filaments of polymer extended onto the substrate near the droplet periphery. The width of these filaments is on the order of 10-20 nm and their height is less than 4 nm, indicating they consist of a small number of individual molecules. Because of the fact that the repeat distance along the chain backbone is 3.2 nm, it is likely that the chains are oriented parallel to the substrate. Also shown in Figure 5 (b) is a TEM micrograph of a showing that similar phenomena can be seen near PMDA-ODA droplet edges in bright field imaging. Because of the low contrast of the TEM image this information was originally ignored until corroborated by the AFM. This result raises the interesting possiblity of quantifying small amounts of bright-field TEM contrast from known fluctuations in sample thickness determined by the AFM.

Figure 6 (a) shows the edge of a genetically engineered silk-like protein (SLPF) droplet on an amorphous carbon film. This AFM image was obtained in the deflection mode. While larger structures are clearly observable on the ring there is a lack of detail inside the ring and at the edge. Figure 6 (b) is a TEM photomicrograph of a section of the same SLPF ring shown in Figure 6 (a). Fine separated whiskers are observed near the edge of the droplet. Similar detail is seen in the interior of the droplet. In TEM subunit whiskers are observable and give rise to the structures noted in the AFM but apparently are masked by amorphous material, which renders them less visible to the AFM. Evidence for the amorphous fascia on the whiskers can be seen near the edge of the TEM droplet.

Since the AFM gives information only about the sample surface but the TEM provides insight about the projection of the structure in the direction of the beam, the combination of AFM and TEM provides a powerful means for determining which features are present only at the surface (by AFM) and which are present through the thickness of the film (by TEM).

CONCLUSIONS

The AFM has been and will continue to provide useful information about the surface microstructure of polymers. However, as we demonstrate, there remain serious concerns about the reliable interpretation of high resolution information due to the problems associated with the interaction of the probe with the sample surface.

Our analysis indicates that frequency artifacts can be induced simply due to the geometry of the interaction between the sample and probe. These artifacts in the image seem to occur primarily at harmonics of the fundamental periodicities in the object. Further work is necessary to explore the nature of these artifacts as a function of probe geometries, and to incorporate more realistic models of the probe-sample interaction. This will include efforts to reduce the force of the interaction. However, this approach will not be able to overcome the purely geometrical nature AFM imaging artifacts. Efforts to alter the strength

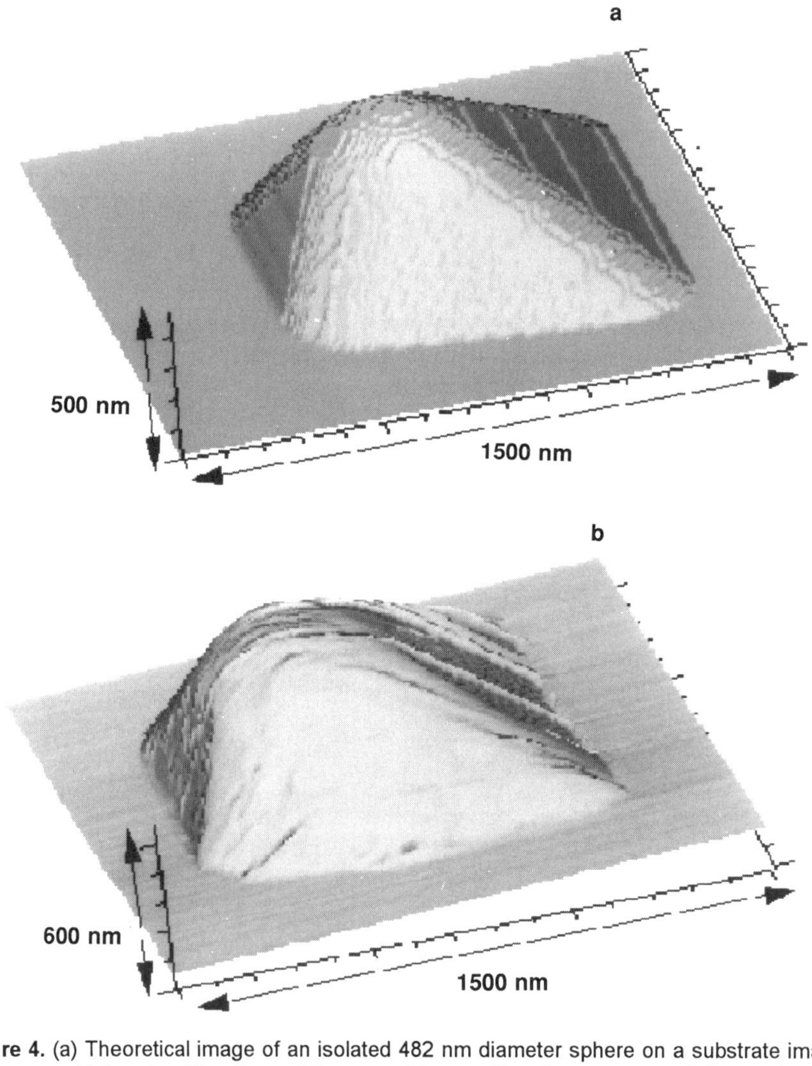

Figure 4. (a) Theoretical image of an isolated 482 nm diameter sphere on a substrate imaged with a pyramidal probe. The slope of the pyramid, as well as the presence of sharp edges on the pyramid are all manifest in the image. (b) Experimental image of 482 nm diameter poly(styrene) latex spheres on a silicon substrate. The agreement with the theoretical image is remarkable. It is evident that imaging spheres with the AFM can be used to extract information about the size and shape of the probe tip.

of interaction between the probe and the sample should reduce problems associated with these artifacts associated with deformation, but it will not be possible to overcome purely geometrical constraints without manipulating the shape of the probe tip. High resolution AFM imaging will require considerably more attention to the size and shape of the probe tip.

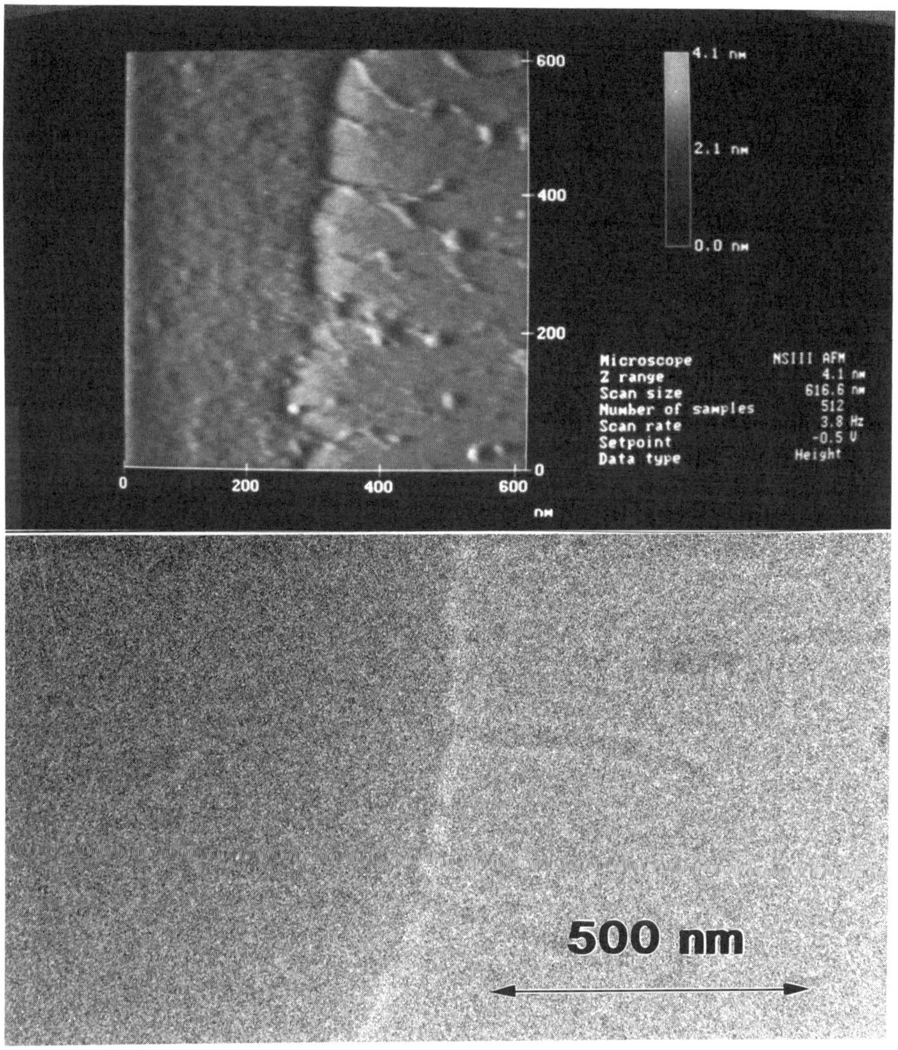

a

b

Figure 5. (a) AFM image of the edge of a PMDA-ODA poly(imide) droplet. (b) TEM image of a similar region. The fingering phenomena of the polymer into individual filaments which is evident in the AFM image can also be seen upon close inspection in TEM. The width of the filaments is on the order of 10-20 nm, and their height is less than 4 nm. The filaments apparently arise during the evaporation of the solvent during sample preparation. Since the repeat distance along the PMDA-ODA chain backbone is 3.2 nm, boundary conditions suggest that the molecules are oriented parallel to the substrate.

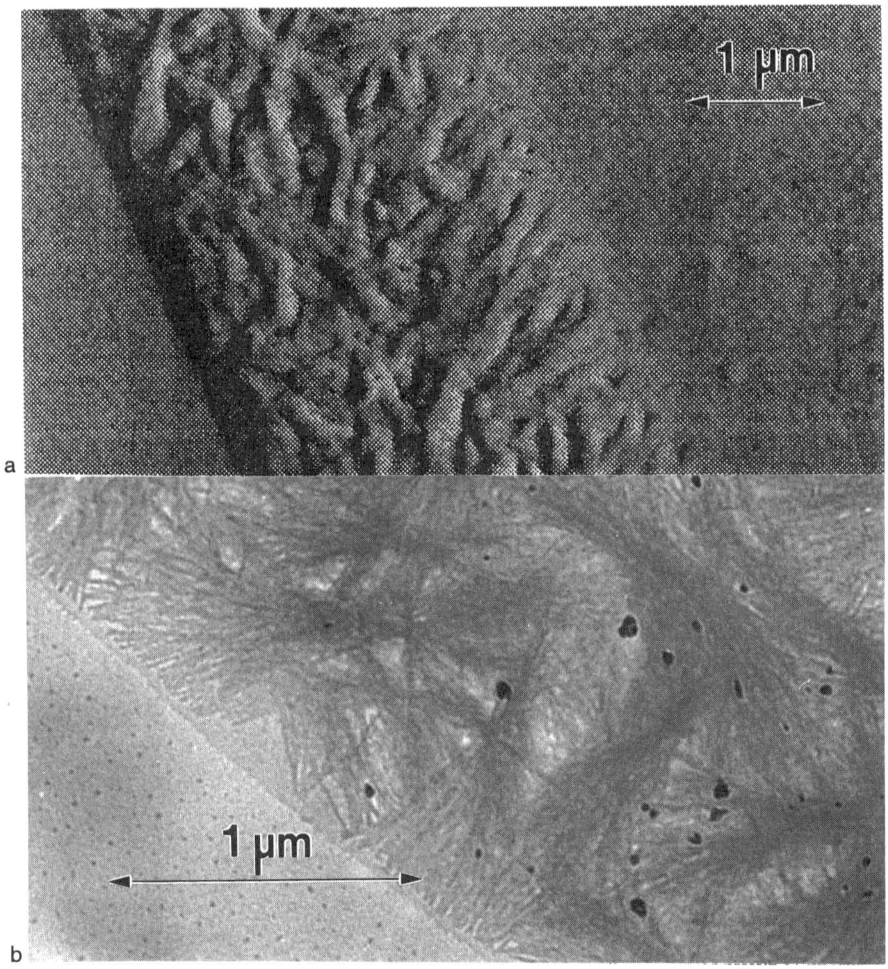

Figure 6. (a) AFM image of the edge of a droplet of a crystalline genetically engineered protein SLPF (silk-like protein with fibronectin functionality). (b) TEM image of of a similar region. The TEM shows well-defined whisker crystallites. The AFM shows the surface roughness but the individual crystallites are not well resolved. The contrast in AFM can be improved after TEM imaging, apparently because a thin surface layer can be etched away by the electron beam. Evidence for a surface layer can also be discerned in the TEM images. By combining AFM with TEM it is therefore possible to extract information about surface structure as well as the organization within the film thickness.

REFERENCES

1. M. Radmacher, R.W. Tillman, M. Fritz and H.E. Gaub, From Molecules to Cells: Imaging Soft Samples with the Atomic Force Microscope, *Science*, 257, 1900 (1992).
2. D.C. Martin, L.L. Berger, and K.H. Gardner, Structural Evolution of a Model Poly(imide): Organization Near Surfaces, *Macromolecules*, 24, 3921 (1991).
3. G.S. Pingali and R. Jain, Imaging models and surface recovery methods for scanning probe microscopy, *in* "Proceedings of the IEEE Workshop on Applications of Computer Vision," Palm Springs, CA, 282 (1992).
4. G.S. Pingali and R. Jain, Probe shape recovery in scanning probe microscopy, in "Proceeding of the MVA '92 IAPR Conference on Machine Vision Applications," Tokyo, Japan, 639 (1992).
5. J.E. Griffith, D.A. Grigg, M.J. Vasile, P.E. Russell, and E.A. Fitzgerald, Scanning probe metrology, *J. Vac. Sci.Technol.*, A10(4), 674 (1992).
6. D.A. Grigg, J.E. Griffith, G.P. Kochanski, M.J. Vasile, and P.E. Russell, Scanning Probe Metrology, *in* "Proceedings of the SPIE Symposium on Microlithography," 1674, 557 (1992).
7. E. Meyer, Atomic force microscopy, *Progress in surface science*, 41, 3-49 (1992).
8. D. W. Schwark, D. L. Vezie, J. R. Reffner, E. L. Thomas, and B. K. Annis, *J. Mat. Sci. Lett.*, 11(6), 352-355 (1992).
9. B. K. Annis, J. R. Reffner, and B. Wunderlich, Atomic force microscopy of extended-chain crystals of polyethylene, *J. Poly. Sci., B: Poly. Phys. Ed.*, 31, 93-97 (1993).
10. O. M. Leung and M. C. Goh, Orientational ordering of polymers by atomic force microcopy tip-surface interaction, *Science*, 255, 64 (1992).
11. G. F. Meyers, B. M. DeKoven, and J. T. Seitz, Is the molecular surface of polystyrene really glassy? *Langmuir*, 8, 2330 (1992).
12. C. R. Giardina and E. R. Dougherty, Morphological Methods in Image and Signal Processing, Prentice Hall (1988).

SCANNING TUNNELING MICROSCOPY AND ATOMIC FORCE MICROSCOPY STUDIES OF CONDUCTING POLYMER FILMS

Timothy L. Porter

Department of Physics
Northern Arizona University
Flagstaff, AZ 86011

Abstract: The techniques of scanning tunneling microscopy, tunneling spectroscopy, and atomic force microscopy have been used to study the nanometer-scale morphology of conducting polymer films. Scanning tunneling microscopy and atomic force microscopy allow for images to be obtained with molecular or atomic resolution on these surfaces. The technique of tunneling spectroscopy allows for the study of the local electronic structure on surfaces. In this report, we have investigated the surface structure of aniline, hydroxyaniline and borate functionalized aniline thin films.

INTRODUCTION

Since Shirakawa and coworkers first demonstrated that the conductivity of polyacetylene increases by several orders of magnitude when exposed to halogen vapors,[1] there has been considerable interest in the synthesis and study of this and other "conducting plastics". Proposed applications for these synthetic metals include electrode materials in new types of electrochemical sensors, including gas sensors,[2-4] ion sensors,[2,4] enzyme biosensors, and nonenzyme biosensors.[5-8] Conducting polymers that have shown promise as materials for this new generation of sensors include polypyrrole and functionalized polypyrroles, polyaniline and functionalized polyanilines, and polythiophene and functionalized polythiophenes. The ability to add functional groups to these standard polymeric materials allows one to tailor the sensing material to promote the desired reaction and opens the door for an almost limitless number of sensing applications.

Gas sensors based on polypyrrole, polyaniline or polythiophene have demonstrated sensitivities for SO_2, NO_2, I_2, NH_3, H_2S, H_2O, alcohols, acetonitrile, and various other oxidizing gases. Biosensitivities for amine, phenols, ascorbic acid, flavinmononucleotide, glucose, ceruloplasmin, and pyruvate and lactate have also been studied. It may someday be possible, using this synthetic chemistry, to "custom design" sensors for virtually any sensing application desired. Detailed studies of these new polymer-based sensors are still in their early stages, and much remains to be learned about the polymers themselves as well as their sensing abilities.

While conducting polymer materials show great promise in new applications, such as gas and biosensing, there exists a great need for fundamental studies on the mechanisms for conduction (and thus sensing) in these materials. Only when these charge transport mechanisms are understood on the molecular level may we begin to model accurately and predict the behavior of a given material in some particular application. The physical structure of these conductive polymers has been shown to be highly complex and is still not well understood on the molecular level. The electronic properties of these materials are highly dependent upon the physical structure or morphology, and this complicated structural-electrical interplay is only beginning to be studied.

The techniques of scanning tunneling microscopy (STM), tunneling spectroscopy (TS) and atomic force microscopy (AFM) are ideal tools to be used in an investigation of both the molecular structure of these materials and their electronic structures. The generally amorphous nature of thin, conducting polymer films makes local, microscopic-scale structural and electronic analysis using conventional techniques, such as X-ray diffraction, electron spin resonance, nuclear magnetic resonance, or X-ray photoelectron spectroscopy, difficult if not impossible. The techniques of STM and AFM permit local, structural analysis on the molecular (or in some cases atomic) scale. The technique of TS gives nanometer-scale, spatially resolved electronic structure information on surfaces being studied. Studies on conducting polymer films using these techniques can help to increase our knowledge of the fundamental conduction mechanisms inherent in these materials and thus lead eventually to greater understanding and control of devices fabricated from conducting polymers.

MACROSCOPIC CONDUCTIVITY OF CONDUCTING POLYMER FILMS

One interesting family of conducting polymers with applications as sensor materials is the polyaniline family (including functionalized anilines). When the emeraldine base form is fully protonated, the formation of spinless bipolarons occurs. The energetically favorable splitting of these bipolarons into polarons then results in the formation of a polaronic metal.[9] This metallic state of aniline has been observed using the technique of ESR with spin-lattice relaxation time and Δg measurements. Also, Pauli susceptibility and thermopower dependence measurements, as well as optical and theoretical band structure calculations, point to the existence of a metallic polaron lattice.

The measured macroscopic film conductivities, however, are much lower than what one might expect from such a metallic material. Measured macroscopic conductivities are only on the order of about 1 S/cm,[10] which means that the film structural effects are very likely the limiting factor in the resulting conductivity. A number of models based on structural aspects have been proposed in order to explain the conductivity differences. These models include a granular polymeric metal model, Fermi glass model, single chain model and metallic bundle model.[11,12] We have used the techniques of atomic force microscopy (AFM) and scanning tunneling microscopy (STM) in order to investigate the nanometer-scale structural features of poly-hydroxyaniline thin films in an attempt to answer some of these questions.

Cyclic voltammographic aniline and hydroxyaniline films were prepared from solutions of 0.5 M aniline or 3-aminophenol in 1.0 M HCl using standard techniques.[13,14] The substrate used was polished 1.5 mm dia. BAS Pt electrodes for all of the films. The cyclic voltammetric films were cycled between -0.2 and 1.0 V at a rate of 300 mV/s until a film of approximately 5 μm in thickness was formed. A film thickness of 5 μm was chosen for the films in order to mask any substrate structural features. AFM scans were performed using a Nanoscope II from Digital Instruments. A constant force of 5 nN was used for all scans. STM images were taken using a custom UHV compatible instrument. All STM scans were performed in a vacuum of 5×10^{-9} Torr or better, using a tip bias voltage of -0.2 V with the sample at ground. TS scans were also performed under UHV conditions, with the sample at this reduced pressure for at least 48 hours prior to study. All TS scans contained 100 data points and were smoothed using a technique similar to that used by Porter.[14] TS data were acquired at approximately 50 different locations on each sample surface.

In Figure 1 below is a 5 x 5 μm AFM scan of the hydroxyaniline surface. The film is segregated into small bundles, ranging in size from about 50 $Å^2$ to over 2000 $Å^2$ in size.

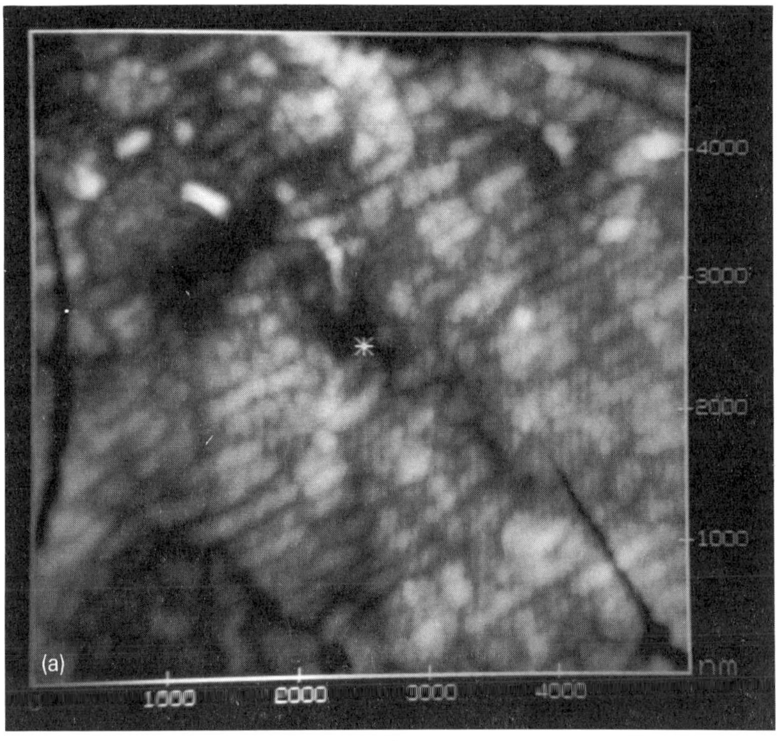

Figure 1. Five μm^2 AFM scan of the hydroxyaniline surface.

Similar scans on the aniline thin films revealed a different surface morphology. Figure 2 below shows a 1 μm^2 STM scan of this surface. The surface is terraced, suggesting a Stranski-Krastanov growth mode for the film. In a previous study on aniline prepared by cyclic voltammographic techniques,[15] a similar layer-by-layer growth mechanism was observed for films grown using a relatively slow sweep rate such as this one.

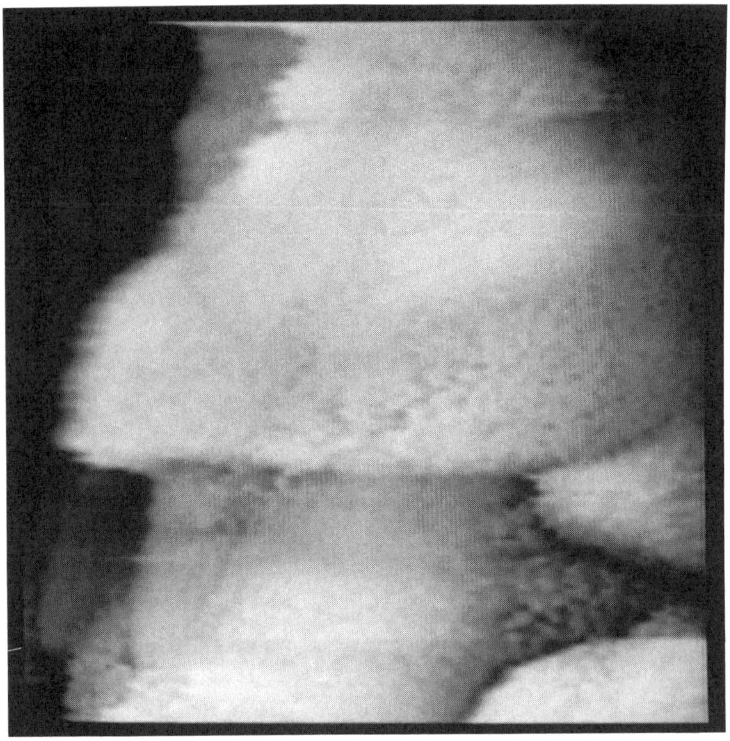

Figure 2. One μm^2 STM scan on the surface of aniline thin film.

TS scans were taken on both surfaces in order to investigate the electronic properties of these two very different surfaces. Figure 3 below shows a (dI/dV)/(I/V) spectrum from the aniline surface. The π^*-π^* bandgap is seen to be about 4 eV. There is a local minimum in the density of states at E_f, although the DOS remains finite. This is the type of DOS expected in a Fermi Glass type of material. No evidence of intragap polaron or bipolaron states is evident on this or any of the TS scans obtained on aniline surfaces.

In Figure 4 we show a TS scan taken from the hydroxyaniline surface. This type of DOS structure was observed on approximately 25% of the surface locations sampled. The remaining 75% of the TS scans were similar to those observed for the aniline films. In this figure, however, we note the presence of a local maximum in the DOS at E_f. The width of this maximum is estimated to be 0.8 eV. Theoretical transport studies for aniline in the metallic state have predicted an approximately 1 eV wide polaron band centered at E_f, while other studies have found evidence for a stable bipolaron structure in this material.[15]

While the TS spectra for both aniline and hydroxyaniline are dominated by a nearly continuous background of states, indicative of a high degree of disorder in the material, the hydroxyaniline film clearly contains regions exhibiting a stable polaron configuration. The DOS for the aniline films more closely resembles the electronic structure expected in a Fermi Glass type of material. The nanometer-scale bundle structure observed in the hydroxyaniline films, coupled with the TS evidence for a stable polaron electronic structure, allows us to conclude that a conducting island model of charge transport is responsible for the observed macroscopic electrical properties of this material. In this model, the material

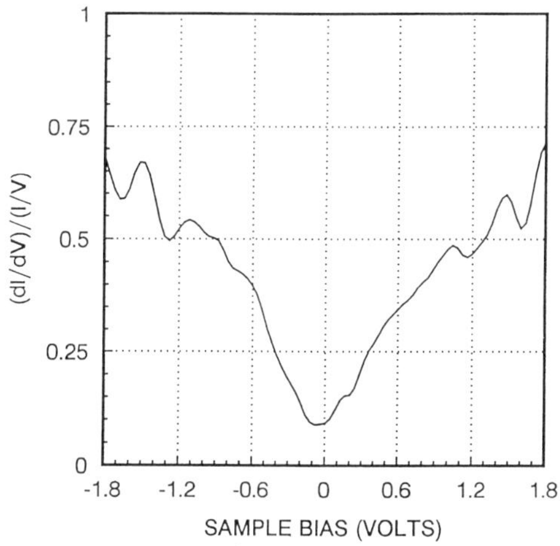

Figure 3. TS scan on the aniline film surface.

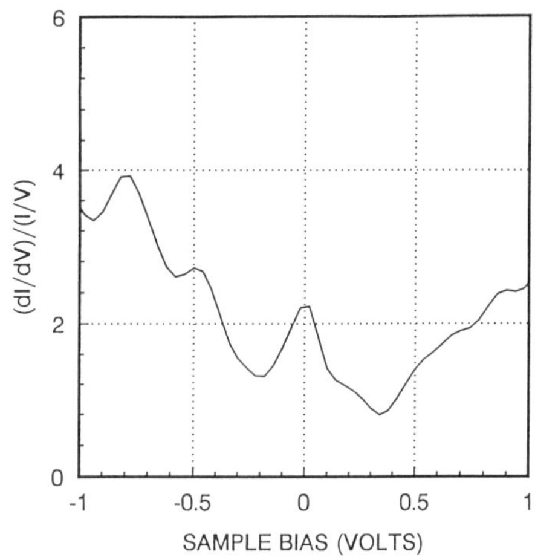

Figure 4. TS spectrum from the hydroxyaniline surface.

is segregated into nanometer-scale islands or bundles. These islands or bundles each exhibit the characteristics expected of a polaronic metal; however, poor coupling between individual bundles reduces the macroscopic conductivity of the film. Conduction between adjacent conducting islands is limited by hopping of charge carriers between the islands, or by charging energy limited tunneling between the islands. The aniline films, however, showed neither the microscopic domain structure nor the polaron configuration expected from the conducting island model. The observed physical and electronic properties of the unsubstituted aniline films were more characteristic of a Fermi Glass type of material, in which the macroscopic conductivity is limited by the overall amorphous nature of the film itself.

MOLECULAR STRUCTURE OF POLYANILINE

The molecular structure of various polyaniline materials has been investigated using the technique of STM. Thick hydroxyaniline films were deposited electrochemically on Pt substrates and imaged with the STM under UHV conditions.[16] Small, amorphous, island-like structures were found to dominate the surface structure. Connecting many of these islands, or lying atop the islands, were numerous single strands of the polymer. Figure 5 shows a 2500 Å by 1200 Å region on this surface. The edge of an amorphous island is present along with three polymer strands.

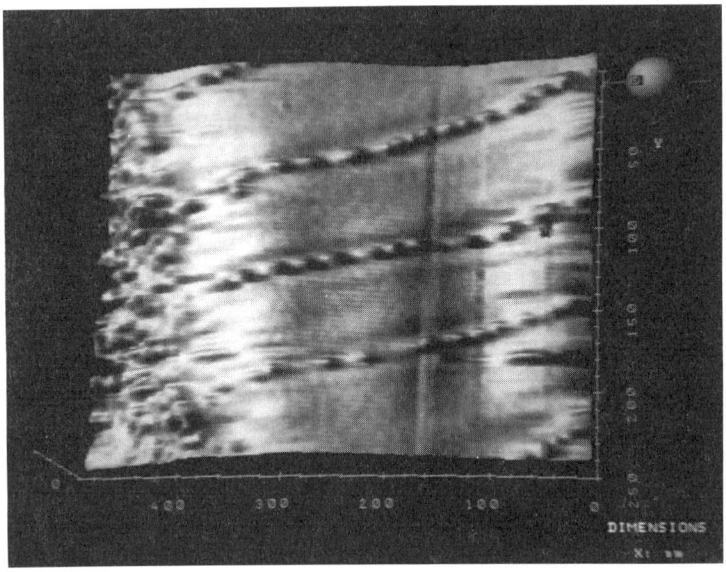

Figure 5. 2500 x 1200 Å STM image on a hydroxyaniline surface.

The average diameter of these strands is 20 Å and their apparent height above the surface averages 10 Å. Similar strands have been observed on surfaces of chemically prepared hydroxyaniline,[16] borate substituted polyaniline,[17] and polythiophene.[18] In Figure 6 below, a high resolution STM image of a single borate substituted polyaniline strand is imaged. The dimensions of this image are 350 Å lengthwise and 220 Å in width. The strand contains individual sections arranged in a zig-zag fashion on the surface. The length of each section of the zig-zag was observed to average about 50 Å. The diameter of these strands, as well as visual evidence of a periodic spacing along the length of these strands, indicates that these are not simply linear polymer strands, but helical or superhelical coils (the expected width of a simple planar polyaniline strand would be only about 4.5 Å). This is an important result, as models attempting to describe the electronic

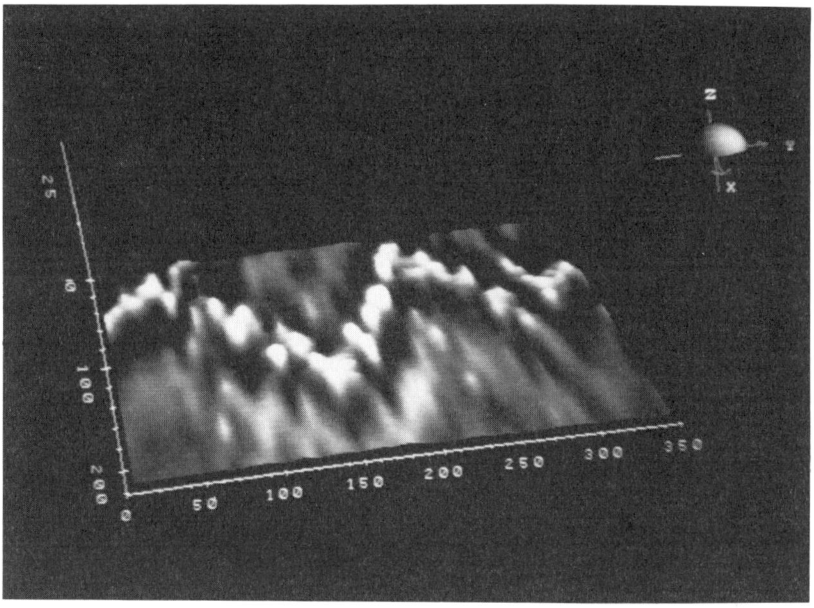

Figure 6. High resolution STM image of single polymer strand on borate-aniline surface.

structure of conducting polymers must begin with a structural picture of the material before the band structure calculations can begin. Studies are now underway in an attempt to prepare and image samples with even greater resolution. The long range goal is to successfully image and study the substructure within individual polymer molecules. This will allow us to use the STM in investigating the molecular interaction of conducting polymers with gasses, bio-molecules and other gaseous or liquid phase materials--an important step toward understanding the uses of these materials as sensors.

CONCLUSION

The techniques of scanning tunneling microscopy, tunneling spectroscopy, and atomic force microscopy can be used to study the both nanometer-scale physical, and local nanometer-scale electronic properties of conducting polymer films. In addition, individual molecular structures can also be resolved in some cases. These techniques allow us to probe and study these new, complicated, materials on a scale heretofore impossible. The further development of these materials as sensing materials depends greatly on a higher lever of understanding of the electronic, nanometer-scale physical and molecular structures. Studies of the molecular scale interactions of these materials with various gasses and biomolecules are also underway using these techniques.

REFERENCES

1. E. Genies, A. Boyle, M. Lapkowski, C. Tsintavis, Polyaniline: a historical survey, *Synth. Met.*36:139-182 (1990).
2. G. Zotti, Electrochemical sensors based on polyconjugated conducting polymers, *Synth. Met.* 51:373-382 (1992).

3. T. Hanawa, H. Hashimoto, S. Kuwabata, H. Yoneyama, Gas sensitivities of electropolymerized polythiophene films, *Synth. Met.* 30:173-181 (1989).
4. A. Boyle, E. Genies, M. Lapkowski, Application of electronic conducting polymers as sensors: polyaniline in the solid state for detection of solvent vapors and polypyrrole for detection of biological ions in solution, *Synth. Met.* 28:C769-C774 (1989).
5. T. Skotheim, P. Hale, H. Karan, Y. Okamoto, L. Samuelson, S. Tripathy, Derivatized Polypyrrole langmuir-blodgett films, applications to bioelectronics, *Synth. Met.* 41-43:1433-1437 (1991).
6. L. Couves, S. Porter, Polypyrrole as a potentiometric glucose sensor, *Synth. Met.* 28:C761-C768 (1989).
7. C. Koopal, R. Czajka, M. Feiters, R. Nolte, B. Ruiter, R. Schasfoort, H. Van Kempen, Polypyrrole microtubles and their use in the construction of a third generation biosensor, *Synth. Met.* 51:397-405 (1992).
8. B. Wheeler, K. Cantrell, G. Caple, J. Francis, S. Grey, S. Vogel, Electrochemical amine sensors using carboxylate functionalized polythiophent films, *J. Electrochem. Soc.* 136:2769-2770 (1989).
9. J. Ginder, A. Epstein, A. MacDiarmid, A. Richter, Insulator to metal transition in polyaniline. Solid State Commun. 63:97-101 (1987).
10. J. Travers, F. Genoud, C. Menardo, M. Nechtschein, Polyaniline: a material still under discussion, *Synth. Met.* 35:159-168 (1990).
11. F. Zuo, M. Angelopoulos, A. Epstein, A. MacDiarmid, Transport studies of protonated emeraldine polymer: a granular polymeric metal system, *Phys. Rev. B.* 36:3475-3478 (1987).
12. K. Mizoguchi, C. Menardo, M. Nechtschein, J. Travers, Spin dynamics in the conducting polymer polyaniline, *Phys. Rev. Lett.* 63:66-69 (1989).
13. T. Porter, G. Caple, C. Lee, P. Oden, Surface structural study of poly-hydroxyaniline and aniline-(3-aminophenylboronic acid) co-polymer films, *J. Vac. Sci. Technol.* A10(4):606-610 (1992).
14. T. Porter, G. Caple, C. Lee, B. Wheeler, Scanning tunneling microscopy studies of substituted polyaniline thin films, *J. Vac. Sci. Technol.* A9(3):1452-1456 (1991).
15. D. Bonnell, M. Angelopoulos, Spatially localized electronic structure in polyaniline by scanning tunneling spectroscopy, *Synth. Met.* 33:301-310 (1989).
16. T. Porter, G. Caple, P. Oden, Surface structural study of poly-hydroxyaniline, *Surf. Sci.* 259:221-230 (1991).
17. T. Porter, G. Caple, R. Dillingham, T. Jones, C. Lee, B. Wheeler, Scanning tunneling microscopy and x-ray photoelectron spectroscopy studies of borate substituted polyaniline, *Synth. Met.* 40:187-196 (1991).
18. T. Porter, G. Caple, S. Jeffers, R. Swift, B. Wheeler, Scanning tunneling microscopy studies of polythiophene and poly-3-bromothiophene films, *Surf. Sci.* 238:L433-L438 (1990).

MORPHOLOGICAL FEATURES OF POLYETHYLENE AND POLYIMIDES BY ATOMIC FORCE MICROSCOPY

Rajkumari Patil and D. H. Reneker

The Maurice Morton Institute of Polymer Science
The University of Akron
Akron, OH-44325-3909

Abstract: Lamellar crystals of linear polyethylene grown from dilute solution in xylene were observed by atomic force microscopy (AFM). Effects of crystallization conditions, annealing and deformation were studied. Features such as lamellar thickness, fold domain boundaries, wrinkles, and pleats were observed. A temperature growth step, produced due to sudden change in temperature during crystallization, was measured accurately.

Patches, not previously studied, on the fold surfaces of the crystals exist in a range of diameters. Patches were 3-70 nm in diameter and only 0.2-1.5 nm high. Localized changes in thickness of lamella and other morphological changes due to annealing at temperatures from 115° to 135° C were observed in great detail. Shearing of lamellae at overlaps and "accidental" tensile deformations were found.

Surfaces of fibers spun from (BPDA-PMB)* were imaged from atomic to micrometer scale. The free surfaces of semicrystalline BTDA** films were observed. A single polyimide fiber (BPDA-PMB) was imbedded in polyamic acid of BTDA, which was thermally imidized at 250° C and then was annealed at 280° C. Orientation of the BTDA by the BPDA-PMB persists through many molecular layers. Surfaces of BPDA-PMB fiber observed with the AFM, showed features at the molecular scale.

* regularly repeating sequences synthesized from 3,3',4,4'-tetracarboxylicbiphenyl dianhydride (BPDA) and 2,2'-bis(methylmethyl)-4,4'-diaminobiphenyl (PMB)

**1,3-bis(4-aminophenoxy)-2,2'dimethylpropane and 3,3',4,4'-benzophenonetetracarboxylic dianhydride (BTDA).

CHEMICAL STRUCTURE OF POLYIMIDES

*

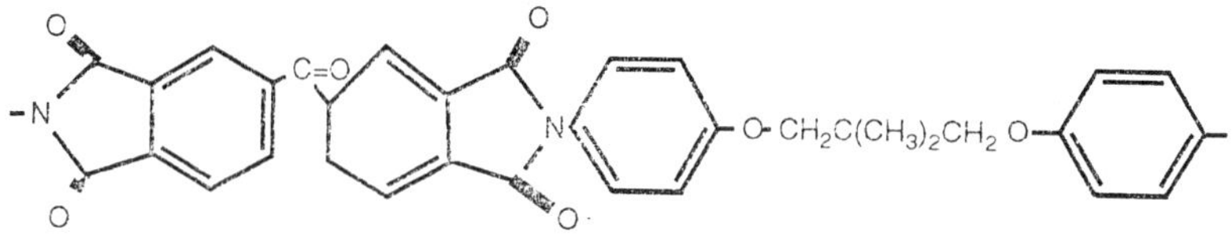

* *

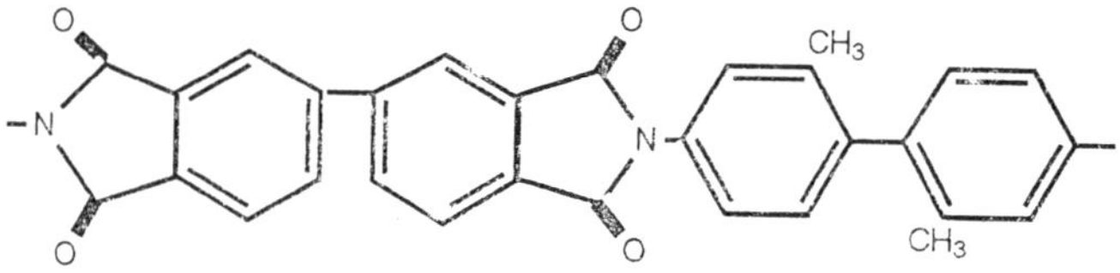

INTRODUCTION

A great deal of information about details of lamellar crystals of polyethylene[1-4] is already known, so the main goal of this paper is to present new results that were obtained with atomic force microscopy. Polyimides are used to form films and fibers due to their excellent physical and mechanical properties[5,6]. BTDA is a candidate for the matrix, and BPDA-PMB is a candidate for the fiber of polyimide composite. The interaction forces at the fiber-matrix interface play an important role in the performance of such composites in many applications, particularly those involving compressive stress. Local molecular chain ordering at the surface strongly affects the mechanical properties of the matrix polymer in the interface region. Ordering of the polyimide molecules at surfaces of BPDA-PFMB (a fluoro analog of BPDA-PMB) fibers and BTDA films is already described[7]. Observation of surface morphology in the molecular to microscale range on BPDA-PMB fibers and BTDA films is shown here. Orientation of the BTDA is observed on the surface of BPDA-PMB single fiber.

EXPERIMENTAL

Linear polyethylene, Standard Reference Material 1475, was obtained from the National Institute of Standards and Technology, Gaithersburg, MD 20899, USA. Dilute solutions of polyethylene, about 0.1% concentration, were prepared by first completely dissolving 20 mg of polyethylene in 20 mL of xylene at 125° C. A self-seeding growth procedure was used. The solution was rapidly cooled to room temperature, then placed in a preheated bath at 98° C and kept there until the solution became transparent. The solution was then placed in a preheated bath at 85° C and kept there for a day while lamellar crystals grew. The solution with suspended crystals was then cooled to room temperature. A portion of the suspension was sprayed onto freshly cleaved mica, which was scanned under ambient conditions by an atomic force microscope (Nanoscope II, Digital Instruments, Inc., Santa Barbara) using a silicon nitride tip at low forces.

Highly oriented fibers of BPDA-PMB were spun from an isotropic solution and then drawn 7 times their original length in the laboratory of Prof. S. Cheng. These fibers were examined with the atomic force microscope. Films of BTDA were prepared by casting solutions of polyamic acid on a glass substrate, thermally imidizing at 250° C, and then annealing at 280° C. A single BPDA-PMB fiber was imbedded in the polyamic acid of BTDA thermally imidized at 250° C, and then annealed at 280° C. The surfaces of all polyimide samples were observed at ambient conditions with the Atomic Force Microscope (AFM), Nanoscope II.

Low pass filtering was done on AFM data to remove some recognizable artifacts. Two-dimensional correlation functions G(x,y) were used to measure periodicities in the AFM data.

RESULTS AND DISCUSSION

Typical single crystals of polyethylene are pyramidal in shape and bounded by growth faces parallel to {110} and {200} planes. A top view image, showing a scan with an area 3 x 3 μm square, is shown in Figure 1. The image reveals truncated pyramidal crystals with {200} edges, grown at high crystallization temperatures. These are the examples of collapsed truncated pyramidal crystals which are obtained by using the self-seeding technique as described in the experimental section above. The crystals are 2 to 3 μm in size and about 10 nm thick. Fold domain boundaries are also observed. Sometimes fold domain boundaries are misaligned. The (110) boundary on the left crystal in this image is an example of such a case. Pleats running parallel to {200} growth face are present.

The thickness of the lamella depends on the crystallization temperature. Thus, if the temperature is abruptly lowered during crystallization, a small step appears on the surface of the lamella (Dosiere[8]). A step of this kind, which runs parallel to the growth edges, is observed in Figure 1. After enlarging a part of the upper left sector of the left crystal, the temperature step is shown in Figure 2 at higher magnification. Figure 2 is rotated by 90° for better visibility. The temperature step is 1.5 nm deep and 88 nm wide. The ability of the AFM to make accurate measurements on the temperature step can improve measurements of the linear growth rates of polymer crystals by Dosiere's method.

The high resolution perpendicular to the lamella obtained by using atomic force microscopy makes possible the observation of low broad patches on the surfaces of collapsed lamellar crystals. Surfaces of crystals grown under different crystallization conditions were observed: and all were found to be covered with patches. Patches are visible in Figure 1 but are dramatically evident in Figure 2 as hills scattered over the surface. In Figure 2, the vertical magnification is much higher than the lateral magnification, thus distorting the shape of the low broad patches and giving the appearance of hill-like features. Patches exist in a range of diameters varying from 3 to 70 nm. The heights of the patches vary from 0.2 to 1.5 nm.

The patches change during annealing. The nature of the change depends on the annealing temperature. Crystals were grown by the usual self-seeding technique. The crystals, suspended in xylene were then sprayed on to pieces of freshly cleaved mica. Each piece of mica with polyethylene crystals deposited on it was then annealed for a few hours. Crystals annealed at 115° C, 120° C, 125° C, 130° C, 132° C and 135° C were scanned with the atomic force microscope. The edges of the crystals were observed to thicken before other changes occurred. The amount of thickening depended on the annealing temperature. The edge thickening process started at 115° C. As the annealing temperature was increased, the lamella thickness near the edges of the crystals increased. A part of a lamellar crystal, annealed at 115° C, is shown in Figure 3.

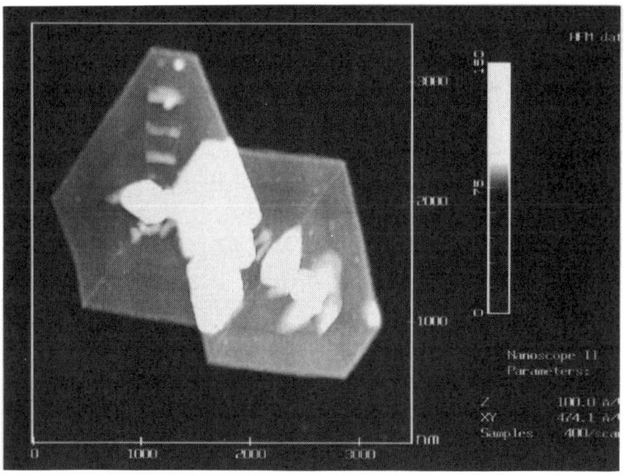

Figure 1. Two overlapping lamellar crystals of polyethylene.

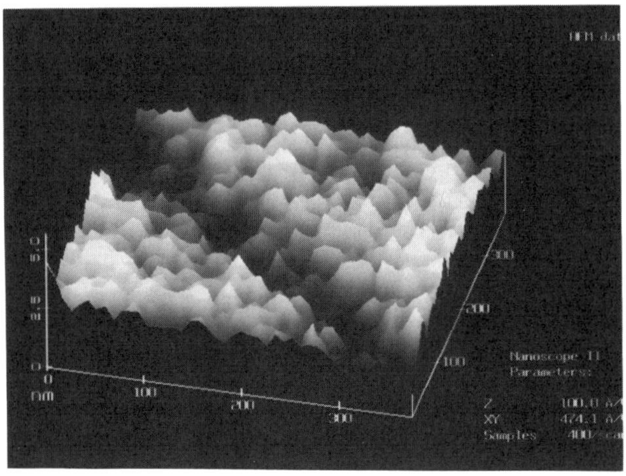

Figure 2. Enlargement of the upper left sector of the left crystal in Figure 1.

The edges of the crystals thickened to 14.5 nm whereas the inner part of the lamella remained at the initial lamellar thickness of 11.5 nm. Above 115° C the thickness of the interior also increased, but by a smaller amount than the edges. The edges also show a slight jaggedness. Fold domain boundaries, pleats and wrinkles are still present. The number of patches, with a diameter of about 10 nm, increases. They are arranged in ordered rows. The crystals annealed at 120° C show a further increase in thickness near the edges, as shown in Figure 4. The edges have thickened to 15.8 nm. Fold domain boundaries and pleats are observed in Figure 4. A higher magnification scan of the left-most sector of the lamella in Figure 4 shows a distinct arrangement of rows of patches in Figure 5. The rows of patches are separated by about 10 nm and the patches are up to 10 nm in diameter. A few big patches, from 40-70 nm in diameter, are also present on the surface.

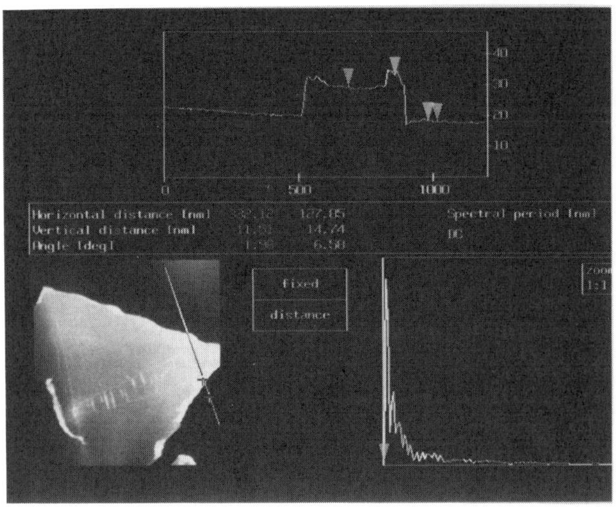

Figure 3. A line section on a lamellar crystal of polyethylene annealed at 115° C.

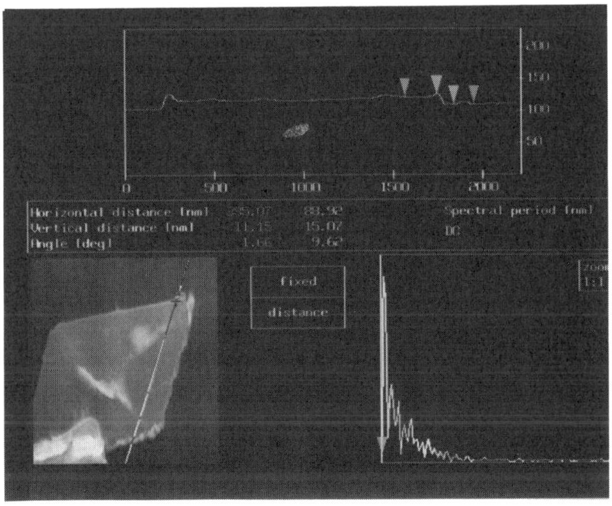

Figure 4. A line section on a lamellar crystal of polyethylene annealed at 120° C.

The first appearance of holes was observed in crystals that were annealed at 130° C. At this annealing temperature, holes develop first near the edges of the crystals. The crystals retain features such as fold domain boundaries, pleats and well developed growth spirals. The growth faces of the crystals retain their general shapes although they are more jagged. The region near the edges has become 19.9 nm thick. Patches are also present on the higher layers of the growth spirals. Smaller patches of up to 10 nm diameter are found in the central part of the crystal. Quasi-periodic rows of patches still persist on the crystals which were heated up to 132° C. Annealing the crystals at 135° C leads to disappearance of smaller patches. Lamellar crystals annealed at 135° C have big holes, as shown in Figure 6. The remaining part of the basal lamella became 21.5 nm thick. The fold domain boundaries are no longer evident. Growth spirals disappeared into unrecognizable heaps when crystals were annealed at 135° C.

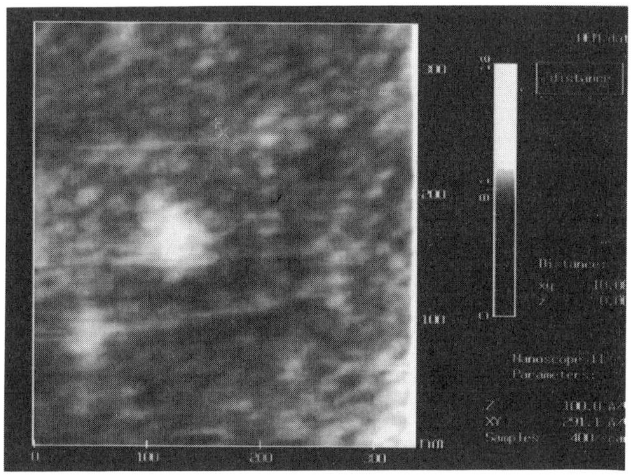

Figure 5. Enlargement of the upper left sector of the lamellar crystal in Figure 4 shows the patches arranged in rows about 10 nm wide.

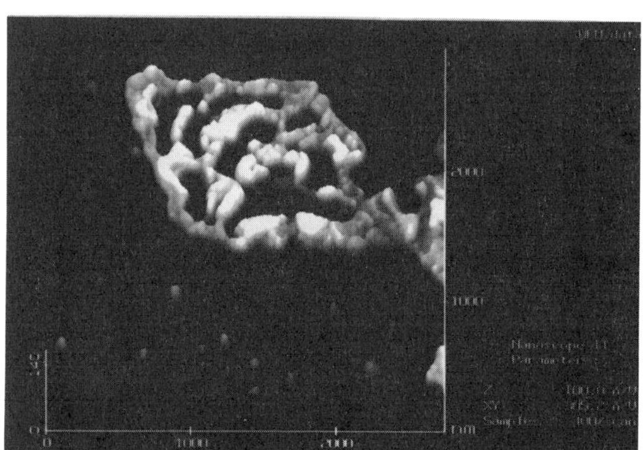

Figure 6: A surface view of a single polyethylene crystal annealed at 135° C.

A number of different polyethylene crystals, grown at various crystallization temperatures, were scanned by atomic force microscopy. Fold domain boundaries were seen on all the crystals examined. In a collapsed pyramidal crystal, the region near a fold domain boundary would be expected to be in compression on one side of the crystal and in tension on the other. The landing of crystals onto the mica substrate can take place with either side up. The number of times crystals land on the sides on which the region near the fold domain boundary is in compression is observed to be greater than the number of times they land on the other side. The fold domain boundaries are thus observed more often with the region around them in compression. These are the usual type of fold domain boundaries. The region compressed by collapse is about 1.5 nm higher than the adjacent fold surface and around 50 nm wide. In a few cases, inverted fold domain boundaries, that is the fold domain boundaries with the region near them in tension, were also observed.

One such case is shown in Figure 7. A top view of two truncated pyramidal crystals is shown in this image. These crystals have edges thicker than the interior, indicating an increase in crystallization temperature after the crystals grew to a certain size. Two overlapped crystals are shown in Figure 7. Pleats running parallel to {200} growth edges are present on the surfaces of both the crystals. Wrinkles are also present on the surfaces as a result of the collapse process. Typical patches are evident on the surfaces of both the crystals. The left crystal shows a thick pleat running across the b-axis. The crystal on the right has an inverted fold domain boundary running along the right end of the b-axis. An enlarged image of the inverted fold domain boundary, in the right part of the right crystal, is shown in Figure 8.

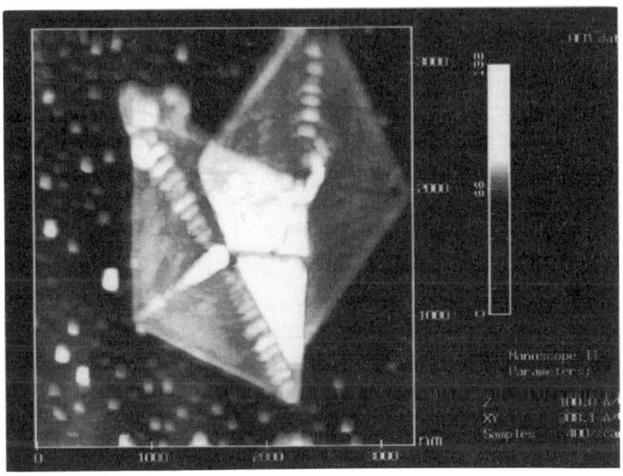

Figure 7. Two truncated lamellar crystals. The crystal on the right shows an inverted fold domain boundary along the right end of the b-axis.

Patches are also present on the surface and are in the range of diameters varying from 20 to 70 nm. An area 400 x 400 nm square containing a part of the inverted fold domain boundary in Figure 8 was scanned at a higher magnification and is shown in Figure 9. The image is rotated by 90° so that the inverted fold domain boundary runs vertically on the page. A line section across the inverted fold domain boundary was made. A crevice is observed at this inverted fold domain boundary. At its top the crevice is 45 nm wide; measured depth is 3.5 nm, but uncertainty about the shape and depth is introduced by the unknown shape of the tip.

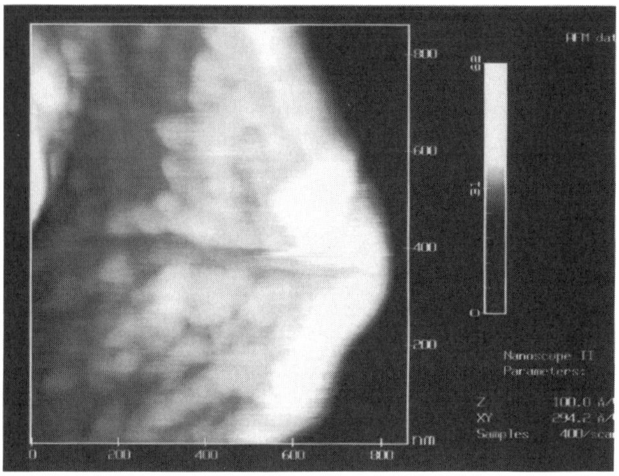

Figure 8. Inverted fold domain boundary at higher magnification.

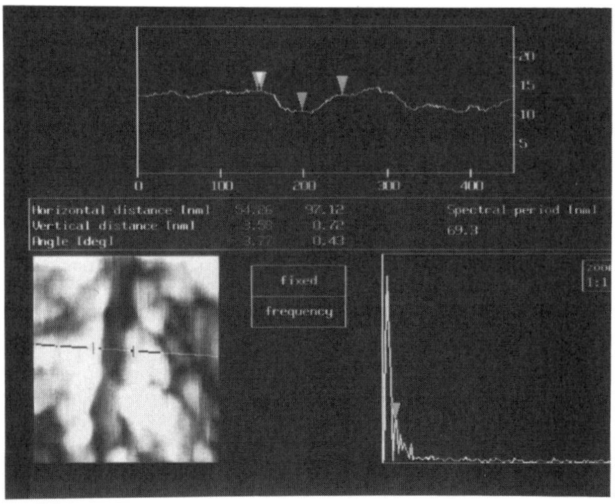

Figure 9. The fold domain boundary along the b-axis of the right lamellar crystal of Figure 8, at higher magnification, is rotated by 90° for a better view. The fold domain boundary is about 3.5 nm deep.

Dendritic crystals were grown from dilute solution in xylene by cooling the solution without using the self-seeding procedure. A part of a large and complicated dendritic crystal with overlapping lamellar crystals near its edge is shown in Figure 10. Two lamellae near the coordinates x = 1700 nm, y = 800 nm in Figure 10 are enlarged in Figure 11. Relative motion, in a sideways direction, of these two lamellae after they came into contact appears to have drawn the material between them into short fibers. Two other overlapping lamellae, near the coordinates x = 2200 nm, y = 1000 nm in Figure 10 are enlarged in Figure 12. In this case, the top lamella sheared in a direction perpendicular to the substrate, leaving a lamellar "island" of polyethylene on top of an apparently continuous basal lamella. If lamellae are sheared in this way during crystal growth, spiral terraces are likely to form.

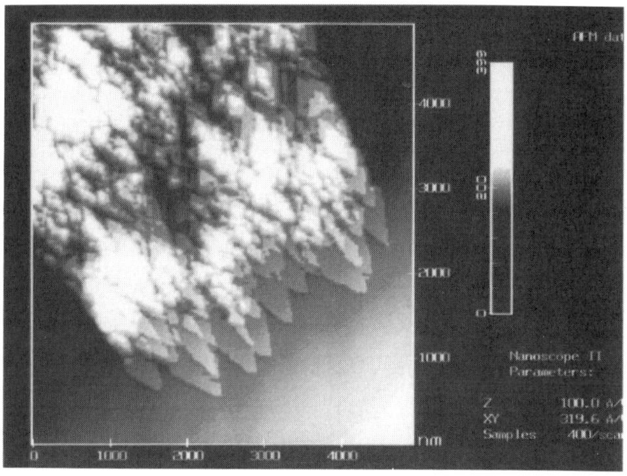

Figure 10. A top-view showing a part of a dendrite.

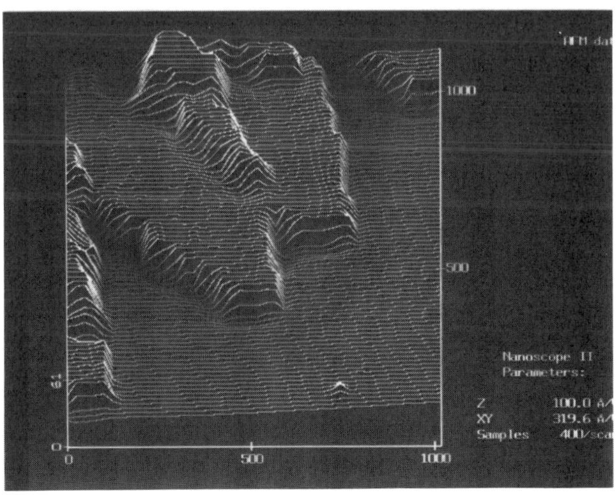

Figure 11. Sideways relative motion of lamellae results in the formation of fibers.

Microfacets are observed at the edges of the lamellae in the large dendritic crystal shown in Figures 10, 11 and 12. Microfacets are observed even at the sheared edge (the edge facing the upper right corner) of the "island" in Figure 12. Surfaces of polyimides in the form of films and fibers were observed by atomic force microscopy. A film of BTDA was prepared by spreading the polyamic acid (precursor) of BTDA on to a glass slide and thermally imidizing it at 250° C followed by annealing at 280° C. An area of about 300 x 300 nm square is shown as a surface view in Figure 13.

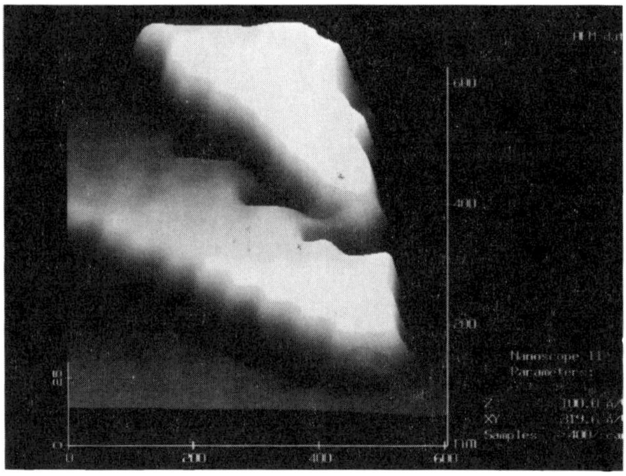

Figure 12. A line-plot of overlapped and sheared lamellae.

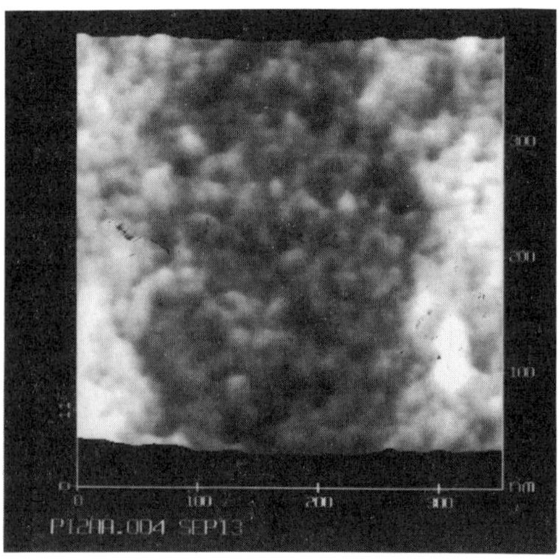

Figure 13. A surface view showing distribution of platelets on the free surface of BTDA.

The image shows a homogenous distribution of platelets. The width of the platelets varies from 15 nm to 50 nm and they are about 9 nm high.

The surface of a single fiber of BPDA-PMB with a draw ratio of seven times was observed by atomic force microscopy. An area of about 300 x 300 nm square on top of the single fiber was scanned. The resulting image is shown in Figure 14. While mounting the fiber onto the sample holder we noted the position and the orientation of the fiber axis with respect to the scan direction so as to keep track of the fiber direction in all the images. The fiber axis in Figure 14 is vertical on the page. The surface is rough on the 250 nm scale shown in the image. The ridges and grooves are running parallel to the fiber axis.

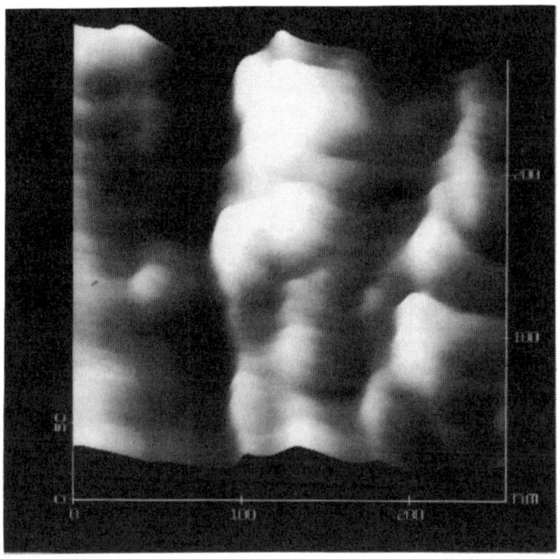

Figure 14. Fibrils running parallel to the fiber axis, which is the vertical direction in this image, observed on the surface of a single BPDA-PMB fiber.

A single fiber of BPDA-PMB was immersed in the polyamic acid precursor of BTDA, which was then thermally imidized at 250° C. The single fiber and the BTDA film were then annealed at 280° C. Images with area 350 x 350 nm square on top of the fiber coated with the BTDA film were examined. An example of such an image is shown as a surface view in Figure 15. The image shows platelets as well as fibrils running parallel to the fiber axis.

The free surface of a single fiber of BPDA-PMB, which was drawn to seven times its original length, was scanned at various places chosen at random. The surface has ridges and valleys, which run parallel to the fiber axis as is shown in Figure 14. Various sites with areas of about 20 x 20 nm square, on the highest plateaus of the ridges, were chosen and scanned to get high resolution images at the molecular scale. One such image is shown as a top view in Figure 16. The surface shows a clear periodic arrangement of features, even though the image is noisy with lines appearing parallel to the scan direction. Two-dimensional correlation functions, which were calculated by self-convolution of the images, were used to measure "average weighted" periodicites. The main feature of morphology of the fiber surface observed at the molecular scale is the presence of periodic arrangement of lines running in the general direction of the fiber axis. The periodicity is 2.0 nm along the direction perpendicular to the fiber axis, as shown in Figure 17.

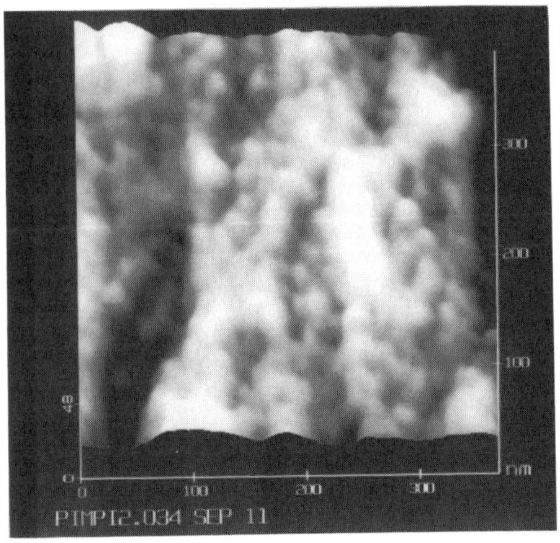

Figure 15. An area of 300 x 300 nm square on top of BPDA-PMB fiber covered by many molecular layers of BTDA crystals. The vertical direction in this image is the fiber axis.

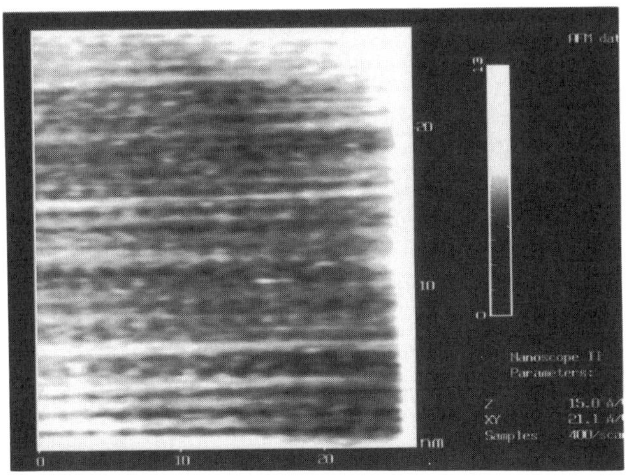

Figure 16. An area of 25 x 25 nm square on the high point of a ridge of a fibril of BPDA-PMB fiber.

CONCLUSIONS

Morphological features on the surface of solid polymer samples can be observed with micro-to-nanometer resolution using atomic force microscopy. Raised patches are conspicuous in all the data obtained from the fold surfaces of collapsed lamellar crystals of polyethylene. Patches are present on the fold domain boundaries. The observation of raised patches opens a new field of inquiry. The contour of the thickness step produced by a sudden change in temperature during crystallization of lamellar crystals can be measured

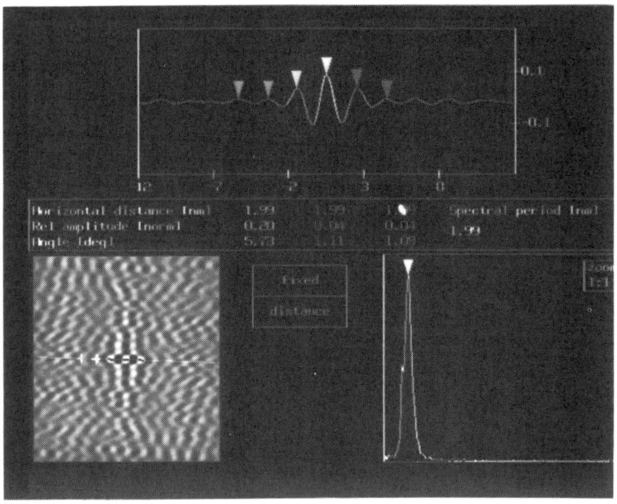

Figure 17. Autocorrelation image of Figure 16 with a line section.

accurately, making possible better measurements of growth rate by the method of Dosiere. Mechanical deformation can be observed at a nanometer scale.

The surfaces of polyimides can be studied at the molecular scale with atomic force microscopy. Orientation of BTDA by BPDA-PMB persists through many molecular layers. Observation at a molecular scale, of drawn polyimide fibers, shows two-dimensional arrays of ordered backbones.

ACKNOWLEDGMENTS

This work was supported by the NSF center for Molecular and Microstructure of Composites and by EPIC, the Edison Polymer Innovation Corporation.

REFERENCES

1. D.H. Reneker, P.H. Geil, Morphology of polymer single crystals, *J. Appl. Phys.* 31: 1916-25 (1960).
2. A. Keller, Growth and Perfection of Crystals, Wiley, New York: 498-532 (1958).
3. D.C. Bassett, A. Keller, On the habits of polyethylene crystals, *Phil. Mag.* 7: 1353-1384 (1962).
4. R. Patil, S.J. Kim, E. Smith, D.H. Reneker, Atomic force microscopy of dendritic crystals of polyethylene, *Polymer Communications*, 31: 455-457 (1990).
5. S. Cheng, Z. Wu, M. Eashoo, S. Hsu, F. Harris, A high-performance aromatic polyimide fiber: 1. Structure, properties and mechanical history dependence, *Polymer* 32: 1801-1810 (1991).
6. S. Cheng, F. Arnold, A. Zhang, S. Hsu, F. Harris, Organosoluble, segmented rigid-rod polyimide films. 1. Structure formation, *Macromolecules*. 24: 5856-5862 (1991).
7. R. Patil, V.V. Tsukruk, D.H. Reneker; Molecular packing at the surfaces of oriented polyimide fiber and film observed by atomic force microscopy, *Polymer Bull.*, 29: 557-563 (1992).
8. M. Dosiere, C.H. Colet, J.J. Point, An isochronous decoration method for measuring linear growth rates in polymer crystals, *J. Polymer Sci.:Polymer Physics Edition*, 24: 345-356 (1986).

STUDIES OF HIGH PERFORMANCE FIBERS BY ATOMIC FORCE MICROSCOPY AND MOLECULAR SIMULATION

G.C. Rutledge,[1] D. Snétivy,[2] and G.J. Vancso[2]

[1]Department of Chemical Engineering
Massachusetts Institute of Technology
Cambridge, MA 02139
[2]Department of Chemistry
University of Toronto
Toronto, Ont, M5S 1A1 Canada

Abstract: Atomic force microscopy (AFM) images obtained from cleaved surfaces of both as-spun and annealed fibers of poly(p-phenylene terephthalamide) (PPTA) have been interpreted with the aid of results from molecular simulations and wide angle X-ray scattering (WAXS). Angstrom-scale periodicities observed in AFM images of the annealed fiber are in accord with those reported in the literature based on WAXS data. For the as-spun fiber, AFM confirms the presence of at least two polymorphs. Among these, a new lattice structure is observed, which has eluded deconvolution from experimental X-ray data. This new lattice exhibits chain conformations and hydrogen bonding structure which is distinct from that seen in either of the two previously reported crystal modifications of PPTA. Interpretation of the AFM images in this case is greatly facilitated by the independent data from molecular simulations. The new crystal facet observed is fully consistent with a third crystal modification previously suggested by molecular simulation.

INTRODUCTION

High polymers containing rigid mesogenic groups in the chain backbone have been developed over the past 20 years for their unusual inherent properties and for the exceptional ease with which these properties may be exploited in processing and engineering of fibers, films, and other end-use articles. Aromatic polyamides (i.e. aramids), aromatic polyesters and copolyesters, and heterocyclic polyazoles are examples of different polymer chemistries which share the features of high molecular weight, main chain mesogens which promote the realization of a lyotropic or thermotropic liquid crystal phase, and significant degrees of aromaticity in the main chain, which imparts thermal stability.

These so-called rigid rod polymers are "high performers" in that near-theoretical degrees of mechanical strength and stiffness can actually be achieved in real articles in combination with better than average stability both to degradation and to dimensional variation with temperature. The advantageous mechanical properties are attributable in large part not to the chemical architecture of the chain per se, but to the high degrees of molecular alignment and order, which are attained during simple processing operations originated from the liquid crystalline phase. By contrast, more flexible chain architectures require considerably greater engineering effort and ingenuity to coax the chains into such macroscopically ordered states. Polyethylene, for example, has a very high theoretical modulus (ca. 300 GPa[1]), but unusual processing methods must be invoked to realize such inherent properties; high pressure die drawing and gel spinning techniques are examples of this.

The exceptional rigidity which is so advantageous to the formation of readily processable liquid crystalline dopes is also characterized by very low mobility of the chains in the solid and severe kinetic constraints on relaxation of the processed material into a thermodynamically equilibrated morphology. As a result, studies of solid state morphology in this class of materials is considerably complicated by the wide range of processing histories used to prepare different samples and the high degree of local heterogeneity that commonly results. Heat treatment is a common post-processing step, which is used to impart partial mobility to the processed material in order to speed up relaxation and to induce a general shift in the solid state morphology towards a more homogeneous, thermodynamically favored form, without going so far as to erase the processing history entirely. However, it remains a very difficult problem to differentiate fully the effects of processing forces imparted during formation from those due to thermodynamic driving forces. Part of the solution to this complexity is the analysis of these high performance materials at various stages in their formation and end-use history.

Atomic force microscopy (AFM) presents a new avenue of investigation of potentially heterogeneous materials at a very fine scale. For studies at the atomic scale AFM is particularly relevant as a complementary technique to wide angle X-ray scattering (WAXS) and computer simulation. WAXS detects structure in reciprocal space and is particularly sensitive to morphologies with more perfect local order and periodicity over longer ranges. Used in transmission mode, it reveals averaged information about the bulk material. However, it is relatively insensitive to local heterogeneity, which manifests itself largely in lower signal resolution. Computer simulation provides reliable information about local energetics and thermodynamic stability, but is limited by the investigator's ability to accurately represent all the details of the energetics of the real material. AFM detects structure in real space, and is thus capable of resolving atomic scale detail even of heterogeneous morphologies, but does so primarily in only two dimensions in space, the third dimension being the intensity of forces associated with a microscopic tip coming in contact with a sample surface, and with the problem that surface morphology in polymers may differ from that in the bulk. In this work we take advantage of the differing strengths of these methods to decipher the complex morphology of poly(p-phenylene terephthalamide) (PPTA), a rigid rod aramid, both before and after post-spin heat treatment of fibers.

EXPERIMENTAL

PPTA fibers (Kevlar™, DuPont) were resolubilized in 100% H_2SO_4 and dry-jet wet-spun through a 1-cm air gap into an aqueous coagulation bath and subsequently collected on a wind-up roller rotating at a velocity sufficient to ensure a low (approximately 1:1

spinerette velocity:wind-up velocity) draw ratio. Of the samples discussed in this work, the larger diameter fiber was spun at 80° C and 20 m/min from a 17.5% by weight polymer-in-acid solution into a 7° C aqueous bath using a 200 μm diameter monofilament spinerette; the smaller diameter fiber was spun at 30° C and 40 m/min from a 10% by weight solution into a 25° C bath using a 54 μm diameter monofilament spinerette. It has been our observation that such differences in solution concentration, spin rate and spinnerette diameter, draw ratio and coagulation conditions result in changes in degree but not in kind in the fundamental morphological features of these fibers. Further details of the spinning process are described elsewhere.[2]

The smaller diameter fiber sample was twice annealed in an argon atmosphere at 450° C for a total of 40 s. The diameters for the as-spun and annealed fiber samples, as determined by laser diffraction, were 103 $\pm$ 2 μm and 23.0 $\pm$ 0.5 μm, respectively. The WAXS studies were conducted using a Siemens D-500 X-ray diffractometer equipped with a graphite monochromator and a scintillation counter. Because of their larger diameters, variation in coagulation rate with radial position and lack of postspin annealing, the as-spun fibers are expected to exhibit greater cross-sectional heterogeneity and lower average degree of molecular orientation. These larger diameter fibers are especially useful as AFM specimens due to the resulting ease of preparation of the samples for microscopy (see below). However, the poor resolution of WAXS patterns made crystal structure determinations in these fibers impractical.

AFM images were taken in air using a NanoScope II (Digital Instruments) with D- and A-type scan heads (the latter was used for molecular resolutions) utilizing NanoProbe 100-μm triangular Si_3N_4 microcantilevers with thick legs; the effective spring constant as specified by the supplier was 0.58 N/m. In the terminology of Digital Instruments, imaging was usually performed in the "height mode"; however, some images were captured in the "force mode". ("Height mode" means that the force between the AFM tip and the sample surface is held constant by the electronic feedback loop. The vertical displacement of the sample, which is needed to maintain the chosen force, is then recorded as the tip is scanned over the sample surface.) For imaging by AFM, the PPTA fibers were first embedded in epoxy resin (Araldit Standard, Ciba-Geigy) and cleaved in the fiber direction using a Sorval MT6000 ultramicrotome with glass knives. On these images differences in the gray tones correspond to variations in the measured displacement. Raw data of images with molecular resolution were obtained with the low-pass filter set to 1 and the high-pass filter to 4; however, for some of the images all filters were disengaged. The images were very stable and reproducible. Electrostatic charges sometimes caused problems, especially after sample cleaving; in such cases good images were obtained after a 24-h waiting period. Graphite and mica images with angstrom resolution were used for distance calibration of the lateral position of the piezo controller. These and additional results from the AFM studies have been presented in greater detail elsewhere.[3] In this paper, we concentrate on conclusions relevant to molecular conformation and crystalline arrangement.

DISCUSSION

Wide Angle X-ray

Immediately after processing, neutralization of residual acid and drying, the fibers exhibit a high degree of molecular orientation parallel to the fiber axis, as evidenced by narrow azimuthal peak widths among the meridional reflections in conventional fiber WAXS photographs. This is attributable to the rotation or reconstruction of nematic domains in the spin dope as it is extended in the spinnerette and downstream in the air gap and is more pronounced in the small diameter, high draw ratio fibers. The spacing of reflections in 2θ along the meridion is consistent with fully extended chains in both as-spun and annealed fibers. The lateral packing of the rodlike molecules as-spun appears to be essentially disordered, resulting in an equatorial trace not unlike an amorphous halo. Superposed on this halo are peaks suggesting the presence of more than one possible packing arrangement, as evidenced by lattice spacings characteristic of both Modification I and of Modification II of PPTA reported previously.[4,5] The lattice parameters for these two WAXS-determined polymorphs are provided in Table I. Other polymorphic forms showing similar lattice spacings cannot be precluded due to the low resolution of the equatorial trace and the virtual absence of any higher order Bragg reflections. Deformed morphologies produced by radial and hoop stresses probably introduced during coagulation and formation of a skin layer on the fiber are not resolved by our WAXS studies.

Table 1. Crystallographic unit cells for Modifications I and II of PPTA, as determined by WAXS (from Ref. 4 and 5).

	Modification I	Modification II
a (Å)	7.87	8.0
b (Å)	5.18	5.2
c (Å)	12.9	12.9
α (deg)	90	90
β (deg)	90	90
γ (deg)	90	90
chain locations (ab projection)	[0,0] [1/2,1/2]	[0,0] [1/2,0]
density (g/cm^3)	1.50	1.50
phenylene ring rotation out of amide plane (deg.)	+30 -38	opposite direction

Subsequent to annealing, the morphology is shifted consistently and irreversibly in favor of a three dimensional lattice characteristic of Modification I. This is evidenced by the disappearance of a broad amorphous component on the equator and of the equatorial peak at 18° in 2θ characteristic of Modification II. In addition, higher order reflections characteristic of Modification I begin to appear. One may conclude that Modification I is the kinetically, and possibly thermodynamically, most favored structure for PPTA under these conditions of processing history. However, one is left with little information about the intermediate morphology present prior to annealing or any clues for controlling subsequent structural transformations.

Atomic Force Microscopy

Atomic force microscopy reveals differences in the structure and heterogeneity of the as-spun and annealed fiber samples. The process of image acquisition and analysis is comparable to a crude sampling of the ensemble of structures present at the surface of each sample. Two factors are likely to shift this sampling with respect to that observed by WAXS: (a) The method of sample preparation by microtomy produces a bias towards surfaces of low fracture energy, and the selection of images for continued analysis based on a subjective decision of acceptable resolution further biases this sampling away from regions destroyed by the knife edge. (b) Only surface features are observed, resulting in an effective bias in the orientation of the morphologies observed. Only structures which are stable to exposure to an unconstrained surface remain unaltered from the bulk. Thus surface reconstruction could lead to irreconcilable differences in interatomic distances between surface and bulk (e.g. WAXS) studies. However, the results presented here are consistent with little or no surface reconstruction, so we may postulate that this possibility has little effect on our studies of PPTA. This is reasonable in light of the kinetic constraints on the rigid rod chains and the strong intramolecular and intermolecular interactions holding the molecules in place in PPTA.

A typical image from the smaller diameter annealed fibers exhibiting resolution at the molecular level in both the x and y directions is shown in Figure 1. During imaging the fiber sample axis lay at an inclination of 25 degrees counterclockwise from the y axis of the image; the high resolution of the chain orientation from WAXS allows us to identify the molecular axis conservatively to within ±15 degrees of this sample axis, and thereby assign features to the molecular chain, indicated by the arrows. Periodicities of the short-range order both perpendicular and along a line inclined with respect to the chain direction (denoted the axial and lateral periodicities, respectively) and the angle between these two directions (denoted the angle of registration; a value of 90 degrees indicates neighboring chains in perfect axial register) were obtained by analyzing the two-dimensional autocorrelation pattern (2-D AP) of the nanographs. The 2-D AP of the raw data in Figure 1 is displayed in Figure 2, with the lines of axial and lateral periodicities highlighted. Analysis of ca. 50 images with molecular resolution from annealed fibers suggested an axial periodicity of 12.7 ± 7 Å, lateral periodicity of 4.8 ± 2 Å, and an angle of registration of 88 ± 2 degrees. Comparison with Table 1 indicates that these are consistent with a hydrogen-bonded facet (h00) of either Modification I or Modification II. In fact, given the similar chain conformations suggested for these two polymorphs, one cannot expect to distinguish these two by surface observation unless one can expose a (0k0) or (hk0) facet. Given that molecular resolution has to date only been accomplished for images consistent with an (h00) facet, one may conclude that the bias produced during sample preparation as described here precludes a significant representation at the surface of facets other than the hydrogen-bonded sheet. However, one might envision other sample preparation methods that would allow exposure of other facets not represented here.

Further analysis of the set of images from the annealed fiber suggest that in addition to the observed periodicities, the offset of features along the chain direction is consistent with phenylene rings in the PPTA chain rotated to different degrees about the chain axis. Modifications I and II are both proposed to contain molecular conformations wherein successive phenylene rings are rotated by alternating ±30-40 degrees out of the plane of the amide group. In an AFM scan the cantilever deflection in the contact-mode imaging (which is dominated by repulsive interatomic forces) will be stronger if the atoms at the apex of the tip are above the "protruding" edges of the aromatic rings.

Figure 3 shows an image after Fourier reconstruction (noise filtering) with an overlay plot of the bc–facet of Modification I (or equivalently Modification II) on the same scale. Bright areas on this AFM image are consistent with higher contact forces resulting from the interaction of the tip with the alternating edges of the aromatic groups in the PPTA chains. In addition, hydrogen bonds perpendicular to the backbone direction result in a

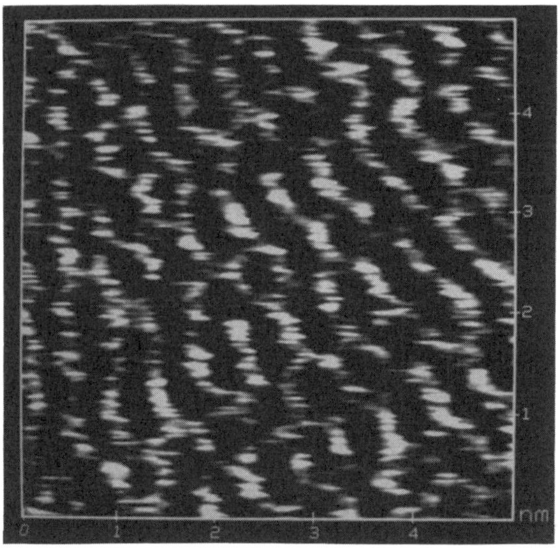

Figure 1. Typical AFM nanograph (image size 5 x 5 nm) of an annealed PPTA microfibril of Modification I (or II) with molecular resolution. The arrows indicate the chain direction. (Reprinted with permission of the American Chemical Society.)

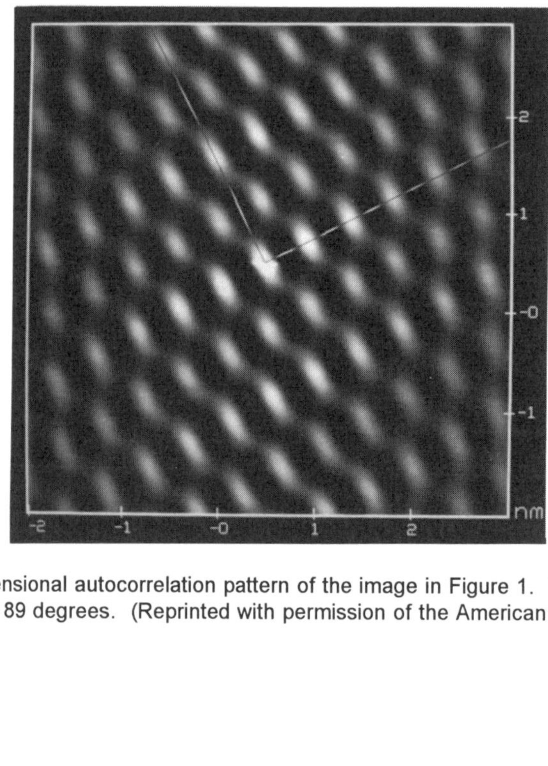

Figure 2. Two-dimensional autocorrelation pattern of the image in Figure 1. The value of the registration angle is 89 degrees. (Reprinted with permission of the American Chemical Society.)

relatively high contact force between tip and sample and can be spotted clearly in the image. This observation of the alternating orientation of the phenylene groups with respect to the plane of the hydrogen-bonded sheet demonstrates the possible use of AFM to study conformation in the solid state. Based on this close agreement between WAXS and AFM results, we assign a *c*-axis lattice spacing to the axial periodicity and a *b*–axis lattice spacing to the lateral periodicity, thus identifying the registration angle as the crystallographic angle α.

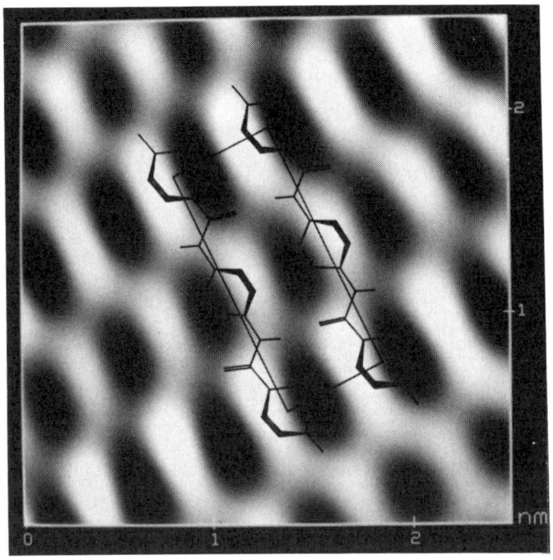

Figure 3. Fourier reconstruction of a typical AFM image of annealed PPTA; the overlay plot shows the corresponding *bc*-facet of the unit cell. (Reprinted with permission of the American Chemical Society.)

Similar analysis of images from the as-spun fibers revealed two sets of images, which enabled us to shed some light on the heterogeneous structure implied by the WAXS data. The majority of images from this material were similar to those already described for the annealed fiber and may similarly be ascribed to the *bc*-facet of either Modification I or II (either or both of which may be present according to WAXS results). The resulting lattice parameters so determined are reproduced in Table 2 along with those for the annealed fiber. In addition, a second set of images were recorded, which were inconsistent with either Modification I or II.

A representative nanograph is illustrated in Figure 4. Once again, the chain axis is readily identified based on WAXS knowledge and is indicated by arrows in Figure 4. The 2D-AP of Figure 4 is shown in Figure 5, with the lines of axial and lateral periodicity indicated. A quantitative analysis averaged over numerous images of this new structure indicated an axial (crystallographic *c*-axis) periodicity of 12.8 ± 0.4 Å, a lateral (crystallographic *b*–axis) periodicity of 4.5 ± 0.2 Å, and a registration angle of 80 ± 2 degrees, in significant disagreement with any previously reported polymorph of PPTA. Additionally, the new structure, which we denote Modification III, does not exhibit the features characteristic of opposite rotation of successive phenylene rings along the chain. Rather, a molecular conformation characterized by near-coplanar rings would account for the alignment of axial features in AFM images of Modification III. As described below, such alteration of the *bc*-facet and of the molecular conformation were both anticipated in computer simulated structures of PPTA based on a thermodynamic criterion.

Table 2. Lattice parameters of the bc-crystallographic facet of PPTA, as determined by AFM

	Annealed, Modification I or II	As-spun Modification I or II	As-spun, Modification III
b (Å)	4.8 ± 0.2	5.1 ± 0.2	4.5 ± 0.2
c (Å)	12.7 ± 0.7	12.7 ± 0.4	12.8 ± 0.6
α (deg)	88 ± 2	91 ± 2	80 ± 2
phenylene ring rotations out of amide plane (deg.)	opposite direction	opposite direction	same direction

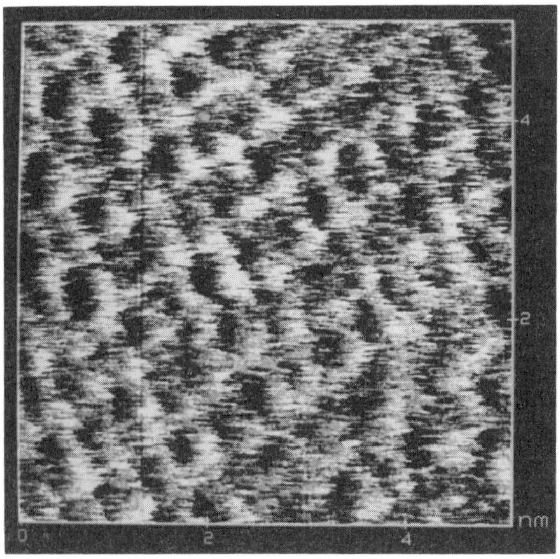

Figure 4. AFM nanograph (image size 5 x 5 nm) of an as-spun PPTA microfibril of Modification III. The arrows indicate the chain direction. (Reprinted with permission of the American Chemical Society.)

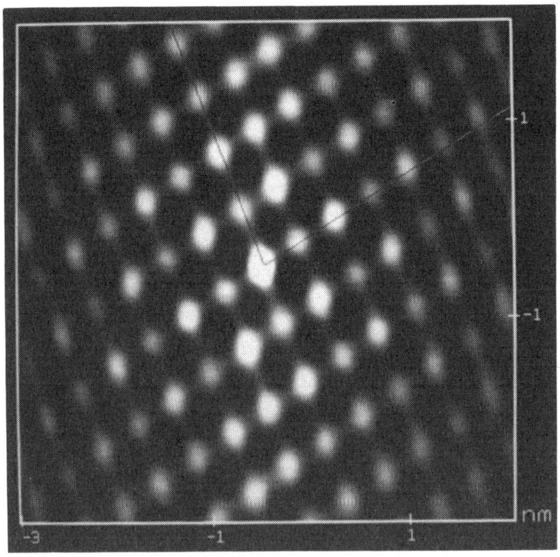

Figure 5. Two-dimensional autocorrelation pattern of the image in Figure 4. The value of the registration angle is 77 degrees. (Reprinted with permission of the American Chemical Society.)

Computer Simulation

The construction of a detailed atomistic model for PPTA has been described in detail elsewhere.[6] Considerable attention has been paid to the development of an appropriate force field, which captures the complex nature of a hydrogen bond-capable aromatic polyamide. The simulation is constructed to produce structures of minimum potential energy, subject only to the constraint of helical periodicity (but not necessarily translational periodicity) along the chain and lattice-like translational periodicity lateral to the chain orientation direction. To this extent, the simulation imposes a considerable amount of order on the molecules under study, but stops short of requiring crystalline order. One unusual aspect of this study in the case of PPTA was the considerable computational effort expended in searching all of the parameter space for viable local energy minima. More commonly, molecular simulation studies are initiated from a point in the vicinity of the experimentally determined structure, from which the closest local minimum energy polymorph is determined. As a result, one usually has little idea of the thermodynamic uniqueness of the local minimum so determined. In fact, the experimental record suggests that numerous polymeric materials exhibit multiple stable crystalline polymorphs.[7] A comprehensive search of parameter space remains a necessary step to establish any sense of uniqueness or universality of a minimum energy structure.

From this study, eight polymorphs of comparable potential energy were identified for PPTA. Of these, three are of relevance to the current discussion. The important structural features are tabulated in Table 3. The computed minimum energy unit cells of these three structures are illustrated in Figure 6. The polymorph denoted Structure #3 in previous reports is readily identified as closely allied with Modification I reported from WAXS measurements. This agreement lends credence to the assertion that Modification I is indeed a thermodynamically stable form, and that the driving force influencing the kinetics of the annealing process is a reduction in potential energy. Comparison of the *bc*-facet of this structure with that observed in both as-spun and annealed fibers and attributed to

Modification I or II also shows good agreement. All three methods discussed here are consistent in suggesting that this polymorph is present in both as-spun and annealed fibers, although to a much more exclusive extent in the latter.

Structure #5 is that in closest agreement with Modification II reported in the literature. However, in this instance the correlation is much less exact. The edge-centered unit cell mirrors that of Modification II and leads to observations of equatorial WAXS traces similar

Table 3. Selected unit cells of PPTA of minimum thermodynamic potential energy determined by molecular simulation

	Structure #3	Structure #4	Structure #5
(Å)	8.3	8.4	8.5
b (Å)	5.0	4.9	4.6
c (Å)	13.1	13.1	13.1
α (deg)	90	80	79
β (deg)	90	89	94
γ (deg)	92	90	93
chain locations (ab projection)	[0,0] [1/2,1/2]	[0,0] [1/2,0]	[0,0] [1/2,0]
density (g/cm^3)	1.45	1.50	1.57
phenylene ring rotations out of amide plane (deg.)	+43 -26	+40 -43	+39 +38

to those reported by Haraguchi et al.[4] and consistent with the 18 degree 2θ peak in the as-spun fiber. The α crystallographic angle of 80 degrees, however, is in disagreement and may possibly be attributed to a difference between a structure of minimum potential energy (simulation) and one of minimum free energy (experimental observation). As noted previously, the formation of hydrogen bonds between chains is a consistent feature of all of the simulated polymorphs, and in Structure #3 this sheet structure is in good agreement with the WAXS measurements. Structure #5 however, exhibits features of both Modification I (by WAXS) and Modification III (by AFM), namely that the chain conformation in Structure #5 shows phenylene rings with alternating rotation directions out of the amide plane (characterstic of Modification I) and an α angle of 80 degrees (characteristic of Modification III).

Structure #6, finally, exhibits three features never before resolved by WAXS but which are demonstrated by AFM to be present in the as-spun PPTA fiber. The shift in the crystallographic angle α from 90 degrees to 80 degrees occurs in both Modification III and Structure #6 (relative to Modification I and its corresponding simulation equivalent Structure #3) and is accompanied by a reduction in the *b*–axis periodicity from 5.0 to 4.6 Å. In addition, this change in lattice structure is concurrent with a change in the molecular conformation of the chain from phenylene rings of opposite rotation direction to rings of like rotation direction. Figure 7 shows the *bc*–facet of Structure #6 superposed on a Fourier reconstruction image of Modification III determined by AFM. Energetically

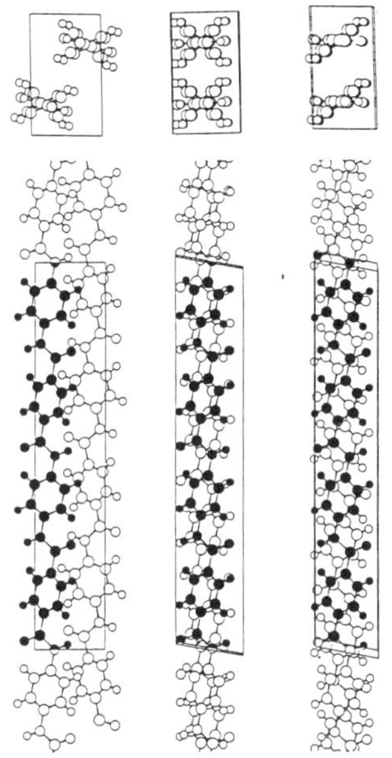

Figure 6. Perspective views of three simulated unit cells of PPTA representing local minima in the potential energy surface. Top view is the *ab*-facet. Side view is the *bc*-facet. (a) Structure #3 (b) Structure #5 (c) Structure #6.

these two chain conformations are nearly indistinguishable; to date we have only been able to suppose that the packing energetics that favor edge-to-face orientation of phenylene rings in adjacent chains somehow preclude the presence of both conformations. AFM convincingly demonstrates that, in unannealed material at least, both conformations are indeed present, and that it is the particular conversion upon annealing to the unit cell of Modification I that disfavors the conformation having coplanar ring groups.

CONCLUSION

The synergism between WAXS, AFM and molecular simulation has proven instrumental in our studies of polymorphism in PPTA fibers. Each enables us to extract a different aspect of structure from the heterogeneous morphology. The annealed fiber appears to offer a fairly consistent picture of near-complete conversion to Modification I. The as-spun fiber, on the other hand, appears to consist of at least two polymorphs, and very likely contains portions of many other low-energy packing arrangements, escape from which is too slow at room temperature due to kinetic constraints on the rigid chains.

We report evidence for a new modification of PPTA. This modification is present in relatively small amount in as-spun material and may also exist only as very small crystallites, thus accounting for the fact that it is very difficult to resolve by X-ray scattering measurements. Molecular simulation, like AFM, is sensitive to crystals even of very small dimensions. For this reason it may be very instrumental in helping to explain features observed in AFM imaging experiments. Where surface reconstruction does not occur, conventional bulk polymer simulation can provide excellent corroboration between theory and experiment. Where surface reconstruction does occur, simulation of such surfaces may ultimately be the only means by which the contrast features of AFM images can be anticipated and deciphered.

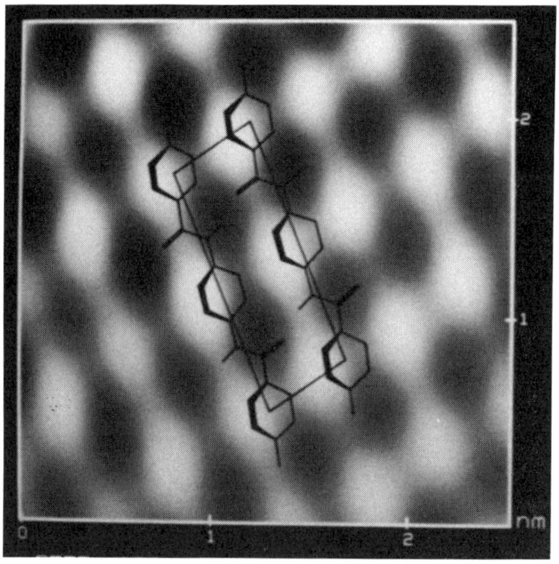

Figure 7. Fourier reconstruction of an AFM image of as-spun PPTA indicative of Modification III; the overlay plot shows the corresponding *bc*-facet of the simulated Structure #6.

ACKNOWLEDGMENTS

Financial support by the Ontario Centre for Materials Research is gratefully acknowledged. The authors wish to thank the DuPont Company for donation of the original Kevlar staple and Dr. B. Glomm and the Swiss Federal Institute of Technology (ETH Zürich) for their assistance and support in fiber preparation.

REFERENCES

1. K. Tashiro, M. Kobayashi, and H. Tadokoro, Calculation of three dimensional elastic constants of polymer crystals. 2. Application to Orthorhombic Polyethylene and Poly(vinyl alcohol), *Macromolecules* 11: 914-918 (1978).
2. G.C. Rutledge, U.W. Suter, and C.D. Papaspyrides, Analysis of structure and polymorphism in poly(p-phenylene terephthalamide) through correlation of simulation and experiment, *Macromolecules* 24: 1934-1943.

3. D. Snétivy, G.J. Vancso, and G.C. Rutledge, Atomic force microscopy of polymer crystals. 6. Molecular imaging and study of polymorphism in poly(p-phenylene terephthalamide) Fibers, *Macromolecules*, 25, 7037-7042 (1992).
4. M.G. Northolt, X-ray diffraction study of poly(p-phenylene terephthalamide) fibres, *Eur. Polym. J.* 10: 799-804 (1974).
5. K. Haraguchi, T. Kajiyama, and M. Takayanagi, Effect of coagulation condition on crystal modification of poly(p-phenylene terephthalamide), *J. Appl. Polym. Sci.* 23, 915-926 (1979).
6. G.C. Rutledge, and U.W. Suter, Detailed atomistic simulation of oriented pseudocrystalline polymers and application to a stiff chain aramid, *Macromolecules*, 24: 1921-1933 (1991).
7. See, for example, P. Corradini, and G. Guerra, *Adv. Polym. Sci.*, 100, 183 (1992).

ATOMIC FORCE AND ELECTRON MICROSCOPIC INVESTIGATIONS OF LEAD SELENIDE CRYSTALS GROWN UNDER MONOLAYERS

Janos H. Fendler and Jianping Yang

Department of Chemistry
Syracuse University
Syracuse, New York 13244-4100

Abstract: Lead selenide (PbSe) particulate films, composed of highly oriented, equilateral triangular crystals, have been *in situ* generated by the exposure of arachidic-acid-(AA) monolayer-coated aqueous lead nitrate [$Pb(NO_3)_2$] solutions to hydrogen selenide (H_2Se). The AA-coated PbSe films, at different stages of their growth, were transferred to solid substrates and characterized by transmission electron microscopy (TEM), electron diffraction, and atomic force microscopy (AFM). The epitaxial growth of PbSe crystals has been rationalized in terms of matching the {111} plane of crystalline PbSe and the (100) plane of the hexagonally close-packed AA monolayer. The parallel alignment of the $\langle\bar{1}10\rangle$ and $\langle 00\bar{1}\rangle$ axes and the perpendicular alignment of the <110> axis to the monolayer have been suggested to be responsible for the observed formation of rod-like PbSe particles. The presence of a negatively charged monolayer has been found to be an essential requirement for the oriented growth of PbSe crystals.

INTRODUCTION

Highly selective *in vivo* crystallization, often referred to as biomineralization,[1-4] is one of nature's common and, until recently largely unnoticed, accomplishments. Biomineralization plays also an important role in bone, teeth, and shell structure formation.[1-4]

Oriented nucleation at specific cell sites and subsequent epitaxial growth are the key factors which govern biomineralization. Elucidation of the chemical species responsible for oriented crystallization is not only inherently interesting, but it may also lead to new methodologies for advanced material construction. Indeed, organized surfactant vesicles, reversed micelles, monolayers, polymers, and proteins have been shown to mediate oriented nucleation of inorganic crystallites in controllable dimensions and morphologies.[5]

Electron (EM) and atomic force (AFM) microscopic characterization of lead selenide (PbSe) particulate films, grown under AA monolayers, is the subject of the present report.

Atomic Force Microscopy/Scanning Tunneling Microscopy
Edited by S.H. Cohen *et al.*, Plenum Press, New York, 1994

METHOD

Spreading arachidic acid over aqueous lead nitrate resulted in characteristic Langmuir isotherms.[6,7] Exposure of an AA-monolayer coated $Pb(NO_3)_2$ solution to H_2Se resulted in the formation of reasonably uniform crystals, which constituted the particulate film. The PbSe particulate film formed under the monolayer was transferred via an amorphous-carbon-coated, 200-mesh copper grid for electron microscopy.

Information on the morphology of PbSe crystals was obtained by atomic force microscopy (AFM). A typical three-dimensional 470 nm by 470 nm image of PbSe crystallites is shown in Figure 1. Interestingly, the equilateral triangular PbSe crystals were not uniform; along their corners they were somewhat thicker than in their middle sections.

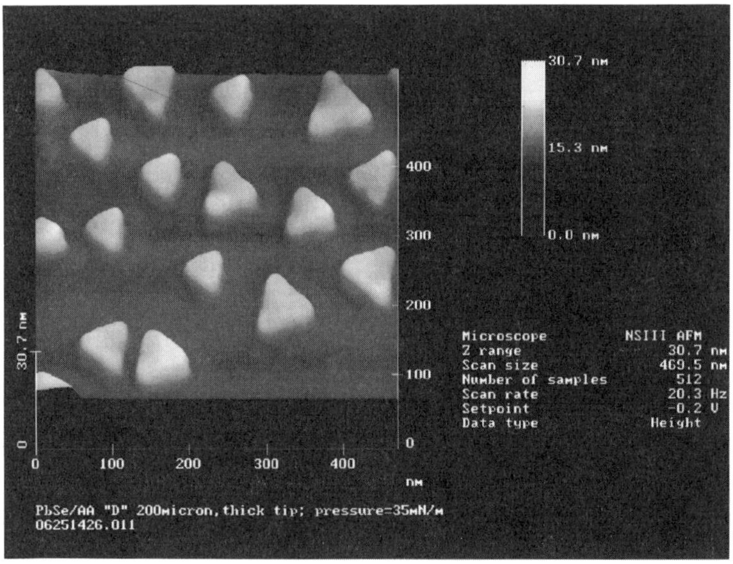

Figure 1. Three dimensional side view of a PbSe particulate film imaged on a 470 nm by 470 nm section by AFM.

RESULTS

At high monolayer surface pressures, formation of rod-like PbSe particles was observed. Typically, the thickness of the rods was 10 nm or less and the length was in the order of 100 nm. In addition to the rods, there were many additional dot-like particles in the 10-nm size range. Electron diffraction resulted in a pseudo six-fold symmetrical pattern caused by the preferred alignment of crystals with their <110> direction parallel to the electron beam. In addition, some of them are being rotated 60° and 120° around the [110] axis. The single crystal diffraction pattern, originating in the alignment of the crystals with their [110] direction parallel to the electron beam, would only give a two-fold symmetry.

Monolayers are shown, in the present work, to have a crucial role in mediating semiconductor crystallite nucleation. Infusion of H_2Se over the aqueous metal ion solution, in the absence of a monolayer, led only to the formation of cubic, irregular, and polydispersed PbSe particles that precipitated and settled at the bottom of the trough.

Furthermore, no PbSe formation could be observed upon the infusion of H_2Se over a positively charged monolayer (dioctadecyldimethylammonium bromide, for example) floating on an aqueous solution of $Pb(NO_3)_2$. Surface charge and charge density on the monolayer head group are clearly important. Negatively charged arachidate ions, formed in the ionization of the carboxyl head groups of AA, attract positively charged lead ions from the bulk phase. This electrostatic attraction is responsible for the very high (up to several molar) lead-ion concentration at the monolayer interface where nucleation of PbSe crystallites is initiated. The high local reagent concentration at the interface is also responsible for the preferential two-dimensional growth of the PbSe clusters.

In monolayers compressed to their solid states, the alkyl chains of AA, fully extended in the air in a planar zig-zag conformation, are oriented approximately normal to the surface in a triangular lattice of hexagonal close packing with a lattice constant of a = 4.85 Å and d_{100} spacing of 4.20 Å.[8] Combined synchrotron X-ray reflection and diffraction data established a structural model for AA monolayers at air-water interfaces.[9-11] The model required the hydrocarbon chains to be packed in a pseudohexagonal lattice and tilted toward their nearest neighbor.[10,12,13]

PbSe is known to crystallize in a face-centered cubic crystalline lattice with a lattice constant of a = 6.1255±0.0005 Å.[14] Distances between the closest Pb-Pb and Se-Se atoms, 4.33 Å, fit the d_{100} network spacing of the AA monolayer. This fit implies the alignment of PbSe along its (111) plane to the hexagonally close-packed AA monolayer. Epitaxial growth of PbSe under AA monolayers is explicable, therefore, in terms of the geometric complementarity between the PbSe and AA headgroups.

A comparison of the interatomic Pb to Pb distance of the (111) plane of the PbSe crystal (4.33 Å) with d_{100} spacing of the AA monolayers (4.16 Å)[15] revealed a mismatch of 4% between the template and the crystal. This should be compared with the less than 1% mismatch between the same template and PbS crystals.[15] The slightly greater intrinsic crystal lattice of PbSe (a = 6.125 Å) than that of PbS (a = 5.94 Å) resulted in a somewhat poorer match of the selenide than the sulfide crystals with the AA monolayer and in a poorer reproducibility in forming equilateral triangular PbSe than PbS crystals.

Appearance of rod-like PbSe particles is rationalized by the good fit between the Pb^{2+} ions in the {110} plane the hexagonally close-packed AA. Thus, the preferred orientation of the [110] axis in PbSe is perpendicular to the monolayer, while the [$\bar{1}$10] and [00$\bar{1}$] axes are parallel to the monolayer and are related to the short and long axes of the PbSe rod, respectively. The overlap of Pb^{2+} ions with the unstrained crystal lattice of arachidic acid is schematically shown in Figure 2(a). Using values a_{PbSe} = 6.125 Å[14] and a_{AA} = 4.81 Å and $d_{100}(AA)$ = 4.16 Å[15] for the intrinsic PbSe and AA crystal lattices led to 2% and 10% mismatch along the vertical [001] and the horizontal [110] axes between PbSe and AA. Stretching the PbSe crystal lattice to a_{PbSe} = 6.27 Å (experimentally determined in the present system) led to an improved overlap between PbSe and AA along the vertical (0.5% mismatch) and the horizontal (8% mismatch) axes (Figure 2(b)). The better overlap in the [00$\bar{1}$] than in the [$\bar{1}$10] direction favors, of course, the growth of PbSe crystallites along the vertical direction and leads to the formation of the observed rod-like structures. The overlap between the AA template and PbSe crystals was analyzed in terms of the observed $d_{100}(AA) = 2a_{PbSe}/3$ relationship.

Generation of rod-like semiconductor particles under monolayers provides a potential method for nanofabrication. Structured materials in the nanometer to submicrometer range are particularly promising and are expected to have unique properties.[16,17] They are size quantized; that is to say that their dimensions are comparable to the length of the de Broglie electron, the wavelength of phonons, or the mean free paths of excitations.[18-20] Exciton confinement, in length of PbSe rods, could lead to two-dimensional size quantization (that is, to the formation, in the language of physicists, of quantum wires). Quantum

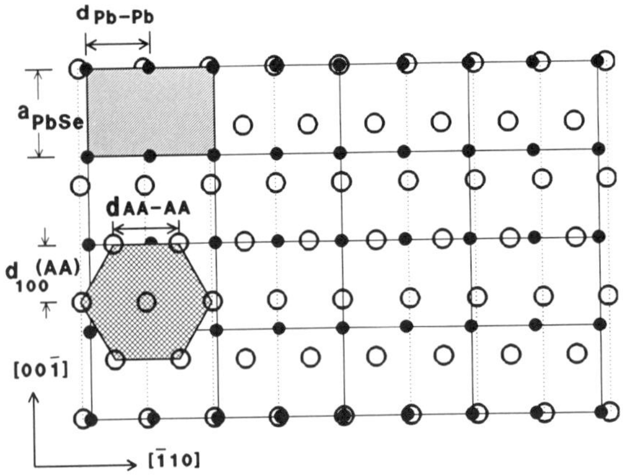

Figure 2. (a) Schematic two-dimensional representation of Pb^{2+} (•) in the (110) plane of unstrained lattice (a = 6.125 Å) and hexagonally close-packed AA (○) headgroups. Two directions are indicated as [001] and [110] in the PbSe lattice. Both unit cells of AA headgroups and Pb^{2+} ions are highlighted by hexagonal and rectangular shaded areas, respectively. The mismatch between PbSe and AA crystal lattices is evident.

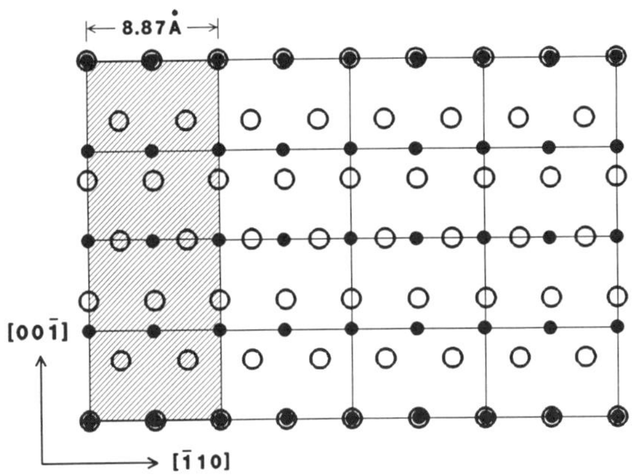

Figure 2. (b) Schematic two-dimensional representation of the improved matching achieved by stretching PbSe lattice along [110] direction. The strained PbSe crystal lattice was experimentally determined as a = 6.27 Å.

wires are expected to have substantially altered mechanical, chemical, electrical, optical, magnetic, electro-optical, and magneto-optical properties. This, in turn, may lead to novel applications and devices.[16,17] Electrochemical formation of epitaxially oriented CdSe quantum dots has already been reported.[21]

ACKNOWLEDGMENT

Support of this research by a grant from the National Science Foundation is gratefully acknowledged.

REFERENCES

1. P.Westbroek and E.W. de Jong, "Biomineralization and Biological Metal Accumulation," Reidel, Dordrecht, Holland (1983).
2. K. Simkiss and K.M. Wilbur, "Biomineralization," Academic Press, San Diego (1989).
3. H.A. Lowenstam and S. Weiner, "On Biomineralization," Oxford University Press, New York (1989).
4. L. Addadi, and S. Weiner, Control and design principles in biological mineralization, *Angew. Chem. Int. Ed. Engl.*, 31:153-169 (1992).
5. J. H. Fendler, Atomic and molecular clusters in membrane mimetic chemistry, *Chem. Rev.*, 87:877-899 (1987).
6. G.L.Gaines, Jr., "Insoluble Monolayer at Liquid-Gas Interfaces," Interscience: New York (1966).
7. A. Ulman, "An Introduction to Ultrathin Organic Films. From Langmuir-Blodgett to Self-Assembly," Academic Press: New York (1991).
8. K. Kjaer, J. Als-Nielsen, C.A. Helm, P. Tippmann-Krayer, H. Möhwald, Synchrotron X-ray diffraction and reflection studies of arachidic acid monolayers at the air-water interface, *J. Phys. Chem.*, 93:3200-3206 (1989).
9. J. Als-Nielsen and H. Möhwald, Synchrotron X-ray scattering studies of Langmuir films, *In*: "Handbook on Synchrotron Radiation," Vol. 4, S. Ebashi, M. Koch, and E. Rubinstein, eds. Elsevier Science Publishers, B.V., 1-53 (1991).
10. D. Jacquemain, S.G. Wolf, F. Leveiller, M. Deutch, K. Kjaer, J. Als-Nielsen, M. Lahav, and L. Leiserowitz, Two-dimensional crystallography of amphiphilic molecules at the air-water interface. *Angew. Chem. Int. Ed. Engl.*, 31:130-152 (1992).
11. D. Möbius and H. Möhwald, Structural characterization of monolayers at the air-water interfaces, 3:19-24 (1991).
12. D. Jacquemain, F. Leveiller S.P. Weinbach, L. Leiserowitz, K. Kjaer, and J. Als-Nielsen, Crystal structures of self-aggregates of insoluble aliphatic amphiphatic molecules at the air-water interface. An X-ray synchrotron study, *J. Am. Chem. Soc.*, 113:7684-7691 (1991).
13. P. Tippmann-Krayer and H. Möhwald, Precise determination of tilt angles by X-ray diffraction and reflection with arachidic acid monolayers, *Langmuir*, 7:2303-2307 (1991).
14. "Handbook of Chemistry and Physics (1983-1984)," CRC Press, Boca Raton, Florida, B-203.
15. X.K. Zhao, J. Yang, L.D. McCormick, and J.H. Fendler, Epitaxial formation of PbS crystals under arachidic acid monolayers. *J. Phys. Chem.*, in press.
16. F. Capasso, Bandgap and interface engineering for advanced electronic and photonic devices, *MRS Bull.*, 23-29 (1991).
17. R.P. Andres, R.S. Averback, W.L. Brown, L.E. Brus, W.A. Goddard, A. Kaldor, S.G. Louie, M. Moskovits, P.S. Percy, S.J. Riley, R.W. Siegel, F. Spaepen, and Y. Wang, Research opportunities on cluster and cluster assembled materials, *J. Mater. Res.*, 4:704-736 (1989).
18. M.G. Bawendi, M.L. Steigerwald, and L.E. Brus, Quantum mechanics of large semiconductor clusters, *Ann. Rev. Phys. Chem.*, 41:477-496 (1990).
19. L.E. Brus, Semiconductor crystallites, *Acc. Chem. Res.*, 23:183-188 (1990).
20. Y. Wang, Nonlinear optical properties of nanometer sized semiconductor clusters, *Acc. Chem. Res.*, 24:133-139 (1991).
21. Y. Golan, L. Margulis, I. Rubinstein, and G. Hodes, Epitaxial electrodeposition of CdSe nanocrystals on gold, *Langmuir*, 8:749-752 (1992).

ATOMIC FORCE MICROSCOPY STUDIES OF ULTRA-THIN FILMS OF CADMIUM PHOSPHIDE NANOCLUSTERS ON MICA

Tomasz Kowalewski, Michael A. Matchett, and William E. Buhro

Department of Chemistry
Washington University
St. Louis, Missouri 63130

Abstract: In this study we demonstrate the use of contact AFM to determine the size of semiconductor nanoclusters with diameters of order of 30 Å. This goal is achieved by taking advantage of tip-induced sample modifications, which are usually the major obstacle in AFM imaging. Nanocluster size is deduced from the depth of holes micromachined with a tip in monolayers and from row spacings in 2-D oriented arrays of nanoclusters produced by prolonged scanning. Ease and accuracy of micromachining of nanocluster films point to their potential applications in nanotechnology.

INTRODUCTION

Interest in nanoscale objects and structures stems from the novel properties of such systems, which are the consequence of quantum confinement due to their finite small size.[1] Both physical and chemical methods are being pursued in an effort of obtaining nanostructured systems.[1,2] Nanophysics methods resort to techniques like chemical vapor deposition, molecular beam epitaxy and various types of nanolithography.[1]

Typical nanochemistry methodology involves carrying out the synthesis in "nanobeaker"- like voids in nanoporous solids[2] or micelles in microheterogeneous suspensions.[3] An example of the latter is the reversed-micelle procedure (water in heptane) which has been used for the preparation of II-VI semiconductor nanoclusters (CdS, CdSe, etc.).[3,4] The sizes of II-VI nanoclusters were indirectly determined by the size of the micelles, which posed the major difficulty in assuring the narrow size distribution desirable in this type of material. Recently we have adapted this procedure to a member of II_3-V_2 family, cadmium phosphide (Cd_3P_2), using titration as the means of size control.[5] It allowed us the isolation of narrowly dispersed, surface-derivatized Cd_3P_2 nanoclusters. We describe in this paper the use of AFM for size characterization and for manipulation of arenethiolate-capped cadmium phosphide nanoclusters obtained by this method.

MATERIALS AND METHODS

Nanocluster Preparation

The Cd_3P_2 nanoclusters were synthesized according to eq. (1)

$$3Cd^{2+} + 2PH_3 + 6OH^- \rightleftharpoons \text{nano-}Cd_3P_2 + 6H_2O \qquad (1)$$

by NaOH titration of the reversed micellar solution prepared by combining heptane, the surfactant AOT, aqueous $Cd(ClO_4)$[5], and stoichiometric amount of the PH_3 precursor $P(SiMe_3)_3$. Successive titration steps of the solution were accompanied with sequential shifting of the UV-visible absorption onset to lower energies (change in solution color from light yellow to red), indicating the stepwise increase in nanocluster size. The nanocluster surfaces were then derivatized with arenethiols of the general formula p-XC_6H_4SH (X=H,Cl,MeO,NH_2), which induced their precipitation. The arenethiolate-capped Cd_3P_2 nanoclusters were isolated as THF and pyridine soluble powders. In this paper we report the studies involving two fractions of such prepared nanoclusters, which will be referred to as yellow and red particles based on the color of the solutions (see above).

Atomic Force Microscopy

Samples for AFM studies were prepared by spin-coating THF solutions of nanoclusters on freshly cleaved mica mounted on AFM sample stages. Preliminary trials indicated that the most uniform coating was obtained when the solution was applied dropwise (6-8 drops) on the substrate rotating at 8000 rpm. Throughout this study the coating conditions were not changed and the only variable in sample preparation was the solution concentration, which ranged from .2 mg/ml to 5 mg/ml. The samples were then immediately imaged at ambient conditions with the aid of Nanoscope III AFM[6], using 15 μm by 15 μm piezoelectric scanner with vertical resolution 0.5 nm. Imaging was carried out with 200 μm wide-legged silicon nitride cantilevers (spring constant 0.12 N/m) in contact mode, with typical contact forces of the order of 10 nN. Most of the large area (≥ 1 μm by 1μm) scans were performed in constant force mode. For small area scans, approaching molecular resolution, constant height and lateral force mode (LFM) were also used, with the latter providing the best contrast as will be discussed later.

RESULTS AND DISCUSSION

Observations in Constant Force Mode: Film Damage and Controlled Micromachining

Early in the study we have observed that scanning at the conditions facilitating very close surface tracking (scanning areas 10 x 10 μm at frequency below 5 Hz or scanning areas 1 x 1 μm and below) is accompanied by extensive film reorganization leading to damage and exposure of the mica substrate. At advanced stages of film damage we have observed patterns similar to those reported for thin polymer films.[7] Typical sequence of images illustrating gradual damage of a film of "yellow" particles is shown in Figure 1 (a) - (f). Figure 1(a) is a large area scan of the film obtained immediately after tip engagement. Subsequently an attempt was made to zoom-in at the smaller area (1μm by 1μm); however, the image was very unstable, with the indications that film disruption was taking place.

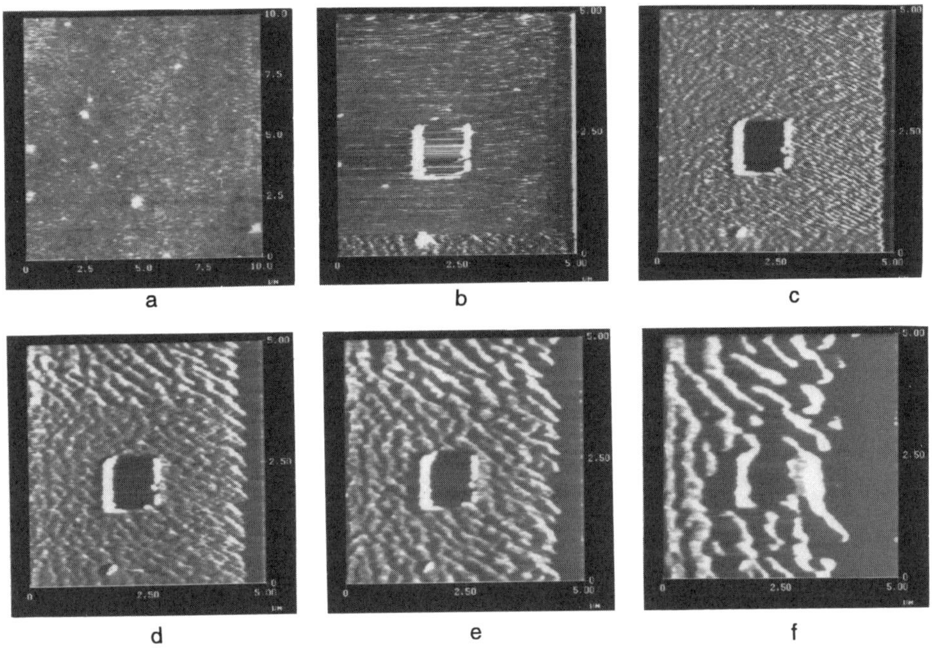

Figure 1. Progress of structural changes during AFM imaging of a monolayer of "yellow" particles. (a) 10 μm by 10 μm first scan image of a film (unaltered); (b) -(f) consecutive scans in 5 μm by 5 μm area leading to gradual formation of bands of coalesced material and exposure of mica. The 1 μm by 1 μm hole in the film in the center of those images has been formed in a 1 μm by 1 μm scan preceding those depicted in images (b)-(f).

Figure 1(b) shows the image obtained after zooming-out to 5 μm by 5 μm, which confirms this observation, showing the square of damaged film in previous scan area. Notice also the difference in film appearance starting from about 0.5 μm from the bottom (slow scan direction is from bottom to top in this image). Whereas the first 0.5 μm shows fairly clear features, the image becomes blurred and multiple horizontal lines appear above this point. This indicates that extensive film damage, accompanied by lateral displacements of material, started after the first 0.5 μm of scan.

Images in Figure 1 (c) - (f) illustrate subsequent stages of film damage, leading to the exposure of the mica substrate and formation of characteristic bands of material at some angle between 45 and 90 degrees to the fast scan direction. Similar patterns have been observed in experiments with thin films of polymers.[7] Images from scans following the noisy one (Figure 1 (b)) are clear and structure changes are more gradual, involving coalescence and growth of bands of material. Such behavior resembles a surface tension driven attempt to decrease the surface to volume ratio.

Tip-sample interactions in contact mode prevented us from observing intact films at magnifications large enough to allow molecular resolution in the film plane. However, exposure of the mica in the course of film breakdown made it possible to take advantage of the subnanometer vertical resolution of an AFM and to measure accurately the film thickness. By changing the concentration of the nanocluster solution used in spin coating from 5 mg/mL down to 0.2 mg/mL we were able to prepare films of various thickness, and identify concentrations resulting in a monolayer of nanoclusters. (Films obtained from concentrations below one corresponding to the monolayer were no longer continuous, and were much more easily destroyed during scanning). The three-dimensional image of a portion of a monolayer film with the exposed square of mica in the center is shown in Figure 2 (a) (this image is actually obtained in the same series as ones from Figure 1, illustrating the progress of film disruption leading to exposure of mica). The image has been corrected for nonplanarity resulting mainly from the nonplanarity of XY displacements of a single tube piezoelectric element. The profile of the cross-section of this film, passing through the areas of the intact film and fully exposed mica is shown in Figure 2 (b).

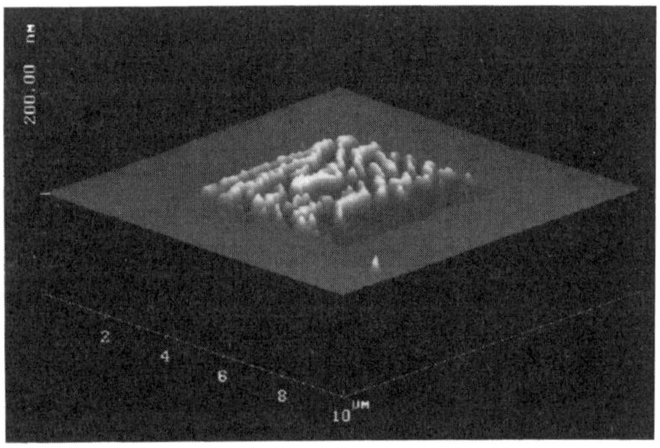

Figure 2. (a) AFM image of the monolayer of "yellow" particles with the central square machined off with the tip

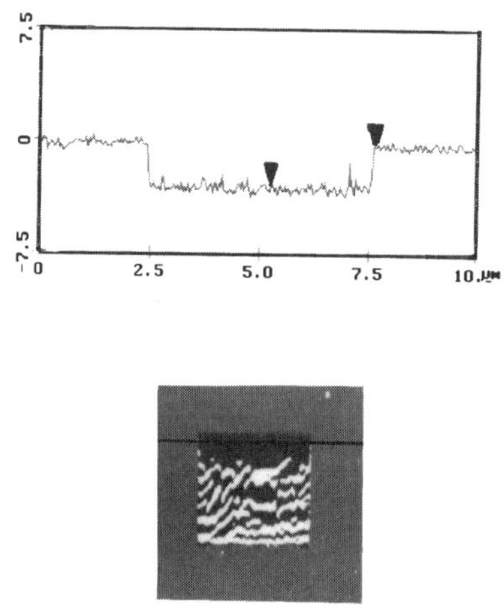

Figure 2. (b) Example of the profile of the cross-section of the micromachined area along the line marked on the miniature image below. Vertical scale on the profile is in nanometers, distance between the arrows, corresponding to film thickness is equal to 2.8 nm.

Film thickness measured from a series of such profiles has been found to be equal to 2.8 ± 0.3 nm in good agreement with earlier transmission electron microscopy observations.[5] Images like one in Figure 2 (b) are also a convincing proof of very narrow distribution of sizes of "yellow" nanoclusters.

Exposure of mica due to film micromachining with a tip, and thus film thickness measurements were possible for films of a thickness up to about 20 nm (about eight nanocluster layers). Progress of micromachining of thicker films differed however from one observed for a monolayer. In most cases, when film disruption occurred after zooming-in from larger area (>5 x 5 μm), it started from the edges perpendicular to the scan direction, and through the stage involving formation of highly mobile, semiliquid layer (Figure 3) led to formation of a stable mound of material somewhere in the middle of a trough micromachined in the film (Figure 4).

Observations of Tip-Induced Order in 2-D Arrays of Nanoclusters in Lateral Force Mode

Multiple experiments indicated reproducibly that prolonged scanning following zooming-in to smaller area of mobile, semiliquid layer as one shown in Figure 3 resulted in formation of equally spaced rows of particles, parallel to the scanning direction. We have determined, that best contrast in imaging of oriented patterns was obtained in lateral force mode (LFM). This observation can be explained as a consequence of the fact, that the size

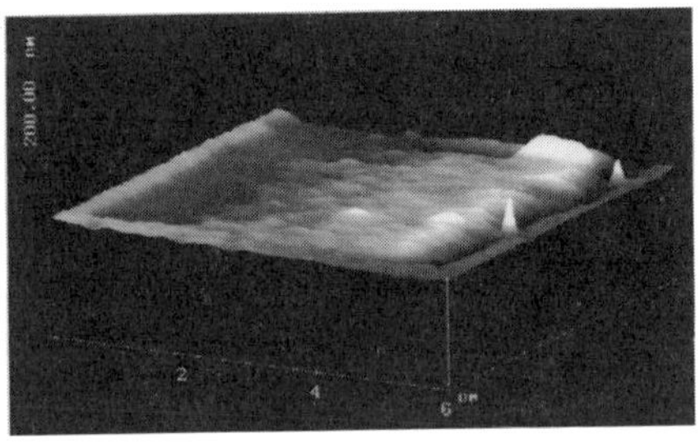

Figure 3. AFM image of an early stage of micromachining of four monolayer thick film of "yellow" particles. Micromachining was carried out in the central square of the image, subsequent observation was performed after zooming-out to 15 μm by 15 μm area.

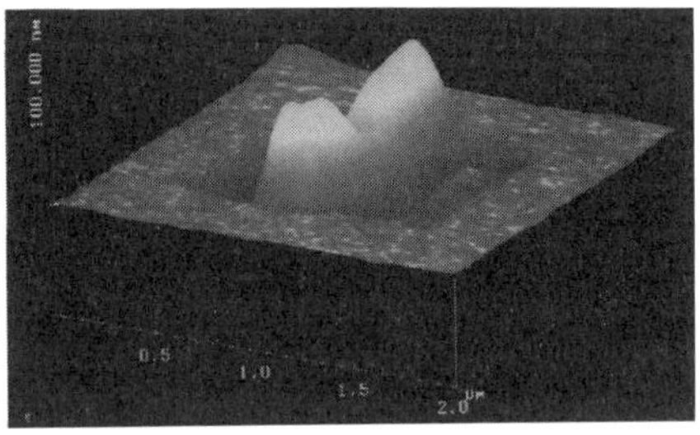

Figure 4. AFM image of final result of micromachining of four monolayer thick film of "yellow" particles (see text).

of surface corrugations, reflecting the nanocluster diameter, is comparable with the tip radius.

Under such circumstances the surface profile cannot be mapped accurately with molecular resolution in constant force or height modes due to two major reasons:

(I) the tip does not penetrate deep enough into the depressions between the nanoclusters;

(II) the character of tip-sample contact is topography dependent - on the crests of surface protrusions number of contact points is smaller than in the areas corresponding to depressions. Effectively - various features of the surface are imaged with different sectors of the tip which imposes limitations on contrast and resolution.

Topography dependence of tip-sample contact provides, however, a contrast mechanism for the lateral force (microfriction) mode. Due to fewer contact points microfriction will be smaller on the crests of sample protrusions than in the valleys. Thus, translating the LFM images into surface topography, one has to take into account that areas of high signal correspond to depressions and vice versa (e.g., by digitally inverting the images). Digitally inverted LFM image of a typical oriented pattern of nanoclusters and its 2-D Fourier transform are shown in Figure 5 (a) and (b). The position of arcs in Fourier transform in Figure 5 (b) corresponds to row spacing of 2.7 ± 0.3 nm in good agreement with diameter of nanoclusters determined from monolayer thickness measurements. The presence of well-defined features in Fourier transform allowed us to carry out meaningful patterns restricted to the arcs. Surface plot of such digitally filtered image is shown in Figure 5 (c).

Oriented patterns were stable enough to be imaged at some angle with respect to initial orientation, leading to images, like one in Figure 6 (a), obtained for scanning angle of 60 degrees. The corresponding 2-D Fourier transform pattern and digitally filtered image are shown respectively in Figure 6 (b) and Figure 6 (c). The row spacing determined from arc position is equal to 2.6 ± 0.3 nm. As can be seen from Figure 6 (a), imaging at some angle to the initial orientation resulted in improvement of resolution along the rows, allowing to identify single particles. This may indicate higher lateral mobility of nanoclusters along the row direction.

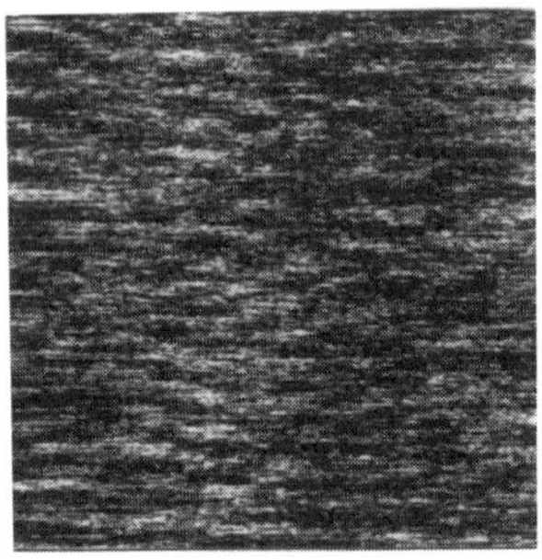

Figure 5. (a) LFM image of 45 nm by 45 nm area of 2-D oriented array of "yellow" nanoclusters. Orientation was produced by prolonged scanning of the mobile layer of nanoclusters as one shown in Figure 3 (see text)

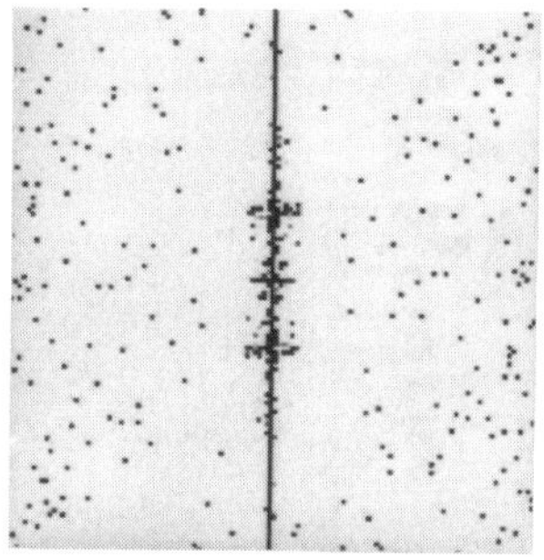

Figure 5. (b) 2-D Fourier transform of an image from Figure 5(a). Position of arcs corresponds to row spacing of 2.7 ± .3 nm.

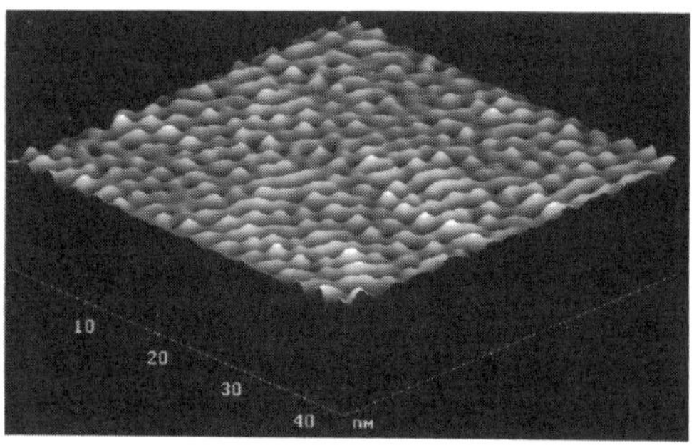

Figure 5. (c) Surface plot of digitally filtered image from Figure 5 (a).

Extensive study of micro- and nanomachining phenomena described above showed that machining effectiveness of the tip decreased with the increase of the angle between the scanning direction and long axis of the cantilever. At a 90 degree angle, "cutting down" to mica in four-particle-layer film required usually several times more scans than for 0 degree scans. We were also unable to produce oriented patterns in 90 degree scans.

Such anisotropy of micromachining properties of pyramid shaped Si_3N_4 tips cannot be explained based just on the tip shape. The possible explanations may invoke the anisotropy of mechanical properties of the triangular cantilever and/or the tilted positioning of the cantilever with respect to the sample surface (both factors make various cutting directions inequivalent).

The ability to produce oriented patterns was also film-thickness dependent. No induced orientation could be obtained for monolayers or films thicker than several layers, apparently

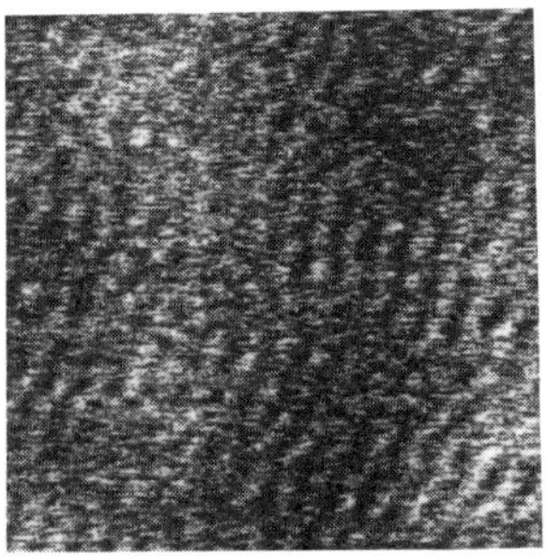

Figure 6. (a) LFM image of 45 nm by 45 nm area of 2-D oriented array of "yellow" nanoclusters obtained for scanning angle of 60 degrees with respect to row direction.

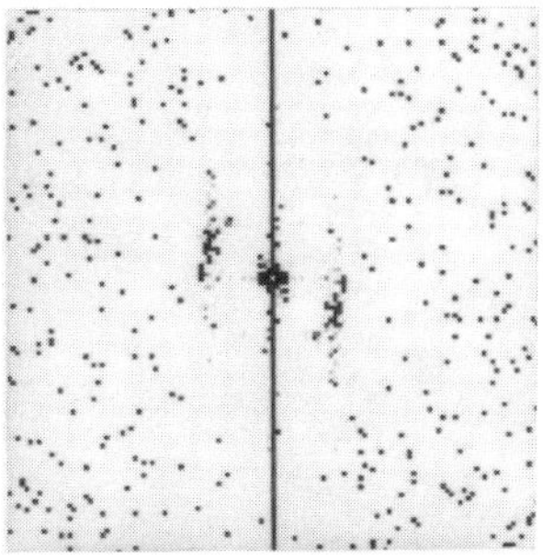

Figure 6. (b) 2-D Fourier transform of an image from Figure 6 (a). Position of arc corresponds to row spacing of 2.6±0.3 nm.

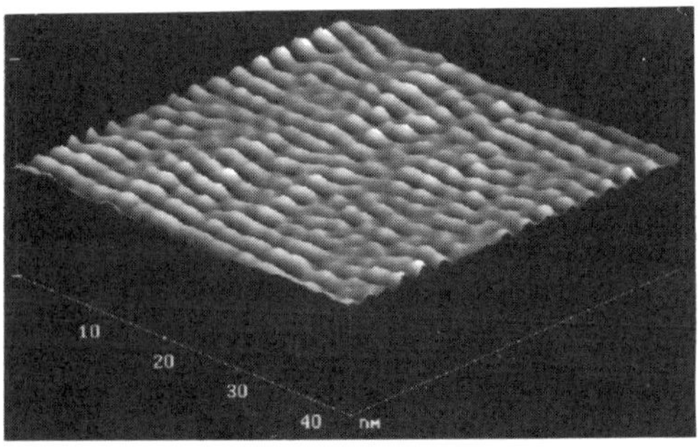

Figure 6. (c) Surface plot of digitally filtered image from Fig. 6a.

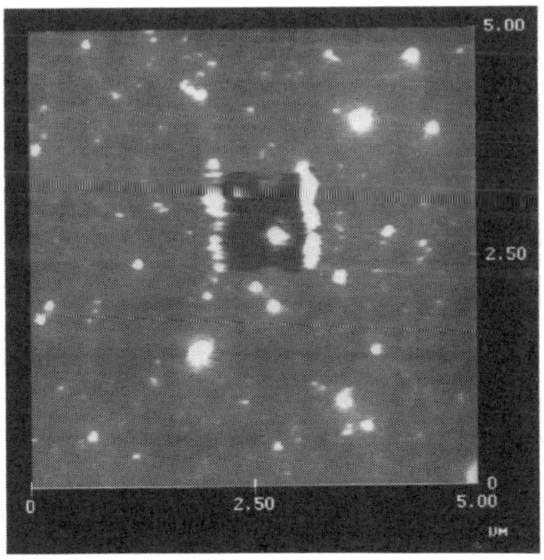

Figure 7. AFM image of a 5 μm by 5 μm area of a monolayer of "red" particles with the central square machined off with the tip. Compare with Figure 2 (a) ("yellow" particles) to see the effect of broad size distribution of "red" particles on film morphology.

the optimum film thickness was about three-four layers of nanoclusters. The inability to introduce orientation in thinner films may be related to surface tension effects leading to different film behavior after disruption (compare Figure 1 with Figure 3). Another important factor may be the need for lateral confinement of the pool of nanoclusters during the orientation, which is easily satisfied if the thickness of the intact film surrounding the hole is equal to several nanocluster diameters. The lack of tip-induced orientation in thicker films is most likely related to an inability to micromachine down to mica surface, thus indicating indirectly the need for surface flatness in producing the 2-D ordered arrays of nanoclusters.

"Red" Particles

A similar set of experiments has been carried out with "red" particles. Both micromaching and (to a lesser extent) nanomachining effects were observed, indicating that, contrary to "yellow" particles, the "red" ones were characterized by a very broad size distribution. An image of the film spin-coated from solution of concentration below which continuous films could no longer be formed is shown in Figure 7.

The smallest film thickness measured in this case was similar to one observed for "yellow" particles. However, as can be seen, the film was very nonuniform, indicating the presence of particles as large as 20 nm. We have also attempted to produce oriented patterns under conditions identical as ones for "yellow" particles. In general, producing the oriented patterns of "red" particles proved to be much more difficult, and orientation, if any, could be detected only by examination of 2-D Fourier transforms of images. An example of this can be seen in Figures 8 (a) and (b) showing respectively the LFM image of 90 nm by 90 nm area of the red particles film after attempted orientation, and its 2-D Fourier transform. The position of the broad maxima in the transformed image corresponds to a nanocluster size range from 3 nm to 15 nm, in agreement both with monolayer observations and earlier TEM experiments.[5]

Figure 8. (a) LFM image of the 90 nm by 90 nm portion of "red" particles film for which orientation was attempted under the conditions analogous to ones leading to oriented patterns of "yellow" particles as shown in Figure 5 (a).

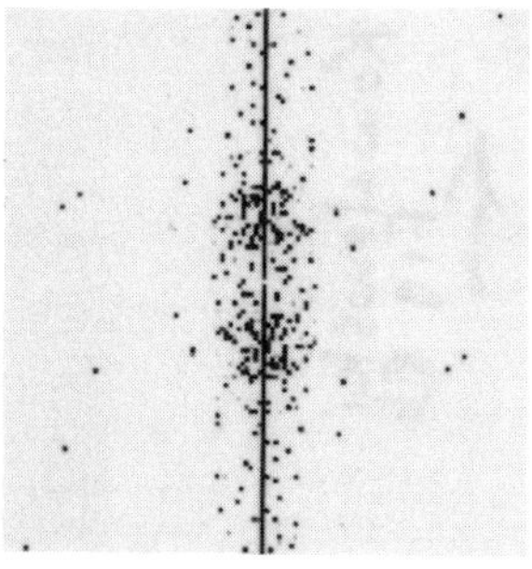

Figure 8. (b) 2-D Fourier transform of an image from Figure 8 (a). The pattern corresponds to broad distribution of spacings between the features parallel to scanning direction, reflecting the width of size distribution of "red" particles.

CONCLUSIONS

The primary goal of this work - determination of the size of semiconductor nanoclusters - falls into the challenging class of AFM problems of imaging "delicate" samples with corrugations of the size comparable with the tip radius. The major obstacles one has to overcome in such situation are tip-induced structural modifications of the sample and topography dependence of image contrast and resolution. In this study we have demonstrated how these factors can be used to the investigator's advantage. Micromachining effects resulting in exposure of the mica surface allowed us to fully utilize the vertical resolution of AFM and to obtain the nanocluster diameter from monolayer thickness measurements. Formation of ordered 2-D arrays of nanoclusters during prolonged scanning over semiliquid layers of nanoclusters formed after film disruption made possible molecularly resolved images in lateral force mode.

Using this approach we were able to confirm that titration-controlled reversed micelle method can be used to prepare narrow size distribution of semiconductor nanoclusters. At early titration stages ("yellow" particles) the size of nanoclusters depends on the amount of base necessary for reaction progress. Narrow size distributions observed in such case imply that this amount does not depend on the micelle size. When sufficient base is added to generate red-orange solutions, flocculation and precipitation occurs as the reversed micelles are apparently disrupted. Size control is lost at this point.

Results regarding film forming capabilities of nanoclusters and use of AFM for micromachining such films point to interesting possibilities of developing nanolithography and other nanotechnology applications. Monodisperse, spherical particles seem to be ideal building blocks for nanoscale structures, allowing, for example, micromachining sharp edged features of sizes well below 100 nm. Additional possibilities arise from the relative ease in varying capping agents for the nanoclusters, which should make it possible to control the cohesive and adhesive properties of the films and to alter them photo- or thermo-chemically. Such capabilities are especially desired in prospective nanolithographic applications.

ACKNOWLEDGMENTS

This work was supported by the Office of Naval Research (T.K.). Funding was provided to W.E.B. through an NSF Presidential Young Investigator award supported by the Monsanto Co. and the Eastman Kodak Co.

REFERENCES

1. M.A. Reed and W.P. Kirk, eds., Proc. Int. Symp.,"Nanostructure Physics and Fabrication," College station, TX, (1989).
2. G.A. Ozin, Nanochemistry: synthesis in diminishing dimensions, *Advanced Mater.* 4:612-649 (1992).
3. M.L. Steigerwald, A.P. Alivisatos, J. M. Gibson, T.D. Harris, R. Kortan, A.J. Muller, A.M. Thayer, T.M. Duncan, D.C. Douglass, and L.E. Brus, Surface derivatization and isolation of semiconductor cluster molecules, *J.Am.Chem.Soc.* 110:3046-3050 (1988).
4. J.H. Fendler, Atomic and molecular clusters in membrane mimetic chemistry, *Chem.Rev.* 87:877-899.
5. M.A. Matchett, T. Kowalewski, A.M. Viano, P.C. Gibbons, N.L. Adolphi, M.S. Conradi, W. E. Buhro, Reversed-micelle synthesis of arenethiolate-capped cadmium phosphide nanoclusters, *Chem. Mater.* (in review).
6. Obtained from Digital Instruments, Santa Barbara, CA.
7. G.F. Meyers, B.M. DeKoven, J.T. Seitz, Is the molecular surface of polystyrene really glassy? *Langmuir,* 8:2330-2335 (1992).

COMPARATIVE STUDY OF THE SURFACE ROUGHNESS OF OXIDE THIN FILMS

Karl H. Guenther,[1] Jean M. Bennett,[2] Frank K. Urban, III,[3] Milad F. Tabet,[3] Marc von Gunten,[4] Ron Bevis,[4] and Mohammad M. Tehrani[5]

[1]Department of Electrical and Computer Engineering
Center for Research in Electro-Optics and Lasers
University of Central Florida, Orlando FL 32826-3271 ; deceased, 1994
[2]Research Department, Naval Air Warfare Center
China Lake, CA 93555
[3]Department of Electrical and Computer Engineering
Florida International University
Miami, FL 33199
[4]Spectra Physics Components and Accessories Group
Mountain View, CA 94039-5017
[5]Litton Guidance and Control Systems
Woodland Hills, CA 91367

Abstract: Reactive low voltage ion plating (RLVIP) is based on plasma assisted electron beam evaporation of metallic starting material that is ionized and reacts with ionized and exited oxygen upon arriving at the surface to be coated. In earlier experiments, the roughness of the coating surface remained in the 0.2 to 0.6 nm root mean square (rms) range even for rather thick multilayer coatings when measured with an optical surface profilometer or a contacting stylus instrument. Contrary to this observation, films made with conventional electron beam evaporation have surface roughness increasing with film thickness. Smooth surfaces are of considerable interest for any optical coating because of low light scattering.

We investigatea the surface of one silica and zirconia single layer coating each using atomic force microscopes (AFMs) of the types: WYKO MicroProbe 3D, Park Scientific AutoProbe (at Spectra Physics), and Digital Instruments NanoScope II (at FIU). A contacting stylus-type profiler (Talystep, at NAWC), and a noncontact optical heterodyne surface profilometer (WYKO TOPO 2-D, at NAWC) were used for comparison. First, AFM measurements using an experimental AFM (at WYKO) and the AutoProbe were repeated with the same samples after they were sized to fit the NanoScope II. For the repeating measurements at WYKO, a new MicroProbe 3D was used. The measurements yielded results that were consistent for each individual AFM, but fairly inconsistent in comparison.

Atomic Force Microscopy/Scanning Tunneling Microscopy
Edited by S.H. Cohen *et al.*, Plenum Press, New York, 1994

Talystep measurements compared in part with AFM measurements although different spatial wavelengths are contained in the surface profiles. The WYKO optical profiler averages over areas at least 0.5 μm in diameter and does not resolve short spatial wavelengths, yielding ≈ 0.4 nm rms roughness for the substrate and coating surfaces alike.

In conclusion, we compared classical profiling of smooth RLVIP coatings with AFM measurements using three different instruments. Optical profiling renders coating roughness comparable to that of uncoated substrate surfaces, due to limited spatial resolution. Stylus profiling and AFMs include shorter spatial wavelengths, yielding larger coating roughness. This test showed also that different AFMs give different results.

INTRODUCTION

Surface roughness of optical thin films and substrates is of critical importance in many applications, particularly those involving laser radiation, as it causes light scattering[1] that can be detrimental to the functioning of a particular optical element or system. This importance is reflected in the vast literature on surface finish and its measurement on which a comprehensive review and collection has been compiled recently.[2]

In many optical shops, the surface quality of optical elements is still assessed visually, perhaps with the aid of a microscope (a Nomarski differential interference contrast microscope is particularly useful for this purpose). However, the quantification of this "quality" in terms of certain light scattering criteria[3] or surface roughness values using sensitive instrumental techniques[4] becomes increasingly critical for very low scatter optics and coatings (< 10 ppm of total integrated scattering, or TIS). In particular, the rms roughness δ has become a widely used measure for the quality of surfaces. However, one has to keep in mind that the value of δ depends on the bandwidth limits of the measuring technique.[5]

Thin film coatings, used for a large variety of optical functions such as anti-reflecting, high reflecting, beam splitting, polarizing, etc.,[6] have been applied to optical surfaces mostly by physical vapor deposition techniques, such as electron beam evaporation, or various sputtering processes. Conventional processes of this kind generate thin films with a columnar microstructure whose surface roughness increases with film thickness.[7] In optical multilayers the surface roughness of individual films can be correlated, uncorrelated, or partially correlated, and more or less increases with the number of layers, giving rise to various small angle and large angle light scattering effects.[8]

The cause of this undesirable microstructure and resulting surface roughness is the low energy (≈ 0.1 eV) of atoms and molecules deposited with conventional physical vapor deposition processes. It does not allow the atoms and molecules to fill all available sites completely because of limited surface mobility, resulting in so-called self-shadowing. By contrast, energetic particle deposition, a term used for thin film deposition processes that provide an average particle energy of several eV (the temperature equivalent of some 10,000 K), allows for high surface mobility that is equivalent to the formation of a transient liquid film even for high refractory oxides. As the actual (or physical) substrate temperature of the substrate is a few hundred K at most and thus "cold" with respect to the "temperature" of the transient liquid film, that film gets "frozen" (quenched) instantaneously, preserving the dense amorphous structure of the "liquid" - i.e., becoming a vitreous film.[9]

Processes that provide conditions for the formation of dense, amorphous films include reactive low voltage ion plating (RLVIP),[10,11] ionized cluster beam deposition (ICBD),[12] and ion beam sputtering (IBS), capable of generating thin films with surface roughness $\delta \leq 0.2$ nm rms and, correspondingly, TIS < 10 ppm.[13]

The purpose of this study thus is threefold. First and foremost, we wanted to investigate a presumed smoothing effect[14] expected from the above mentioned transient liquid film formation in energetic particle deposition processes. Secondly, we wanted to compare the results of surface roughness measurements obtained with established surface profilers, an optical heterodyne interferometric profiler (WYKO TOPO 2-D) and a contacting stylus-type profiler (Talystep) with measurements performed with atomic force microscopes (AFMs) at near-atomic resolution. Third, this study aimed at comparing three different AFMs and exchanging experience obtained with measuring smooth optical surfaces.

EXPERIMENTS

Sample Preparation and Sequence of Measurements

Zerodur substrates of about 1" diameter and about ¼" thick were fabricated at Litton Guidance and Control Systems. Such substrates are normally polished extremely well when used for ring laser gyro mirrors. To investigate the presumed smoothing effect mentioned in the Introduction, the polishing process used with the substrates for this investigation was terminated before the final smoothness was reached, to provide for a nominal roughness of about 0.6 nm rms.

The substrates were coated in July 1992 at CREOL using commercially available high vacuum coating equipment of the Balzers BAP 800 type that is specifically equipped for RLVIP.[15] For this initial investigation, we limited the selection of coating materials to SiO_2 and ZrO_2. Both oxides have been used extensively for a variety of coatings in the thin film laboratory at CREOL, and have shown excellent smoothness even for rather thick multilayers (such as a reflector for 1064 nm wavelength consisting of 12 pairs of SiO_2 and ZrO_2 thin films of quarter-wave optical thickness each).[16] For coating the substrates, we used holders that masked half of the surface so as to have both uncoated and coated areas available for surface roughness measurements on the same piece (Figure 1). The samples

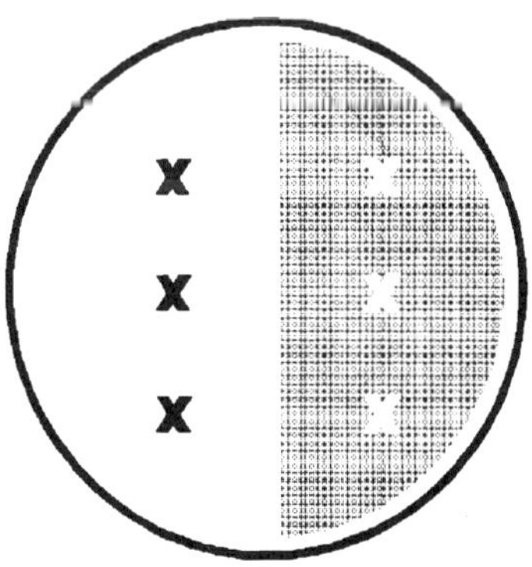

Figure 1. Semicoated sample surface. "x" indicates typical measurement positions for the uncoated (white) and coated areas (gray).

used for this round-robin measurement experiment are listed in Table 1(a). For brevity, they are referred to here as 706 and 716.

Table 1. (a) Samples investigated

Sample ID	Coating	Thickness
920706	SiO_2	350 nm
920716	ZrO_2	248 nm

After coating at CREOL, the samples were sent to the coinvestigators at the participating laboratories for their specific investigations. The "chain of events" is listed in Table 1(b), where the measurement techniques are actually listed in ascending order of vertical and horizontal resolution, i.e., the optical profiler before the mechanical profiler and that before the AFMs (those are listed in chronological order). This listing is consistent with the descriptions of the instruments and the presentation of experimental results.

Table 1. (b) Processing history of samples

Process	Instrument	Laboratory
Substrate manufacturing	Grinder, Polisher	Litton G&CS
Coating	BAP 800	CREOL
Optical profilometry	WYKO TOPO 2D	NAWC
Contacting profilometry	Talystep	NAWC
Atomic Force Microscopy	WYKO Experimental	WYKO
Atomic Force Microscopy	Park Scientific AutoProbe	Spectra Physics
Resizing	Lathe	Spectra Physics
Atomic Force Microscopy	Park Scientific AutoProbe	Spectra Physics
Atomic Force Microscopy	Digital Instruments NanoScope II	ECE/FIU
Atomic Force Microscopy	WYKO MicroProbe 3D	WYKO

For shipping, we used a double-hemisphere type packaging of clear plastic that prevented the surfaces of the samples from touching the packing material. Care was taken during the whole round robin experiment not to contaminate the surfaces, and cleaning them was normally limited to blowing off any loose dust or lint that may have settled there before an actual measurement, except where noted differently.

Surface profiling of the samples started actually with an experimental AFM at WYKO in October 1992, followed by optical and mechanical profiling in January 1993 at NAWC. Next, the samples were measured at Spectra Physics with an AutoProbe AFM. When we learned that the NanoScope II at FIU would not accommodate the samples in their original 1" diameter, ¼" thickness size, the samples were cut down in size at Spectra Physics while protecting the coated surfaces with an acetone soluble strip coating. After removal of the strip coating, each coated surface received a drag wipe with reagent quality solvents (acetone, then methanol). Subsequently, the substrates and coatings were remeasured with he AutoProbe to observe any changes in surface roughness or topography caused by the resizing process.

The samples were then sent to FIU and measured there with a NanoScope II. When these measurements were completed, the coating and substrate surfaces were investigated again at the WYKO Corp., now with a MicroProbe 3D. To be sure that they were clean (after repeated measurements and shipping), the samples were put in an ultrasonic cleaner using acetone for the solvent, and agitated for 3 min. and 1 min. (ZrO_2 and SiO_2, respectively). Finally, the sample surfaces were drag wiped with Kodak lens tissue moistened with methanol.

Instruments Used for Surface Characterization

WYKO TOPO 2D. This optical profiler uses white light illumination to eliminate spurious fringes from stray reflections caused by laser light, and a 1024 element linear imaging array. A variety of objective lenses allows for surface profiles of different length, up to 5 mm. The calculated surface profile is repeatable to 0.01 nm rms under good environmental conditions.

Rank Taylor Hobson Talystep. This contacting stylus-type surface profilometer, manufactured by Rank Taylor Hobson in the UK., has been computer automated to measure optical surface roughness down to the < 0.1 nm range.[17]

Park Scientific AutoProbe™. This AFM is equipped with a 10 x 10 μm^2 scanner and high aspect-ratio tips (Ultralevers™); automated sample navigation and tip approach; autofocus, 3:1 zoom, oblique view optical microscope and integral 4-point pneumatic isolation stage. Sample sizes to 50 mm in diameter and 20 mm in height are accommodated.

Digital Instruments NanoScope II. This AFM has a 12 x 12 μm^2 scanner and automatic tip approach. It has a 45x monocular microscope, a top view AFM optical head that allows on-axis viewing of the probe and sample, and a 2 x 2 mm^2 sample positioning stage, which accommodates samples up to 1.5 cm in diameter.

WYKO MicroProbe 3D. This WYKO AFM is a completely new instrument. It has a lower noise level than the earlier model that was used initially in October 1992, ~ 0.05 nm (0.5 Å rms), and has many automatic features. (The earlier model is not described further in this paper but some of the measurements are included, for comparison, in Table 6 summarizing the results.) The sample can be translated under the probe tip using x and y translation motors. The probe is set on the surface automatically; no manual adjustments are necessary.

Measurement Details

WYKO TOPO 2D. System configuration:

Obj. magnification	41.1x
Profile length	321 µm
No. Average	4 Profiles
Method	Absolute
Wavelength	655.5 nm

Measurement configuration: Fringes were aligned parallel to the linear detector array and broadened such that the zero optical path difference fringe nearly filled the field of view. The roughness analysis is based on the TOPO-2D absolute roughness embedded algorithm. The sample is displaced (more than 25 μm) between initial and final measurement pairs of each profile. Subtraction of the profiles eliminates the contribution of the roughness on the reference surface.

Talystep

Stylus radius	0.8 μm
Stylus loading	1 mg
Profile length	200 μm
Stylus speed	3.070 μm/s
Data points	5323
Sampling interval	0.037 μm
Instrumental noise	0.05 nm (0.5 Å) rms roughness equivalent

Park Scientific AutoProbe™

Stylus	monolithic B-doped silicon cantilever and tip, radius 5 nm
Stylus loading	200 nN
Vertical resolution	~0.04 nm rms
Map areas	5 x 5 μm^2, 2 x 2 μm^2
Data points	512 x 512 for all areas

Digital Instruments NanoScope II

Stylus	silicon nitride tip, radius ≈ 30 - 50 nm
Stylus loading	typically 10 - 100 nN
Vertical resolution	~0.06 nm rms
Map areas	5 x 5 μm^2, 2 x 2 μm^2, 0.5 x 0.5 μm^2
Data points	400 x 400 for all areas

WYKO MicroProbe 3D

Stylus	silicon nitride pyramidal tip, radius ≈ 30-50 nm
Stylus loading	20 nN
Instrumental noise	~0.05 nm rms
Map areas	55 x 55 μm^2 cropped to 50 x 50 μm^2, 5.6 x 5.6 μm^2 cropped to 5 x 5 μm^2, 2.8 x 2.8 m^2 cropped to 2 x 2 μm^2
Data points	512 x 512 for all areas

RESULTS

Roughness Measurements Using the WYKO TOPO 2D Optical Profiler at NAWC

In Table 2 below appear the roughness measurements obtained with the WYKO TOPO 2D profiler at the Naval Air Warfare Center, China Lake, CA.

Surface Profiles Obtained with the Talystep Mechanical Profiler at NAWC

In Figure 2 (a) to (d) appear the surface profiles obtained using the Talystep mechanical profiler at the Naval Air Warfare Center, China Lake, CA.

Table 2. Rms roughness of bare (left) and coated (right) Zerodur sample surfaces

Sample	rms roughness (nm)	Avg	Sample	rms roughness (nm)	Avg
706-bare	0.32, 0.34, 0.4.3	0.37nm	706-SiO_2	0.38, 0.40, 0.39	0.39nm
716-bare	0.44, 0.34, 0.50	0.43nm	716-ZrO_2	0.42, 0.34, 0.37	0.38nm

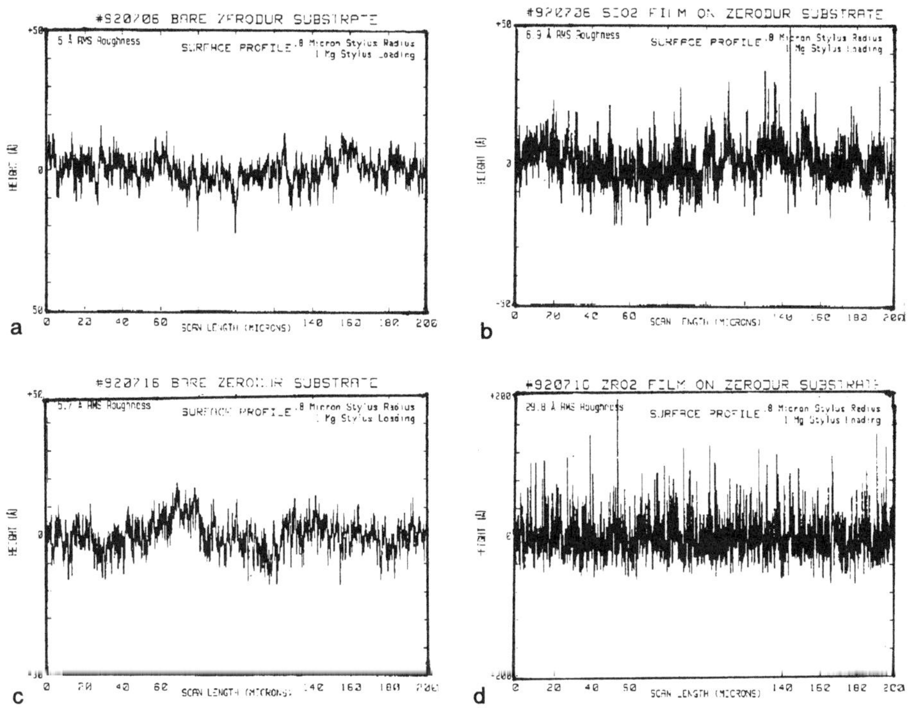

Figure 2. 200 µm long surface profiles for Sample 706 and Sample 716. Sample 706: (a) bare substrate (b) SiO_2 coating; height scale: -5 to +5 nm in both graphs. Sample 716: (c) bare substrate (d) ZrO_2 coating; height scale: - 5 to + 5 nm for the left scan, -20 to + 20 nm for the right scan.

Images and Data Taken with the Park Scientific AutoProbe AFM at Spectra Physics

The images (before resizing) and data taken at the Spectra Physics Components and Accessories Group, Mountain View, CA, appear in Figure 3 (a) to (f). The roughness values (nm) obtained follow in Table 3.

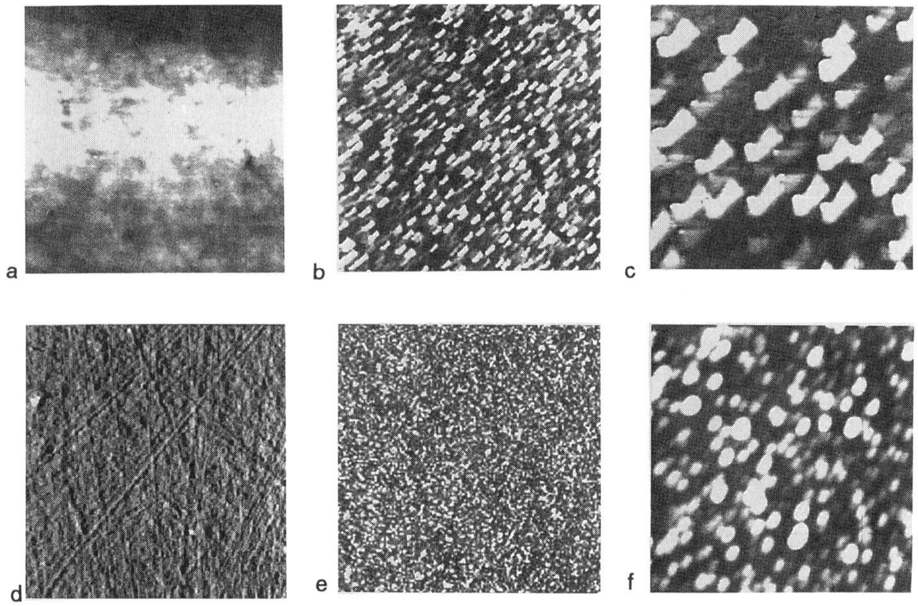

Figure 3. Sample 706: (a) bare substrate 2x2 μm^2 (b) SiO_2 coating 5x5 μm^2 (c) SiO_2 coating 2x2 μm^2. Sample 716: (d) bare substrate 5x5 μm^2 (e) ZrO_2 coating 8x8 μm^2 (f) ZrO_2 coating 2x2 μm^2.

Table 3. Roughness value (nm) obtained from AutoProbe measurements; scan areas are μm^2

Sample	Description	Scan	p-v roughness	rms roughness
Before Resizing	706-bare Substrate	5x5	17.2, 12.3	1.8, 1.9
		2x2	12.3, 11.6	1.5, 1.2
	706-SiO_2Coating	5x5	34.6, 31.8	2.6, 2.6
		2x2	38.4	4.4
	716-bare Substrate	5x5	16.9, 21.2	0.77, 0.78
		2x2	7.7, 5.0	0.74, 0.66
	716-ZrO_2Coating	5x5	106.6, 106	5.0, 5.0
		2x2	90.2	5.4
After Resizing	706-bare Substrate	2x2	12.3	0.93
	706-SiO_2 Coating	5x5	36.4, 27.2	3.7, 3.2
		2x2	31.8	4.6
	716-bare Substrate	2x2	8.9	1.1
	716-ZrO_2 Coating	5x5	44.1	4.8
		2x2	42.2	4.9

Surface Images Taken with the Digital Instruments NanoScope II AFM at FIU

In Figure 4 (a) to (i) appear the images taken at FIU, Miami, FL, with the Digital Instruments NanoScope II. The rms roughness values obtained follow in Table 4.

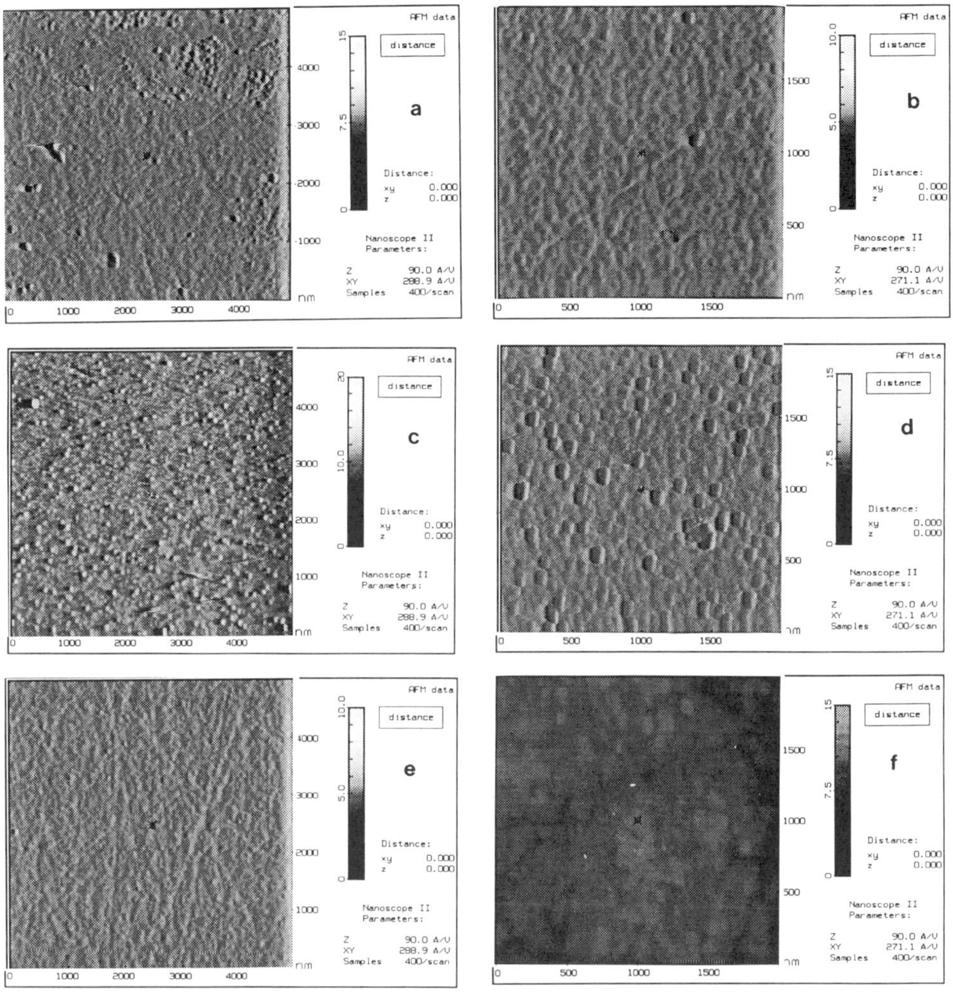

Figure 4. (a) - (f) Surface images taken with Digital Instruments NanoScope II AFM at FIU. Sample 706-bare, substrate surface (a) 5x5 μm^2 (b) 2x2 μm^2: Sample 706-SiO_2, coating surface (c) 5x5 μm^2 (d) 2x2 μm^2. Sample 716-bare, substrate surface (e) 5x5 μm^2 (f) 2x2 μm^2.

Table 4. Rms roughness values (nm) obtained with the NanoScope II

Sample	706-bare	706-SiO_2	716-bare	716-ZrO_2
2x2 μm^2	0.23	0.56	n.a.	2.47
5x5 μm^2	0.71	1.34	0.28	2.99

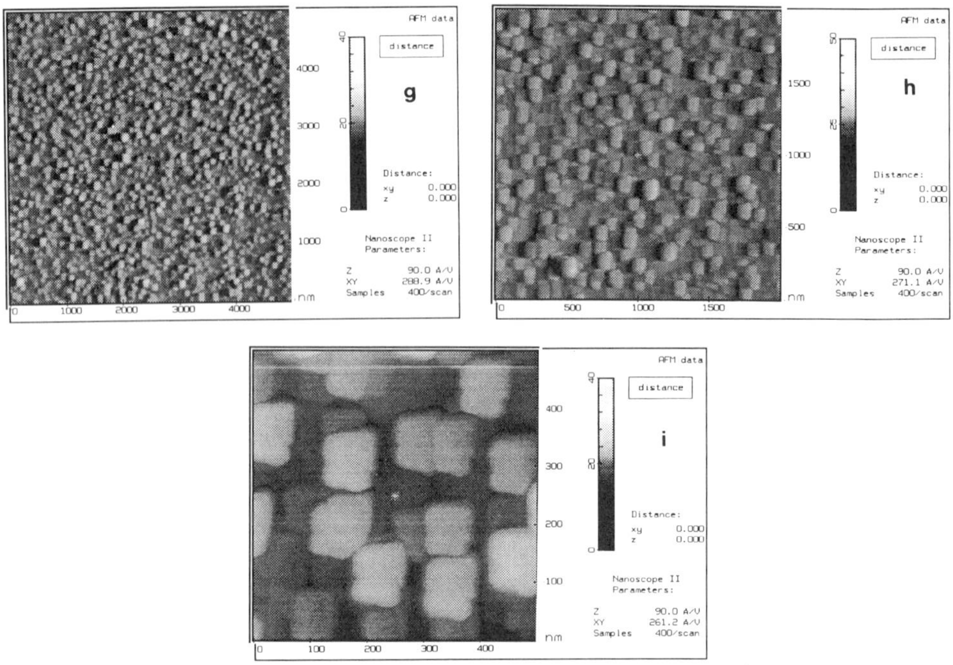

Figure 4. (g)-(i) Sample 716-ZrO_2, coating surface (g) 5x5 μm^2 (h) 2x2 μm^2 (i) 0.5x0.5 μm^2.

Images and Data Obtained with the WYKO MicroProbe 3D AFM at WYKO Corp.

The images in Figure 5 (a) to (h) below and Table 5 data that follow are from the WYKO MicroProbe 3D AFM at WYKO Corp., Tucson, AZ. The following three-dimensional images (Figure 6 (a) to (h)) have been created with the same data as the two-dimensional images in Figure 5.

DISCUSSION

When comparing classical surface profiling techniques with AFM measurements, resolution is obviously the key difference. However, it is interesting also to compare quantities proportional to the pressures exerted on the sample surface by the diamond stylus of the Talystep and by the pyramidal silicon nitride tip of the AFM in contacting mode. A value proportional to the contact force/radius^2 for the MicroProbe 3D is 1.25 mgf/μm^2. The corresponding value for the Talystep is 1.0 mgf/μm^2. Hence, to a first approximation, the ratio of pressure exerted by the styli of the two instruments is about unity.

The samples were first measured in October 1992 at WYKO using an experimental AFM in October of 1992. The last measurements of the sample surfaces were taken there

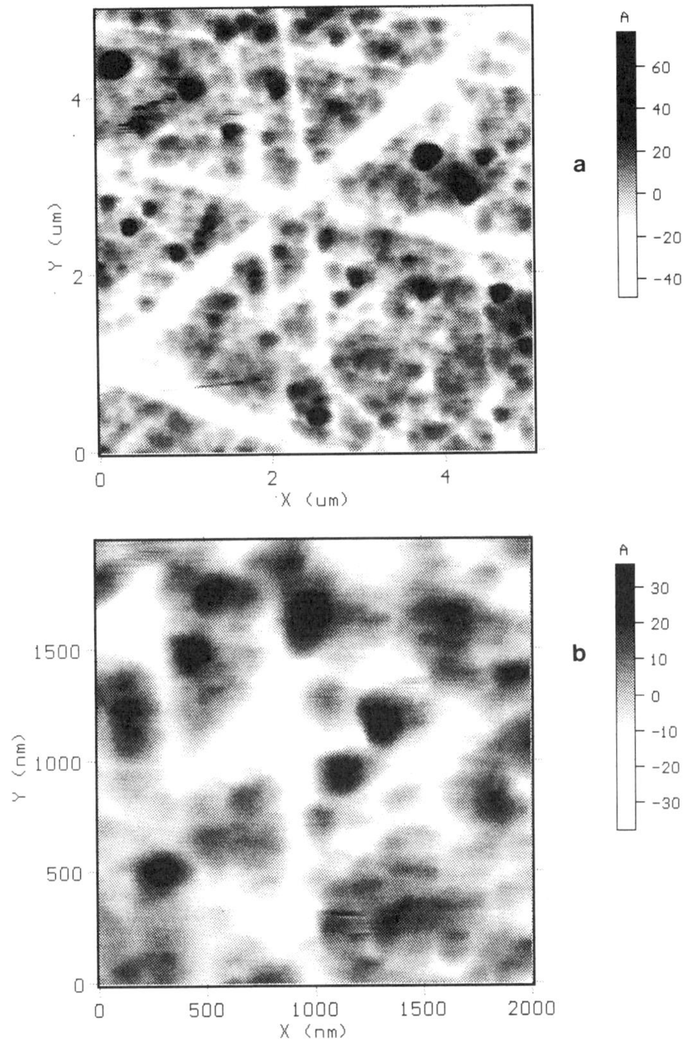

Figure 5. (a) - (h) Images and data obtained with the WYKO Probe 3D AFM at WYKO Corp. (a)-(b) Sample 706-bare, substrate surface: (a) 5x5 μm^2 (b) 2x2 μm^2.

as well, though using a new MicroProbe 3D AFM, after they had gone through that chain of events listed in Table 1 (b). It is particularly interesting to compare those initial and final measurements (Table 6) as they may best reflect any changes the samples might have gone through in that course of time. Note that the first measurements covered larger scan areas (10 x 10 and 50 x 50 μm^2). Three data buffers were taken for all measurements recorded above. They all looked identical and the roughness values calculated were within 0.1 nm rms of each other.

Measuring smaller areas of the samples using the AFMs showed that both of the films appear to be made up of cubes. The cubes in the ZrO_2 film arrange themselves so that some lie above others on the surface. Thus, the roughness measured for the smaller areas can

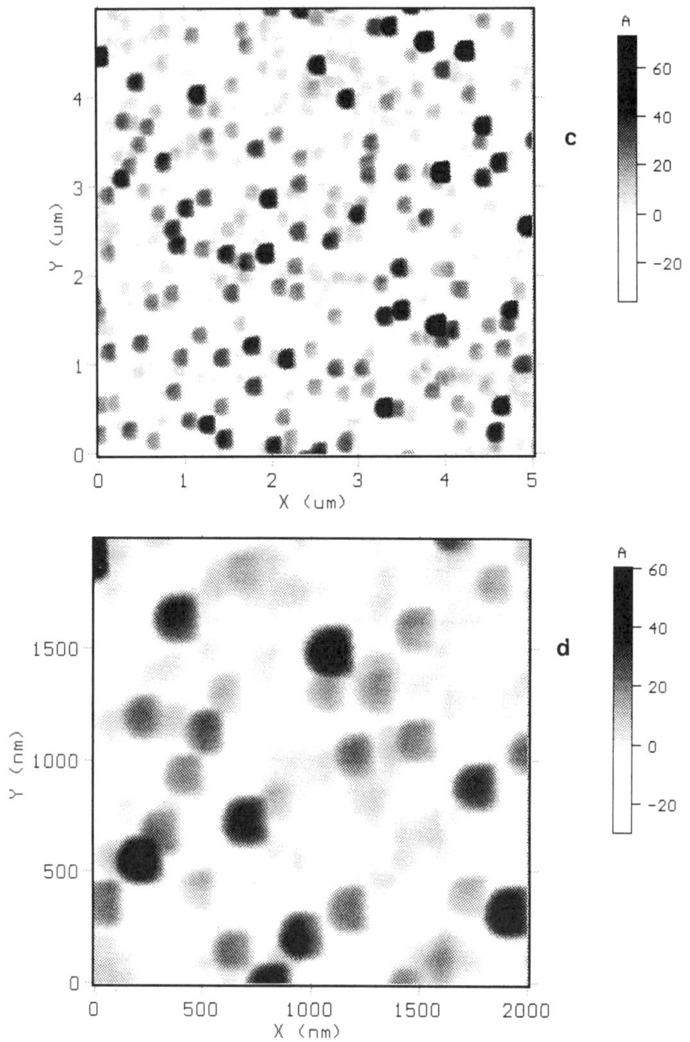

Figure 5. (c)-(d) Sample 706-SiO_2, coating surface: (c) 5x5 μm^2 (d) 2x2 μm^2.

vary considerably depending on which cubes are included in the picture. The effect is less pronounced for the SiO_2 film.

The Zerodur substrate under the SiO_2 film appears to be much rougher than the Zerodur substrate under the ZrO_2 film. It indeed has bumps in it as shown in Figure 5 (a), (b). There were many defects on the substrate that appeared in the 50 x 50 μm^2 scans (not shown in this paper). These were avoided in the smaller area scans. However, the roughness for the 5 x 5 μm^2 areas (1.28 nm rms) was consistent with that measured previously for the 10x10 μm^2 areas (1.16 nm rms). The roughness of the Zerodur substrate

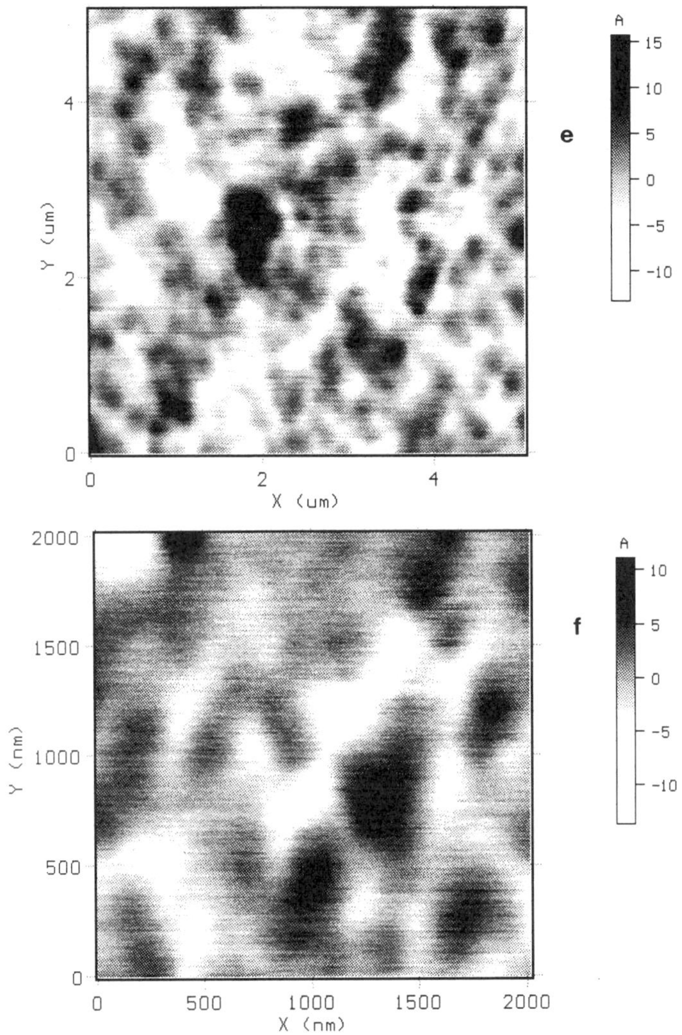

Figure 5. (e)-(f) Sample 716-bare, substrate surface: (e) 5x5 μm^2 (f) 2x2 μm^2.

under the ZrO_2 film for the 5 x 5 μm^2 areas (0.40 nm rms) was consistent with the previous measurements on a 10 x 10 μm^2 area (0.65 nm rms). The surface was also lumpy, as shown in Figure 5 (e)(f).

The roughness of the ZrO_2 coated Zerodur was 4.27 nm rms for a 5 x 5 μm^2 area at the last measurement, as compared to 3.30 nm rms obtained previously for a 10 x 10 μm^2 area. Thus, this film appears rougher than before; possibly there has been some recrystallization. The major disagreement between the two measurements is with the results for the SiO_2 film. In the first measurement, the roughness for a 10 x 10 μm^2 area was 4.71 nm rms, while the last time the roughness for a 5 x 5 μm^2 area is 1.42 nm rms. Also, the character of the film roughness has changed from being irregular pyramidal-shaped lumps to cubes similar to the ZrO_2 film (see Figure 5(c),(d)).

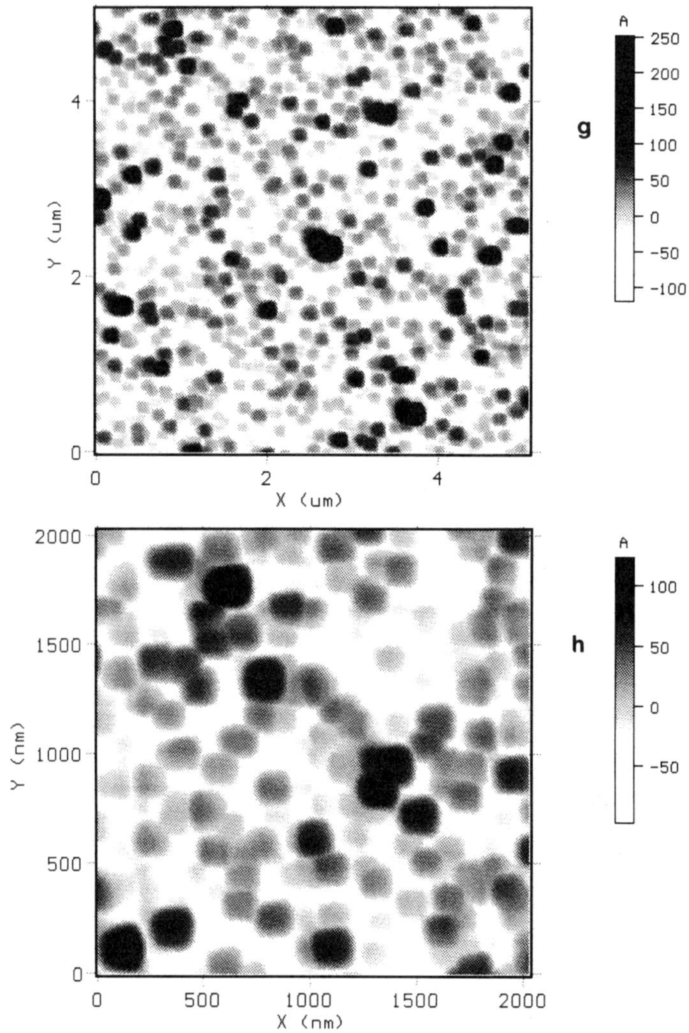

Figure 5. (g)-(h) Sample 716-ZrO_2, coating surface: (g) 5x5 μm^2 (h) 2x2 μm^2.

Table 5. Rms roughness values (nm) obtained with the MicroProbe 3D

Sample	706-bare	706-SiO_2	716-bare	716-ZrO_2
2x2 μm^2	0.91	1.53	0.32	4.36
5x5 μm^2	1.28	1.42	0.40	4.27

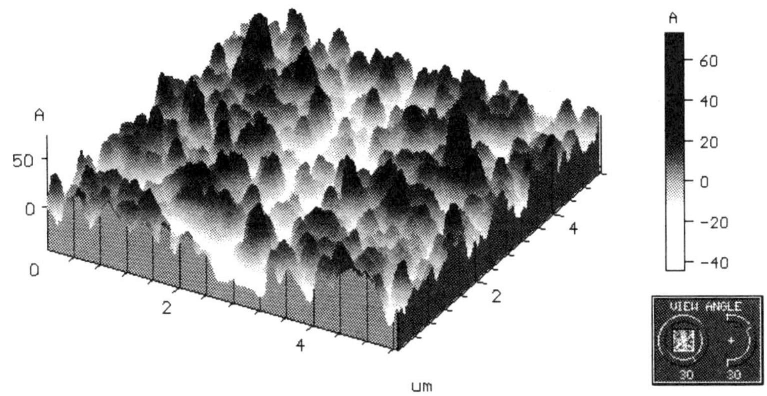

Figure 6. (a)-(h) Three-dimensional images created with same data as in Figure 5. (a) Sample 706-bare, substrate surface: 5.5 μm^2.

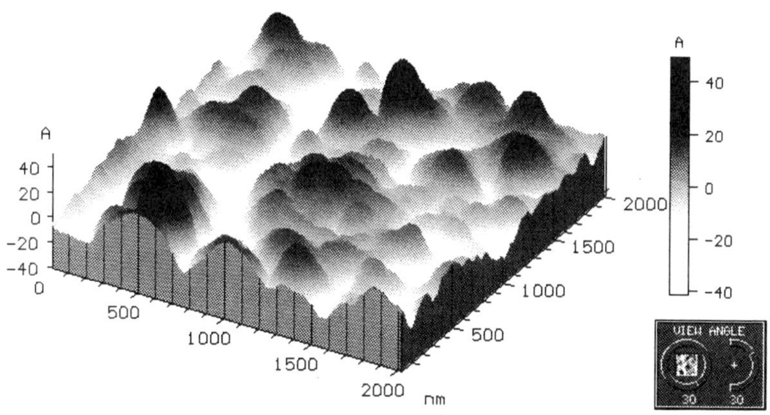

Figure 6. (b) Sample 706-bare, substrate surface: 2x2 μm^2.

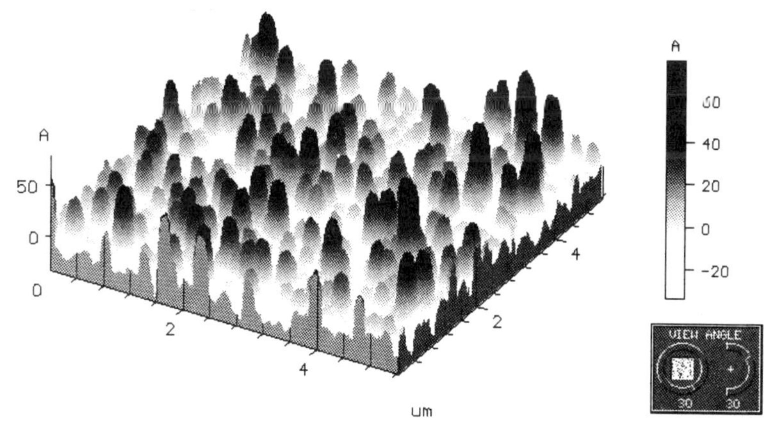

Figure 6. (c) Sample 706-SiO_2, coating surface: 5.5 μm^2.

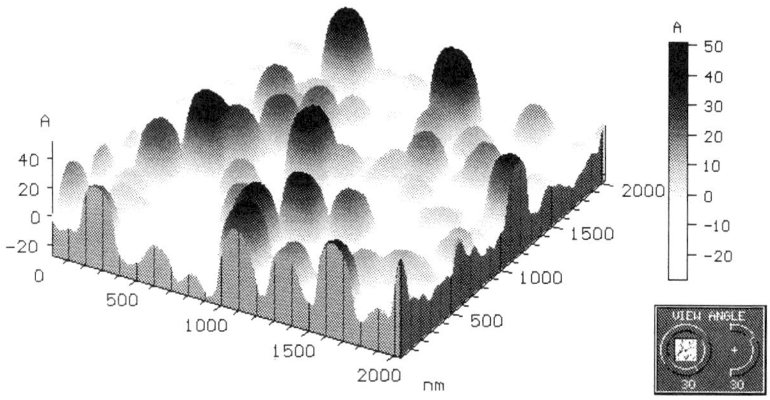

Figure 6. (d) Sample 706-SiO_2, coating surface: 2.2 μm^2.

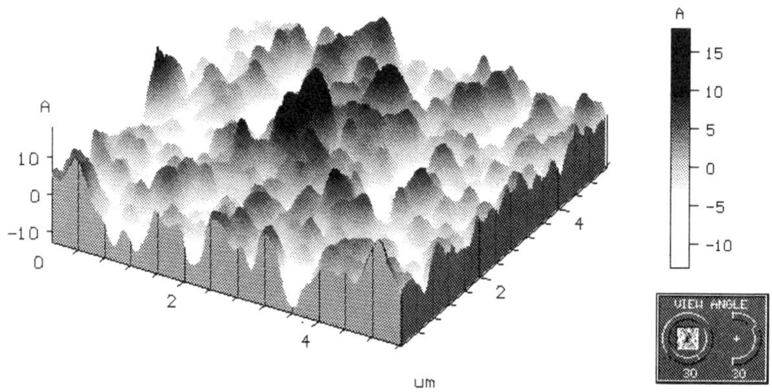

Figure 6. (e) Sample 716-bare, substrate surface: 5.5 μm^2

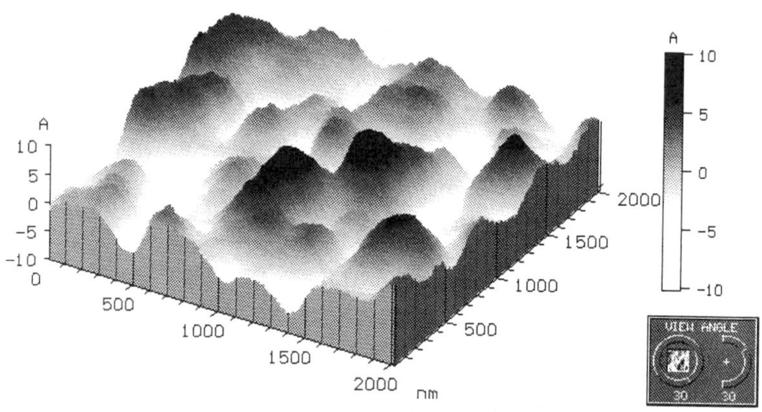

Figure 6. (f) Sample 716-bare, substrate surface: 2.2 μm^2

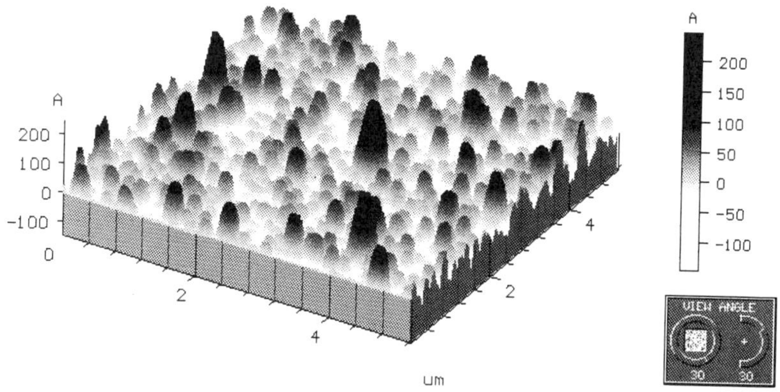

Figure 6. (g) Sample 716-ZrO_2 coating surface: 5x5 μm^2

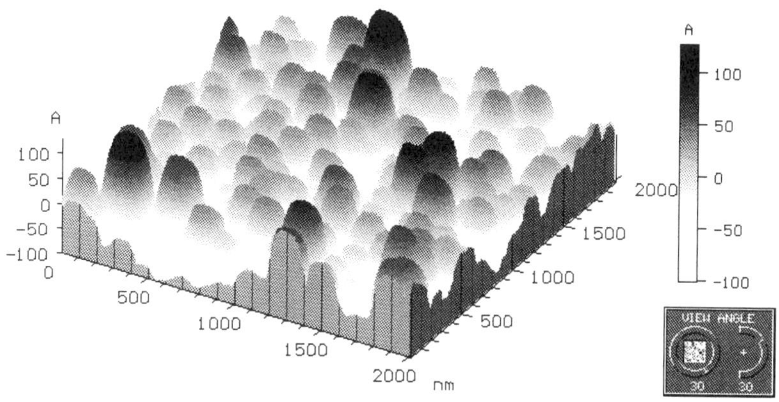

Figure 6. (h) Sample 716-ZrO_2 coating surface: 2x2 μm^2

Table 6. Summary of average rms roughness values (nm) obtained from this study

Instrument	Date of measurement	Scan (μm) or (μm^2)	Sample 706 Bare surface	Sample 706 SiO_2 coating	Sample 716 Bare surface	Sample 716 ZrO_2 coating
TOPO 2D	1/8/93	≈ 320	0.37	0.39	0..43	0.38
Talystep	1/6/93	200	0.52	0.67	0.56	2.96
WYKO AFM	10/14-15/92	50x50	0.94	6.02	0.55	2.76
(experimental)		10x10	1.16	4.71	0..65	3.30
AutoProbe, before	3/23/93	5x5	0.18	2.60	0.77	5.00
& after resizing	3/24/93	5x5	0.93	3.45	1.10	4.85
NanoScope	3/27/93	5x5	0.71	1.34	0.28	2.99
MicroProbe 3D	4/8-9/93	50x50	3.77	2.34	0.84	4.69
		5x5	1.28	1.42	0.40	4.27

Although the number of measurements taken with each of the instruments available for this study is small due to time constrains, we can say that roughness values (rms) obtained by the three different AFMs vary by about a factor of two to four for the same sample and scan area. Measurements taken with any AFM were reasonably self-consistent for the two scan areas, 2x2 and 5x5 μm^2. Between the NanoScope II and the MicroProbe 3D, the surface features found particularly on the ZrO_2 film look like cubes in the first case while they appear more rounded in the 3D image of the latter. In 2D surface images taken with the MicroProbe 3D (not shown here), they appeared also more like cubes. This surface feature may be indicative of cubic zirconia.

When comparing Talystep surface profiles with AFM measurements that have considerably better lateral resolution, we note that the roughness measured on the Zerodur substrates is primarily longer surface spatial wavelength components for the Talystep profiles, compared to the granular structure measured by the AFMs. The latter structure is not resolved by the Talystep stylus.

The SiO_2 film is apparently slightly "lumpy" in addition to having the large "blocky" structure shown in the AFM 3-D pictures. This latter structure is barely resolved by the Talystep stylus. The spikes on the Talystep profiles are isolated particles deposited on top of the main film and avoided in the AFM measurements.

The ZrO_2 film structure measured by the Talystep profiler is entirely caused by the particles that were avoided or truncated in the AFM data. The closely spaced film structure seen in the AFM 3D pictures was not resolved by the Talystep stylus.

The WYKO TOPO 2D optical profiler averages over areas at least 0.5 μm in diameter and does not resolve short spatial wavelength features, yielding $\approx$ 0.4 nm rms roughness for both the substrate and coated surfaces. This difference in measurement results of surface roughness that clearly depends on the lateral resolution of the profiling instrument raises the question which value is really important for the function of an optical coating or surface. Assuming that it is an interaction with light similar to that occurring in the optical surface profiler, its values are certainly meaningful in this context. Of course, for fundamental investigations into the mechanisms causing surface roughness, the near-atomic resolution as provided by today's AFMs and other scanning probe microscopes is most desirable.

CONCLUSIONS

In conclusion, we have measured the roughness of two partially coated Zerodur samples, both on the bare and the coated surfaces, using an optical and a mechanical profilometer and three AFMs. While the optical profiler yielded comparable (small) roughness values for the uncoated and coated surfaces, the AFM measurements do not support the previously made assumption of a smoothing capability of thin films deposited by reactive low voltage ion plating. The AFM measurements are reasonably self-consistent for each instrument but lack good agreement between them.

ACKNOWLEDGMENT

We wish to thank Tami L. Balter, John C. Podlesny, and Jay Jahanmir of the WYKO Corporation for providing the MicroProbe 3D pictures and data shown here.

REFERENCES

1. J.M. Bennett, and L. Mattson, "Introduction to Surface Roughness and Scattering," Opt. Soc. Am., Washington, D.C. (1989).

2. J.M. Bennett, Surface finish and its measurement, "Collected Works in Optics," Vol. 2, Opt.Soc. Am.,Washington, D.C. (1992).
3. K.H.Guenther, J.A. McCandless, and F.D.Orazio, Jr., Correlation of light scattering measurements and visual ranking of optical surfaces, *Appl. Opt.* 32 (1993).
4. K.H. Guenther, P.G. Wierer, and J.M. Bennett, Surface roughness measurement of low scatter mirrors and roughness standards, *Appl. Opt.* 23:3820-3833 (1984).
5. E.L. Church, and P.Z. Takacs, Instrumental effects in surface finish measurement, *In*:"Surface Measurement and Characterization," J. M. Bennett, ed., Proc. SPIE 1009, 46-55 (SPIE - The International Society of Optical Engineering, Bellingham, WA) (1989).
6. ISO 9211, "Optical Coatings," Draft International Standard (International Standardization Organization, Geneva, Switzerland).
7. K.H. Guenther, The microstructure of vapor-deposited optical coatings, *Appl. Opt.* 23:3806-3815 (1984).
8. K.H. Guenther, H.L. Gruber, and H.K. Pulker, Morphology and light scattering of dielectric multilayer systems, *Thin Solid Films* 34:363-367 (1976).
9. K.H. Guenther, Optical thin films deposited by energetic particle processes, *In*: "Thin Films for Optical Systems," K. H. Guenther, ed., Proc. SPIE 1782, 344-355 (SPIE - The International Society of Optical Engineering, Bellingham, WA).
10. K.H. Guenther, Recent advances in optical coating technology: reactive ion plating deposition of Optical coatings, *In:*"Optical Thin Films and Applications," R. Herrmann, ed., Proc. SPIE 1270, 211-221 (SPIE - The International Society of Optical Engineering, Bellingham, WA) (1990).
11. K. H.Guenther, Recent Advances in Reactive Ion Plating Deposition. *In:* Optical Thin Films III: New Developments," I. R. Seddon, ed., Proc. SPIE 1323, 29-39 (SPIE - The International Society of Optical Engineering, Bellingham, WA).
12. T. Takagi, "Ionized Cluster Beam Deposition and Epitaxy," Noyes Publications, Park Ridge, New Jersey (1988).
13. D.T. Wei, Ion beam interference coating for ultralow optical loss, *Appl. Opt.* 28:2813-2819.
14. K.H. Guenther, M.M. Tehrani, and J.M. Bennett, Coating and substrate surface roughness, *In*: "OSA Annual Meeting Technical Digest," *Opt. Soc. Am.*, 58 (1991).
15. K.H. Guenther, D. Boon Loo, D. Burns, J. Edgell, D. Windham, and K.-H. Muller, Microstructure analysis of thin films deposited by reactive evaporation and by reactive ion-plating, *J. Vac. Sci.Tech.* A 7, 1436-1445.
16. F.J. Boero, R.A. Chipman, and K.H. Guenther, Smoothing of optical surfaces with dielectric thin films produced by reactive ion plating deposition, *In:* "Laser Induced Damage in Optical Materials: 1988," H. E. Bennett, A. H. Guenther, B. E. Newnam, and M. J. Soileau, eds., NIST Spec. Publ., U. S. Dept. of Commerce, 775, 339-347 (1989).
17. J.M. Bennett and J.H. Dancy, Stylus profiling instrument for measuring statistical properties of smooth optical surfaces, *Appl. Opt.* 20, 1785-1802 (1981).

METHODOLOGIES AND TECHNIQUES

APPLICATIONS OF A COMBINED SCANNING TUNNELING MICROSCOPE AND QUARTZ MICROBALANCE

Chris Daly and Jacqueline Krim

Physics Department
Northeastern University
Boston, MA 02115

Abstract: It has recently become possible to carry out sliding friction measurements of adsorbed monolayers by means of a quartz crystal microbalance technique. These ongoing studies have revealed a range of applications that might result by combining the shaking action of the quartz microbalance with the imaging and patterning capabilities of a scanning tunneling microscope. We describe here how such an apparatus can be constructed and discuss a number of potential applications.

INTRODUCTION

The advent of new experimental techniques capable of studying the force of friction which results when a finite number of atoms is made to slide on a crystalline substrate has led to the emergence of a new field involving friction studies at nanometer length scales (nanotribology). Atomic force microscopes[1] have been utilized to study friction, adhesion and lubrication for surface contacts as small as a single atom[2,3] and have also been configured for nanoindentation experiments capable of nanonewton force resolution and hundreths of nanometer depth resolution.[4-6] Surface force apparatus has meanwhile been modified to allow studies of the shear forces between two crystalline mica surfaces, either in direct contact or separated by molecularly thin liquids.[7-9]

A quartz crystal microbalance (QCM) technique has recently been developed to measure the sliding friction coefficients of atomically thin adsorbed films,[10-12] and data for Xe monolayers on silver substrates are in quantitative[13] agreement with theoretical prediction based on electronic friction considerations.[14]

Other theoretical studies of atomic-scale friction include a calculation of phonon-induced friction levels for steady-state shearing of layered compounds,[15] and the use of a local density approximation to calculate the total energy of a palladium atom moving on a graphite surface as a function of an externally applied normal force.[16] Recent molecular dynamics simulations[17-19] include studies of slippage at fluid-fluid, fluid-solid, and solid-solid interfaces and contact points.

These research activities have, for the most part, developed independently and spontaneously in terms of both the field of nanotribology and also standard tribology research. As the field develops, it is anticipated that the various experimental, theoretical and computational tools will be used in conjunction to further speed progress. In this spirit, we describe here how a QCM can be constructed so as to operate in conjunction with a scanning tunneling microscope (STM), to combine the shaking action of the QCM with the imaging and patterning capabilities of the STM. A number of applications that may result from this apparatus are then discussed.

SLIDING FRICTION MEASUREMENTS WITH A QCM

A quartz crystal microbalance consists of a single crystal of quartz which oscillates in transverse shear motion with a quality factor Q near 10^5. The driving force (supplied by an external oscillator circuit) typically has constant magnitude and is periodic with frequency f = 4 to 10 MHz, the series resonant frequency of the oscillator. Each major face is plated with a metal electrode which, for our Xe monolayer experiments, is prepared by thermal evaporation in ultra-high vacuum (UHV) and can be characterized *in situ* with Auger electron spectroscopy, and scanning tunneling microscopy (STM).

Sliding friction measurements of Xe monolayers are carried out by transferring the sample, within the ultra-high vacuum chamber, to a tip whose temperature can be regulated between 80 and 155K. At these temperatures, condensation of Xe is in the form of two or more layers followed by the condensation of bulk crystallites.[20,21] Since the van der Waals adsorption energy of the second monolayer is significantly lower than that of the first, the second layer does not begin to form until the first layer is complete.[22] Van der Waals forces perpendicular to the surface are 10^{14} times stronger than any corresponding gravitational forces. "Popping up" out of the plane due to the shaking action of the substrate is therefore not expected to occur as it might in a system governed by gravitational forces. Strong perpendicular forces do not, however, prevent horizontal motion of the film, which is expected to be governed by substrate corrugation effects rather than the absolute magnitude of the van der Waals force.[14]

Adsorption onto the microbalance electrodes produces shifts in both the frequency and the amplitude of vibration, which are simultaneously recorded as a function of pressure. Amplitude shifts are due to frictional shear forces, which are exerted on the surface electrode by the adsorbed film (or alternatively by an STM tip or a three dimensional vapor). Interfacial friction coefficients η_2 are deduced from the relations:

$$\delta(Q^{-1}) = 4\pi\tau\delta f \quad \eta_2 = \rho_2/\tau, \tag{1}$$

where τ is a characteristic film slip time and ρ_2 is the mass per unit area of the adsorbed film (amplitude shifts are converted to quality factor shifts δ(Q-l) through calibration with a gas which does not condense at 77.4 K). Figure 1 displays typical data for Xe adsorption obtained in this manner. Characteristic slip times for monolayer adsorption are on the order of 1-3 x 10^{-9}s, and are in quantitative agreement with theoretical predictions based on electronic friction considerations.

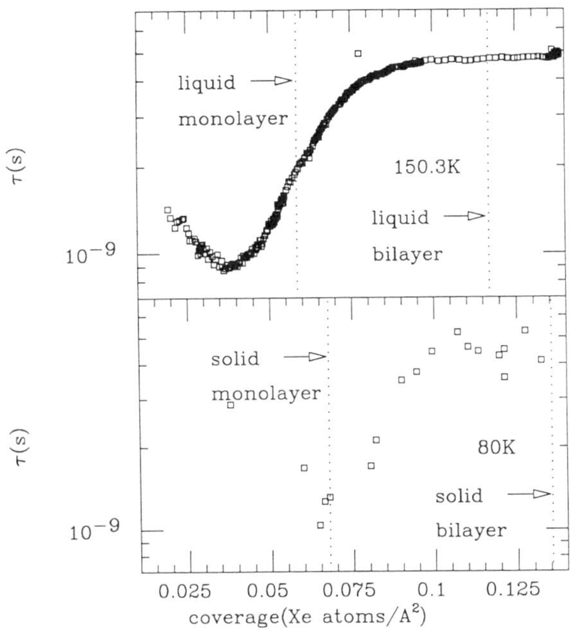

Figure 1. Slip time as a function of coverage for both liquid and solid Xe on a gold substrate. The coverages are shown for films which have the same density as bulk liquid or solid xenon.

QCM IN COMBINATION WITH STM

A quartz crystal microbalance consists of a single crystal of quartz which has metal film electrodes deposited onto each of its two major faces. The crystal oscillates in transverse shear motion, driven at its own resonance frequency by means of an external oscillator circuit, which applies an ac voltage between the two electrodes of the microbalance.

A scanning tunneling microscope consists of a scan tip, which is electrically biased with respect to the sample and held close enough so as to produce a tunneling current on the order of nanoamps. Imaging is performed by monitoring changes in the tip height while scanning at constant tunneling current, or by monitoring changes in the tunneling current while scanning at constant tip height.

Figure 2 shows a photograph of the most basic arrangement in which a QCM can be combined with a STM. The quartz crystal, a thin disk with diameter ≈ 0.75 cm, is mounted horizontally, directly below the scan tip of a Digital Instruments, Nanoscope II STM. The crystal is supported by two metal spring clips, each of which are in electrical contact with one of the electrodes present on each major face of the crystal. The wire connected to the upper electrode of the crystal is connected with the bias voltage of the STM, while the lower contact wire exits the microscope to an external oscillator circuit (Figure 3).[23] The oscillator circuit for this arrangement is selected on account of the fact that one electrode of the quartz oscillator is held at electrical ground. To operate the crystal in conjunction with the STM, we hold this electrode at the desired bias level for STM operation and scan either the upper electrode itself or alternatively a sample mounted on top of it. The oscillator's transistor is meanwhile biased at 18 volts positive to the STM bias voltage.

Figure 2. Digital instruments Nanoscope II STM with QCM mounted horizontally below the scan tip.

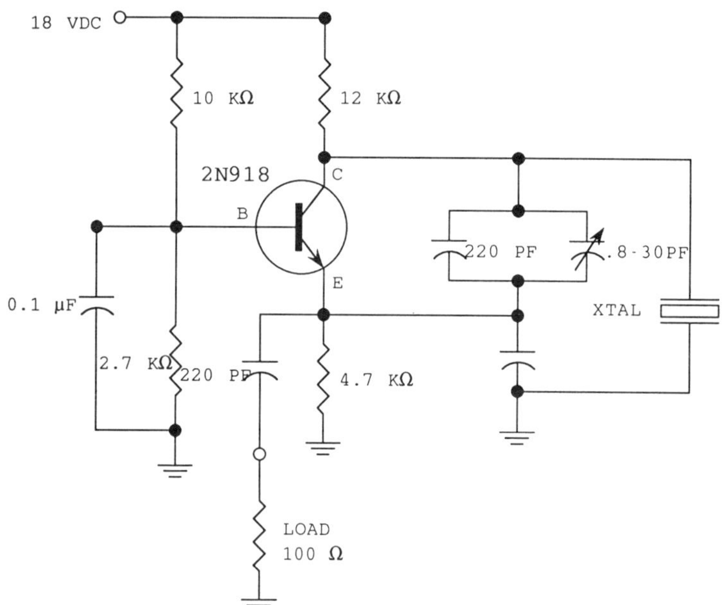

Figure 3. Schematic diagram of the Clapp oscillator circuit.

This arrangement allows the QCM and STM to be operated either independently or simultaneously. The microbalance can be utilized in its original function to detect monolayer or submonolayer quantities of adsorbed films on the sample surface, while the STM can be used in its original function for imaging the surface. Alternatively, the STM tip can be used to machine or alter the sample, while the QCM can be used to monitor frictional drag forces of the STM tip on the surface of the QCM, whether it be in the presence or absence of adsorbed films on the surface. The flexibility of this apparatus allows the possibility of a number of applications ranging in scale from the nanometer to the macroscopic. We discuss here several of those which appear to be the most promising.

APPLICATIONS

Atomic-Scale Application: Sliding Friction of Atoms

To date, our studies of atomic-scale friction have focussed on fundamental issues such as the nature of the energy dissipative mechanism as atoms slide to a stop on a surface. Nonetheless, there are a variety of practical applications which may result from these studies. While the link between these studies and macroscopically observed friction levels has yet to be established, it is important to note that the scale of that which is considered "practical" has progressively been shrinking. A new field of nanoscale technology is emerging, wherein atomic-scale friction measurements may be directly applicable. It is now experimentally possible to slide single xenon atoms with an STM tip, consistent with our QCM observation of the ease with which Xe atoms can be translated on the shaking surfaces of the microbalance. QCM measurements which we have carried out on water molecules, which are physisorbed at room temperature,[24] have also indicated high mobility on the part of these molecules.

Mesoscopic Application: Nanolithography

Our experiments demonstrate that atoms can be moved by shaking the substrate below them with the quartz crystal microbalance. We therefore ask whether it is not possible to assemble atoms or particles into patterned nanostructures by a method which utilizes the shaking action of the QCM to position them into place. In our conception of this technique, we would utilize the STM tip for machining and scanning purposes and the QCM for shaking and microweighng purposes.

One possible sequence would be the following: The STM tip is used to image the surface and to select an appropriate place on the surface for "machining" a trough or hole in the sample. The machining could take place by directly scratching the sample surface with the STM tip, or through a field emission process in which the tip makes no contact with the sample. The features produced by such processes range from the nanometer to micron in length scale. A source material, destined to fill the troughs, would be predeposited in small amounts on the sample's surface, using the microbalance to monitor the amount deposited. By increasing the temperature of the substrate, the tip-to-surface mobility, as well as the trans-surface mobility, of the source atoms could be increased as well. By shaking the substrate on the surface of the QCM, the source atoms might slide into place along the trough. The system could then be cooled down in order to fix the atoms in place. The STM tip could be used throughout in order to image the process.

Recent work on the properties of C60 mark it as an excellent candidate for such a process. It is a potentially unreactive molecule that is expected to be an excellent lubricant, and its electrical properties allow it to be imaged clearly by an STM. Alternatively, a metal source material with a low melting point such as indium might be attempted in order to fill the troughs so as to create wires.

Macroscopic Applications: Molecular Conveyor Belts

The sliding about of adsorbates on a QCM raises the possibility that these molecules could be made to slide differentially in a preferred direction. This conveyor action could be achieved in several different ways. One would be to drive the QCM so as to cause it to shake with a sawtooth oscillation. Another is to create sawtooth, or "ratchet", surface features on the substrate itself. Such a ratchet geometry can be produced by depositing the metallic film at off-normal incidence, or alternatively by ion-bombarding a surface so as to erode it at an angle.[25]

Such "molecular conveyor belts" could conceivably be used to transport molecules in a particular direction at a macroscopic scale. We estimate the typical distance a molecule moves per oscillation for our Xe experiment is on the order of 0.1 angstroms for the relatively rough surfaces. The QCM typically oscillates at a frequency of around 5 MHz, so one could expect to achieve average velocities of around 50 micrometers per second, or one millimeter per 20 seconds.

ACKNOWLEDGMENTS

This work has been supported by the ONR, Grant No. N00014-89-J-1853, and NSF grant DMR-86-57211.

REFERENCES

1. J.B. Pethica, Comment on "Interatomic Forces in Scanning Tunneling microscopy: Giant Corrugations of the Graphite Surface," *Phys. Rev. Lett.* 57:3235-3239 (1986).
2. C.M. Mate, G. McClelland, R. Erlandsson, S. Chiang, Atomic-scale friction of a tungsten tip on a graphite surface. *Phys. Rev. Lett.* 59:1942-1945 (1987).
3. R. Erlandsson, G. Hadziioannou, C.M. Mate, G. McClelland, S. Chiang, Atomic-scale friction between the muscovite mica cleavage plane and a tungsten tip, *J. Chem. Phys.* 89:5190-5193 (1988).
4. N.A. Burnham, D.D. Dominguez, R.L. Mowery, R.J. Colton, Probing the surface forces of monolayer films with an atomic force microscope, *Phys. Rev. Lett.* 64:1931-1934 (1990).
5. U. Landman, W.D. Luedtke, N.A. Burnham, R.J. Colton, Atomistic mechanisms and dynamics of adhesion, nanoindentation, and fracture, *Science* 248:454-461(1990).
6. N.A. Burnham, R.J. Colton, Measuring the nanomechanical properties and surface forces of materials using an atomic force microscope. *J. Vac. Sci. Technol.* A7:2906-2913 (1990).
7. J.N. Israelachvili, P.M. McGuiggan, A.M. Homola, Dynamic properties of molecularly thin films. *Science*. 240:189-191(1988).
8 .J. Van Alsten, S. Granick, Molecular tribometry of ultrathin liquid films, *Phys. Rev. Lett.* 61: 570-2573 (1988.).
9. M. Hirano, K. Shinjo, R. Kaneko, Y. Murata, Anisotropy of frictional forces in muscovite mica. *Phys. Rev. Lett.* 67:2642-2645 (1991)
10. J. Krim. R. Chiarello, D.H Solnina, X-ray reflectivity and adsorption isotherm study of fractal scaling in vapor-deposited films, *Phys. Rev. Lett.* 66:181-186 (1991).
11. E.T. Watts, J. Krim, A. Widom, Experimental observation of interfacial slippage at the boundary of molecularly thin films with gold substrates, *Phys. Rev.* 41:3466-3472 (1990).
12. J. Krim, A. Widom, Damping of a crystal oscillator by an adsorbed monolayer, *Phys. Rev.* B 38: 12184-12191 (1988).
13. J. Krim, R. Chiarello, Sliding friction measurements of molecularly thin films, *J. Vac. Sci. Technol.* A9: 2566-2570 (1991).
14. B.N.J. Persson, D. Schumacher, D. Otto, Surface resistivity and vibrational damping in adsorbed layers, *Chem. Phys. Lett.* 178: 204-212 (1991)
15. J.B. Sokoloff, Theory of energy dissipation in sliding crystal surfaces, *Phys. Rev.* B 42:760-765 (1990).
16. W. Zhong, D. Tomanek, First-principles theory of atomic scale friction. *Phys. Rev. Lett.* 64: 3054-3057 (1990).

17. U. Landman, W.D Luetke, M.W. Ribarsky, Structural and dynamical Consequences of Interactions in Interfacial Systems, *J. Vac. Sci. Technol.* A7:2928 (1990).
18. U. Landman, W.D. Luedtke, E.M. Ringer, Atomistic mechanisms of adhesive contact formation and interfacial proceses, *Wear,* 150-165 (1991).
19. P.A. Thompson, M.O. Robbins, Simulations of contact-line motion: slip and the dynamic contact angle, *Phys. Rev. Lett.* 63:766-769 (1989); Shear flow near solids: epitaxial order and flow boundary conditions, *Phys. Rev.* A 41:6830-6837 (1990).
20. D. Fargues, P. Dolle, M. Alnot, J.J. Ehrhardt, Adsorption of xenon on Ni(111) studied by photoemission and LEED, *Surf. Sci.* 214: 187-196 (1989).
21. J. Krim, J. Suzanne, J.G. Dash, Triple point wetting of molecularly thin films, *Phys. Rev. Lett.* 52:640-644 (1984).
22. L.W. Bruch, J.M. Gay, J. Krim On the limit of compression of a physisorbed monolayer, *Jour. de Phys.* (Paris) 46:425-433 (1984).
23. M.E. Frerking, "Crystal Oscillator Design and Temperature Compensation," Van Nostrand Reinhold, New York, 88 (1978).
24. J. Dayo, C. Daly, and J. Krim, private communication.
25. J. Krim, and I. Heyvaert, private communication.

SURFACE SCIENCE AT THE NANOSCALE:

MOLECULAR IMAGING AND SURFACE FORCES

Richard J. Colton,[1] Eric I. Altman,[2] and Steven M. Hues[1]

[1]Chemistry Division, Code 6177
Naval Research Laboratory
Washington, DC 20375-5342
[2]Chemical Engineering Department
Yale University
New Haven, CT 06520

Abstract: This paper presents a general description of the use of the ne tip-based proximal probes--scanning tunneling microscopy (STM) and atomic force microscopy (AFM)--to study the nanoscale properties of surfaces. Using STM, we have examined the nucleation and growth of C_{60} films on Au(111) and Ag(111). In addition to imaging, we have measured the charge transfer between the metal and C_{60} using STM in the spectroscopic mode. AFM studies which focus on developing quantitative methods for measuring surface forces and the mechanical properties of surfaces are also presented.

INTRODUCTION

The new tip-based proximal probes--scanning tunneling microscopy (STM) and atomic force microscopy (AFM)--have stimulated a revolution in surface science. These techniques can image the surface of many materials with atomic resolution and provide information about the structure and organization of atomic and molecular adsorbates on surfaces. The tip and its associated force or field can also be used to manipulate atoms and molecules to form unique structures. However, the STM and AFM imaging mechanisms of molecular systems are not well understood. Because STM images correspond to a convolution of the tip and surface electron density-of-states near the Fermi level, molecules with deeper lying states are not expected to be readily imaged even though several examples have been reported. STM can also be used in a spectroscopic mode to probe the electronic structure of molecular adsorbates and their interaction with the substrate.

The interaction force between the tip and sample can also be used to form an image by AFM. In its simplest form, the atomic force microscope can be configured as a contact-mode surface profilometer capable of "atomic resolution." The imaging mechanism is not well known, however, because the surface forces can actually cause appreciable tip-sample deformation at contact. Under higher applied loads, the microscope becomes an indentor capable of measuring the mechanical properties of materials as thin as a monolayer. In the noncontact mode, the microscope is a probe of long-range force interactions between surfaces.

SCANNING TUNNELING MICROSCOPY AND SPECTROSCOPY

Scanning tunneling microscopy (STM) is based on the observation that electrons, because of their wave-like nature, can tunnel through the potential barrier established when two surfaces under appropriate bias are positioned about a nanometer from each other.[1,2] The microscope consists of a sharp tip and flat surface. Electrons tunnel between the tip and the surface when a small bias voltage is applied. An image of the surface is acquired by scanning the tip or surface using piezoelectric actuators. As the tip approaches a high (or low) spot on the surface, the tunneling current increases (or decreases) in response to the change in separation. The extreme sensitivity of the technique is based on the exponential dependence of the tunneling current on the tip-sample separation.

In addition to imaging, STM can be used in a spectroscopic mode where the tip is held over a specific location and the tunneling current-voltage characteristic measured. When the tip is biased negative, the tunneling electrons originate in filled states of the tip and end in empty states of the surface (and vice versa); current-versus-bias data therefore yield a convolution of the tip and surface electron density-of-states.

To illustrate the utility of both the imaging and spectroscopic modes of STM, we summarize some of our recent work on the nucleation and growth of bucky ball films on Au(111) and Ag(111) surfaces in ultra-high vacuum.[3-6] (Au(111) and Ag (111) were chosen because their surfaces are relatively inert, and (111) surfaces could be grown with large terrace widths separated by monatomic steps. In addition, Au(111) undergoes an unusual $23\times\sqrt{3}$ reconstruction that acts as an ordered array of nucleation sites for the growth of Ni, Fe, and Co films.) On Au(111), C_{60} grows in a layer-by-layer manner with nucleation at the step edges. Although no LEED pattern due to adsorbed C_{60} is observed, STM indicates that two ordered adsorbate phases predominate. One structure (shown in Figure 1(a)) grows along the crystallographic directions of the substrate but with a large lattice mismatch. Here, the Au(111) step kinetically defines the orientation of the overlayer. The second structure (shown in Figure 1(b)) is rotated 30° with respect to the substrate and the periodicity is $2\sqrt{3}\times2\sqrt{3}$ R30°. The lattice matching is nearly perfect in this orientation; all molecules can be in equivalent adsorption sites. Therefore, this structure is thermodynamically favored. The adsorption of C_{60} is found also to lift the $23\times\sqrt{3}$ reconstruction of the Au(111) surface. Because ordinary hydrocarbons do not lift the reconstruction, C_{60} shows an unusually strong interaction with Au(111). The thermal stability of C_{60} monolayers on Au(111) also indicates a strong interaction. Multilayers could be desorbed at 300°C while first layer molecules remain until 500°C. Because Au is considered catalytically inert, stronger interactions are expected with other metals, suggesting that the reactivity of fullerenes can be controlled through transition metal catalysis.

STM images of pure C_{60} adsorbed on Au(111) reveal an apparently "inhomogeneous" monolayer with some molecules appearing brighter than others as shown in Figure 1.[4] The "inhomogeneity" is not due solely to differences in adsorption site geometry since contrast variations are also observed in $2\sqrt{3}\times2\sqrt{3}$ R30° domains (see Figure 1(b)). Time-lapsed STM images, intramolecular contrast, and scanning tunneling spectroscopy (STS) show that the

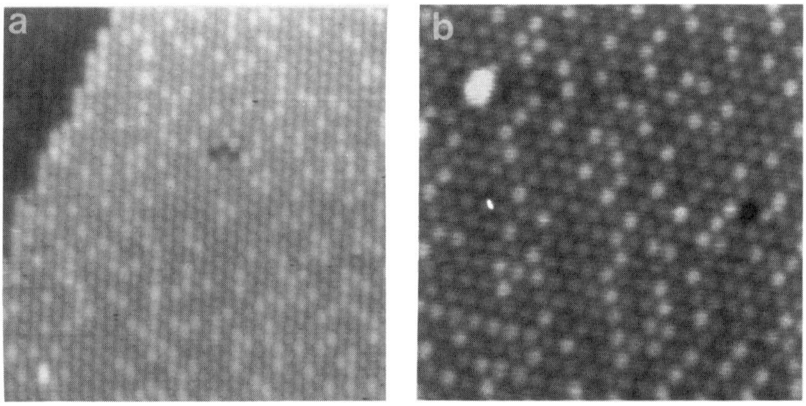

Figure 1. STM images of C_{60} monolayer film on Au (111) showing (a) an in-phase 38 x 38 C_{60} domain (29 x 29 nm) and (b) a 2√3x2√3 R30° C_{60} domain (19 x 19 nm).

apparent height variations can be attributed to electronic differences between molecules bound in different rotational orientations on the surface. The tunneling spectra in Figure 2 show that bonding variations are due to differences in the efficiency of coupling between the C_{60} LUMO and the Au and Ag surfaces. The shift in the unoccupied state density towards the Fermi level for the first layer is attributed to charge transfer from the metal to the molecule. The barrier height over the adsorbed molecules increases compared to bare Au, consistent with charge transfer from the substrate to the C_{60} molecule.

The interaction of C_{60} with Ag(111) was also studied and compared with Au(111).[5] Ag has nearly the same lattice constant and electronic structure as Au. However, the Ag(111) surface does not reconstruct and should be more chemically reactive than Au. On Ag(111), growth proceeds also in a layer-by-layer manner, but difference in growth of the first layer due to the Au reconstruction was observed. However, the 2√3x2√3 R30° structure occurs almost exclusively. STM/S results are similar to those for Au, again showing an inhomogeneous layer due to variations in adsorbate bonding. STM images of the second and third layer exhibited no contrast variations, consistent with contrast variations due to the interaction of C_{60} with the metal substrate.

For both Au(111) and Ag(111), STM images of the monolayer revealed a variety of internal structures. These structures are a convolution of the structure of C_{60} and the underlying metal substrate. Variations in internal structure for 2√3x2√3 R30° domains in which all molecules can be in equivalent surface sites, again indicates that the appearance of adsorbed C_{60} in STM depends on the rotational orientation of the molecule on the surface. (No internal structure of second layer molecules is observed, consistent with their free rotation in the solid at room temperature.) Occupied state images were found to reflect the adsorption site symmetry while unoccupied state images reflected the molecular symmetry.[6] The observed internal structure of adsorbed C_{60} was assigned to an on-top site with the five-membered ring on the metal surface.

ATOMIC FORCE MICROSCOPY AND SPECTROSCOPY

Atomic force microscopy (AFM) was invented in 1986 by Binnig et al.[7] as "a new type of microscope capable of investigating surfaces of insulators on an atomic scale."

Stemming from developments in STM, it is now possible to image insulators--organic and biological molecules, salts, glasses, and metal oxides--some with atomic resolution. Being able to obtain high-resolution, topographical images of insulators under a variety of conditions--at ambient pressure, in aqueous or cryogenic liquids, etc.--has created many applications, some of which are illustrated in these Proceedings. However, beginning in 1987, Mate et al.[8] introduced a new direction for AFM where atomic-scale frictional forces could be measured. Likewise, in 1989, Burnham and Colton[9] used AFM to measure the surface forces and nanomechanical properties of materials. Today, there are many examples of using AFM as a high-resolution profilometer, surface force probe and nanoindentor. AFM is evolving from a qualitative imaging tool to a quantitative probe of surface forces such as van der Waals, capillary, electrostatic, capacitive, double-layer, magnetic or adhesive forces.

A force microscope consists of a sensor that responds to a force and a detector that measures the sensor's response. The sensor--a cantilever beam with an effective spring constant, k--moves in response to the forces acting on its tip. The detector measures the cantilever's position, which can be converted to force using Hooke's Law, $F = -kz$, where z is the cantilever displacement. (The displacement can be measured by tunneling, capacitive or optical instruments.) If the cantilever beam is either microfabricated or a thin wire, then small spring constants (0.1-100 N/m) are possible. For 0.01 nm displacement sensitivity, 10^{-8}-10^{-12} N forces are measurable. These forces are comparable to the forces associated with chemical bonding, e.g., ~10^{-7} N for an ionic bond and ~10^{-11} N for a hydrogen bond.

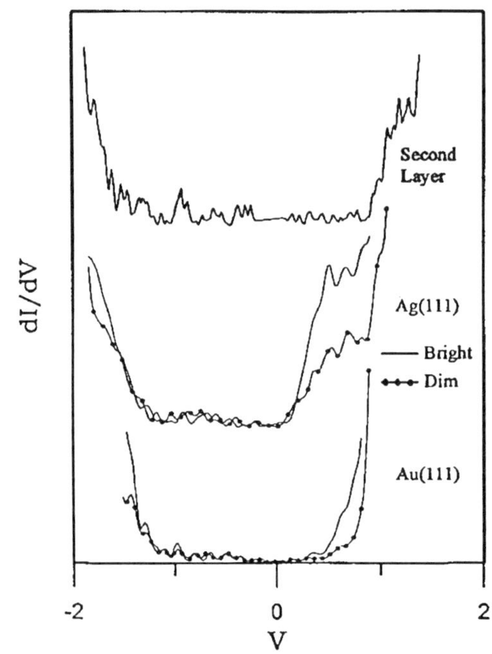

Figure 2. Tunneling spectra for second layer C_{60} molecules on Ag(111) and first layer molecules on Au(111) and Ag(111). Bright and dim refer to spectra recorded over molecules that appear bright or dim in topographic images.

A plot of the force interaction between two surfaces (typically the tip and flat surface) as a function of relative tip-sample separation constitutes a *force curve*, and such measurements have been termed "force spectroscopy." We have done extensive work[9-15] describing how to interpret force curves so as to gain information about i) the point of tip-sample contact, ii) the magnitude and functional dependence of adhesive and long-range attractive forces, iii) possible mechanisms of long-range force interactions (surface layers, fixed dipoles, patch charges), iv) tip-sample contact area, which relates to the imaging mechanism in contact-mode force microscopy, and v) the elastic modulus and plasticity of thin and thick films.

Figure 3 shows typical force curves in schematic form.[12] The cycle starts (at the left) with the sample far away and the cantilever in its rest position. As the separation decreases, the cantilever bends towards the sample (below the baseline) responding to an

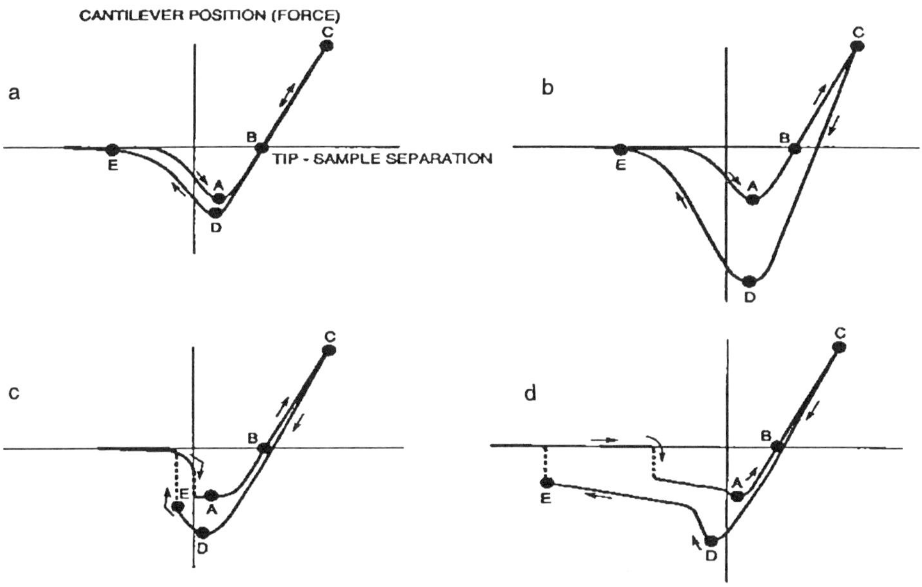

Figure 3. Types of force curves (schematic): (a) long-range attractive force, no instabilities, capillary, or plastic deformation; (b) with plastic deformation; (c) with cantilever instability; (d) with thin liquid film.

attractive force. The *attractive region* of the curve extends over several tens of nanometers. *Contact* may be defined as the point where repulsion is first detected. A *transition or deformation region* extends from contact, through the *pull-on force* or maximum negative load during loading, to some point where the attraction becomes negligible in comparison with the positive applied load. Continuing to push the sample in the direction of the cantilever after contact until a predetermined load is reached causes the cantilever to deflect through its original rest position. At higher loads we have *Hertzian* behavior provided that there is no bulk plastic deformation. The slope of the curve in this repulsive force region (above the baseline) is a function of the nanomechanical properties and local geometry of the tip and sample. The steeper the slope, the higher the modulus of the material. On unloading, some *hysteresis* is almost always seen, the *pull-off force* or *adhesion* being greater than the pull-on force.

The repulsive part of the curve is easiest to explain qualitatively. Accordingly to the classical relationships derived by Hertz (see, for example, ref. 13), the slope of the curve plotting the force versus the contact area or penetration depth is dependent on the geometry of the indentor and on the modulus of the material. Figure 4 clearly shows this trend where the slope increases with the modulus of the material. The radius of the contact area predicted by the Hertzian relation is an estimate of the minimum radius as it ignores any tip-sample deformation caused by local surface forces.

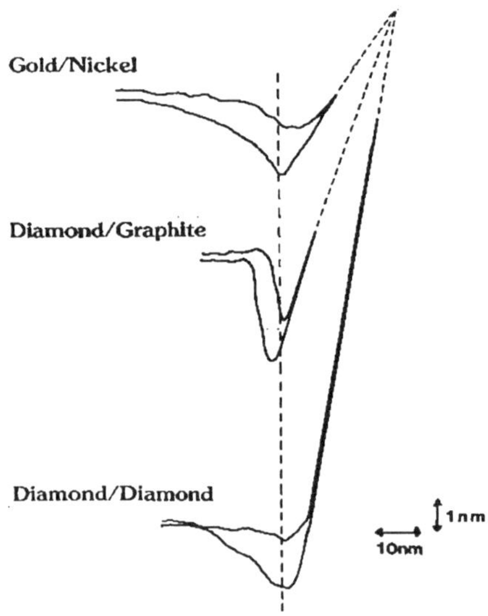

Figure 4. Force-separation curves for different tip/sample materials. Vertical axis is cantilever deflection (1 nm ≈ 260 nN). Horizontal axis is sample displacement (nm).

The attractive part of the curve should likewise be easy to explain given the extent of our understanding of intermolecular and surface forces based on van der Waals theories. However, unlike experiments done with the surface force apparatus[16] with clean mica surfaces in which a $1/D^2$ dependence for the nonretarding dispersive force is measured over a distance of 10 nm, the attractive force measured by AFM is typically much longer ranged. While van der Waals forces certainly exist and should be readily measurable by AFM, their relative magnitude and distance dependence are masked by other forces due to capillary and patch charge effects (see below) caused by adsorbed films and contaminants. By controlling the cleanliness of these surfaces (as can be done in an ultra-high vacuum environment), the van der Waals forces should become the dominant attractive force between uncharged, nonmagnetic surfaces. In solution, other forces associated with double-layer and hydration need also to be considered.

As an example of our current work dealing with attractive forces, we have developed a model based on work function anisotropies (and their associated patch charges) as an origin of long-range surface forces.[14] The work function is very sensitive to perturbations at the surface of a material. Surface preparation, uneven distribution of adsorbates, crystallographic orientation, the presence of surface steps, hillocks, pits or defects can all influence the work function leading to regions on the surface with slightly different electrostatic potentials or patch charges. We have recently modelled the interaction between patch charges on the tip and sample and obtained the result shown in Figure 5. Clearly the shapes of the theoretical and experimental curves for a diamond tip and graphite surface are similar, and the distance dependence of the model appears to be correct. A van der Waals force curve is also shown for comparison.

The deformation part of the curve where tip and sample make contact, while perhaps the most interesting part, is the most difficult to interpret. In the Hertzian model, when two elastic spheres make contact, they meet at a point. When the spheres are loaded, they deform causing the contact area to increase. Removing the load causes the contact area to decrease until at zero applied load the surfaces are once again in point contact. At this point, the surface can be separated with zero pull-off force. The shortcoming of the Hertzian model is that it neglects the attractive forces between the two surfaces in and around the contact region. The local attractive forces are typically strong enough to cause

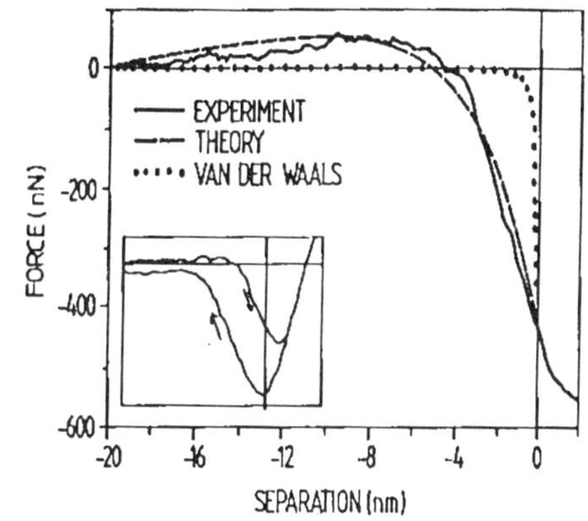

Figure 5. AFM experimental data (solid line), theoretical fit (dashed line), and van der Waals force (dotted line), where force is shown as a function of tip-sample separation. Inset shows full force curve.

the materials to deform thereby increasing the area of contact. There are two classical models which describe the contact deformation of elastic materials: the Johnson-Kendall-Roberts-Sperling (JKRS) theory and the Derjaguin-Muller-Toporov (DMT) theory (see for example refs. 12 and 13). The JKRS theory modifies the Hertzian model by adding a short-range, attractive force in the contact area only, forming a connective neck between the materials. This model works well for materials with low moduli, high work of adhesion and large radii (e.g., Ni/Au). The DMT theory adds a long-range attractive force outside the contact area. The theory works well for materials with high moduli, low work of adhesion and small radii (e.g., diamond/diamond). Each theory predicts a finite pull-off (or adhesive) force to break contact at zero applied load and reduces to the Hertzian model at high loads.

An interim goal of our research is to describe the shape of the force curve analytically.[12] We have recently derived an expression using the relationships described above. With this expression we can fit the shape of the force curve, particularly in the deformation region, using moduli and contact area as fitting parameters. From these curve fitting routines it then becomes possible to determine the mechanical properties of materials regardless of their thickness (i.e., the mechanical properties of monolayer films and adsorbates can be studied). The remaining goals of the research involve a comparison of the classical models with atomistic models as described by Landman et al.[17] and Harrison et al.[18-22], and the development of a full atomistic mechanism of the nanomechanical properties of materials including adhesion.

ACKNOWLEDGMENT

We would like to acknowledge the support of our other collaborators--Don Brenner, Nancy Burnham (ISI-KFA, Jülich, FRG), Dan DiLella, Charles Draper, Barbara Gans, Judith Harrison, Gil Lee, Ken Lee, and Hubert Pollock (U. Lancaster, UK). Funding support from the Office of Naval Research is gratefully acknowledged.

REFERENCES

1. "Scanning Tunneling Microscopy and Spectroscopy: Theory, Techniques and Applications 1993," D.A. Bonnell, ed. VCH Publishers, Inc., New York, NY.
2. "Methods of Experimental Physics, Scanning Tunneling Microscopy," J.A. Stroscio and J.W. Kaiser, eds., vol. 27, Academic Press, Inc., Boston, MA (1993).
3. E.I. Altman and R.J. Colton, Nucleation, growth and structure of fullerene films on Au(111), *Surf. Sci.* 279, 49-67 (1992).
4. E.I. Altman and R.J. Colton, Characterization of the interaction of C_{60} with Au(111), *in* "Atomic and Nanometer Scale Modification of Materials: Fundamentals and Applications," Ph. Avouris, ed., NATO ASI Series, Kluwer Academic Press, London 303-314 (1993).
5. E.I. Altman and R.J. Colton, The interaction of C_{60} with noble metal surfaces, *Surf. Sci.*, 295:13-33 (1993).
6. E.I. Altman, and R.J. Colton, Determination of the orientation of C_{60} adsorbed on Au(111) and Ag(111), *Phys. Rev.* B48:18244-9 (1993).
7. G. Binnig, C.F. Quate, and Ch. Gerber, Atomic force microscope, *Phys. Rev. Lett.* 56:930 (1986).
8. C.M. Mate, G.M. McClelland, R. Erlandsson, and S. Chiang, Atomic-scale friction of a tungsten tip on a graphite surface, *Phys. Rev. Lett.* 59:1942 (1987).
9. N.A. Burnham and R.J. Colton, Measuring the nanomechanical properties and surface forces of materials using an atomic force microscope, *J. Vac. Sci. Technol.* A7: 2906-13 (1989).
10. N.A.Burnham, D.D Dominguez, R.L Mowery, and R.J. Colton, Probing the surface forces of monolayer films with an atomic force microscope, *Phys. Rev. Lett.* 64:1931-4 (1990).
11. N.A. Burnham, R.J. Colton, and H.M Pollock, Interpretation issues in force microscopy, *J. Vac.Sci. Technol.* A9:2548-56 (1991).
12. N.A. Burnham, R.J. Colton, and H.M. Pollock," Interpretation of Force Curves in Force Microscopy, Nanotechnology," vol. 4, 64-80 (1993)
13. N.A. Burnham and R.J. Colton, Force microscopy, *in* "Scanning Tunneling Microscopy and Spectroscopy: Theory, Techniques and Applications," D.A. Bonnell, ed., VCH Publishers, Inc., New York, NY, 191-250 (1993).
14. N.A. Burnham, R.J. Colton, and H.M. Pollock, "Work Function Anisotropies as an Origin of Long-Range Surface Forces," *Phys. Rev. Lett.* 69:144-47 (1992).
15. S.M. Hues, R.J. Colton, E. Meyer, and H-J Güntherodt, Scanning probe microscopy of thin films, MRS Bulletin, 18:41-49 (1993).
16. J.N. Israelachvili, "Intermolecular and Surface Forces," Academic Press, Inc., New York, NY, 153 (1986).
17. U. Landman, W.D. Luedtke, N.A. Burnham, and R.J. Colton, Atomistic mechanisms and dynamics of adhesion, nanoindentation and fracture, *Science* 248:454-61(1990).
18. J.A. Harrison, D.W. Brenner, C.T. White, and R.J. Colton, Atomistic Mechanisms of Adhesion and Compression of Diamond Surfaces, *Thin Solid Films* 206:213-19 (1991).

19. J.A. Harrison, R.J. Colton, C.T. White, and D.W. Brenner, Atomistic simulation of the Nanoindentation of diamond and graphite surfaces, *Mat. Res. Soc. Symp. Proc.* 239:573-7 (1992).
20. J.A. Harrison, C.T. White, R.J. Colton, and D.W. Brenner, Nanoscale investigation of indentation, adhesion, and fracture of diamond (111) surfaces, *Surf. Sci.* 271:57-67 (1992).
21. J.A. Harrison, C.T. White, R.J. Colton, and D.W. Brenner, Molecular dynamics simulations of atomic-scale friction of diamond surfaces, *Phys. Rev.* B 46:9700-8 (1992).
22. J.A. Harrison, C.T. White, R.J. Colton, and D.W. Brenner, Atomistic simulations of friction at sliding diamond interfaces, *MRS Bulletin* 18:50-3 (1993).

LINEARITY AND CALIBRATION OF SCANNING PROBE MICROSCOPE IMAGES

Paul E. West, Renee Jobe, and Tim Van Slambrouck

TopoMetrix Corporation
Santa Clara, CA
95054-1162

Abstract: Scanning probe microscopes rely on piezoelectric ceramics for scanning a probe across a sample surface. Artifacts in SPM images are a result of the nonlinearity, hysteresis, and creep in piezoelectric ceramics. Examples of the artifacts in SPM images created by piezoelectric scanners and methods for correcting these artifacts are presented.

INTRODUCTION

The ideal microscope generates a direct map of a sample surface, such that the measurements that are made are an exact magnification of the actual sample. In order to make repeatable, quantitative measurements, a scanning probe microscope (SPM) must be calibrated in three axes. Methods for calibration of SPM instruments are essential for this tool to become accepted as a quantitative analytical instrument. The protocol for calibration depends on the application of the instrument.

Factors that limit the ability to calibrate SPM images include the probe geometries, the scanner nonlinearities, the probe-sample angle, and the errors introduced by data processing. Calibrating a scanning probe instrument requires the systematic standardization of each of these subcomponents.

Verification of calibration in an SPM instrument requires standards with horizontal and vertical features ranging from the angstrom to the μm range. Additionally, calibration must be established by comparison of SPM images with images measured by other types of microscopes.

The objective of this presentation is to exhibit the factors that must be considered in order to measure calibrated images with a scanning probe microscope.

PROBES

SPM images are a combination of the sample surface and the probe that is used to image the surface. Many types of probes are available. An understanding of the interaction between probes and samples is essential to the calibration of SPM images. Probe related factors that can create calibration errors include: not selecting the optimal probe for the application, errors in probe geometry, and dynamics of the probe/cantilever.

Table 1 lists several types of probes commonly available for the atomic force microscope. Each type of probe is best suited for making calibrated measurements on specific types of samples.

Table 1. Cantilever/probes used in the scanning force microscope

Cantilever Tip	Super Tip	Pyramidal	Nanosensor
Cantilever Comp. Geometry	SiN2 V-Shape 200μm long arms 18 μm wide 1μm thick	SiN2 V-Shape 200 m long arms 18 μm wide 1 μm thick	Silicon (100) I-Shape 450 m long arms 35 m wide 5 - 7 μm thick
Force Constant	0.064 N/m	0.064 N/m	> 3.51 N/m
Tip Composition Geometry	Diamond-Like Long rod 1.5μm long	SiN2 Pyramidal 4μm base 4 μm high	Silicon (100) Long cone 10 - 20 μm long
Aspect Ratio Tip Radius	> 10 to 1 < 100 Å	1 to 1 < 500Å	3 to 1 < 200 Å

As an example, the images illustrated in Figure 1 (a) and 1 (b) were measured with a supertip and a pyramidal tip, respectively. The pyramidal tip is unable to reach the bottom of the trench, causing considerable error in the measurement of the vertical and horizontal dimensions of a trench.

Errors in probe geometries are created by the manufacturing process used to fabricate the probe. For example, pyramidal silicon nitride probes are fabricated with 1 μm photolithography processes and thus have a 0.1 μm uncertainty. Probes manufactured by ion beam[1] and electron beam techniques[2] have tolerances equivalent to the resolution of the electron or ion beam.

SCANNERS

Calibration errors created by scanners utilized in SPM instruments are reduced with scan linearity correction techniques. The primary source of the nonlinearities in an SPM scanner is the piezoelectric materials used in the scanner. This is because piezoelectric ceramics expand in a nonlinear manner with an applied voltage. Corrections may be made with open or closed loop techniques. Correction methods in the horizontal (XY) axis are different from the correction methods used for the vertical (z) axis.

In open loop methods, the SPM system is programmed so that nonlinear behavior of ceramics is corrected either during scanning or after the image is made[3]. Using this method, a different correction is required for each scan range, rotation and scan speed. Open loop methods are capable of imaging samples with an error that is less than 1% of the scan range in the x-y axis. Errors in the z axis with open loop techniques are dependent on the type of piezoelectric material used in the scanner.

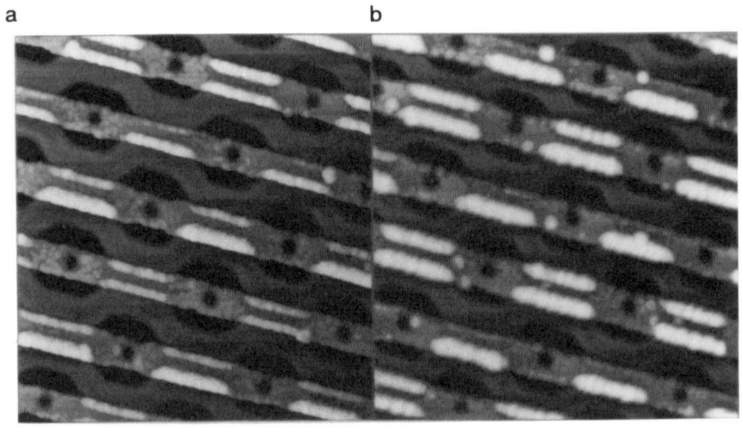

Figure 1. Semiconductor test pattern imaged with a supertip (a) and a pyramidal tip (b).

Closed loop methods use a sensing device to measure the actual scanner motion in the x and y axes and drive the ceramic to produce linear scan displacement. Methods commonly used for closed-loop corrections include optical sensors,[4] capacitance sensors,[5] and strain gauge sensors.[6] Sensors can be used to directly measure the motion in the z axis.

PROBE-SAMPLE ANGLE

Several calibration errors are derived from a nonperpendicular angle between the probe and the sample. Measurements on extremely flat samples and patterned samples are prone to errors as a result of improper probe-sample angles. Many of the errors are reduced by accurately controlling the angles between the probe and the sample.

Often, when measuring micro-roughness of flat samples such as optics surfaces or silicon wafers, bowing in the images is observed. The observed bowing in the images is caused by the nonlinear response of the piezoelectric scanner as it moves the probe across the flat surface. The bowing is reduced or eliminated when the probe sample scanning axis is perpendicular to the probe axis. Meeting this condition is easily achieved with probe scanning microscopes in comparison to sample scanning microscopes.

DATA PROCESSING

Computer processing is routinely applied to images generated with a scanning probe microscope. Errors in calibration may be created by the data processing algorithms. The most common processing technique that creates loss of calibration is filtering, either bandpass or Fourier filtering (see Figure 2).

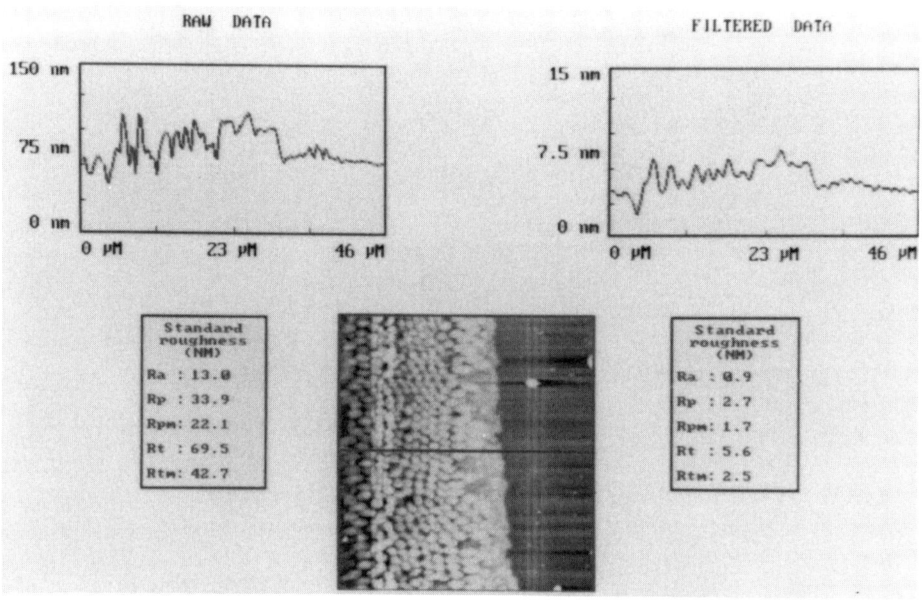

Figure 2. Change in line profile of a surface caused by bandpass filtering.

Filtering is often used to remove low frequency components such as bowing from an image. Such filtering can inadvertently remove slowly varying surface features from an image or change the height of surface features with relatively small width compared to the scan range. Errors from filtering are minimized by applying the same filter to all images.

STANDARDS

Verification of calibration in a scanning probe microscope requires a standard with a well-defined surface topography. Topography must be defined in the x, y and z axes. There are three primary ranges for calibration standards; atomic scale (dim.<100 angstroms), intermediate range (100 angstroms<dim.<1 μm), and large scale (dim.>1 μm), see Table 2.

Table 2. Range of calibration standards required for scanning probe microscopes

SPM CALIBRATION STANDARDS

	X - Y	**Z**
D< 100 Å	Crystal Lattice	None
100 Å < D < 1 μm	None	Profiler Standards
D > 1 μm	Patterned Wafers	Profiler Standards

At the atomic scale the lattice spacing of crystalline materials serves as a standards in the x-y axis. However, there are no calibration standards for vertical dimensions below 100 angstroms. At this scale, long-term stability with respect to chemical reactions is a critical factor in identifying standards.

At intermediate ranges, between 0.001 and 1 μm, there are no standard for calibrating the x-y scan ranges. In the z axis, standards are developed for surface profilers. For scan ranges larger than 1 μm, patterned semiconductor samples serve as standards for the x,y and z axes.

CONCLUSION

Scanning probe microscopes are capable of making routine quantitative measurements. However, appropriate calibration methods and standards must be identified. Several subcomponents including the probes, scanners, stage and processing software must be calibrated to achieve reliable quantitative data.

REFERENCES

1. H. Ximen, P. Russell, Microfabrication of AFM tips using focused ion and electron beam techniques, *Ultramicroscopy*, 42:1526-1530 (1992).
2. D. Keller, D. Deputy, A. Aldvino, K Luo, Sharp vertical walled tips for SFM imaging of steep or soft samples, *Ultramicroscopy* (in press).
3. R. W. Basedow, T.D. Cocks, Piezoelectric ceramic displacement characteristics at low frequencies and their consequences in Fabry Perot interferometry, *Phys. E. Sci. Instrum.*, 13:840-844 (1980)
4. R.C. Barrett, C.F. Quate, Optical scan-correction system applied to atomic force microscopy, *Rev. Sci. Instrum.*, 62 (6):1393-1399 (1991).
5. J.E. Griffith, G.L. Miller, C.A. Green, D.A. Grigg, D.A., P.E. Russel, A scanning tunneling microscope with a capacitance-based position monitor, *J. Vac. Sci. Technol.* B, 8(6):2023-2027 (1990).
6. K. Mizutani, T. Kawano, Y. Tanaka, A piezoelectric-drive table and its applications to microgrinding of ceramic materials, *Precision Engineering*, 1 12 (4):219-226 (1990).

SAMPLE HOLDERS FOR IMAGING INTACT PARTICLES WITH THE SCANNING FORCE MICROSCOPE

A.A. Garcia[1] and M. Edwards[2]

[1]Department of Chemical, Bio and Materials Engineering
Arizona State University
Tempe, Arizona 85287-6006
[2]Tempe Union High School
Tempe, Arizona 85287

Abstract: We have fabricated and tested sample holders for scanning force microscopy (SFM) imaging of individual, intact particles in the range of 70-500 micrometers in diameter. The sample holders keep particles rigidly in place at the end of a micropipet during scanning (sample stage is vibrated by the piezo tube) through the use of flexible tubing connected to a simple laboratory aspirator. This design is shown to permit high resolution imaging of intact polymer adsorbent particles. The sample holders were tested by imaging adsorbent particles with the trade designations of Amberlite XAD-4, XAD-71, XE-348 and IRA-900 (Rohm & Haas Co., Springhouse PA). Polymer adsorbents were imaged in either the dry or hydrated state without immersing the SFM probe in solution.

INTRODUCTION

Scanning probe microscopy was born with the invention of the scanning tunneling microscope (STM) by Binnig and Rohrer.[1] The scanning force (also called atomic force) microscope (SFM) invented by Binnig, Quate and Gerber[2] combined the idea of using a probe to come into direct contact with the surface and the raster scanning ability demonstrated by STM so that atomic level imaging of electrically insulating materials could be performed. Since scanning force microscopes have become commercially available, there has been a dramatic increase in applications, such as biological studies, nanolithography, quality control in semiconductor wafer manufacturing, measurement of local properties, and characterization of materials. An excellent review of articles on scanning probe microscopy is provided by Snyder and White.[3]

Atomic Force Microscopy/Scanning Tunneling Microscopy
Edited by S.H. Cohen *et al.*, Plenum Press, New York, 1994

The basic operation of a commercial SFM is illustrated in Figure 1, where the working probe is a small pyramid of silicon nitride at the end of a V-shaped cantilever. A raster scan (similar to the scanning performed by the electron gun of a TV set) is performed by

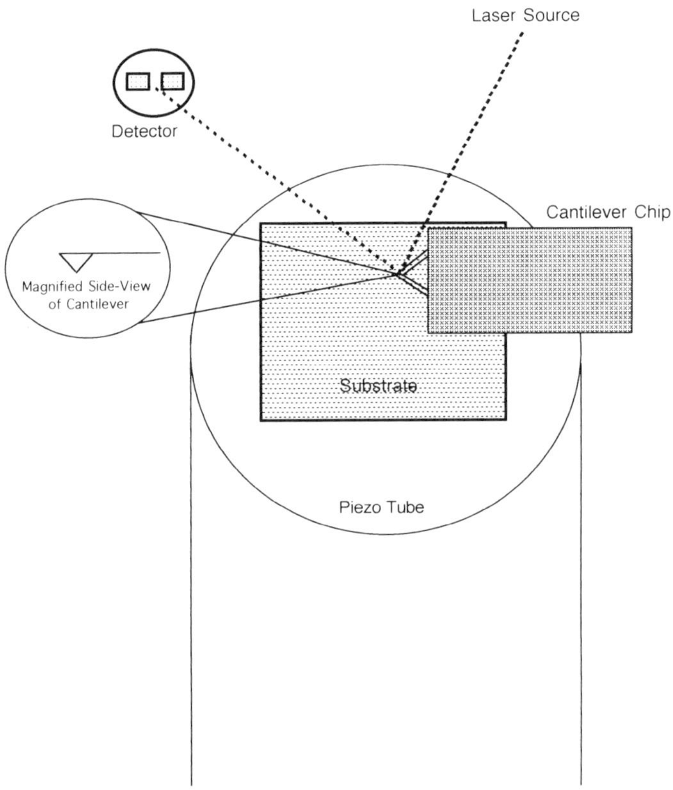

Figure 1. Illustration of the basic operation of a commercial scanning force microscope.

applying voltages to deflect a tube made of piezoelectronic material. The image is acquired by either vibrating the cantilever over a surface or by vibrating the sample stage under a fixed cantilever. The cantilever is coated with a thin layer of gold so that a laser beam can be aimed at the back of the cantilever with the beam being reflected into a photodiode array detector. As the probe goes over a raised feature, the cantilever deflection is then detected by the photodiode array since the laser spot on the detector shifts during this event. Usually scanning is performed in "constant contact force" mode. A computer-controlled feedback system maintains constant force by moving the sample stage up or down (i.e., changing the voltage applied to the z-piezoelectronic element) during scanning to keep the laser spot on the photodiode array from shifting its position.

Many improvements aimed at increasing the utility of SFM have been recently reported and made commercially available, such as, for example, improvements in probe design/shape and the development of liquid sample stages and electrochemical cells. We became interested in developing a sample stage for imaging single particles down to colloidal size (micrometer-scale) after learning of the work by Horber, et al.[4] who developed a micropipet sample stage for SFM imaging of biological cells in aqueous solution. The microscope used by Horber and colleagues scanned the surface by vibrating the cantilever and laser assembly while keeping the micropipet stationary.

Since the scanning force microscopes at Arizona State University operate by vibrating the sample while keeping the cantilever stationary, we realized that our sample stage designs must be sufficiently rigid so that the images of the particle surface would not be distorted by any "whipping" action caused by micropipet movement. In order to verify that our particle sample stages could in fact generate accurate images of a particle surface, we used polymer adsorbents which have been studied by scanning electron microscopy. The polymer adsorbents also provided a source of spheres with a relatively wide range of sizes. Material properties, such as macroreticularity (well defined pore structure) and rigidity were also beneficial in helping to discern surface features and in the general operability of the scanning force microscope.

EXPERIMENTAL METHODS

Particle sample stages were based on two different designs. Figure 2 is a photograph showing the sample stage design consisting of a Kimax™ soft glass capillary tube with an outer diameter of approximately 1 mm encased in modeling plastic (Edmund Scientific, Barrington, NJ). This design is subsequently referred to as the "elbow design". The capillary is bent by partially melting the joint with a bunsen burner flame and bending by hand so that a short stub of about 0.5-1 cm is directed vertically upwards.

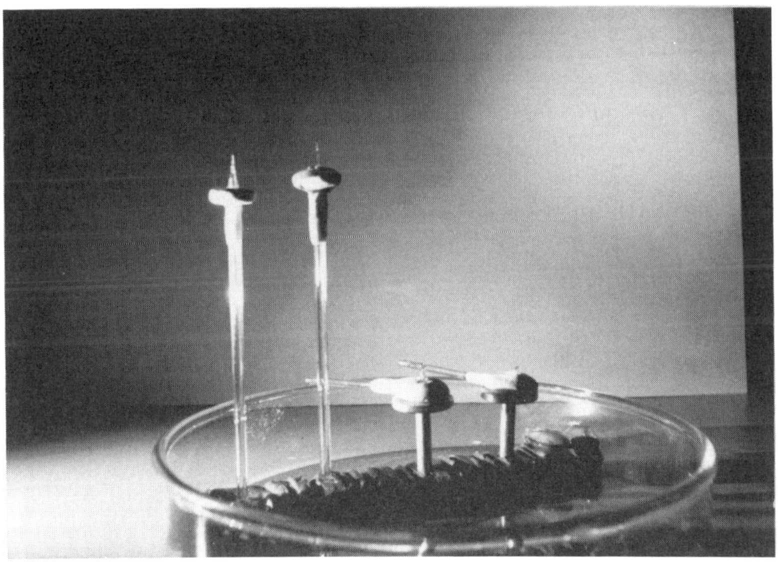

Figure 2. Particle sample stages used for imaging with the Park Scientific Ambient AFM. Both the "elbow" and "volcano" designs are shown.

By placing the tip of the vertical stub in the outer cone of the Bunsen burner flame, the hole can be reduced in size. Care is taken to prevent the capillary from sealing at the tip or along the neck near the tip. Some difficulties inherent with this technique are: (1) the hole size cannot be easily predicted since many of the factors dictating hole size are not well controlled; and (2) the shape of the tip opening is far from circular. The important advantages of the "elbow design" are simplicity and low cost. By encasing the capillary in modeling plastic, the sample holder can then be easily attached to the AFM (Park Scientific Instruments, Sunnyvale, CA) sample stub using a superglue. The superglue bonding can be broken using a razor blade allowing for reuse of the sample stub and holder.

Figure 2 also shows the second sample stage design consisting of a borosilicate glass capillary with a "volcano" shape tip formed by a capillary puller (Narishige, Japan). Since bending the tip in a fashion similar to the "elbow design" proved difficult, this sample stage was used by inserting the capillary through the hollow piezo element of the Park Scientific Ambient AFM. Modelling plastic was used to form a collar around the microcapillary - taking the place of the SFM sample stub. The capillary was secured to the piezo tube by using the allen screw provided by the SFM manufacturer to hold the sample stub in place. The "volcano design" has a significant advantage over the "elbow design" in that the size of the opening at the tip can be controlled by varying parameters, such as the tension, the rate of heating, and the coil temperature in the capillary puller. A major concern in this design is that extreme care must be taken to prevent moisture from coming into contact with the high voltage wires attached to the piezoelectric element.

Surprisingly to us, we determined that attaching long (2 m) flexible tubing to the capillary tube in both sample stage designs and using a simple laboratory aspirator to create a slight vacuum allowed for generally good quality images of particle surfaces. We hypothesize that the flexible tubing dampens much of the vibration induced by the running water. It is clear that due to the relatively low mass of the particles, a high vacuum is unnecessary. Further, by a judicious choice of capillary tip opening size (using an opening slightly smaller than the diameter of the particle) the particle can be held tightly in place--also improving image resolution by preventing the particles from "rocking" during scanning. Further vibration isolation is introduced by suspending the scanning force microscope on bungee cords from a camera tripod.

The SFM used in these experiments was a Park Scientific Ambient AFM. The SFM probes used were 4.5 μm conical tip silicon cantilevers (Park Scientific Co., Sunnyvale, CA). For experiments with the sample stage shown in Figure 2, a scan tube with a maximum x-y scan range of 250 μm and a maximum z-extension of 14 μm was used. Also, a scan tube of 75 μm with a maximum z-extension of 5 μm was used with the "volcano design." Images were collected in the constant force mode. All the imaging data shown in this paper have not been image processed.

RESULTS AND DISCUSSION

To illustrate the use of these sample stages, a photograph of the volcano design is shown in Figure 3 (a) (sideview) and 3 (b) (topview) where an Amberlite XAD-71 adsorbent particle (nominal diameter between 50-80 μm) is held in place by aspirator vacuum. The V-shaped (180 μm thin leg) AFM cantilever is seen in contact with the adsorbent particle. The laser has been turned off in order to obtain good quality photographs. Positioning of the cantilever on the particle is relatively easy with the Park Scientific Instrument AFM since a 400X video microscope for overhead viewing and a 50X monocular field microscope for side viewing is provided. The two most important constraints dictating the operability of particle imaging involve electrostatic charge build-up and the z-piezo range of the scanning head.

Figure 3. (a) Sideview of an Amberlite XAD-71 adsorbent particle held in place by vacuum using the "volcano" stage design. A scanning force microscope cantilever is seen in contact with the adsorbent particle.

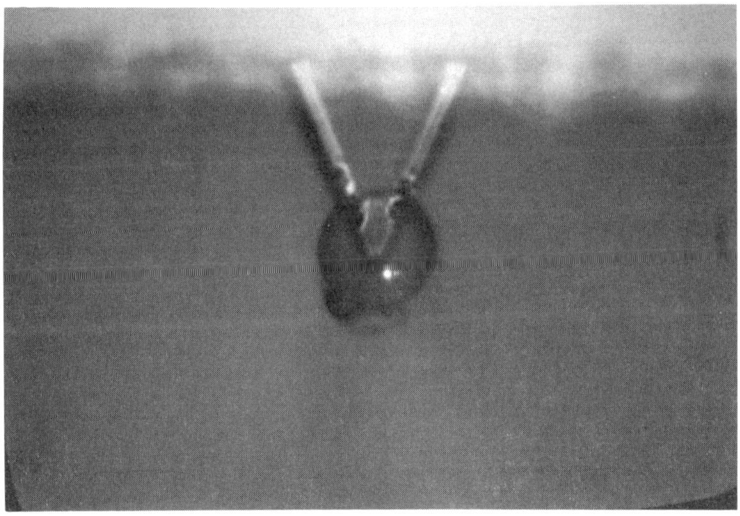

Figure 3. (b) Topview of the setup described in the caption of Figure 3 (a) where the circular objective is the adsorbent particle and the V-shaped object is the cantilever. The cantilever used here has a leg length of approximately 180 μm.

Electrostatic charge, which accumulates on the particle, can cause the cantilever to fluctuate wildly as the cantilever comes in close proximity to the particle. Both sample holder designs do not allow for electrical grounding. Fortunately, in our experiments we have been infrequently disturbed by electrostatic charge build-up because of the use of antistatic carpeting and humidity control.

Particle imaging creates another limitation in that particle curvature limits the area that can be scanned without withdrawing the SFM probe and repositioning. The z-piezo element extension normally used to follow the surface features caused by microscopic roughness is principally needed in particle imaging to maintain force contact on the curved surface. Assuming a spherical particle and using the law of cosines for triangles, the vertical distance traveled by a probe as a function of the scan length and the radius of the particle is given by:

$$z = \sqrt{2}\,r\left(1-\cos\left(\frac{s}{r}\right)\right)^{\frac{1}{2}}\sin\left(\frac{s}{2r}\right)$$

where r and s are the radius of a great circle of the sphere and the arc length (which corresponds to the scan length).

Using the maximum z-range of 14 micrometers available on the Park Scientific Ambient AFM, this equation yields the maximum scan length and the percent of the hemisphere area that can be scanned as a function of particle radius (see Figures 4(a) and 4(b)). Since the curves in Figures 4(a) and 4(b) represent the entire z-piezo element extension, these are the maximum possible values. For reasonable microscope operability, not all of the z-piezo extension should be devoted to maintaining force contact with the particle. About 1-2 μm of z-piezo extension should be reserved for microscopic surface roughness. It is also important to remember that the entire hemisphere of the particle (assuming that half of the particle protrudes from the microcapillary tip) can be scanned by repositioning the cantilever when the z-piezo limit is exceeded. However, reconstructing the entire particle topography would require careful mapping of the entire surface.

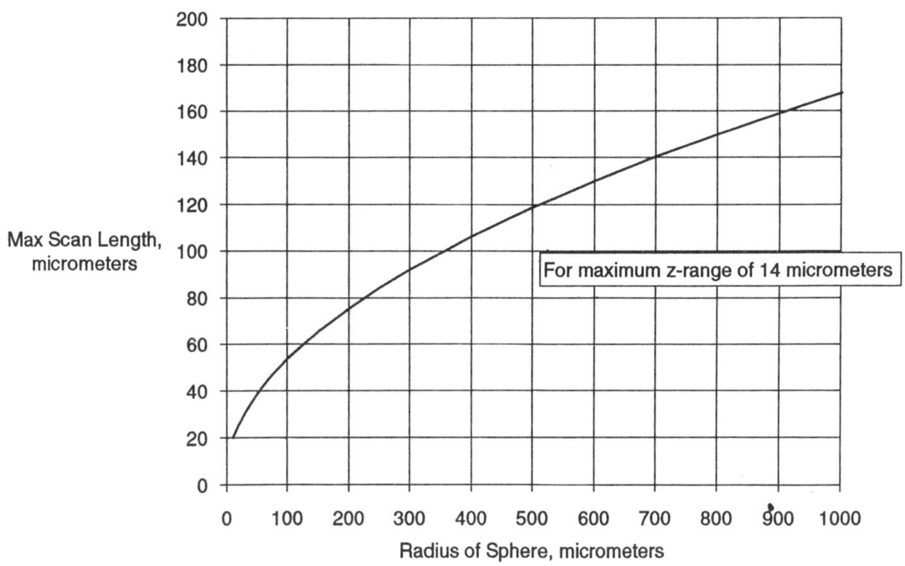

Figure 4. (a) The maximum scan length as a function of the radius of a spherical particle for scanners with a maximum z-piezo element range of 14 micrometers.

Figure 5 to Figure 8 are unfiltered and uncropped AFM images of hydrated and dry adsorbents obtained using the particle sample holders. All of the adsorbents are manufactured by Rohm & Haas. Ambersorb XE-348 is a carbonaceous adsorbent made by sulfonation followed by pyrolysis of macroreticular polystyrene-divinylbenzene adsorbents[5] such as the Amberlite XAD-4 polymer adsorbent. All adsorbents except Ambersorb XE-348 were imaged in their hydrated state. The hydrated particles were prepared for imaging by placing a drop of adsorbent suspension on top of the microcapillary while

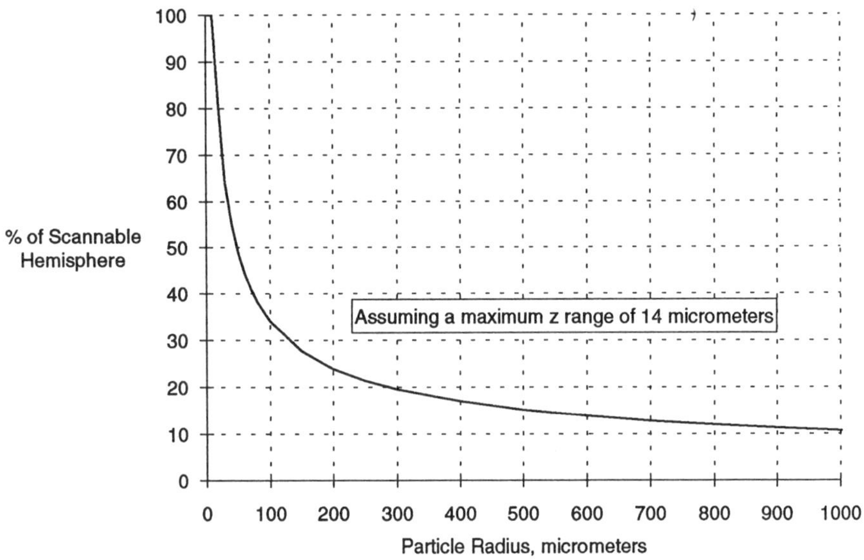

Figure 4. (b) The percentage of the spherical particle surface that can be scanned as a function of the radius of a spherical particle for scanners with a maximum z-piezo element range of 14 micrometers.

drawing a vacuum. Room humidity was sufficiently high so that we did not expect a high rate of particle drying. Also, the water vapor pressure above the adsorbent micropores is lower than the normal vapor pressure.[6] Because of these reasons we assumed that the hydrated particles did not lose a significant amount of water and that hydrated particles can be imaged without using a liquid sample cell and immersing the cantilever in solution.

Horizontal streaks decrease image quality (see Figure 8). This can be caused by vibrational disturbances, the probe "sticking" to the surface, or by weakly bound impurities being swept around by the probe. Image streaking can be removed by postimage processing techniques, such as averaging using nearest neighbor pixels or using pixels

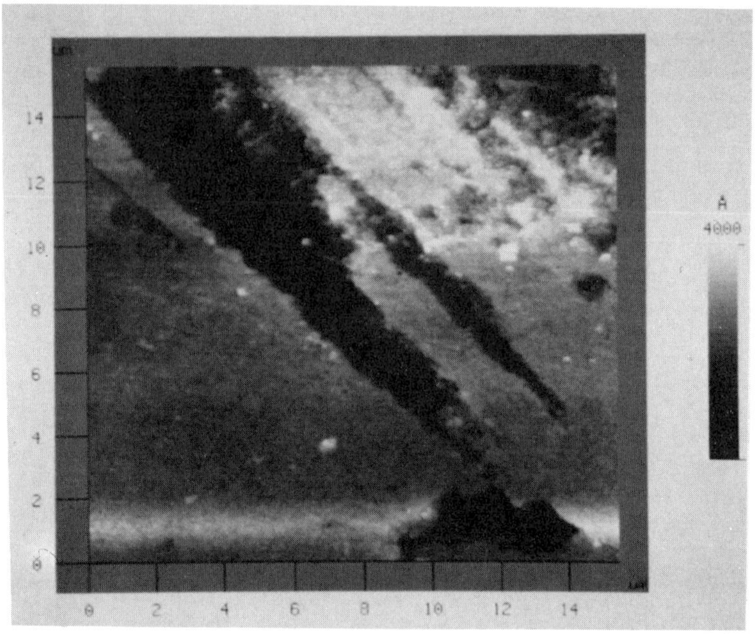

Figure 5. An SFM image of a hydrated Amberlite XAD-4 adsorbent particle obtained using the single particle holder.

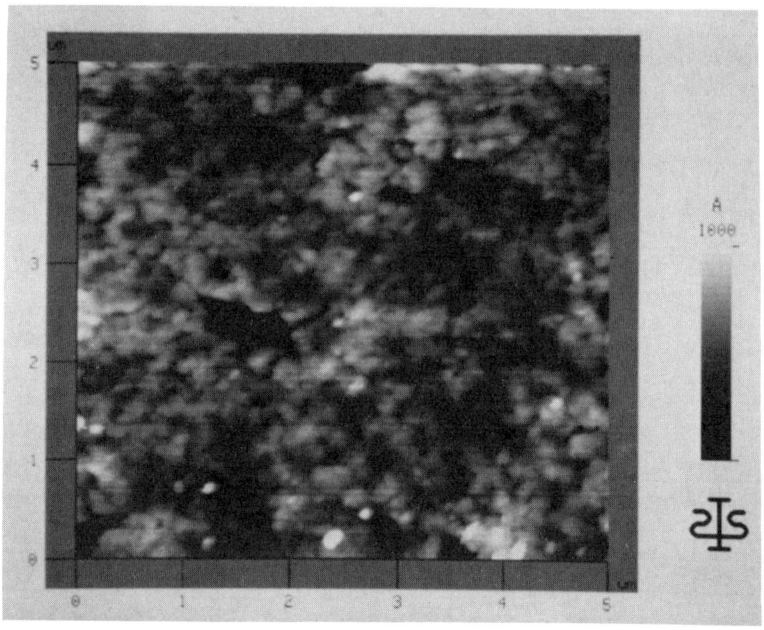

Figure 6. An SFM image of a hydrated Amberlite IRA-900 ion exchange particle obtained using the single particle holder.

in an entire line scan. However, some of the image data would then be approximated. A general observation that can be made of these images is that the surface of these adsorbents can be characterized as consisting of submicrometer spheroidal particle agglomerates, which appear to be sintered. Macroreticular resins are known to be formed by the aggregation of submicrometer particles that precipitate out of the suspension during polymerization.[7] Scanning electron micrographs of Amberlite XAD-4[7] and Ambersorb XE-348[8] also show a granular-like structure for the interior of these particles.

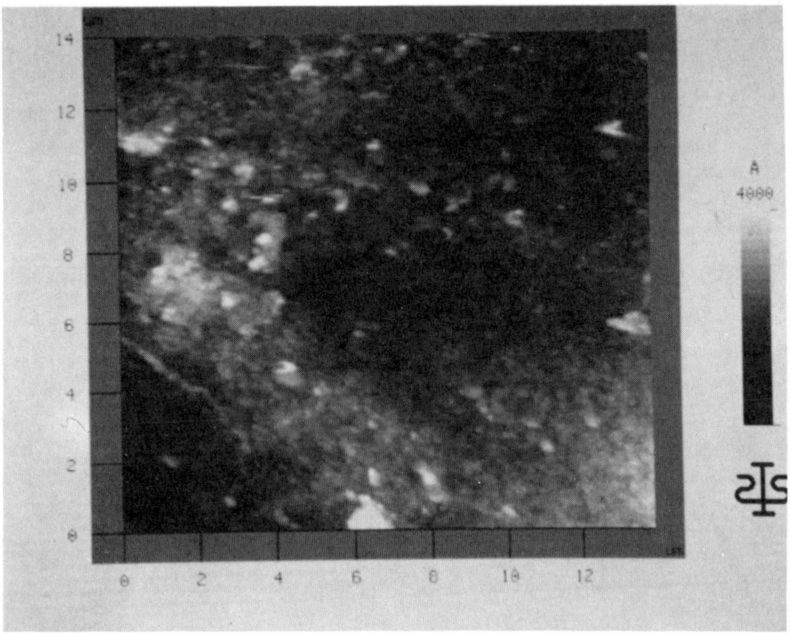

Figure 7. An SFM image of a hydrated Amberlite XAD-71 adsorbent particle obtained using the single particle holder.

The resolution of the SFM images shown in Figures 5-8 can be calculated by dividing the scan length by 512 since 512 scan lines were collected to form these images. Thus, some of the images have a resolution as high as 10 nanometers. The z or "depth" resolution of SFM can be as high as 0.01 nanometers, although many factors such as the surface feature geometry, servo response gain, and thermal fluctuations can contribute to decreasing the resolution. An advantage that SFM imaging holds over scanning electron microscopy (SEM) is that the intact particle can be imaged. In SEM the particles are embedded in a matrix and sliced for mounting onto the SEM stage.

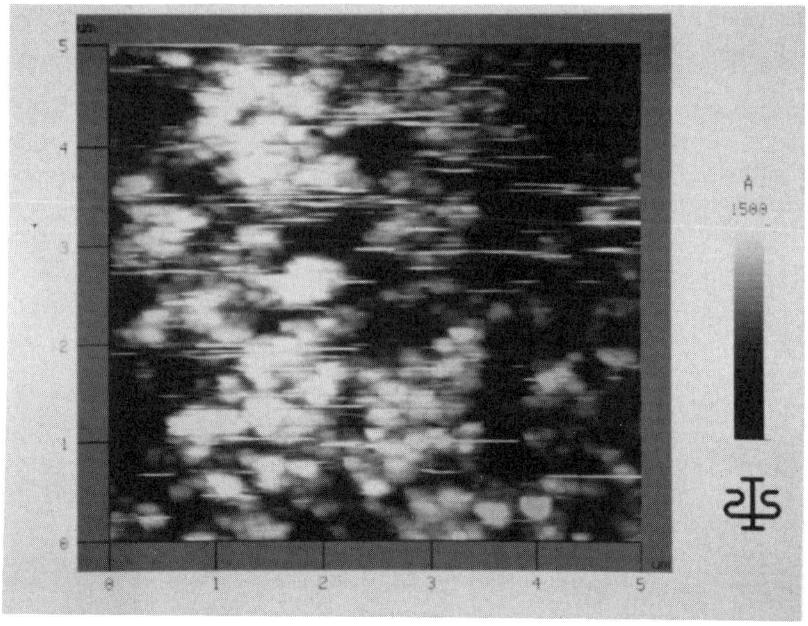

Figure 8. An SFM image of a dry Ambersorb XEN-575 adsorbent particle obtained using the single particle holder.

SFM images can be used to obtain information on particle surface porosity and roughness. Figures 9 and 10 give morphological data for Amberlite XAD-4 and Ambersorb XE-348 showing that for these images the distribution of pore entrance diameters is more uniform and the surface roughness is lower for the unpyrolyzed macroreticular polymer adsorbent. However, Figure 5 shows that the image in Figure 9 is an image of a smaller scan area on the XAD-4 particle. The large crack on the surface of Amberlite XAD-4 in Figure 5 indicates that this adsorbent can also have very large (several micrometers in width) pores. The data in Figures 9 and 10 are given here as an illustration of how the SFM topographical data can be used to characterize particle morphology. A definitive study of these adsorbents would require more images of several particles in order to accurately portray their surface porosities.

SUMMARY

A new technique for imaging micrometer-scale particles using the scanning force microscope holds promise as a versatile method for characterizing surface morphology. In order to test the stability of imaging while immobilizing the particles using a slight vacuum, imaging experiments were carried out with macroreticular adsorbents with particle diameters ranging from 70-500 micrometers. Two particle sample holder designs that utilize glass microcapillaries were tested with particles in the dry or hydrated state. These sample holders can be used with scanning force microscopes that vibrate the sample or the cantilever during scanning and do not require the use of a liquid cell to image hydrated particles.

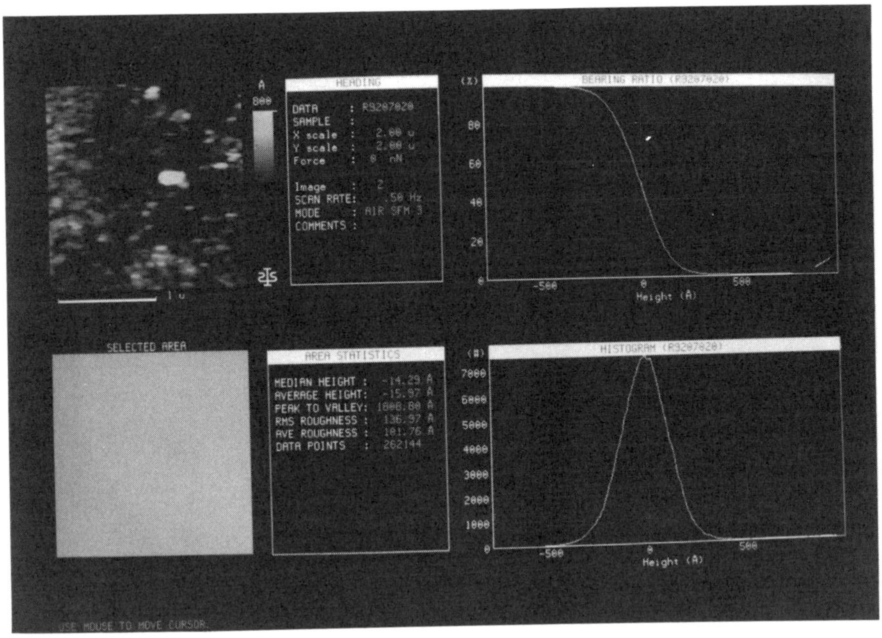

Figure 9. An analysis of the surface topography information obtained on Amberlite XAD-4. Area statistics such as average height and roughness are given. Also, a bearing ratio defined as the percent of the image that has a height greater than a certain value. The derivative of the bearing ratio yields a histogram that is a measure of the pore width distribution.

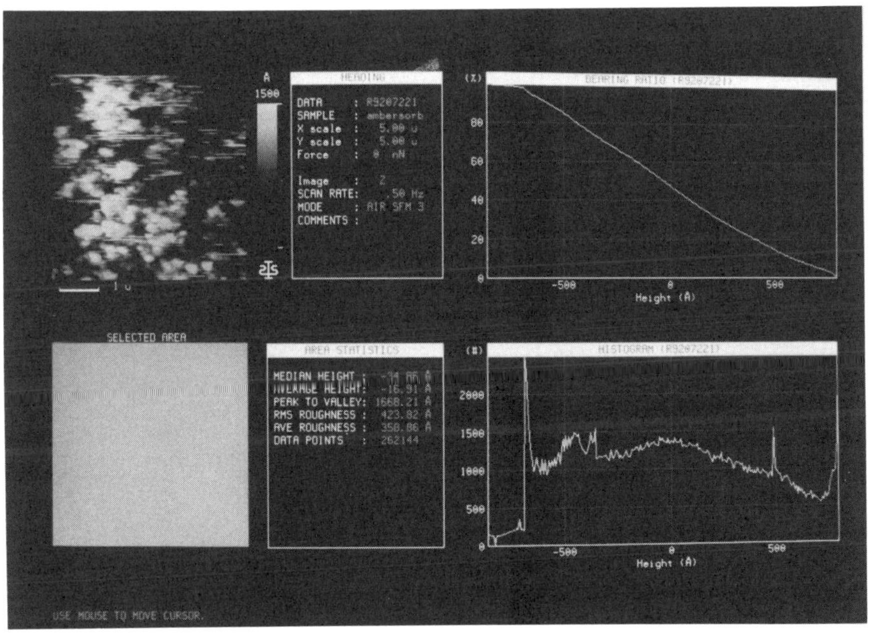

Figure 10. The same analysis (explained in the caption of Figure 9) for Ambersorb XE-348.

ACKNOWLEDGMENTS

Marcus Edwards was funded by the American Chemical Society under Project SEED. Funding was also provided by the Whitaker Foundation. The authors are indebted to Mr. Y. Li (Arizona State University Physics department) for the original idea of the "elbow design" and to John Graham (Park Scientific, Sunnyvale CA) for his valuable insight and creativity towards the "volcano design". Richard Satterlie (Arizona State University Zoology department) kindly gave us access to the microcapillary puller in his laboratory. We also are thankful to Horber and colleagues for suggesting this work by the presentation of their research on micropipet imaging of cells at the 1992 Scanning Microscopy International meeting in Chicago. We also thank Rohm & Haas and Supelco for providing the adsorbent samples.

REFERENCES

1. G. Binnig, H. Rohrer, Ch. Gerber, and E. Weibel, Surface study by scanning tunneling microscopy, *Phys. Rev. Lett.* 49:57-60 (1982).
2. G. Binnig, C. Quate, and Ch. Gerber, Atomic force microscope, *Phys. Rev. Lett.* 56:930-933 (1986).
3. S. Snyder and H. White, Scanning tunneling microscopy, atomic force microscopy and related techniques, Anal. Chem. 64:116R-133R (1992).
4. J. Horber, W. Haberle, F. Ohnesorge, G. Binnig, H. Liebich, H. Mahnel, C. Cerny, and A. Mayr, Investigation of living cells in the nanometer regime with the scanning force microscope, *Scanning Microscopy* 6(4):919-925 (1992).
5. J. Neely and E. Isacoff, "Carbonaceous Adsorbents For The Treatment of Ground And Surface Waters," Marcel Dekker, Inc., New York (1982).
6. A. Adamson, "Physical Chemistry of Surfaces," 4th Edition, John Wiley & Sons, New York (1982).
7. F. Slejko, "Adsorption Technology," Marcel Dekker, Inc, New York (1985).
8. S.H. Cohen and R.A. Segars, Thin sectioning of carbonacious adsorbent spheres for visualization by light microscopy and scanning electron microscopy, *Scanning Microscopy*, 5(2):363-367 (1991).

SCANNING TUNNELING MICROSCOPY OF FREEZE FRACTURE REPLICAS OF BIOMATERIALS

John T. Woodward[1] and Joseph A. Zasadzinski[2]

[1]Department of Physics
[2]Department of Chemical and Nuclear Engineering
University of California, Santa Barbara
Santa Barbara, California 93106

Abstract: Under ambient conditions STM measurements of feature heights on biological and other soft materials are often much larger than expected and vary from point to point on the surface: STM imaging of these samples is also accompanied by deformations of the surface that would not be expected from the conventional picture of noncontact STM imaging through a vacuum gap. To explain these observations we have developed a two spring model for the interaction between the tip and sample that suggests that a fluid meniscus couples the tip to the sample leading to large height amplifications and the possibility of damage to the surface. To test this theory we imaged platinum carbon replicas of cadmium arachidate multilayer Langmuir-Blodgett films under a dry nitrogen environment, exposed to humid air, and returned to a nitrogen environment. Feature heights increased significantly in the humid environment, but were reversible upon return to a dry nitrogen environment.

INTRODUCTION

With the advent of the scanning tunneling microscope (STM) in 1982 it became possible to image solid, conducting samples with atomic resolution.[1] The subsequent invention of the atomic force microscope (AFM) made it possible to routinely image nonconducting surfaces as well.[2] However, the local pressure applied by these probes to the sample surface during imaging has limited their usefulness in imaging soft samples. As many biological structures are too soft for direct imaging, we have modified the technique of freeze fracture replication for use with the STM.

Freeze fracture replication was originally developed as a way to observe structures with the transmission electron microscope (TEM) that could not tolerate the high vacuum inside the TEM. This includes many of the hydrated biological systems that at present are incompatible with the scanning probe microscopes. The preparation consists of rapidly freezing the sample so as to lock in the structure or phase of interest. The sample is then fractured under high vacuum and a thin metal film is deposited on the surface. The original sample is then dissolved and the metal replica of the surface is observed in the microscope. For observation in the TEM, the metal film is deposited at an angle and structural information is obtained from the resultant shadowing.[3] For the STM the metal is deposited vertically onto the sample in order to form a continuous metal replica of the surface for imaging.

It should be noted that this technique represents a sacrifice of what is often considered the prime motivation for using a scanning probe microscope: atomic resolution. At best we could expect each atom on the surface of the replica to have been in contact with the sample surface. In reality the replica is composed of tiny crystallites of about two nm in diameter. It is these crystallites that limit the resolution of STM on freeze fracture replicas. However, in return for our sacrifice of resolution we are able to image a class of samples that are otherwise incompatible with scanning probe microscopes. For many of these samples it is the macromolecular structure, not the atomic scale detail, that is of interest.

We present here the techniques we have developed as a result of attempting to image two organic structures. The first is a reorganized Langmuir-Blodgett film of cadmium arachidate on mica. The reorganized films have regions that are 1, 3, 5 and 7 layers thick and have been well characterized using the AFM.[4] As the thickness of these layers is well known from X-ray scattering[5,6] they are ideal for calibrating the z direction of scanning heads and determining the reproducibility of height measurements. In addition there are large (>10 μm^2) flat areas between the steps that allow the smoothness of the replica to be assessed. The second sample is the ripple or $P_{\beta'}$ phase of dimyristoyl phosphatidylcholine (DMPC). While the ripple wavelength has been measured using X-ray scattering[7-10] as well as freeze fracture TEM[11-13] there are unresolved questions about the ripple amplitude, waveform, and the possibility of a smaller ripple running roughly perpendicular to the main ripple.

In our first attempts to image freeze fracture replicas with the STM we prepared the replicas in a manner similar to that used for the TEM. The replicas were 2.5 nm of platinum carbon (Pt/C) with a backing layer of 30 nm of carbon added for structural support. The replicas were mounted on 500 mesh electron microscope grids.[14] While we were able to obtain many stable images of the ripple phase of DMPC we occasionally found images with ripple amplitudes that were several times the usual height and on a few occasions we obtained amplitudes over 100 times the typical value. Amplitudes could also change dramatically when the same area was imaged for several hours. We examined the possibility that the scanning tip might be deforming the replica, and found that if a liquid meniscus forms between the tip and replica it is possible for deformations to occur which could substantially enlarge the measured height of a surface feature.[15] Since all of our imaging was done in air we consider it likely that a meniscus of water or other contaminant forms between the tip and sample.

We developed a model that suggests three main methods for reducing the amplifications. The first two of these are relatively simple and obvious: make the replica thicker and mount it on a substrate with smaller pores. The amplifications are reduced, but not eliminated, by adding a 100-200 nm backing of silver or gold to the replica and mounting on a fine mesh silver membrane with pores 1-3 μm in diameter.[15] The third suggestion was to image in an inert environment. A liquid meniscus will not form between the tip and sample at a low relative humidity. To this end we built a chamber for the STM out of a belljar and a base plate that allows us to evacuate the chamber and backfill with nitrogen. Thus, we can routinely and easily image samples in an inert environment without the expense or maintenance required for an ultra high vacuum STM.

EXPERIMENTAL

To test whether we were able to eliminate the height amplifications, we wanted to use a sample of known height that we could replicate. We chose multilayer films of cadmium arachidate prepared by Langmuir-Blodgett deposition. By holding the films in the subphase prior to the deposition of the final monolayer the films reorganize creating a terraced surface with most levels differing by a bilayer or multiple bilayers in height.[4] To calibrate the STM head we used both a three-layer film on mica with a half-hour wait between the second and third layers and a five-layer film with a half hour wait between the fourth and fifth layers. Both were prepared as described in Ref. 4 and coated with Pt/C using the following procedure. The film was loaded into a Balzers BAF 400K/CL freeze fracture device and the chamber was evacuated to 10^{-7} mbar. We then deposited 0.8 nm of Pt/C at a 45° angle on the stationary sample and another 0.8 nm with the sample rotating. This guaranteed a good conducting path between layers and enough conductivity on the initially shadowed side to prevent tip crashes. The deposition was performed at room temperature.

We then imaged the Pt/C coated films with a Nanoscope II STM in air with several different cut platinum-iridium tips. Over 40 step heights were measured from 18 different images. The height was measured by taking bearing plots that straddled the steps using the Nanoscope software. The step height was taken to be the distance between peaks in the bearing plot with an uncertainty relative to the peak widths. The weighted mean was set equal to the known value for a bilayer step of 5.5 nm.[4,5] Seventy percent of the measured values fell within eight percent of the mean. This spread of values is only slightly larger than the spread found using the AFM to measure the height of the steps directly.

We also made a replica of a five-layer film with half-hour wait between the fourth and fifth layers. The film was loaded into the Balzers freeze fracture unit and pumped down to 10^{-7} mbar and held at room temperature. The replica was made by depositing 2.0 nm of Pt/C with the sample rotating and then backing with 35 nm of carbon. An additional backing of 200 nm of silver was added in an evaporation chamber. The replica was removed by floating the sample on a 5% hydrofluoric acid solution overnight. The replica was then mounted on a silver membrane as described in Ref. 15. The replica was positioned in the STM and was placed under the belljar. The bell jar was pumped out using a mechanical pump with a cold trap on the line to about 0.1 torr for over an hour and backfilled with dry nitrogen. A total of 10 different images taken with three different tips gave 28 bilayer step heights. These measurements gave a weighted mean of 5.1 nm for a bilayer step, and once again 70 percent of the values fell within 8 percent of the mean. On this basis we find that there is no measurable amplification of height measurements made with the STM when imaging in an inert environment.

There has recently been some controversy over whether a meniscus that formed between a tip and sample that was covered by a continuous liquid film could apply a force to the sample.[10,17] In order to demonstrate more clearly that liquid condensation is responsible for the deformations that we have observed we endeavored to change the environment in which the STM was operating while it scanned over the same object. We began by imaging the replica of the cadmium arachidate film in an inert environment as described above. A suitable feature was found and imaged for 15 to 30 minutes to ensure that it was stable. Air was then admitted to the chamber while imaging. After about 15 minutes, the chamber was flushed with nitrogen to determine if the changes observed in air were reversible.

We followed this procedure six times and found that three of the runs showed no amplification, two of the runs showed small amplifications (less than the 8% spread in height measurements), and one run showed some dramatic changes. This is entirely consistent with our model which predicts amplifications only when imaging over a pore and no amplification when imaging over an area that is in contact with the silver membrane. Three images and cross-sections from the run that had large amplifications are shown in Figure 1. Figure 1 a was taken after fifteen minutes of a thirty minute continuous scan under nitrogen. Figure 1(b) was taken less than five minutes after exposure to air. The island resembling Australia has increased in height by 17% and the difference between the highest and lowest points on the image has doubled. Figure 1 (c) was taken less than five minutes after we began flushing the chamber with nitrogen. The height of the island is only 8.5% above its original value. The island returned to its original height after less than half an hour of flushing with nitrogen. We attribute these changes to the formation of a meniscus between the tip and sample due to condensation from the air and the subsequent evaporation of the condensation upon return to the nitrogen environment. We therefore feel that we are able reproducibly to image replicas with an STM by imaging in an inert environment. We can make height measurements on the order of 5 nm with an accuracy of better than 10%.

The limiting factor in resolution of images of replicas is now the graininess of the replica itself. The images in Figure 1 clearly show that the replica is composed of small crystallites. This replica was deposited at room temperature and has crystallites that are about 5 nm in diameter. Figure 2 is a replica of the ripple phase of DMPC that was prepared similarly to the replicas in Figure 1 except that the sample was at -170 °C when the replica was deposited. This method formed crystallites about 1-2 nm in diameter.

Several researchers have investigated the optimal conditions for depositing replicas of Pt/C as well as tantalum-tungsten (Ta/W) for imaging in the TEM. Ruben has shown that vertically deposited Pt/C films can be made with grains as small as 0.5 nm if the films are less than 1.0 nm thick.[18] However, these films are generally not continuous as is required for the STM. Increasing the thickness to form a continuous film increases the grain size as well. Gross et al. have made continuous films of Pt/C that are essentially grainless, but this required a freeze fracture machine capable of obtaining temperatures of -260 °C and pressures less than 10^{-9} mbar both of which are beyond the capabilities of standard freeze fracture systems.[19]

ONGOING RESEARCH

We have begun to explore the use of amorphous or 'glassy' metals to try to produce continuous, grainless films with a standard liquid nitrogen cooled freeze fracture unit. Glass-forming metals are typically binary alloys that have a deep eutectic at which the melting point is substantially less than either of the pure components. When rapidly cooled from a vapor or liquid state, as they are when deposited onto a cooled substrate during freeze fracture replication, crystallization is inhibited and they retain the amorphous structure of the liquid state. One general class of glass-forming metals is alloys of approximately 80 atomic percent transition metal and 20 atomic percent silicon or phosphorus. For example, platinum-silicon has a eutectic near 78% Pt/22% Si and palladium-silicon near 84% Pd/16% Si. Our first attempt to make an amorphous film was with an 88% titanium/12% silicon alloy. However, we were not able to get reliably stable images with the STM. We suspect that this is due to the silicon oxidizing and forming an insulating layer on the surface. The alloy also gave poor contrast in the TEM, making it difficult to assess the continuity and crystallinity of the sample using that technique. We have therefore turned our attention to the higher atomic number alloys so that we may use the STM and TEM to help optimize the quality of freeze fracture replicas.

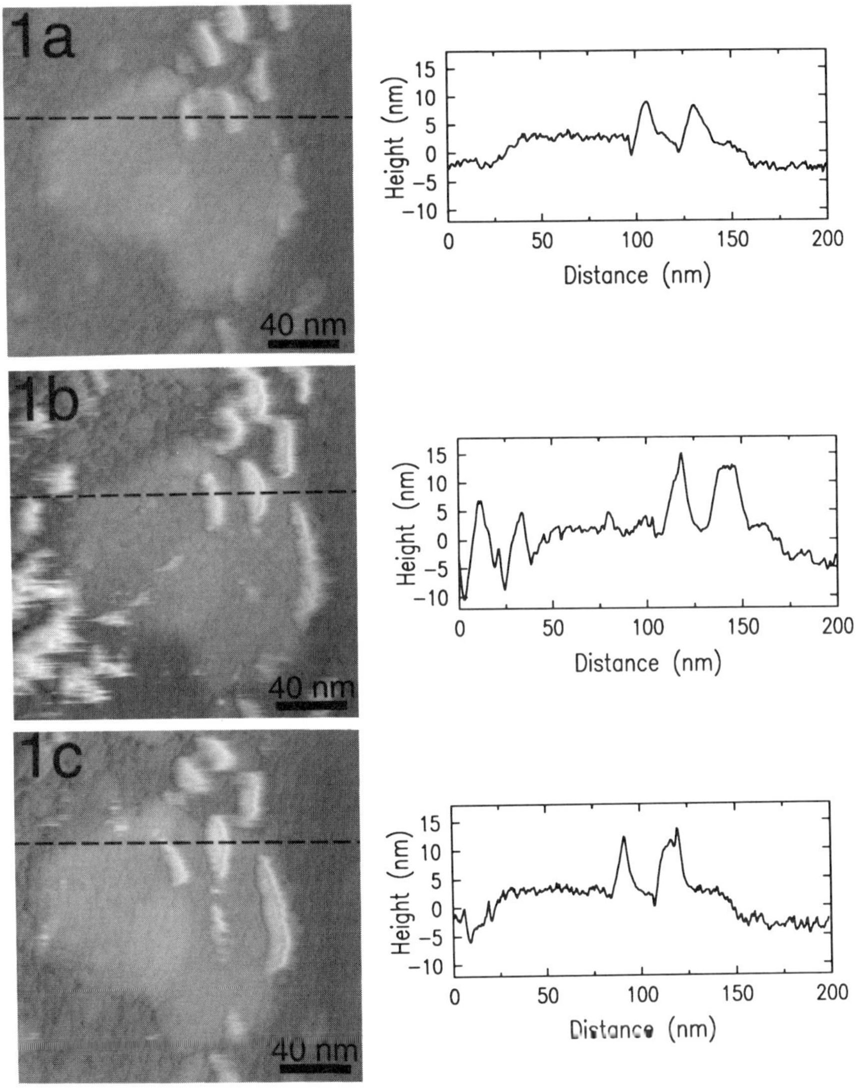

Figure 1. (a) An STM image taken under nitrogen of a replica of a reorganized cadmium arachidate film and the height profile along the dashed line. (b) The same area after less than five minutes exposure to air. The step heights have increased and additional features are visible. (c) The same area less than five minutes after being returned to a nitrogen environment. The amplification is reduced and the additional features have disappeared.

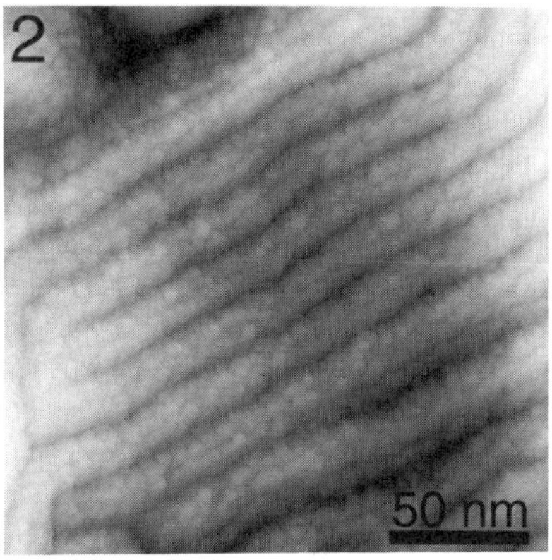

Figure 2. An STM image of a freeze fracture replica of a 30% DMPC/70% water solution at 21 °C. The image shows the long wavelength (22 nm) Λ conformation of the ripple phase. This phase has a characteristic 'M' shaped ripple conformation. The individual crystallites that compose the replica are clearly resolved. The shallow valleys along the top of each ripple, however, are somewhat obscured by the crystallites.

CONCLUSION

In conclusion, we have modified the method of producing and imaging freeze fracture replicas with the STM to enable accurate and reproducible imaging. Our experiments have shown that by making replicas with thicker backing films and imaging them in an inert environment we are able to eliminate distortions due to the formation of a fluid meniscus between the tip and sample. With these improvements the current limitation on the resolution of this technique is the quality of the replicating film itself, which is an area that we are actively exploring.

ACKNOWLEDGEMENTS

We would like to acknowledge Ravi Viswanathan for his preparation of the Langmuir-Blodgett films. We are also indebted to Paul Hansma, Jacob Israelachvili, and Suzi Steinberg for helpful discussions. This work was supported by a Whitaker Foundation Biomedical Engineering Grant, the Office of Naval Research grant #N00014-90-J-1551, and National Science Foundation grant #CTS9212790.

REFERENCES

1. G. Binnig, H. Rohrer, Ch. Gerber, and E. Weibel, Surface study by scanning tunneling microscopy, *Phys. Rev. Lett.* 49:57-60 (1982).
2. G. Binnig, C. Quate, and Ch. Gerber, Atomic force microscope, *Phys. Rev. Lett.* 56:930-933 (1986).
3. J. Zasadzinski and S. Bailey, Applications of freeze fracture replication to problems in materials and colloid science, *J. Electron Microsc. Tech.* 13:309-344 (1989). This issue is an excellent reference for freeze fracture technique.

4. D. Schwartz, J. Garnaes, R. Viswanathan, and J. Zasadzinski, Surface order and stability of Langmuir-Blodgett films. Science 257:508-511 (1992).
5. P. Tippman-Krayer, H. Mohwald, and Yu L'vov, Structural changes before and during desorption of Langmuir-Blodgett films, *Langmuir* 7:2298-2302 (1991).
6. Y. Sasanuma, Y. Kitano, A. Ishitani, H. Nakahara, and K. Fukuda, Characterization of long-periodic layered structures by x-ray diffraction III: structure of a Langmuir-Blodgett film of cadmium arachidate at elevated temperatures, *Thin Solid Films* 199:359-365 (1991).
7. A. Tardieu, V. Luzzati, and F. Reman, Structure and polymorphism of the hydrocarbon chains of lipids: a study of lecithin-water phases, *J. Mol. Bio.* 75:711-733 (1973).
8. M. Janiak, D. Small, and G. Shipley Temperature and compositional dependence of the structure of hydrated dimyristoyl lecithin, *J. Bio. Chem.* 254:6068-6078 (1979).
9. G. Smith, C. Safinya, D. Roux, and N. Clark, X-ray study of freely suspended films of a multilamellar lipid system, *Mol. Cryst. Liq. Cryst.* 144:235-255 (1987).
10. D. Wack and W. Webb, Synchrotron x-ray study of the modulated lamellar phase $P_{\beta 1}$ in the lecithin-water system, *Phys. Rev.* A 40:2712-2730 (1989).
11. E. Luna and H. McConnell, The intermediate monoclinic phase of phosphatidylcholines, *Biochim. et Biophys. Acta.* 466:381-392 (1977).
12. D. Ruppel and E. Sackman, On defects in different phases of two-dimensional lipid bilayers, *J. Phys.* (Paris) 44:1025-1034 (1983).
13. J. Zasadzinski and M. Schneider, Ripple wavelength, amplitude and configuration in lyotropic liquid crystals as a function of effective headgroup size, *J. Phys.* (Paris) 48:2001-2011 (1987).
14. J. Zasadzinski, J. Schneir, J. Gurley, V. Eilings, and P. Hansrrla, Scanning tunneling microscopy of freeze-fracture replicas of biomembranes, *Science* 239:1013-1015 (1988).
15. J. Woodward, P. Hansma, and J. Zasadzinski, Precision height measurements of freeze fracture replicas using the scanning tunneling microscope, *J. Vac. Sci. Technol.* B 9:1231-1235 (1991).
16. J.-Y. Yuan, Z. Shao, and C. Gao, Alternative method of imaging surface topologies of nonconducting bulk specimens by scanning tunneling microscopy, *Phys. Rev. Lett.* 67:863-866 (1991). J.-Y. Yuan, Z Shao, C. Gao, Yuan, Shao, and Gao reply, *Phys. Rev. Lett.* 68 : 2564 (1992).
17. J. Woodward, J. Zasadzinski, and D. Schwartz, Comment on 'Alternate method of imaging surface topologies of nonconducting bulk specimens by scanning tunneling microscopy', *Phys. Rev. Lett.* 68:2563 (1992).
18. G. Ruben Ultrathin (1 nm) vertically shadowed platinum-carbon replicas for imaging individual molecules in freeze-etched biological DNA and material science metal and plastic specimens, *J. Electron Microsc. Tech.* 13:335-354 (1989).
19. H. Gross, T. Muller, I. Wildhaber, H. Winkler, High resolution metal replication, quantified by image processing of periodic test specimens, *Ultramicroscopy* 16: 287-304 (1985).

THE SCANNING PROBE MICROSCOPE AS A METROLOGY TOOL

Rebecca S. Howland

Park Scientific Instruments, Sunnyvale, CA 94089; Tencor Instruments, Mountain View, CA 94043

Abstract: Linearity of the piezoelectric scanner is critical to using the scanning probe microscope as a metrology tool. In this paper, the inherent nonlinearities associated with piezoelectric materials and scanners synthesized from them are described both macroscopically and microscopically. Various corrections to the nonlinearities are presented and compared.

INTRODUCTION

The first scanning probe microscope (SPM) was used to produce atomic-resolution images of the reconstructed surface of silicon over 10 years ago. New applications have driven the SPM from an imaging tool for surface science research in the direction of an instrument for quantitative measurements. Several barriers stand in the way of making this general purpose research instrument into a production worthy metrology tool. The two most problematic areas are tip integrity and scanner reproducibility. Progress in achieving precise, repeatable scanner control will be addressed in this article.

THE PIEZOELECTRIC SCANNER

In virtually all scanning probe microscopes piezoelectric scanners are used as extremely fine positioning stages to move the probe relative to the sample. The scanner is driven in a kind of raster pattern by the SPM electronics (Figure 1). The scanner moves back and forth across each line, stepping in the perpendicular direction (*slow scan direction*) to the next successive scan line. SPM data are collected in only one direction. This approach differs from a traditional raster pattern (e.g. on a standard television set) in which alternating lines of data are taken in opposite directions. The single-direction sampling minimizes line-to-line registration errors that result from scanner hysteresis.

The image data are sampled digitally at equally spaced intervals as the scanner moves across each scan line. For instance, the data recorded could be the height of the scanner in

z for constant-force AFM or constant-current STM. It could also be the deflection of the cantilever or the tunneling current for constant-height AFM or STM. The spacing between each measuring point (the step size) is determined by the full scan size divided by the number of data points per line. Typical SPM scan sizes range from tens of angstroms to hundreds of micrometers, and usually contain either 256 or 512 data points per line. The number of data points per line usually equals the number of lines. The result is, ideally, a dense, square grid of measurements. This file structure has several benefits in the ease of data handling, image processing and display. The difficulties in achieving a perfectly square measurement grid--and the consequences of having a distorted measurement grid--are described in this paper.

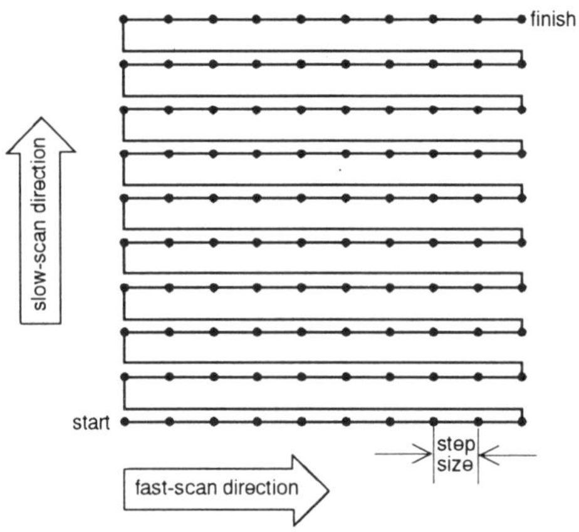

Figure 1. Scanner motion during SPM data acquisition.

DESIGN, FABRICATION AND OPERATION

Scanners are designed to move in three orthogonal directions. They move by changing dimensions in response to an applied voltage, a fundamental property of piezoelectric materials. Piezoelectrics also develop an electrical potential in response to the application of mechanical pressure. This property is exploited in the sensing element of a phonograph arm.

The scanners are usually fabricated from lead zirconium titanate, or PZT, and then doped to create specific material properties. First, a powder is pressed together and sintered, resulting in a polycrystalline solid. Each of the crystals in this piezoelectric material has its own electric dipole moment, the basis of the scanner's ability to move in response to an applied voltage. These dipole moments are randomly aligned after sintering. A process called "poling" must be used to realign them before the scanner will be able to move. In this process, the scanners are heated to about 200 °C to free the dipoles, and then submitted to a DC voltage. After a few hours the dipoles have rotated into alignment and the scanner is cooled rapidly to "freeze" the dipoles into place. The scanner can then reliably respond to an applied voltage by extending and contracting in the desired manner.

Orientation of the dipoles (polarization) is maintained by consistent use of the scanner. Misaligned dipoles are realigned every time voltage is applied to initiate scanning motion.

If the scanner is not used regularly, a significant fraction of the dipoles will wander (depolarize or de-pole) over a period of weeks. De-poling is significantly accelerated if the scanner is subjected to temperatures over 150 °C. This means that an SPM with a heated stage would require adequate thermal isolation from the scanner. (Note: the Curie temperature for PZT materials is about 300 °C).

Many SPMs use variations of the hollow tube design depicted in Figure 2. Electrodes are mounted on the outside of the tube, electrically segmenting it into vertical quarters for $+x$, $+y$, $-x$ and $-y$ travel. An electrode is also mounted to the center of the tube to allow travel in the z direction. Alternating voltages applied the $+x$, $-x$ electrodes induce

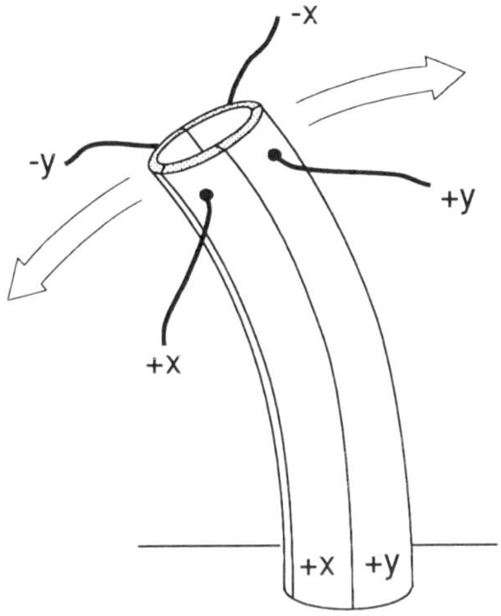

Figure 2. Piezoelectric scanner, with segmented tube design.

mechanical strain on the tube, and cause it to bend back and forth in the x direction. Likewise, voltages applied to the z electrode cause the scanner to extend or contract vertically. For constant-force AFM or constant-current STM the voltage applied to the z electrode of the scanner at each measurement site usually comprises the data set, although in some cases an external sensor is used to measure the scanner height directly. The following discussion assumes the data set consists of voltage applied to the z electrode in the absence of external sensors.

SPMs typically use scanners with a lateral scan range of tens of angstroms to over 100 μm. Height differences from the subangstrom range to about 10 μm can be distinguished. Maximum scan size is determined by the length, diameter, wall thickness and strain coefficients of the particular piezoelectric ceramic from which the scanner is fabricated.

Compactness, subangstrom resolution, and high-speed response are what make piezoelectric scanners so valued in SPM operation. Unfortunately, these very properties bring some challenges as well.

CHALLENGES OF PIEZOELECTRIC SCANNERS

As a first approximation, the strain in a piezoelectric scanner varies linearly with applied voltage. (Strain is the change in length divided by the original length, $\Delta l/l$). The following equation describes the ideal relationship between the strain and an applied electric field:

$$s = d\,E$$

where s is the strain in Å/m, E is the electric field in V/m, and d is the strain coefficient in Å/V. The strain coefficient is characteristic of a given piezoelectric material and determines the size of the voltage that must be applied to move the scanner a given distance.

In practice, the behavior of piezoelectric scanners is not so simple. The relationship between strain and electric field diverges from ideal linear behavior. If we start from zero applied voltage and gradually increase the voltage, the resulting plot of the extension of the piezoelectric material as a function of this voltage will appear, not as a straight line, but as an s-shaped curve (see Figure 3). Quantitatively, the **intrinsic nonlinearity** of a piezoelectric material is the ratio of the maximum deviation Δy from linear behavior to the ideal linear extension y at that voltage. Simply, the nonlinearity is $\Delta y/y$, expressed as a percentage. The piezoelectric materials used in SPM systems usually carry between 2% to 20% intrinsic nonlinearity.

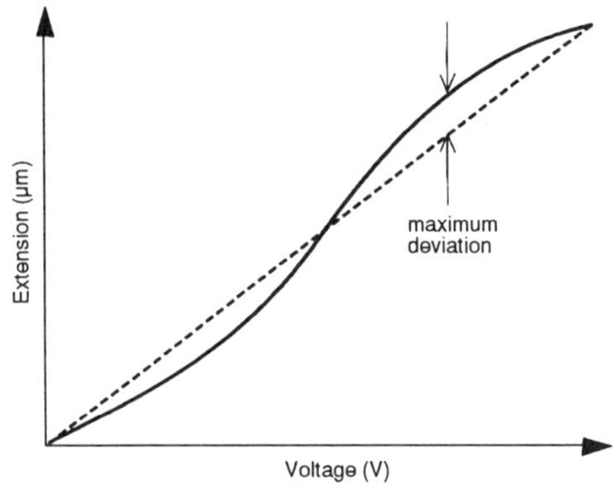

Figure 3. Intrinsic nonlinearity in a piezoelectric scanner: a linear voltage produces a nonlinear extension of the scanner.

The effect of intrinsic nonlinearity in the plane of the sample surface is a distortion of the square measurement grid (see Figure 1). Since the scanner does not move linearly with applied voltage, the measurements points will not be equally spaced. An SPM image of a surface with periodic structures will therefore show nonuniform spacings and curvature of linear structures. Less regular surfaces may not show recognizable distortion, even though distortion is present.

Intrinsic nonlinearity will also cause errors in height measurements (in the z direction perpendicular to the plane of the sample surface). A sample with a known step height is usually used to calibrate the SPM. A reading of the voltage applied to the z electrode as it traverses the step of known height gives the z component of the strain coefficient. However, if that strain coefficient is directly applied to measure a feature different from the calibrated height, the intrinsic nonlinearity of the scanner will result in errors.

The situation is complicated by the **hysteresis** inherent in piezoelectric ceramics. Suppose we start with zero voltage applied to the scanner, gradually increase it and then bring it back down to zero (assume the voltages are changing very slowly). If the extension of the scanner as a function of the applied voltage is plotted, the descending and ascending curves differ, as shown in Figure 4.

Quantitatively, the hysteresis of a piezoelectric scanner is the ratio between the maximum displacement between the two curves and the maximum extension which a voltage can create in the scanner, $\Delta Y/Y_{max}$. Hysteresis can be as high as 20% in piezoelectric materials.

Recall that SPM data are usually collected in one direction to minimize registration errors caused by such scanner hysteresis. Figure 4 indicates that data collected in the return direction would be shifted slightly, distorting the measurement grid of Figure 1. Since most SPMs have the capability to take scans in any fast-scan direction, one can demonstrate hysteresis by comparing data sets taken in opposing fast-scan directions. For example, compare a data set taken left-to-right to one taken right-to-left. Likewise, hysteresis in the direction perpendicular to the sample plane will cause erroneous step-height profiles (Figure 5). Figure 4 tells us that as the scanner contracts to trace up the step in Figure 5, a specific

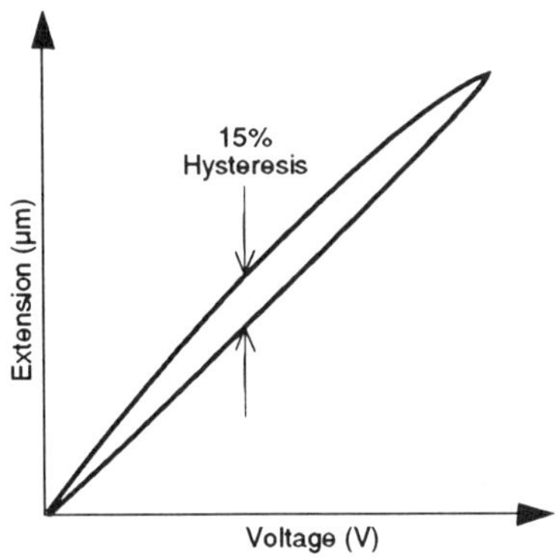

Figure 4. Hysteresis in a piezoelectric scanner: the deflection of the scanner differs depending upon whether it is extending or contracting.

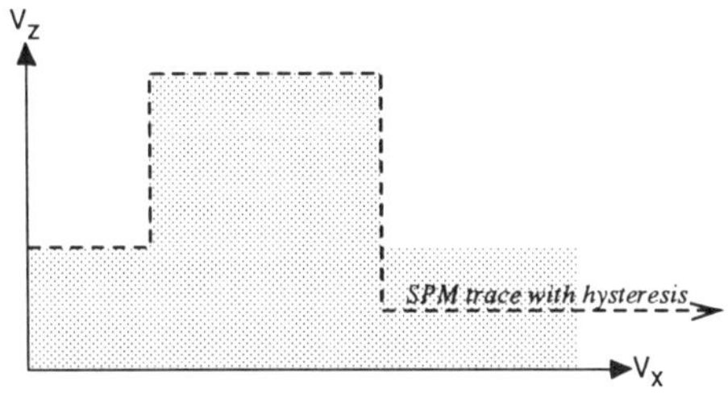

Figure 5. The effects of hysteresis on an SPM image of a step.

voltage is applied. However, to go back down the same step the scanner must extend, which takes more voltage than contraction for the same displacement. Since the SPM image is represented by the voltage applied to the scanner, the image profile would look like that in Figure 5 (assuming the strain coefficient in *z* was calibrated for a contracting scanner).

An abrupt voltage change first produces an abrupt change in scanner position, but then produces a slow, undesirable "creeping" motion in the scanner. Instead of changing dimensions all at once, the scanner moves by two processes: the first occurring in less than a millisecond, the second at a much slower rate. Both processes are in the same direction. The second step (Δx_c in Figure 6) is commonly referred to as **creep**.

Quantitatively, creep is the ratio of the second dimensional change to the first, $\Delta x_c/\Delta x$.

Creep is expressed as a percentage and usually quoted with the characteristic time interval T_{cr} over which creep occurs. Typical values of creep range from 1% to 20%, over times ranging from 10 to 100 seconds.

The times involved during typical scans place the lateral motion of the scanner in the curved portion of the response curve represented in Figure 6. As a result, two scans recorded at different speeds show slightly different length scales (magnifications) when creep is present. When it is, the only reliable measurement is one taken at the scan speed used to calibrate the SPM.

Creep can affect throughput when the SPM is asked to "zoom-in" on an interesting feature. To characterize a defect on a sample surface, the user usually takes a large scan of the defect. If it is located in the corner of this large scan, the user would want to take a much smaller, higher resolution scan centered on this defect. To move the scanner to its new center position in preparation for the scan, the user applies an offset voltage. But because this action involves applying a sudden voltage step, in this case an offset voltage

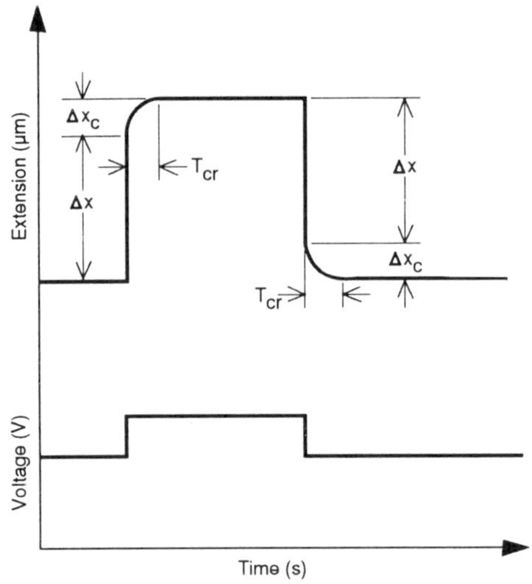

Figure 6. Creep in a piezoelectric scanner: delayed response of the scanner to an abrupt change in voltage.

in the plane of the sample, creep will cause the scanner to miss the target.

The effects of creep in the z direction are illustrated in Figure 7. As the tip traverses the step from bottom to top, the scanner immediately contracts with a voltage correspond-

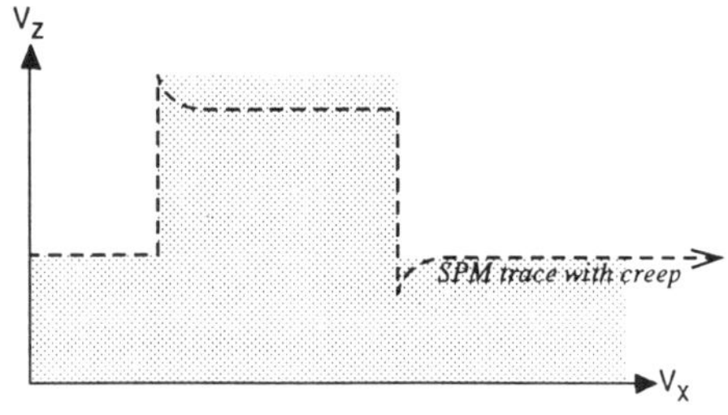

Figure 7. The effects of creep on an SPM image of a step.

ing to the full step height. However, creep will cause the scanner to continue to contract slowly. To maintain tip-to-sample contact, the SPM must apply a voltage in the opposite direction to counteract the creep. The same compensating voltage must be applied when the tip traverses the step from top to bottom.

More realistic demonstrations of the effect of creep on the profile of a step are seen whenever the SPM is used to scan any sample having steep slopes. When imaging defects, grains or other steep surface features, you may notice ridges and shadows appearing on either side of the feature in the SPM image. If the shadows occur where the tip has traveled down a steep slope, or ridges where it has traveled up a steep slope, creep may be at work. Reversing the fast-scan direction and taking the same image will help the user separate creep artifacts from true ridges and trenches.

Figure 8 graphs the results of simply using or not using a piezoelectric scanner, phenomena grouped together under the heading of **aging** effects. The strain coeffficient, d, of the piezoelectric material changes exponentially with time and use. The aging rate is the change in strain coefficient per decade of time. The deflection achieved for a given voltage

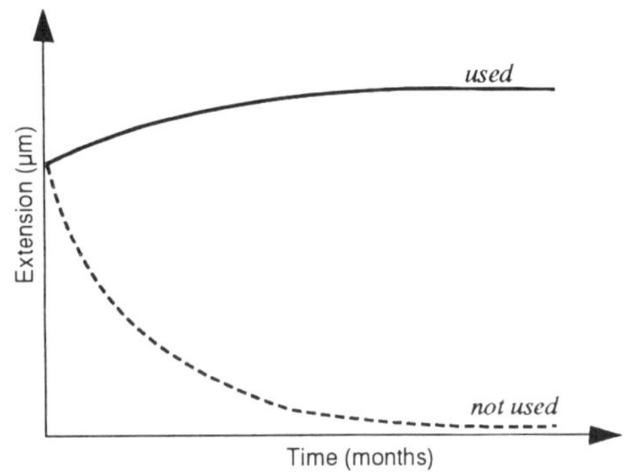

Figure 8. Aging of a piezoelectric scanner with use or left idle: maximum deflection of scanner, or deflection for a given voltage, changes over time.

gradually decreases when the scanner is not used. Conversely, the deflection achieved for a given voltage increases slowly with time when the scanner is used regularly. The aging rate of an idle SPM scanner can produce a decrease in lateral strain coefficients (and therefore, an error in length measurements taken from SPM images) on the order of 10% over months.

Time and usage effects are really part of the same process. As mentioned above, piezoelectric materials are polycrystalline ceramics. Each of the tiny crystals that comprise the scanner has its own dipole moment. Repeated application of voltage in the same direction--as would occur during scanning--causes more and more of the dipoles to align along the axis of the scanner. The amount of deflection achieved for a given voltage depends upon how many dipoles are aligned. Thus, the more the scanner is used, the further the scanner will travel.

On the other hand, if the scanner is not used the dipole moments of the crystals will gradually become randomly oriented again. As a result, fewer and fewer dipoles will contribute to the deflection of the scanner if it is left idle. When an SPM is delivered the scanners have already been "poled," which means that they have already been exercised to the point where the deflection of the scanner is close to its maximum.

The dependence of the scanner deflection on time and use means that the scanner may not be extending the same distance for a given applied voltage, as it did when it was calibrated. As a result, when a feature on an SPM image is measured, the values of lateral and vertical dimensions may be in error.

The term **cross coupling** refers to the tendency of the *x*-axis or *y*-axis scanner movement to have a spurious *z*-axis component (Figure 9). It arises from several sources and is fairly complex. The electric field is not uniform across the scanner. The strain fields are actually complex tensors. Some "cross talk" occurs between *x*, *y* and *z* electrodes. But the largest effect is geometric. Geometric cross coupling is based in the way piezoelectric scanners are constructed (usually as segmented tubes or tripods).

To produce the *x-y* motion of a tube scanner, one side shrinks and the other side expands. As a result, a piezoelectric tube scanner moves in an arc, not in a plane (see Figure 2). A voltage applied to move the piezoelectric tube along the *x* or *y* axis (parallel to the surface of the sample) necessitates that the scanner extend and contract along the *z* axis (perpendicular to the surface of the sample) to keep the tip in contact with the sample.

A tripod scanner is designed with three mutually perpendicular bars or tubes glued

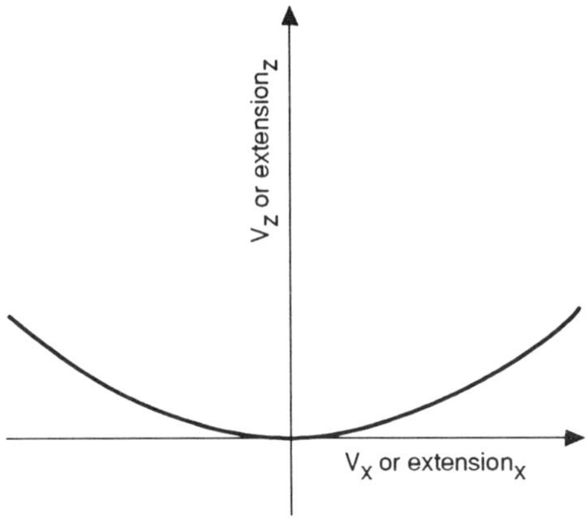

Figure 9. Cross coupling in a piezoelectric scanner: a voltage applied along one axis causes deflections along other axes.

together at one end. This design is also susceptible to cross coupling because the three bars of piezoelectric material are attached to one another. When the x bar extends or contracts, it causes a rotation of the y and z bars.

Cross coupling can cause an SPM to generate a bowl-shaped image of a flat sample. A profile of such an image is shown in Figure 10 with the now familiar example of a step. In interpreting Figure 10, one should keep in mind that the SPM image is the voltage required to compensate for the curvature generated by the arc of the scanner. The bowl shape may not always be evident in the final image because the curved background can be subtracted out using image-processing software. The best way to determine if the scanner is subject to cross coupling is to image a sample with a known radius of curvature, such as a lens. Software corrections can only flatten it or leave it untouched. In the first case, one ends up with a spuriously flat image, and in the second case the curvature of the scanner will be added to the curvature of the lens. The true curvature of a lens can be measured only when cross coupling is eliminated.

In systems which raster the sample underneath a stationary tip, cross coupling can cause another problem, depicted in Figure 11. Unless scan curvature is eliminated, lateral can calibration--and thus measurements in the plane of the sample--will artificially vary as a function of sample thickness. With a typical scanner having a maximum lateral range of 10 µm, replacing a sample 1 mm thick with a sample 10 mm thick will increase the apparent scan length by 65%, with corresponding errors in dimensions of features measured from the SPM image.

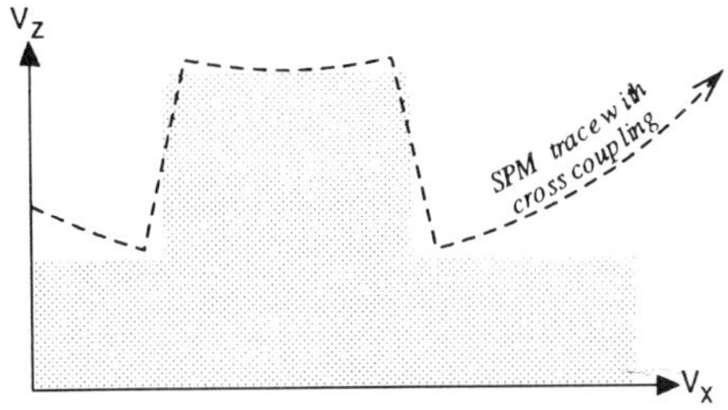

Figure 10. The effects of cross coupling on an SPM image of a step.

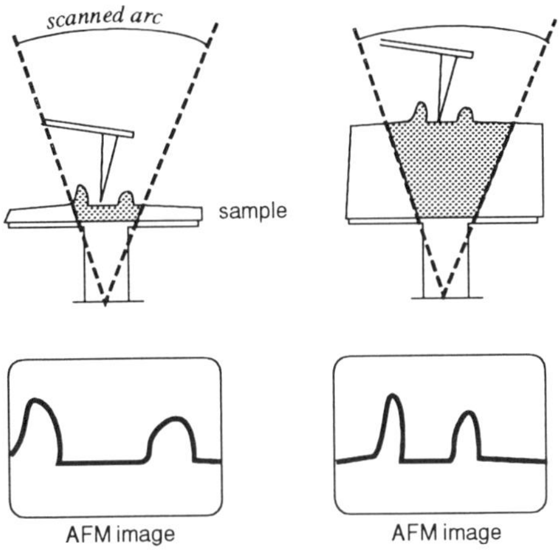

Figure 11. Replacing thin sample with thick sample effectively extends the length of the scanner, as it scans in an arc. If this effect is not corrected, length measurements artificially depend upon sample thickness.

The problem of the lateral calibration depending upon the sample height can be corrected in systems with fully automated tip-to-sample approach. With this design the system can measure the thickness of the sample and apply an appropriate correction.

Throughout this section we have used the example of a single step to demonstrate hysteresis, creep and cross coupling in the vertical direction. To show a single image in the laboratory that illustrates each of these effects in isolation is virtually impossible. Figure 12 shows the sum of the effects of hysteresis, creep and cross coupling on the image of a single step (the aspect ratio of the tip may also contribute to the shape of the sidewalls). An SPM profile with a shape like that of Figure 12 could be reproduced experimentally.

HARDWARE AND SOFTWARE SOLUTIONS

The nonlinear behavior of piezoelectric scanners described above has traditionally been addressed using software corrections. Some new systems on the market use hardware solutions that eliminate most of the nonlinearities instead of correcting them after the fact. These hardware approaches include optical, capacitive, and strain-gauge techniques.

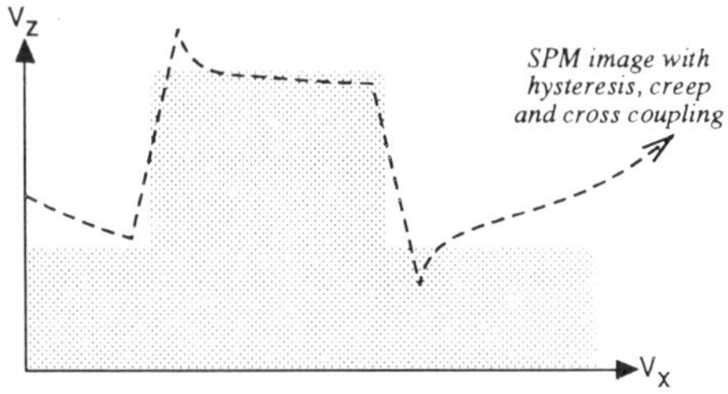

Figure 12. The combined effects of scanner hysteresis, creep and cross coupling on an SPM image of a step.

The traditional software approach uses a straightforward procedure. The first step is to image a calibration grating. The system compares the measured data with the known characteristics of the grating. Then the system determines how to adjust the measured data to conform to the known characteristics. It stores that information in a map, or lookup table. Afterwards the system can compensate for nonlinearity while collecting data, by adjusting the voltage applied to the scanner in accordance with the lookup table (Figure 13).

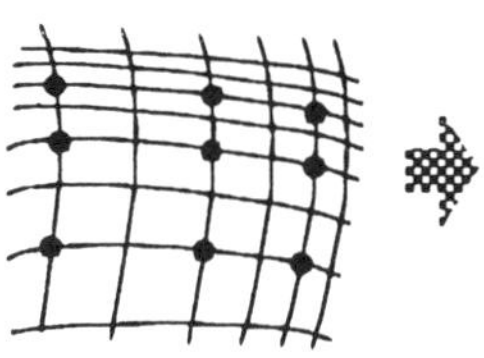

Look-Up Table

32.1	45	98	14
44.5	682.7	532	87
.347	556	.459	3321
128	32	441	.45
3	.34	21	3.68
67	.349	5697	221
2167	29	311	65
.321.2	993	87	239
46.89	27	554	12

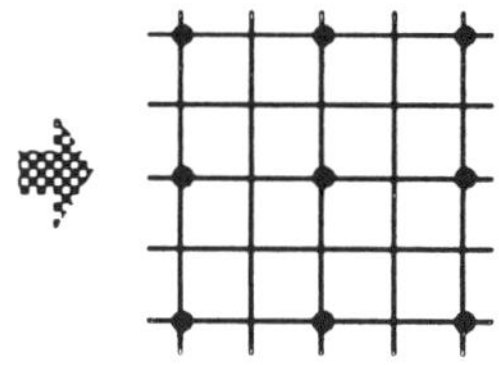

Figure 13. Schematic of generating lookup table for real-time software correction of scanner nonlinearities.

Software approaches are relatively simple and inexpensive to implement. Their main disadvantage is that they compensate only partially for scanner nonlinearities. The values stored in the lookup table are strongly dependent upon the scan speed, scan direction, and whether the scanner was centered within its scan range during calibration. As a result, software corrections are most accurate only for scans that reproduce the conditions under which the calibration was done. In practice, SPM systems that use software corrections typically vary from true linear behavior by as much as 10%. Often this deviation is unnoticed by the SPM user who calibrates and checks the calibration of the SPM using the same procedure every time.

To address the scanner problems with hardware, external sensors are used to measure the scanner's actual position (Figure 14). As the scanner approaches each measurement point, a feedback system reads the scanner's position from the sensors and applies voltage to the scanner to drive it to the desired position before the data point is sampled. In this way the data can be sampled in a square measurement grid. The external sensors themselves must be stable and immune to all sources of nonlinearity, since their purpose is to improve the linearity of the SPM. Fortunately it is much easier to build a linear sensor than a linear scanner.

To address the scanner nonlinearity with an optical solution[2], a "flag" or reflector is attached to the scanner. A light shines on the flag, and a detector responding to either the phase or the position of the light determines the actual position of the scanner. The system then uses feedback to adjust the voltage to the scanner to place it in the proper position to generate a linear scan.

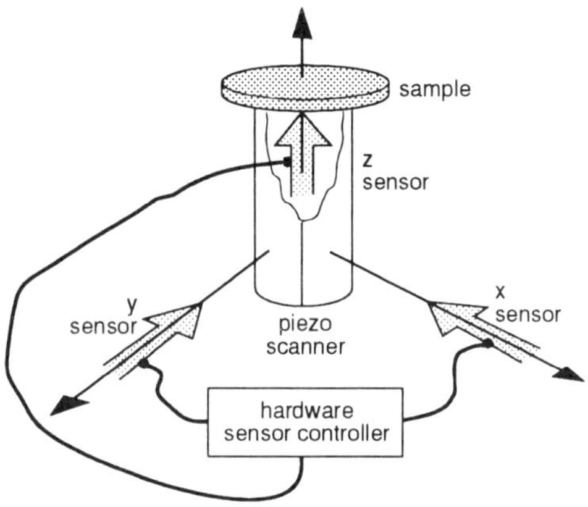

Figure 14. Schematic of piezoelectric scanner with external position-sensitive detectors.

In capacitive approaches[3], a metal electrode is placed on the scanner and another is mounted nearby. An electrical circuit then measures the capacitance between the electrodes, which varies with their separation. Having located the scanner in this way, the system uses feedback to adjust the scanner's position to compensate for nonlinearity.

Alternatively, a strain gauge could be mounted on the scanner. The active element in a strain gauge is a piece of piezoresistive material--material whose resistance varies when it is under tensile or compressive stress. When the scanner moves, the resistance in the strain gauge changes. The magnitude of the resistance of the strain gauge reflects the amount of bending in the scanner, thereby measuring its position. The SPM uses feedback to adjust the scanner's position and generate a linear scan.

Because the scanner's actual position is measured, systems using hardware sensors compensate for intrinsic nonlinearity, hysteresis, creep, aging and cross coupling as a matter of course. They can reduce the total nonlinearity of the system to better than 1%.

CONCLUSION

Hardware approaches to correct scanner nonlinearities have brought SPM systems closer to measurement tools by addressing the sources of the nonlinearities. With reduction of scanner distortions to less than 1%, images more truly reflect the topography of the sample surface. In addition, measurements taken from the images are more repeatable and precise.

REFERENCES

1. J.E. Griffith, D.A. Grigg, M.J. Vasile, P.E. Russell, E.A. Figeral, Characterization of scanning probe microscope tips for linewidth measurement. *J. Vac. Sci. Technol.* B 9 (6): 3586-3589 (1991).
2. R.C. Barrett, and C.F. Quate, Optical scan-correction system applied to atomic force microscopy, *Rev. Sci. Instrum.* 62 (6):1393 (1991).
3. J.E. Griffith, G.L. Miller, and C.A. Green, A scanning tunneling microscope with a capacitance-based position monitor," *J.Vac. Sci. Technol. B* 8 (6): 2023-2027 (1990).

USE OF ATOMIC FORCE MICROSCOPY IN THE DETERMINATION OF IMAGE CONTRAST IN MICROTOMED SAMPLES OF THERMOTROPIC LIQUID CRYSTALS

Timothy J. Bunning,[1] Deborah L. Vezie,[2] Peter D. Haaland,[3] Hao Jiang,[3] Pamela F. Lloyd,[4] Edwin L. Thomas,[2] and W. Wade Adams[1]

[1] Wright Laboratory/MLPJ, Bldg 651
Wright-Patterson AFB, OH 45433-7702
[2] Dept. of Materials Science and Engineering
Massachusetts Institute of Technology
Cambridge, MA 02139
[3] Lawrence Associates, Inc.
Dayton, OH 45431
[4] UES, Inc.
Dayton, OH 45432

Abstract: Atomic force microscopy (AFM) is used in conjunction with transmission electron microscopy (TEM) and scanning electron microscopy (SEM) to explain for the first time the origin of image contrast in TEM for cholesteric liquid crystals (LC). Samples investigated were side-chain thermotropic siloxane-based cholesteric liquid crystalline materials whose room temperature glassy phase slowed for easy quenching of the LC order. Microtomed samples were prepared by cutting parallel to the helical axis and therefore perpendicular to a thin film cross-section. These sections exhibited a series of alternating dark and light bands whose sizes are related to the periodic structure of the cholesteric helix. The surfaces of the microtomed sections show a corrugated topology whose repeat distance corresponds to one half the pitch. Peak-to-valley heights on the order of 200 Å were observed. Image contrast in the TEM was observed to increase substantially with time and electron dose due to beam damage of the materials. Since $\Delta I/I$ is directly related to $\Delta t/t$, the increase in contrast can be due to general mass loss (decrease in t) or preferential mass loss (increase in Δt). AFM was performed on both unirradiated samples and samples that were exposed to large electron doses to explain the contrast mechanism.

INTRODUCTION

Cholesteric liquid crystals have the structure shown in Figure 1. They are essentially a twisted nematic phase where the twist axis is perpendicular to the director. This twist is caused by an optically active carbon in the mesogenic structure and both left-handed and right-handed helices can form. The distance it takes to rotate the director 360° is called the pitch.

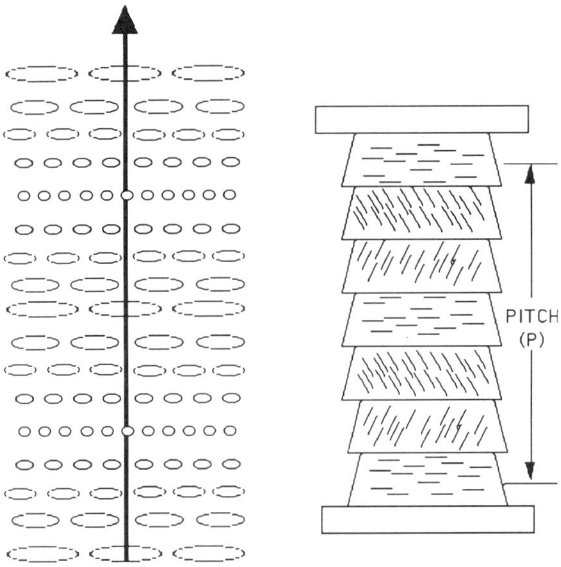

Figure 1. Schematic of a cholesteric liquid crystal geometry.

Cholesteric materials have received much attention due to their helical structure, which exhibits optical activity. Optical properties, such as birefringence, circular dichroism, and optical rotation are observed although the phenomena of selective reflection[1] has been the most widely exploited. Two types of macroscopic packing can occur, focal conic and Grandjean textures as shown in Figure 2. In the former texture, the helices are more or less randomly arranged, which causes the materials to be opaque. If the substrates are rubbed together while the material is in the mesophase region, spontaneous alignment occurs causing the helical axis to reside perpendicular to the substrates. It is in this conformation that the system exhibits selective reflection[1] as shown schematically in Figure 2(c).

Cholesteric liquid crystalline polymers are conducive to electron microscopy techniques as they can be ordered in the mesophase and vitrified into a solid. The solid film maintains the anisotropic ordering of the mesophase and can subsequently be examined using solid state techniques. Examination of low molecular weight liquid crystals is typically performed with polarizing optical microscopy and for systems with large pitches (>2 μm), the helix can be visualized when in the fingerprint texture.[2] At low temperatures, however, these systems crystallize, thereby destroying the helicity.

Most EM studies to date have been performed using scanning electron microscopy after freezing the cholesteric order into room temperature glasses. Fracture of these films parallel or oblique to the helical axis produces a periodic surface topography due to the spatially varying molecular directors (n) which can be imaged on fractured bulk films.[3,4,5] When the fracture surface is exactly parallel to the helical axis, the surface periodicity

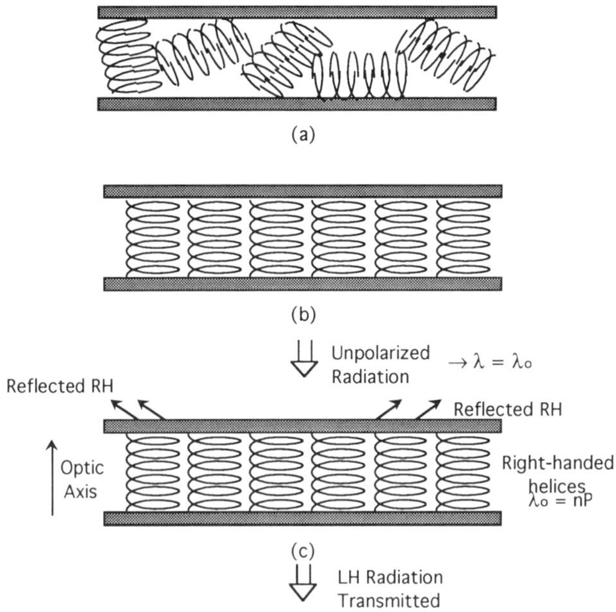

Figure 2. Focal-conic (a) and Grandjean (b) packing structures. Selective reflection (c) from a Grandjean thin film. LH-lefthandedness, RH-right-handedness.

corresponds to one-half the pitch (P) as there is no difference between the n and -n directors. A model, based on preferential crack propagation in spatially varying anisotropic glasses, proposed previously, predicts the contour of the fracture surface relative to the helical direction.[6] Various types of defects can also be observed in these fracture surfaces.

Recently, several groups[7-11] have reported on image contrast observed with bright-field transmission electron microscopy. A series of dark and bright lines whose spatial periodicity corresponded to one-half the pitch value were observed. This contrast was not material dependent as many different materials, including cellulose acetate butyrate, siloxane lcp's, cellulose, and copolyesters exhibited this behavior. This behavior was observed for thin microtomed specimens where the microtoming direction was parallel to the helical planes and TEM accelerating voltages were 50 to 100 keV.

Since the spatial contrast was related to the pitch distance, most groups attributed the source of the contrast to the anisotropy of the molecular directors. Since in our opinion, no adequate explanation for the contrast mechanism had been forwarded, an investigation of possible sources of this contrast using scanning electron microscopy, transmission electron microscopy, and atomic force microscopy was undertaken. A complete discussion of these results can be found elsewhere.[12] The information presented here is a small part of this work primarily concentrating on the atomic force images from thin microtomed sections of a glassy thermotropic cholesteric liquid crystal.

EXPERIMENTAL PROCEDURES

The cholesteric material investigated belongs to a class of cyclic siloxane materials whose room temperature glassy phase allows the mesophase to be vitrified. The compound examined, shown in Figure 3, had equal amounts of biphenyl-4-allyloxybenzoate and cholesteryl-4-allyloxybenzoate mesogens attached to a pentamethylcyclosiloxane core. It exhibits a Tg at 50 °C, a wide cholesteric mesophase, and a clearing temperature of 220 °C. Films were formed by shearing at 150°C to induce the Grandjean texture where the helical axes are perpendicular to the film surfaces. This material exhibits a selective reflection wavelength of 514 nm, which corresponds to a pitch of P = 342 nm. Bulk films were embedded in a room temperature epoxy system and sections were taken parallel to the helical axis. TEM was performed on a JEOL 100CX at 100 keV and images were recorded on Kodak SO163 film. Microtomed sections were 60-80 nm thick. Free surfaces of a Grandjean aligned sample sheared on silicon wafers were also investigated. AFM was performed with a Park Scientific SFM-BD2210 Scanning Force Microprobe running from a SPC-400 Scanning Microprobe Central Control Unit. A standard silicon nitride tip with a radius of curvature of 40 nm was used with mid-range forces in the repulsive mode.

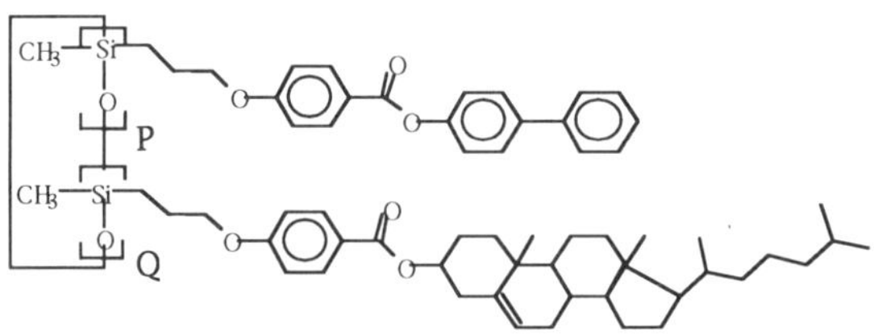

Figure 3. Schematic of cholesteric cyclic siloxane investigated

RESULTS AND DISCUSSION

A typical bright-field transmission electron micrograph from a thin microtomed sample is shown in Figure 4. A strong intensity contrast is evident, which has a periodicity measured corresponding to one-half the pitch of the sample. Although the contrast present in this image is reminiscent to that of "stained" block copolymers, no staining was performed on these samples. The contrast was not initially present when using low dose conditions. The steady state constrast shown in Figure 4 was observed to develop very quickly with increasing dose and thus is attributed to beam damage. Large values of defocus typically enhance the contrast from an in-focus BF image.

Surfaces of both unirradiated and irradiated microtomed sections were examined with AFM. Figure 5 shows an AFM image of an unirradiated sample containing a highly corrugated surface with a periodicity (167 nm) close to one-half the pitch of the sample (171 nm). An edge dislocation with a Burger's vector of P is shown in the middle of the scanned area in Figure 5(b). The three-dimensional structure of the surface topography is shown in Figure 5(c) and height-profile analysis indicates an average peak-to-valley height of 20-25 nm.

The surface topography of these thin films indicates that crack propagation ahead of the knife tip is not flat. When one thinks of "cutting" thin sections, one typically thinks that a film with constant thickness is the result. Instead, these AFM images indicate that crack propagation in these anisotropically varying glasses follows the path of least energetic resistance. Indeed, these fracture surfaces resemble the surface topography observed for bulk fracture surfaces in the SEM. In slow fracture, the fracture path tends to minimize the surface area exposed, but locally the crack tip is preferentially propagated relative to the anisotropy of the molecules.

The surface topography of irradiated samples reveals the same degree of corrugation as the unirradiated samples as shown in Figure 6. The increase in contrast observed with irradiation in the TEM[12] was initially thought to be due to an increase in the peak to valley

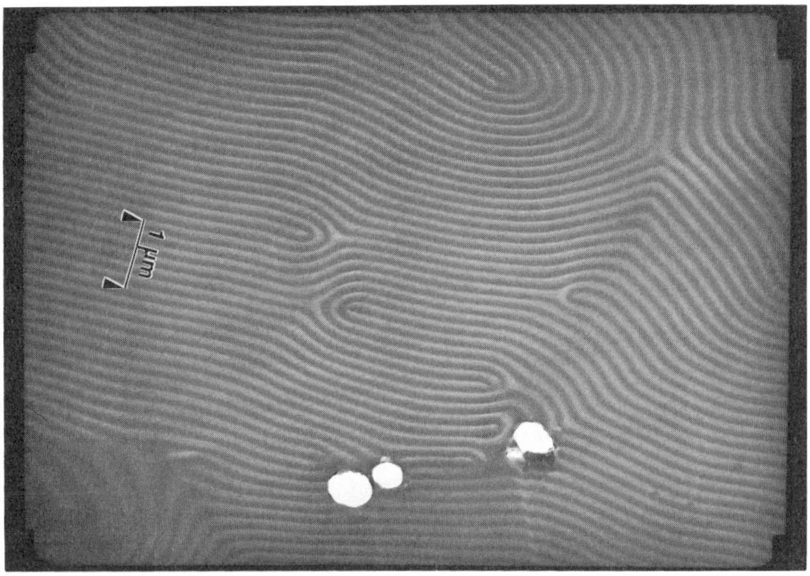

Figure 4. Bright-field TEM image of a cholesteric liquid crystal.

height corrugation in the microtomed section. AFM was also performed on the stub used for microtoming to address whether the surface corrugation was due to anisotropic mechanical relaxation of the thin films. Images from the stub, essentially an infinitely thick film, showed the same type of surface topography, thereby confirming that the spatially periodic topography was induced in the microtoming process.

TEM experiments on the thin films also revealed that the edge orientation of the helices is typically much different than the surface orientation. Near the surface, the helices were observed to rotate 90°, thus the helix direction was more or less parallel to the surface. This conformation of the helices has been predicted previously[13] as one solution to the boundary layer problem in which the twisted structure of the bulk cannot readily adapt to this boundary condition and is instead satisified by an alternate series of s = +1/2 and s = -1/2 singular lines as shown in Figure 7. The free surfaces in planar (Grandjean) samples cause the molecules to be perpendicularly oriented at each surface.

Attempts to view a free surface with scanning electron microscopy failed and thus AFM experiments were attempted. The surface topography was complex although the same spatial periodicity as the microtomed sections was observed, as shown in Figure 8. The peak-to-valley corrugation was, however, an order of magnitude lower at 2-4 nm.

Instead of a surface satisfying packing considerations with +1/2 and -1/2 singularities

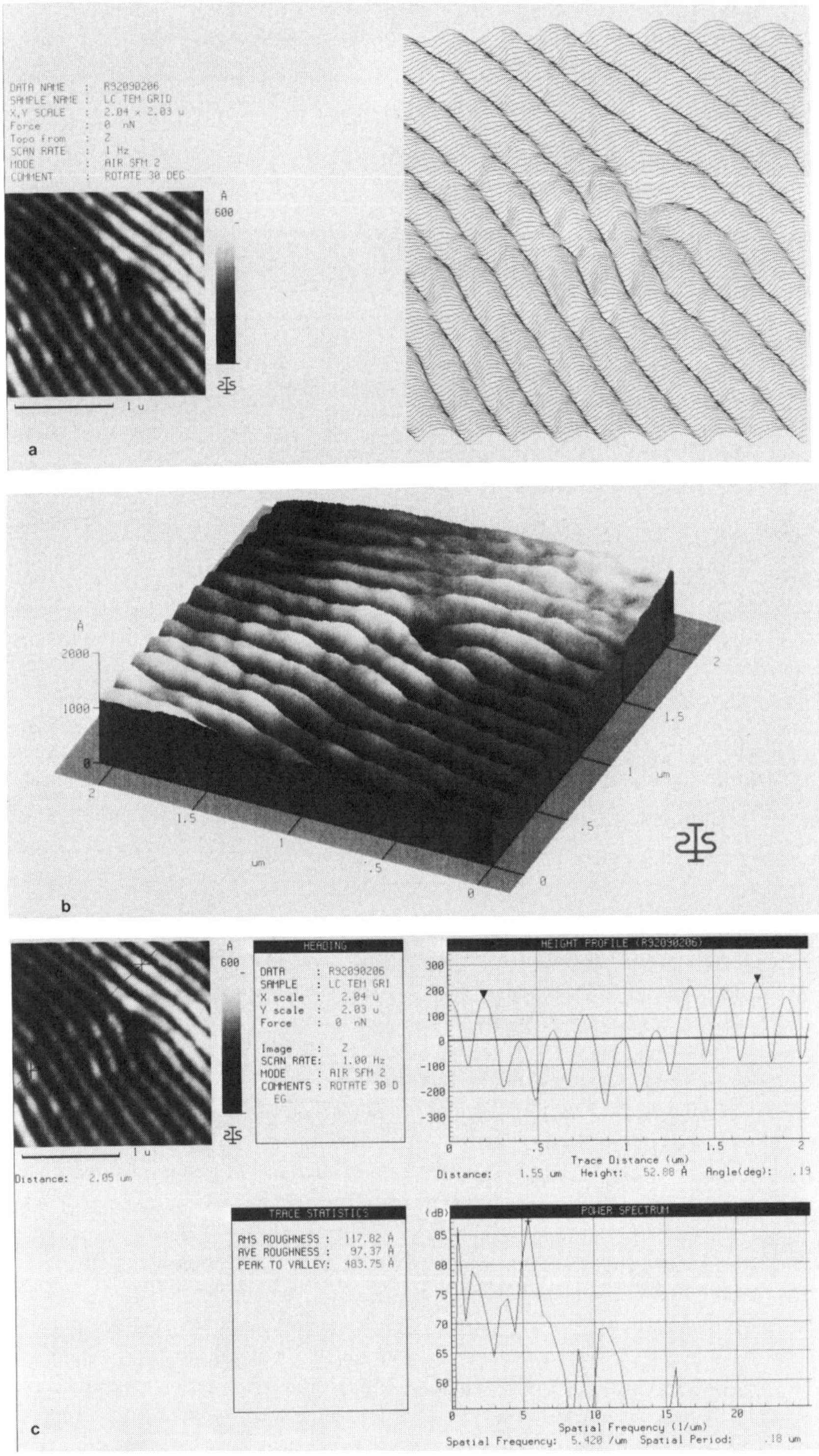

Figure 5. (a) AFM image of surface of unirradiated microtomed sample; (b) the area in three dimensions showing a dislocation in 3-Ds; and (c) a height profile analysis across this section.

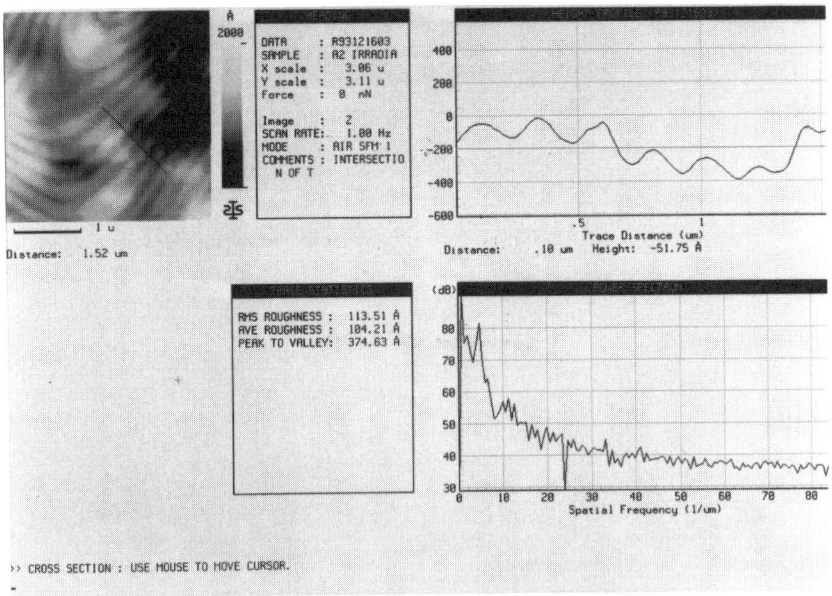

Figure 6. AFM image of irradiated (100 keV) microtomed section.

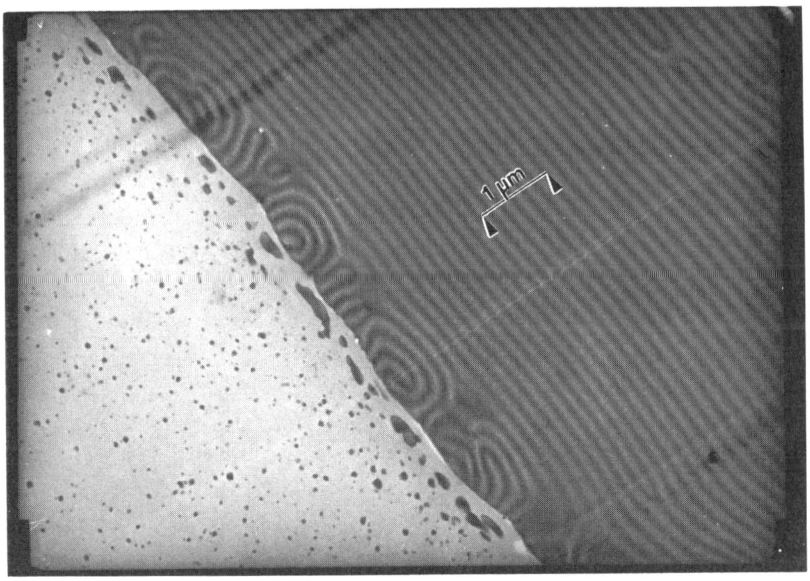

Figure 7. TEM image of the edge (surface) showing different orientation of helices at the surface than in bulk.

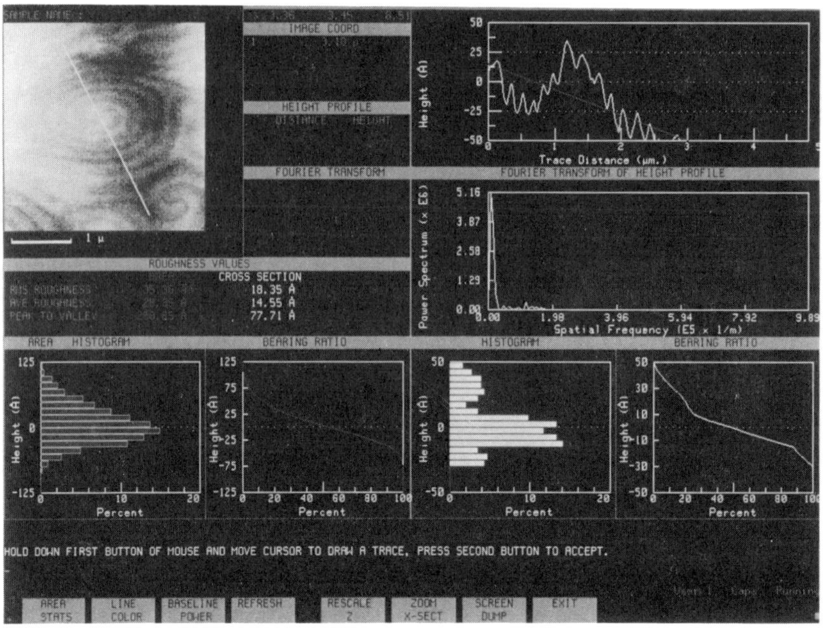

Figure 8. AFM image of surface of Grandjean thin film

as shown in the TEM image, the surface topography is essentially a +1 disclination. The topography is not one-dimensional as explained by the theory but, rather, two-dimensional as exhibited by the complex swirls of periodic topography. The surface topography seen here is manifested in the TEM image as a slight undulation at the surface of the section.

CONCLUSIONS

Although not fully discussed here, AFM experiments contributed to an understanding of the image contrast observed with transmission electron microscopy from glassy cholesteric liquid crystalline materials. Interesting surface topography was observed from microtomed thin sections of the sample. The topography is thought to be due to preferential crack propagation in an anisotropic spatially varying glass. Similar peak to valley height corrugation was also observed on the free surface of Grandjean films although the surface topography was an order of magnitude smaller than the fractured thin films. This surface topography is a consequence of a low energy conformation of a helical axis on a free surface.

REFERENCES

1. H.L. DeVries, Rotatory power and other optical properties of certain liquid crystals, *Acta. Cryst.* 4:219-226 (1951).
2. P.J. Collings, "Liquid Crystals," Princeton University Press, Princeton, N.J. 222 (1990).
3. J. Voss and B. Voss, SEM Studies of cholesteric liquid crystals, *Z. Natureforsch* .31A:1661-1663
4. T.J. Bunning, H.E. Klei, E.T. Samulski, R.L.Crane, and R.J. Linville, Bilayer structures in cholesteric, cyclic siloxane liquid crystals, *Liq. Cryst.* 10:445-456 (1991).
(1976).

5. M.J. Costello, S. Meiboom, and M. Sammon, Electron microscopy of a cholesteric liquid crystal and its blue phase. *Phys. Rev.* A. 29:2957-2959 (1984).
6. D.W. Berreman, S. Meiboom, J.A. Zasadzinski, and M.J. Sammon, Theory and simulation of freeze-fracture in cholesteric liquid crystals, *Phys. Rev. Lett.* 57:1737-1740 (1986).
7. P. Sixou, J.M. Gilli, A. Ten Bosch, F. Fried, P. Maissa, L. Varichon, and M.H. Godinho, Cholesteric mesophases, *Physica Scripta*, T35:47-52 (1991).
8. J.M. Gilli, M. Kamaye, and P. Sixou, Phases bleues "figees" dans un polysiloxane mesomorphe, *J. Phys. France*, 50:2911-2918 (1989).
9. V. Dave, W.G. Glasser, and G. Wilkes, Evidence for cholesteric morphology in films of cellulose acetate butyrate by transmission electron microscopy, *Polym. Bull.* 29:565-570 (1992).
10. J. Giasson, J-F Revol, A.M. Ritcey, and D.G. Gray, Electronic microscopic evidence for cholesteric structure in films of cellulose and cellulose acetate, *Biopolym*, 27:1999-2004 (1988).
11. H. Hara, T. Satoh, T. Toya, S. Iida, S. Orii, and J. Watanabe, Cholesteric liquid crystalline polyesters. 1.) Cholesteric Liquid Crystalline Copolyesters Based on Poly(chloro-1,4-phenylene trans-1,4-cyclohexanedicarboxylate), *Macromolecules*, 2 1:1419 (1988).
12. T.J. Bunning, D.L. Vezie, P.F. Lloyd, P.D. Haaland, E.L. Thomas, and W.W. Adams, Cholesteric liquid crystals: image contrast in the TEM, *Liquid Crystals*, in press.
13. A. Saupe, Disclinations and properties of the director field in nematic and cholesteric liquid crystals, *Mol. Cryst. Liq. Cryst.* 21:211-238 (1973).

PHOTON TUNNELING MICROSCOPY OF POLYMERS

John M. Guerra,[1] Mohan Srinivasarao,[2] and Alex Hsieh[3]

[1]Polaroid Corporation.
Optical Engineering Department.
Cambridge, MA 02139
[2]University of Massachusetts Polymer Science Department
Amherst, MA 01002
[3]U. S. Army Laboratory Command
Materials Technology Laboratory
Watertown, MA 02172-0001

Abstract: The photon tunneling microscope (PTM) can image polymer surface morphology with vertical resolution below a nanometer and lateral resolution of 0.10 μm. PTM produces visual, whole-field tunneling images as well as real-time high resolution three-dimensional images with continuous interactive perspective, ready for stereopsis. In addition, surfaces are imaged without metallization, shadowing, vacuum, electrons, or scanning probes. A brief review of the theory, instrumentation, and features of photon tunneling microscopy is presented. Applications of PTM to such diverse polymeric surfaces as polystyrene sphere structures, single polyethylene crystals, silicon dioxide thin films on polycarbonate, and a real-time dewetting study of polystyrene at the glass transition temperature are presented and briefly discussed.

INTRODUCTION

Photon tunneling microscopy (PTM)[1,2,3] has all the benefits of reflected light microscopy: whole field imaging with direct visual access, and in real-time, with lateral resolution (while less than its scanning probe cousins) benefiting from the super numerical aperture afforded by the tunneling or near-field information that is normally lost in a light microscope. However, PTM also shares the subnanometer vertical resolution and profiling ability of its scanning probe microscopy cousins, and so is proving to be a valuable tool for imaging and quantifying the many kinds of surfaces that arise from polymer morphological processes.

Related to and descended from frustrated total internal reflection microscopy (FTIRM) in the teaching of McCutchen[4] (alternately referred to as evanescent wave or surface contact microscopy when used in the transmission mode taught by Ambrose), PTM is differentiated from its scanning probe relatives at the tunneling interface by a contiguous transducer that acts like a close-packed array of probes; information is gathered in a parallel manner, rather than serially as with scanning methods. This allows more kinds of samples to be imaged in real-time without probe errors or damage. Additionally, its unique three-dimensional analog display of the photometric video image facilitates real-time imaging with continuous and immediate image rotation about all axes, as well as profiling. Most importantly for polymer applications, PTM images dielectric surfaces wholly and directly with no metallization, carbon shadowing, vacuum, or damaging electrons. Further, surface alteration is negligible: while the transducer is in optical proximity to the sample surface, it is most everywhere separated by a tunneling gap (or else it would not function), it moves with the sample, is flexible enough to accommodate particulate, and is of a hardness comparable to most polymers.

Finally, the photon tunneling microscope can be adapted for high and low temperature studies in, for example, the annealing of polymers.

BRIEF DESCRIPTION OF PTM

The photon tunneling microscope, like its forbear, is based on the behavior of light at a total internal reflection (TIR) surface, or the boundary between a medium 1 with index of refraction η_1 and medium 2 with a lesser index of refraction η_2, where total reflection occurs in medium 1 if the angle of incidence is equal to or greater than the critical angle:

$$\sin^{-1}(\eta_2/\eta_1). \tag{1}$$

Probing the rarer medium near the TIR surface (hence "near-field") with a dielectric surface of η_3 greater than η_2 reveals energy coupling from the first medium into the third medium, indicating the presence in the rarer medium gap, when unperturbed, of an electromagnetic wave with the same frequency as its parent wave in the denser medium, and "evanescent" because it does not carry any energy from the TIR surface if unperturbed. The field strength of the evanescent wave decays exponentially as:

$$E_{evanescent} = E_0 \exp(-Z/d_p), \tag{2}$$

where Z is the distance normal to the TIR boundary and d_p is the penetration depth at which the field strength falls to E_0/e:

$$d_p = \frac{\lambda_1}{2\pi(\sin^2\Theta - \eta_{21}{}^2)^{\frac{1}{2}}} \tag{3}$$

where η_2/η_1 is the ratio of refractive indices of medium 1 and 2, respectively, and λ_1 is the illumination wavelength in medium 1.

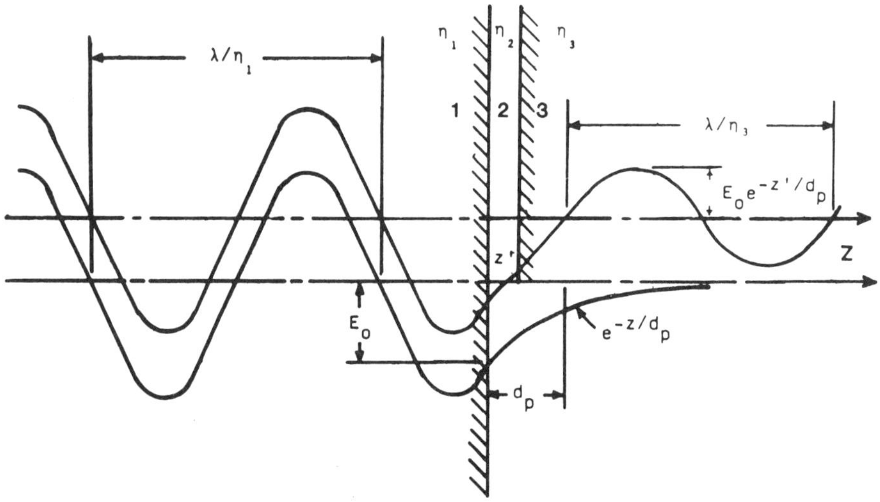

Figure 1. Evanescent field, associated with photons tunneling into medium 2, as a continuation of the parent field in medium 1.

Evanescent waves are fully explained with classical optics[5,6] and one need not invoke quantum theory but doing so adds to the intuitive understanding of the microscopy. The analogy to the electron barrier tunneling experiment has been elegantly made (Coon[7], Zhu[8], and others) and the effect called optical or photon tunneling. In this modality equation (2) describes the probability of finding a photon from medium 1 in the tunneling gap at some distance Z normal to the TIR boundary, and so photon tunneling has three properties of import to microscopy:

First, each tunneling photon serves as a height probe of a third medium with $\eta_3 > \eta_2$, as TIR is locally frustrated to a degree exponentially and inversely dependent on Z. Thus when the third medium is a dielectric sample with homogenous optical properties, the reflected light will be modulated in intensity corresponding spatially to the modulation of the tunneling gap by the sample's surface topography, and an image of the surface topography in grayscale is formed wherein height information is mapped into intensity space.

Secondly, photometry of this tunneling image yields quantified height measurement directly and without ambiguity, as bright is down and dark is up in the intensity-mapped topography. More importantly, the exponential decay of the tunneling probability (potential) with distance Z normal to the TIR boundary frees vertical resolution from wavelength constraints. Vertical resolution is limited only by photometric resolution, determined by the signal to noise ratio of the detector. The "diffraction barrier" of light microscopy is broken, while all the advantages are retained, and height is quantifiable.

Thirdly, a tunneling photon has the same wavelength in the air tunneling gap as in medium 1, so that high lateral resolution is achieved without oil immersion of the sample.

Configuring a reflected light microscope for TIR and photon tunneling (Figure 2) requires an objective with a numerical aperture larger than one, in order to contain the critical angle for the boundary of the objective's distal surface and the tunneling gap in air. For an aqueous tunneling gap, as with *in vivo* biological samples, the NA>1.33. When the distal surface is positioned within the objective's depth of focus and in optical unity, TIR of the epi-illumination is effected.

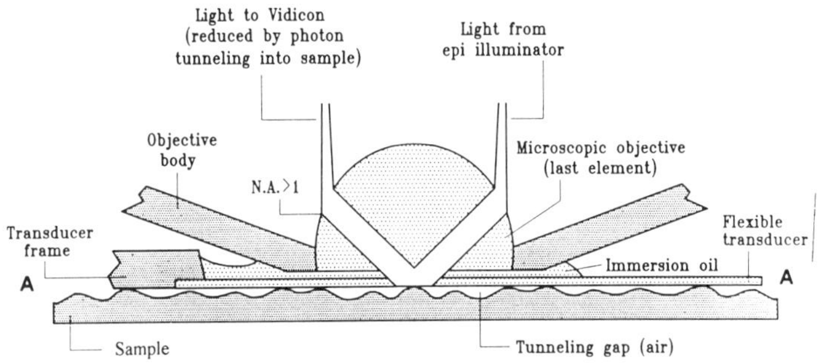

Figure 2. Schematic of PTM objective. Only one numerical aperture is shown for clarity, and the tunneling gap and surface topography are exaggerated for illustration.

In practice, a thin optical window, called a transducer because it converts height modulation into light modulation, is contacted to the sample surface. Then the objective is oil immersed to the transducer (the sample remains dry) and brought to focus on the transducer's distal surface. The immersion oil interface, in addition to providing the required optical unity for TIR (frustrated now by photon tunneling into the sample), also allows the sample and transducer to travel together with respect to the microscope. This eliminates damaging sliding contact, protects the sample from the oil, and allows complete freedom to move about the surface and explore. A recent innovation is the use of a proprietary flexible optical membrane as the transducer: it is stiff over the field of view of the microscope but flexible enough on a larger scale to accommodate curved surfaces and even dirt on the sample.

The vertical tunneling range is about 0.75, or about 0.3 μm for green light. Where the gap grows larger than this, the transducer serves as a soft contact interference reference surface, with only refocussing and suppression of TIR illumination required for interferometry.

Unlike the scanning tunneling microscopes, whether electron or photon, which operate at constant tunneling current, the photon tunneling microscope forms an image with the variations in tunneling referenced to a constant surface (the transducer). The grayscale tunneling image thus formed of the surface is viewed directly through the microscope. The high resolution, contrast and immediacy of this image is astonishing; reproduced images cannot do it justice. With only a little practice, reading the tunneling image with its unambiguous grayscale becomes second nature.

For height quantification, a CCD or vidicon detector views the tunneling image at the microscope's phototube in the normal manner, and this output is readily analyzed and displayed in any number of ways, including three dimensions, by any number of image processing programs now available for PCs. Detector tradeoffs must be considered, in that while the CCD has the better signal to noise ratio and therefore higher vertical resolution and vertical range, the vidicon has higher lateral resolution at lower cost, with real-time response; besides, a pasicon tube yields a vertical resolution of about a nanometer. Additionally, the nonlinear vidicon response to intensity corrects to first order for the tunneling's exponential response to linear height variation, so that in the absence of image processing to correct the grayscale, the three-dimensional display of the topography is not distorted.

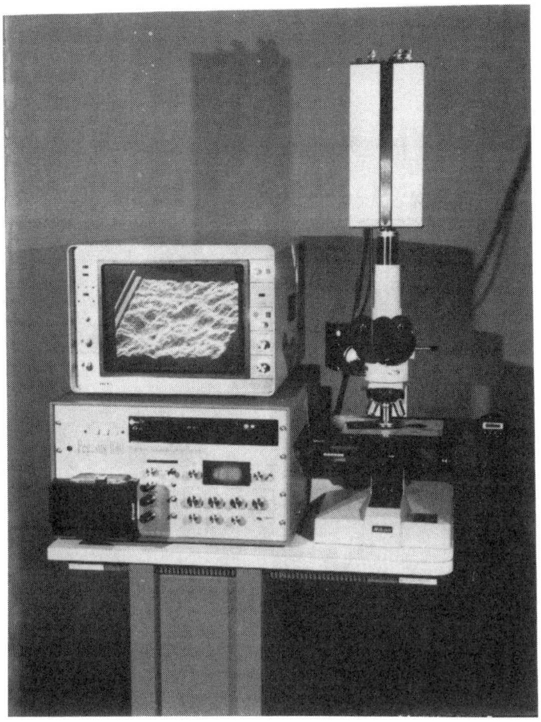

Figure 3. A commercial photon tunneling microscope (courtesy of Dyer Energy Systems, Inc.).

Analog has the advantage in the three-dimensional display as well. A high resolution XY oscilloscope displays each of the video raster lines as amplitude traces, thus electronically mapping intensity back into height for a real-time three-dimensional image. Additional analog circuits to shear or expand and collapse the multiple oscilloscope traces creates an immediate control over three-dimensional perspective, with simple turns of a few knobs. The quality of reality and immediacy this display provides helps to interpret the surface and understand it. A single trace can be isolated anywhere in the image for electronic cross-sectioning and height measurement.

PTM IMAGES OF POLYMERS

The following applications were chosen to demonstrate the diversity of polymeric surfaces to which PTM can be used to advantage, although in some cases it may be argued that other techniques, including traditional light microscopy, might be the method of choice; in these cases we offer the PTM images to show correlation. However, in other cases PTM offers clear advantages in keeping with its aforementioned features.

Also, some of the images beg further investigation with more complete measurement and understanding of the processes and variables; they are included because of, rather than in spite of, this to add value and completeness to this survey of photon tunneling microscopy of polymers.

It is well known that many synthetic polymers can be crystallized from dilute solutions to form what are usually referred to as "polymer single crystals". The discovery of such "single crystals", made almost at the same time by Keller[9], Till, and Fischer in 1957 has had a profound impact on the understanding of crystallization and self-assembly of macromolecules (including recently, proteins). Their electron diffraction data showed that the polymer chains (polyethylene) were normal to crystal platelets formed from dilute solution, and most remarkable was the fact that the platelets were only about 100Å thick. This led Keller and O'Connor to conclude that the molecules must fold sharply on themselves to give rise to lamellar periods of only 100Å. Since then several studies have been carried out on solution grown polymer crystals.

PTM images of single polyethylene crystals from solution are shown in Figure 4 (a) and (b). Lamellae are 70Å high (the vertical magnification is exaggerated in all of these images). These crystals are frequently imaged in the literature by light,[10] electron,[9] and atomic force microscopes[11] and so are a good target for correlation study.

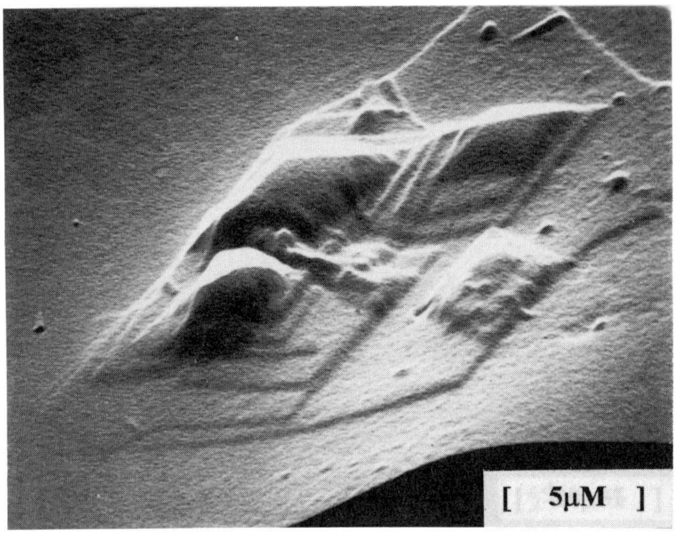

Figure 4. (a) PTM image of a grouping of single polyethylene crystals with 70Å lamellae.

Figure 4 (c) captures a polystyrene smooth film forming beautiful structures as it dewets from a glass substrate at its glass transition temperature. A thermal gradient has been induced to cause the structure. This is an example of PTM's real-time capability, even at high sample temperatures: the entire structure-forming process was recorded on time-lapse video over a span of two hours.

Silicon dioxide SiO_2 coatings on polycarbonate exhibit structure as seen in Figure 4 (d) just prior to crazing, much like a meandering river, complete with oxbows. An image of the same sample with SEM is shown in Figure 4 (e) for comparison.

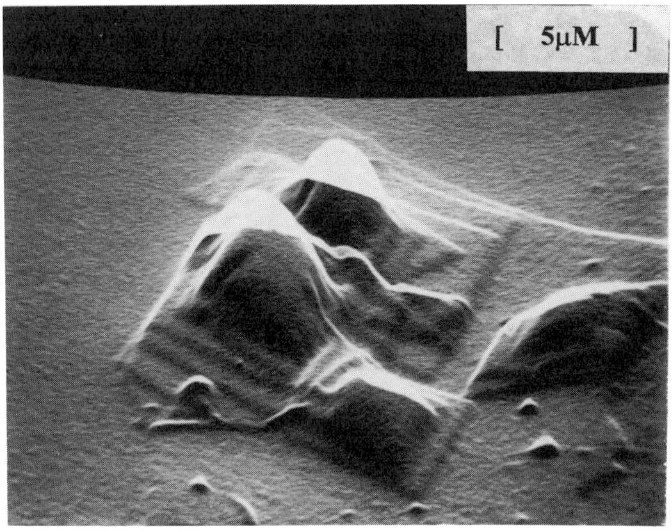

Figure 4. (b) Another PTM perspective of the polyethylene crystals in Figure 4(a).

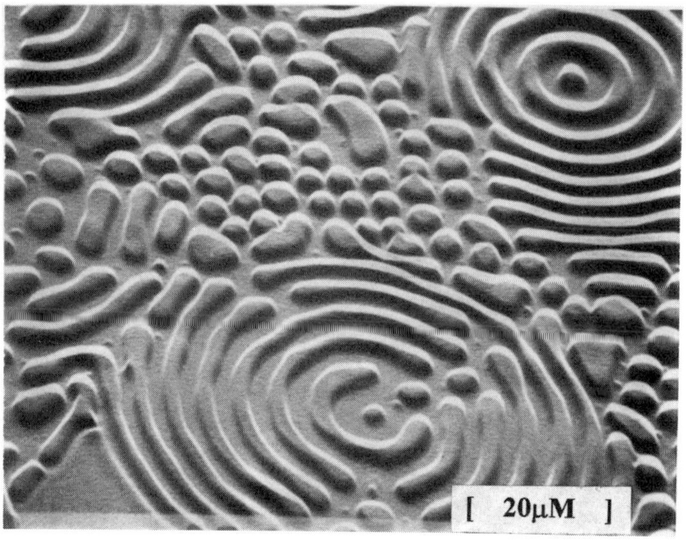

Figure 4. (c) Dewetting polystyrene film with a thermal gradient about T_G. Field of view is 95 μm and structure height, PV, is 0.2 .

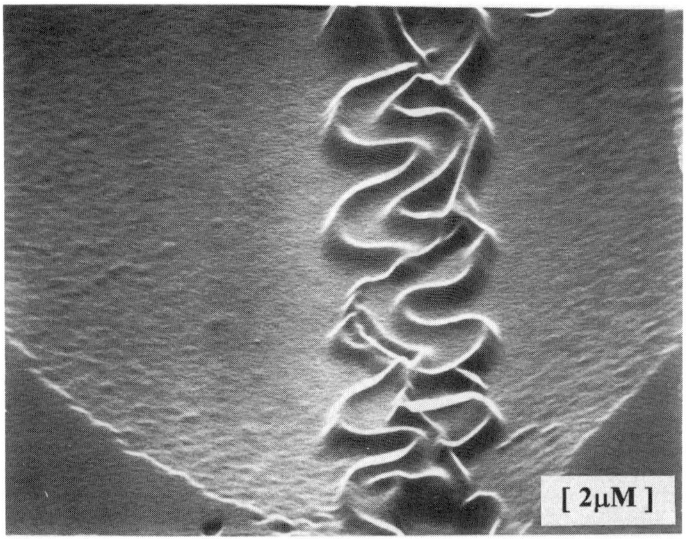

Figure 4. (d) SiO_2 coating crazing precursor

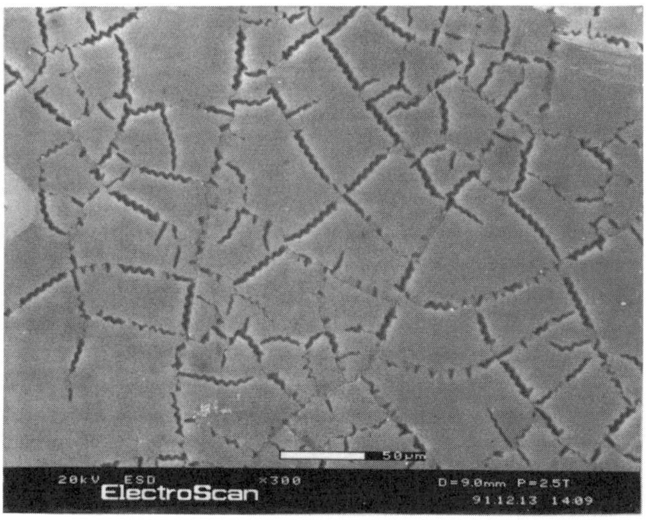

Figure 4. (e) SEM view of same sample (courtesy ElectroScan).

Polystyrene spheres below about 600 nm diameter will form close-pack hexagonal structures if air dried from suspension on a substrate. Figure 4(f) shows such a structure formed from 50 nm spheres. The vertical valleys constitute crazing. The diagonal orthogonal ridges are the side view of the crystal-like organization. In Figure 4(g), Nomarski imaging at 100X magnification reveals a beautiful system of radial and tangential crazing, with the area of each tile remaining constant, requiring that new rows form at discrete radii. The low magnification shows this pattern very well. A higher magnification image from the photon tunneling microscope is seen in Figure 4(h). While these features are also seen in Nomarski at higher magnification, PTM allows their height to be measured (200 nanometers), and shows a shift in the ridges across the crazing that indicates they were once continuous and therefore formed before the crazing (rather like polymer geology). Further, the cusping of the tile edges is made obvious in PTM. The spheres themselves are just beyond PTM's lateral resolution, but structure is indicated.

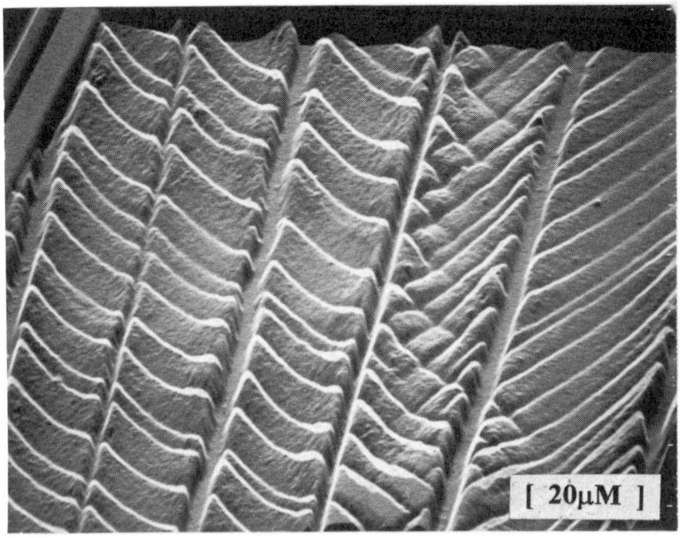

Figure 4. (f) Structure from hexagonal close-packed 50 nm polystyrene spheres.

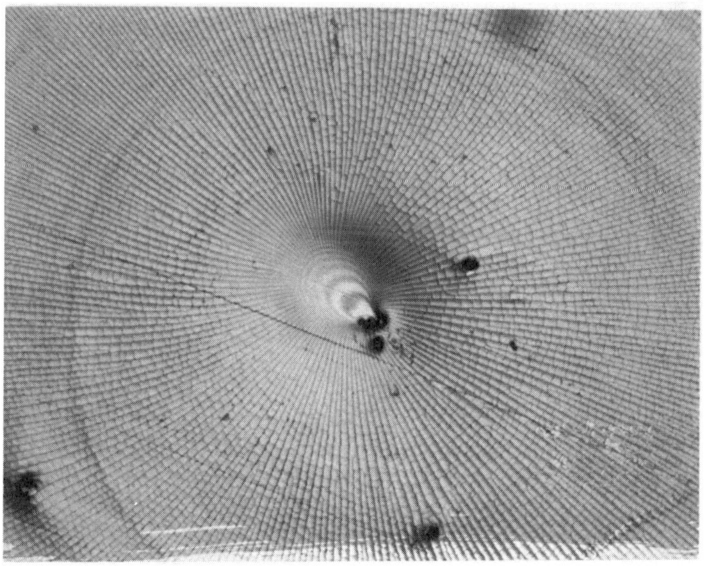

Figure 4. (g) Nomarski image of crazing.

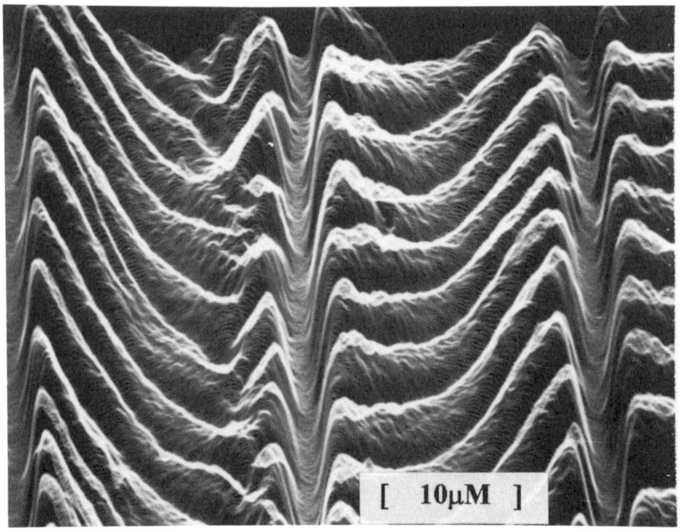

Figure 4. (h) Higher magnification PTM view of above at orthogonal juncture.

Figure 4(i) is an example of a machined polymer surface: a diamond-turned optical surface in acrylic, where the peak-to-valley height of the turning remnant is about 150Å over the 60 μm field of view. This information is needed not only to assure meeting the optical performance specification but is used to modify and improve both the precision lathe, tool, and process parameters as well.

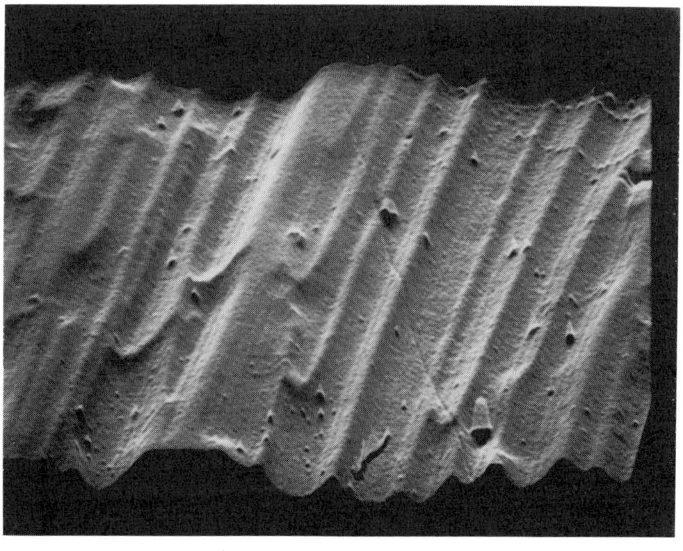

Figure 4. (i) Diamond-turned acrylic optic (courtesy Mr. Don Combs, Polaroid Corp.). Horizon is 60 μm across.

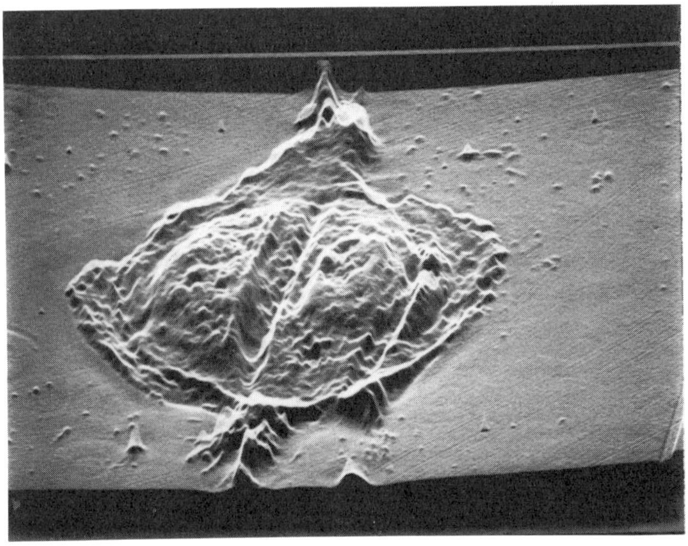

Figure 4. (j) A life science polymer application: a human cell, critical point dried, looking at attachment interface toward transducer. (Courtesy Dr. Sheldon Penman, M.I.T.)

Finally, Figure 4(j) is a human cell (a biological polymer application) that has been critical point dried but not stained, and was prepared directly on the transducer. In this way, cell attachment points can be studied. The inhomogeneity of the optical properties in some biological materials may contribute to a false topography, but *a priori* knowledge or subsequent imaging of the optical properties by flooding the tunneling gap allows correction.

CONCLUSION

The high lateral (0.10 μm) and subnanometer vertical resolution of the photon tunneling microscope, together with its real-time, three-dimensional interactive display and ability to image and profile dielectrics without scanning probes, vacuum, coating, or electrons, make PTM a powerful tool for imaging and measuring polymer surfaces.

ACKNOWLEDGMENTS

The authors wish to thank Polaroid Corp., specifically Dr. W. T. Plummer for supporting this work and Dr. R. Gaudiana for many helpful suggestions, the Department of the Army, and U. Mass., specifically Dr. R. Stein for his valuable support, guidance, and suggestions.

REFERENCES

1. J.M. Guerra, Photon tunneling microscopy, *Appl. Opt.* 29:3741-3752 (1990).
2. J.M. Guerra and W.T. Plummer, Optical proximity imaging method and apparatus, U.S. Patent 4,681,451 (21 July 1987).

3. J.M. Guerra, Photon tunneling microscopy, *Proc. Soc. Photo-Opt. Instrum. Eng.* 1009:254-263 (1988).
4. C.W. McCutchen, Optical systems for observing surface topography by frustrated total internal reflection and by interference, *Rev. Sci. Instrum.* 35:1340-1345 (1964).
5. J. Strong, "Concepts of Classical Optics," Freeman, San Francisco, 124-126, 516-518 (1958).
6. S.G. Lipson, and H. Lipson, "Optical Physics," Cambridge University Press, London, 79-109, 282-285 (1969)
7. D.D. Coon, Counting photons in the optical barrier penetration experiment, *Am. J. Phys.* 34:240-243 (1966).
8. S. Zhu, Frustrated total internal reflection: a demonstration and review, *Am. J. Phys.* 54(7): 601-607 (1986).
9. A. Keller, Regular rotation of growth terraces in polymer single crystals, *Kolloid-Zeitschrift & Zeitschrift Für Polymere*, Band 219, Heft 2: Seite 118-131 (1967).
10. P. Sullivan and B. Wunderlich, The interference microscopy of crystalline linear high polymers, *SPE Transactions,* 4:2, 2-8 (1964).
11. F. Lin, N.V. Gvozdic, and D.J. Meier, Application Note No. 2-1092-001, Topometrix Corp., Santa Clara CA (Oct., 1992).

MORPHOLOGY AND MOLECULAR ORDERING OF LANGMUIR-BLODGETT AND SELF-ORGANIZED FILMS FROM ORGANIC COMPOUNDS

Vladimir V. Tsukruk[1] and Darrell H. Reneker[2]

[1]Department of Engineering Technology
Western Michigan University
Kalamazoo, MI 49008
[2]Institute of Polymer Science
The University of Akron
Akron, OH, 44325

Abstract: The formation of ultrathin ordered molecular films at solid surfaces from various organic compounds, including discotic liquid crystals (LCs), alkylsilanes, ladder polymers, polyglutamates, lipids and dye molecules were explored. The molecular packing and mesoscale morphology of Langmuir-Blodgett (LB) films and self-organized films from these compounds at the surfaces of silicon wafers and glasses were studied by atomic force microscopy. Parameters of molecular ordering in the bulk state as well as thickness of molecular layers at the solid support were obtained from X-ray analysis and used for analysis of AFM data. Molecular modeling, including conformational analysis and molecular packing, provided additional information useful for interpretation of molecular images.

Discotic LCs adsorbed from dilute solution form ordered monolayer films on a silicon substrate. In the monolayers, discotic molecules are in an edge-on orientation. Dense side-by-side packing of the rigid, disc-shaped cores of the molecules is observed with periodicity of 0.35 nm. The columns lie parallel to the solid support. Monolayers are very flat and possess a local microroughness of 0.2 - 0.5 nm. Chemisorption of alkylsilane molecules leads to formation of flat monomolecular films with alkyl tails arranged perpendicular to the surface. The surface of LB films from ladder polymer precursor and stearic acid is characterized by microroughness of 0.6 - 1 nm with 2 nm thick single layers with sharp edges. The surface morphology of composite LB films from polypeptide (polyglutamate) bilayers on cadmium arachidate films is discussed in mesoscale terms (defects, domain structures, orientation, roughness). Molecular scale images are obtained both for the side alkyl chains and rod-like polyglutamate backbones. Surface reconstruction of the aged LB films is detected. Surface morphology of single crystals of dye molecules of one molecule thickness grown on lipid monolayers is discussed. Local mechanical modifications of the selected areas of the polymeric monolayers on the namometric scale produced by controllable scraping by the AFM tip are demonstrated for various films studied.

INTRODUCTION

Inducing of molecular ordering of desirable symmetry in ultrathin films of organic materials at solid surfaces is an important step in creation of "nanostructure" materials with new properties and uses. Depending upon the functionality of the molecules (switchable conformation, nonlinear optical response, selective chemical binding, charge generation, for example) these materials can find various applications in nonlinear optics as optical switches, waveguides, molecular sensors, anisotropic conductive films, and piezochromic films.[1,2] Many functionalized polymeric and low molecular weight compounds possess natural tendencies toward self-organization in a variety of lamellar and uniaxial supramolecular structures, display high mobility and high thermal stability and are used for fabrication of ultrathin ordered molecular films.

There are three main ways to fabricate ultrathin molecular films from organic compounds at solid supports: self-assembling of monolayers by physical or chemical adsorption of the molecules from solution; formation of ordered monolayers at an air-water interface and their transfer onto the solid support (Langmuir-Blodgett (LB) technique); and single crystal growth at ordered surfaces. In the present paper we consider the major features of surface mesoscale and molecular scale structures of several kinds of ultrathin molecular films. The set of ultrathin films discussed includes: self-assembled monolayers (SAM) of silanes and discotic liquid crystals (LCs); one-component and composite LB films from rod-like polyglutamates; amphiphilic ladder polymers; and composite films from lipids and dye molecules.

Experimental scanning tunneling and atomic force microscopy (STM and AFM) techniques, complemented by X-ray data provide quantitative information about molecular ordering in ultrathin molecular films. Molecular modeling including conformational analysis and molecular packing was used for rigorous interpretation of molecular images. AFM provides a unique possibility for direct observation of surface structures at ambient conditions on a wide range of scales from optical (micrometers) to molecular (nanometers).[3-6] Application of this technique to LB mono- and multilayers of several organic compounds has provided structural information about molecular conformation and shape; symmetry and parameters of positional ordering in two-dimensional lattices of molecules; possible structural reorganization of surface structures during phase transitions; and local defects such as holes, edges and dislocations.[6-13]

Images of the surfaces at ambient temperature were obtained with an atomic force microscope, the Nanoscope II (Digital Instruments, Inc.) using a pyramidal Si_3N_4 tip according to well-established procedures.[3,5] Scanners D and A were used for imaging on scales from 15 μm to 1.5 nm with applied forces of some tens of nN. In surface modification experiments, forces of more than 100 nN were used. The scanning rate was as low as 0.13 Hz for large scale images and 10-20 Hz for molecular scale scans. Mesoscale images were obtained with minimal values for the feedback gains and without any additional input filters. Raw AFM data were processed using flattening and, for some images, high-frequency filtering with smooth background subtraction.

All the images presented here were obtained repeatedly and were stable under experimental conditions. Roughness was calculated as standard deviation of heights in a square of appropriate size, excluding macroscopic features. The authenticity of the molecular images was checked using criteria such as: image stability during repeated scans; reproducibility of molecular features and spacings with variation of scale; appropriate rotation of the observed features with rotation of the scanning direction; and reproducibility of the images at different locations on the films. Finally, the average periodicities were obtained from two-dimensional correlation functions and Fourier-transformations of the images. Then the measured periodicities were compared to intermolecular distances expected from X-ray measurements of the same systems in the bulk state, in molecular films, and with intermolecular distances obtained by molecular modeling.

Molecules with various architectures are considered: silanes were represented by hexadecyltrichlorosilane[14]; twin discotic LCs with chemical linking of triphenylene and TNF moieties were observed[15,19]; polyglutamates (PG)[16] and amphiphilic ladder polyheteroarylenes represented rod-like macromolecules[17]; and cyanine dye molecules were grown on the surface of lipid monolayers.[18] Ordered molecular films studied[14-22] include films formed by the chemical adsorption of silane molecules and by physical adsorption of discotic LCs.[14,19] LB films were fabricated from PG bilayers on a solid support or on a cadmium arachidate film.[16,20] LB films were formed from amphiphilic complexes of ladder polyheteroarylenes with stearic acid.[17] Single crystal sheets of dyes were grown at the surface of a charged lipid monolayer.[18,22] Silicon wafers, glass and formvar were used as solid supports.

RESULTS AND DISCUSSION

Self-Organized Films

Alkylsilanes. Many technological processes require the spreading of a liquid on a solid surface and thereby depend upon wetting phenomena.[2,12,23,24] Active modification of surface properties can be done by coating the surface with alkylsilanes, which are able to form stable and smooth monolayers.[2,12] Covering of a silicon wafer by a monolayer of hexadecyltrichlorosilane 14 (Figure 1 (a)) creates a surface with a roughness of 0.1-0.2 nm. Very rarely, one or two times per square μm, holes with a diameter of 50 to 100 nm and a depth of 2.4 to 2.8 nm were observed.

The depth of the holes corresponds to the thickness of the layer formed by the silane molecules with the alkyl chains virtually perpendicular to surface (2.6 nm as estimated from molecular models). Self-assembled monolayers from hexadecyltrichlorosilane on a glass surface are rougher and contain of holes 100 - 300 nm in diameter and one monolayer deep (Figure 1 (b)).

Discotic Liquid Crystals. Ordered molecular LB films were prepared from various amphiphilic discotic LCs based upon derivatives of discotic triphenylene compounds and their mixtures with trinitrofluorenones (TNF).[15,25] These films possess a set of physical properties interesting for application as molecular conductors and optical switches. For example, in-plane, anisotropic photoconductivity was demonstrated recently for LB films fabricated from discotic LC molecular charge transfer complexes.[26,27]. In the bulk LC state, formation of charge transfer complexes produces a thermally stable (up to 260 °C) LC columnar orthorhombic phase with alternating packing of donor and acceptor groups within the columns.[15] We observed the ordered monomolecular films which were formed on a solid support by self-assembling twin discotics.[15] Ordered monomolecular films from discotic LC donor-acceptor compounds were formed from dilute solution by physical adsorption on the surface of silicon wafers.[19] They are built of single columns with the edges of the discotic molecules in contact with the solid surface. Monolayers are molecularly flat in areas 5 - 15 μm across with roughness of 0.2 - 0.5 nm (Figure 2 (a)). On surface of monomolecular films, ridges are observed, as can be seen from Figure 3(a), which have a height of 100 - 200 nm and width of 0.5 - 2 μm. They are aggregates of

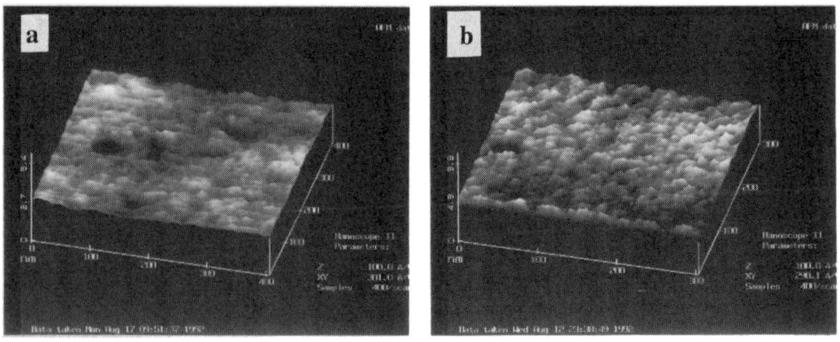

Figure 1. AFM images of SAM from hexadecyltrichlorosilane (a) on a silicon wafer, 400 nm x 400 nm and (b) on glass, 300 nm x 300 nm.

precipitated material. The ridges are weakly attached to the substrate and can be pushed away by applying higher forces during scanning.

Molecular resolution of the individual molecules packed in columns was obtained (Figure 4 (a)). The columns in ordered monolayers lie parallel to the solid support and are arranged with the diagonal planes of the unit cell parallel to the silicon surface. The dense side-by-side packing of the rigid cores of triphenylene and TNF groups is characterized by the closest possible intermolecular distance of 0.35 ± 0.03 nm (Figure 4 (a)). Positional

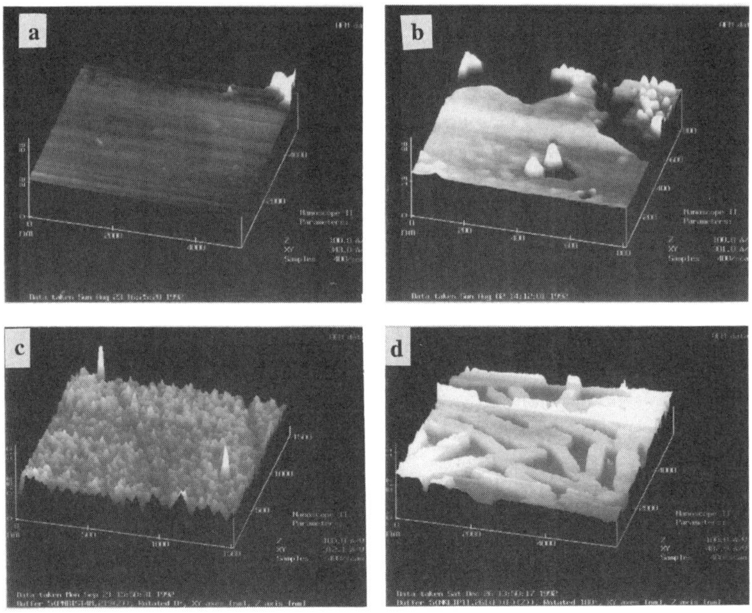

Figure 2. AFM images of molecular films on mesoscale:(a) SAM from discotic LCs, 2.5 μm x 2.5 μm; (b) LB bilayers of PG, 0.8 μm x 0.8 μm;(c) LB films from complexes of ladder polymer and stearic acid, number of layers n = 3, 1.5 μm x 1.5 μm ; (d) single crystals of cyanine dye grown on lipid monolayer, 4 μm x 4 μm. The sharp bumps that are occasionally seen are presently unexplained.

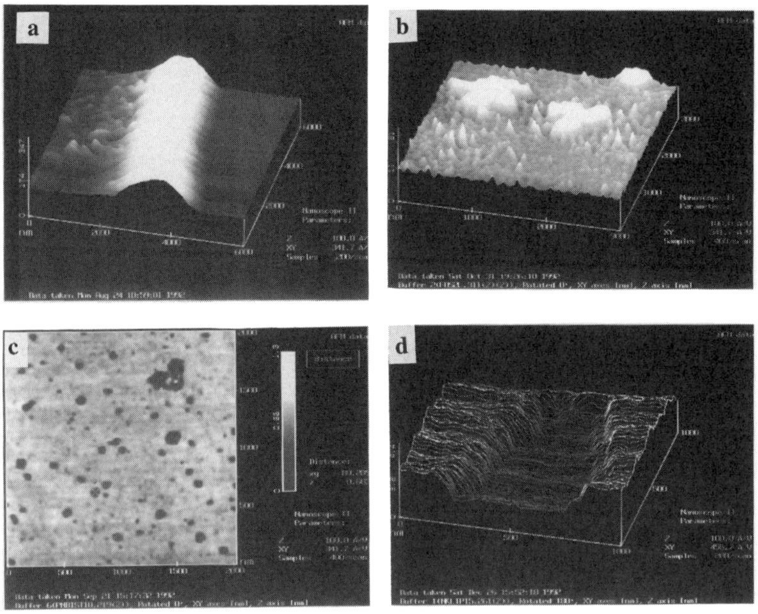

Figure 3. AFM images of surface defects of molecular films: (a) ridge of the bulk material on the surface of SAM from discotic LCs, 6 μm x 6 μm;(b) extra flat layers of PG on the LB bilayers, 3 μm x 3 μm; (c) holes in a LB film made from complexes of ladder polymer and stearic acid, 2 μm x 2 μm (d) hole with stepped edges in a single crystal of dye grown on a lipid monolayer, 0.8 μm x 0.8 μm .

and orientational correlations of the molecules extend over some tens of nearest neighbors. Correlations are much less pronounced for molecules with longer spacers.

Use of the AFM tip to apply high forces during scanning leads to a scraping of the monolayers from the silicon surface in areas some hundreds of nanometers across (Figure 5 (a)). The depth of the holes equals to 1.8 ± 0.5 nm and corresponds to the thickness of single columns of discotic materials. This demonstrates the possibility for controlled modification of soft molecular films from liquid-crystalline materials on a nanometer scale. It also shows that mechanical deformation characteristics can be used to identify layered films.

Langmuir-Blodgett Films

Polyglutamates. Formation of ordered molecular thin films of LB type from rod-like rigid macromolecules possessing a high degree of orientational ordering in the nematic phase was demonstrated for polyglutamates (PG).[7,16] These films can be used for optical waveguides.[1,16] Preferred orientation of the polyglutamate backbones along the dipping direction causes anisotropic birefringence in the plane of LB films. We discuss surface mesoscale morphology, microscopic orientation, mechanical stability and possible modification of PG bilayers transferred onto a silicon wafer by the LB technique.[20]

The typical surface morphology of PG bilayer consisted of flat areas with rarely occurring surface defects (Figure 2(b)). Surface roughness in 0.5 μm x 0.5 μm square regions was estimated to be in the range of 0.1 - 0.2 nm, which is comparable with the roughness of the supporting silicon surface. At higher magnifications the surface structure is characterized by patches 0.2 - 0.4 nm in height and 100-200 nm across. In addition we

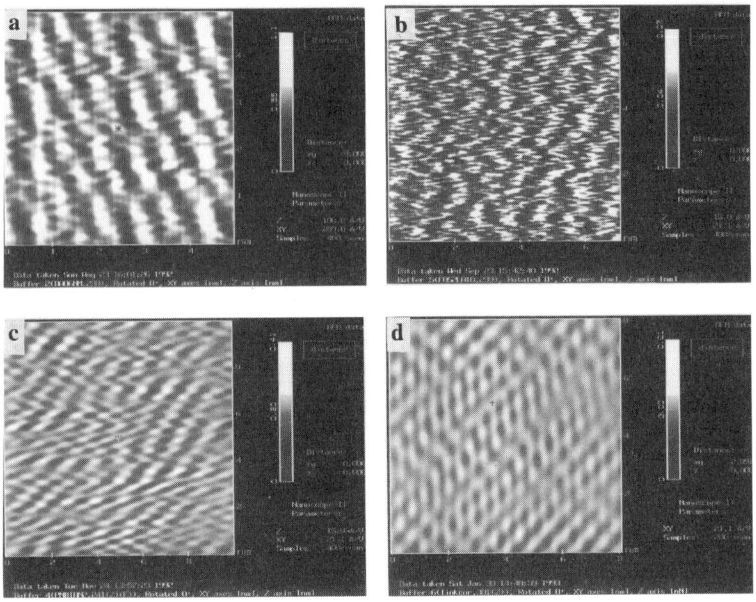

Figure 4. Molecular scale images of the films: (a) columnar ordering of the discs in monomolecular films from discotic LCs, 5 nm x 5 nm; (b) PG backbones in LB bilayers, 7 nm x 7 nm; (c) ladder backbones in cyclized LB film from polyheteroarylene, 12 nm x 12 nm; (d) a x b projection of surface of a single crystal of dye molecules, not covered by lipid, 10 nm x 10 nm.

observed a "cellular" morphology formed by the domains of PG with the lateral sizes of the grains in the range of 2 - 3 μm.

Extra PG layers with lateral dimensions of 0.5 - 1 μm were formed on top of bilayer PG films (Figure 3 (b)). These layers, 1.6 and 3.5 nm high, were formed by single or bilayer packing of PG backbones. Holes with depths of 3 - 4 nm, corresponding to the PG bilayer thickness, were occasionally observed in the samples studied.

Two types of molecular images were obtained from PG bilayer films. Characteristic features in the first image type correspond well with those expected from a layer of alkyl side chains packed in a distorted lattice with weak positional ordering. Molecular scale images of the second type, as shown in Figure 4 (b), consist of wavy ridges with the spacing between neighboring ridges of 1.1 ± 0.2 nm. This intermolecular distance (the distance between rods in one layer) is close to the 1.23 nm distance between the backbones proposed by Watanabe[28] and similar to the bulk rod to rod spacing of about 1.1 nm observed by Tsujita.[29] Along the backbones only a weak modulation of heights with periodicity of 2.5 nm was observed. This value is close to the known pitch of the α-helix, which is 2.7 nm. The observed molecular features are due to the rod-like PG backbones.

The stability of surface structures of soft polymeric materials during interaction with an AFM tip is a crucial point in the correct interpretation of STM and AFM images.[6] Dramatic changes in surface morphology are observed for the soft PG bilayers deposited directly on a silicon wafer. Even very gentle scanning with forces in the nanoNewton range lead to visible changes in surface morphology from scan to scan and to formation of

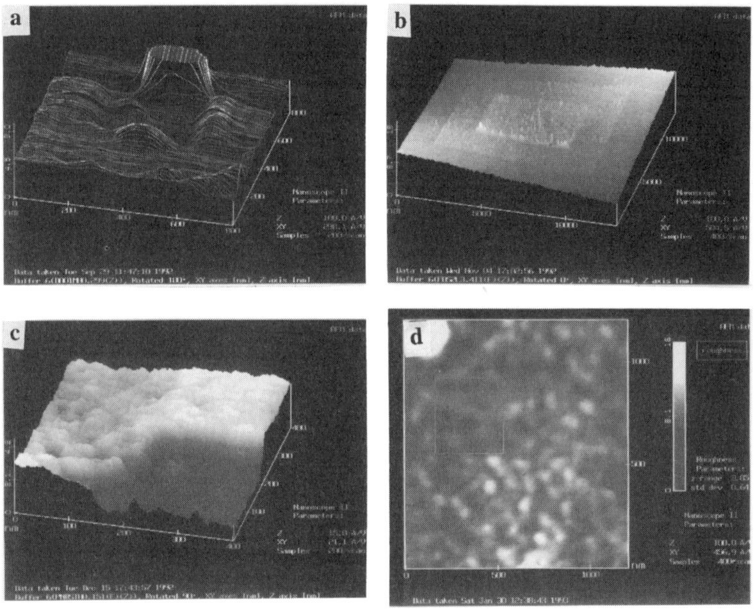

Figure 5. Surface modification of molecular films by the AFM tip: (a) hole 1.8 nm deep and 400 nm x 400 nm across, made in monomolecular film from discotic LCs, 0.8 μm x 0.8 μm scan; (b) random surface distortions by repeated scans with increasing sizes at the surface of LB bilayers of PG (square tracks of first and second scans are visible), 13 μm x 13 μm; (c) edge of a hole that is one layer deep (2.0 nm), made in a LB film from ladder polymer and stearic acid, 400 nm x 400 nm; (d) shallow square depression caused by a scan with high forces, in the upper quarter of image, on a film of cyanine dye covered by diacetylene-containing lipid monolayer, 1.2 μm x 1.2 μm.

ridges, and, finally, after a few scans, to so-called "abrasion" pictures.[7,30] The observed orientation of the ridges is determined by the macroscopic orientation of the rod-like PG macromolecules. Repeated scans lead to an increase in the roughness to 0.9 and 1.2 nm for the second and third scans. Prolonged scanning leads to destruction of the initial molecularly flat bilayer surface and formation of a heterogeneous surface with material concentrated in high ridges aligned along the dipping direction, which is perpendicular to the scan direction.

Scanning of the same PG bilayer in the direction parallel to the dipping direction leads to development of random surface distortions with a roughness of 0.4 - 0.8 nm for the third scan (Figure 5 (b)). Use of a higher force during scanning in a selected sample area produces a local scraping of the PG film. In contrast, PG bilayers deposited on cadmium arachidate multilayers display a greater mechanical stability during scanning.

Ladder polymers. Ladder materials with conjugated backbones possess high third-order nonlinear optical susceptibility, complemented by an extremely high thermal and chemical stability.[31] The anisotropic shape of the semirigid backbones promotes high in-plane ordering of the macromolecules during formation of LB films which can serve as optical waveguides.[31] We observed features of the surface morphology and molecular ordering of LB films from complexes of poly(naphthoylene benzimidazole) precursor (PNBI) and stearic acids. These films were studied in their initial state after deposition as well as after thermocyclization at high temperature.[17,32] The molecular structure of PNBI in the bulk state derived from X-ray data, was described earlier.[17] LB films from complexes of PNBI with stearic acid, consisting of one to five molecular layers, possess relatively flat surface morphology (Figure 2 (c)). The roughness steadily decreased from 1.4 - 1.6 nm for the first layer to 0.3 - 0.5 nm for the film containing five layers. The major type of defect is many small holes 0.6 - 0.9 nm deep and 60 - 200 nm in diameter (Figure 3 (c)).

For LB films with seven to nine layers, a heterogeneous surface domain morphology is developed. The domains are 10 - 15 nm high and 1 - 3 μm across. The thickness of a single layer in these LB films, determined from the height of layer edges is 2.0 nm, which corresponds to the average value derived from X-ray data. Roughness rises to 1 - 1.5 nm as the number of layers increases to nine. The upper layers in these films are very weakly attached and can be easily scraped away by the AFM tip during scanning with higher forces. Holes a single layer deep can be made on the nanometer scale by the AFM tip, by scanning with higher forces (Figure 5 (c)).

Cyclization of LB films from PNBI and stearic acid complexes, by heating to 350 °C, destroys the initial multilayered structures, as judged from X-ray data. Roughness of the film surfaces increases after thermal treatment and reaches 2 - 3 nm for the films consisting of nine layers. Edges of layers with heights in the range of 0.6 - 2 nm are observed for these films. On the molecular scale, we observe wavy ridges aligned along the dipping direction, which corresponds to the direction of the macromolecular backbones (Figure 4 (c)). The average lateral periodicity is 0.85 ± 0.03 nm. This matches quite well to X-ray data for intermolecular distances in the lateral packing of the ladder backbones of PNBI macromolecules in the planes in which phenyl rings lie.

Composite Films of Dye and Lipid

Cyanine dyes can be used for formation of thin molecular Langmuir-Blodgett (LB) films with interesting photosensitive properties and prospects for nonlinear optical applications.[33] The type of molecular ordering (aggregation of cyanine dyes in molecular films) determines their optical response and can be controlled via preparation conditions, e.g., selecting proper surfactant and nucleation parameters.[18,22] We discuss the surface morphology and molecular ordering of single crystal sheets of photosensitive ionic dye molecules grown at the surface of a charged lipid monolayer by adsorption from the subphase. The photosensitive cyanine dye NK-2012 was adsorbed from the water subphase onto monolayers of two different positively charged lipids; dimethyldioctadecyl-ammonium bromide (named DODAB) or dimethylene-containing lipid dimethyl[(-bis-12-hydroxyethyl hexacosa-10,12-diynoate) ammonium bromide] (named DAB). Lipid monolayers with adsorbed NK2012 molecules were transferred onto a 20 nm thick layer of formvar supported on glass (see references 18, 22).

Dye molecules absorbed on a monolayer of the lipid DODAB formed single crystals, as judged from electron diffraction and fluorescence microscopy.[18] The average thickness of this composite monolayer derived from X-ray data is about 3.3 nm.[21] The dye crystals were nearly rectangular with sizes from 2 μm to 100 μm across, depending upon preparation conditions (Figure 2 (d)).

The surface of the by lipid monolayer that covered the dye crystals is smooth with roughness in 1 μm x 1 μm areas in the range of 0.1 - 0.3 nm. The height of the crystals is very uniform everywhere on the film surface and equals to 1.5 ± 0.2 nm, which corresponds to the size of the molecules along the shorter in-plane axis.[18,21] The lipid monolayer that covered the single crystal sheet of dye molecules sometimes had depressions 1.5 nm deep with smooth or stepped edges (Figure 3 (d)). This suggests that the lipid layer sheared into holes between the dye crystals. Frequently, the walls of depressions intersect at the angle of 75° ± 2°, which corresponds to the angle γ for the a x b plane of the unit cell of NK2012.[18]

The crystals of dyes, covered by a monolayer of DAB, have a dendritic morphology. The roughness of lipid monolayer on top of single crystal surface in an area 0.5 µm x 0.5 µm is 0.6 - 1.3 nm. The observed differences of dye molecule aggregation at the surface of monolayers composed of DODAB and DAB can be understood by taking into account the possibility of polymerization of the acetylene-containing DAB lipid chains, promoted by the cyanine dye molecules.[34] At the surface of single crystals, not covered by a lipid monolayer, molecular scale images are observed (Figure 4 (d)). The observed arrangement corresponds to aggregation of dye molecules with unit cell parameters of 2.1 nm x 0.6 nm, close to those determined from electron diffraction.[18]

Scanning with higher forces on the surface of DODAB covered dye crystals scraped holes in the lipid and piled material along the edges of the holes. In contrast, for DAB lipid-based films, only a shallow square depression of deformed monolayer with no scraped material at the edges was observed (Figure 5 (d)). The depth of the depression is only 2 ± 1 nm, and the surface in the depression area is very smooth (roughness is 0.5- 0.7 nm in comparison with the roughness outside the area of 1 - 1.2 nm). Only modest plastic deformation of DAB lipid monolayer occurs due to the higher mechanical stability of lipid alkyl chains that are cross-linked by polymerized diacetylene groups.

CONCLUSIONS

In conclusion, AFM measurements of various ordered molecular films from different classes of organic low molecular weight and polymeric compounds were made. Several aspects of these systems were revealed: their mesoscale surface morphology; typical naturally occurring defects of the surfaces; molecular scale ordering; and surface modification and stability during scanning with the AFM tip.

Self-assembled films from silane studied on a silicon wafer possess very smooth surfaces with occasional holes one monolayer deep. Ordered monomolecular films from discotic LCs can be formed at the surfaces of silicon. These films are molecularly flat over areas that are 5-10 µm across. They are characterized by a roughness of 0.1 - 0.5 nm. The films are formed by dense packing of one layer of columns lying parallel to the solid surface. The parameters of intracolumnar and intercolumnar packing of the molecules correspond well to those reported for the bulk liquid crystal state. The ordering in a column is an alternating packing of the donor and the acceptor moieties of the molecules. Surface modification of ordered monomolecular films by scraping well-defined rectangular areas on a nanometer scale was demonstrated.

The surface of PG bilayers deposited on silicon by the Langmuir-Blodgett technique is extremely smooth. Holes, grooves and "extra" layers are seen only occasionally. Defects in the film include holes through one or two PG layers, or the entire film thickness, as well as "extra layers" covering small areas of the sample. Molecular images observed correspond to the close lateral packing of rod-like polyglutamate backbones aligned along the dipping direction. The stability of bilayer surfaces to repeated scanning was excellent for the PG bilayers on top of a cadmium arachidate multilayer. In contrast, bilayers of PG on a silicon surface developed distinctive, scanning-induced ridges, aligned along the dipping direction for scanning with the AFM tip moving transversely to the dipping direction. Repeated scanning led to distortion of initially flat bilayered films and formation of "abrasion patterns". By use of higher forces on the tip, PG layers can be scraped from selected nanometer scale areas of the surface of a silicon wafer.

LB films from complexes of PNBI with stearic acid consisting of four to five molecular layers possess relatively flat surface morphology with roughness of 0.3 - 0.5 nm for a film containing five layers. A low concentration of hole defects in the films and the edges of layers is typical for these films. The major types of the surface defects are small holes 0.6-0.8 nm deep and 60-100 nm in diameter. The sharp edges of a single layer of 2.0 ± 0.3 nm thick were also observed. The upper layers in thicker LB films (more than six layers) are weakly attached and can be easily scraped away. Holes a single layer deep can be made on nanometer scale with the AFM tip, by scanning with higher forces. For the cyclized films, flat-on arrangement of the backbones with edge-by-edge, in-plane distance of 0.85 nm periodicity was realized.

The crystals of cyanine dye molecules, grown from water solution onto a charged lipid monolayer, form very uniform multiphase films of 3.3 nm thickness, composed of single crystals sheets exactly one molecule thick (1.5 nm). The dye molecules are in an edge-on position and the lipid tails are tilted in their own layer. The covering lipid layer tends to shear into holes in the dye crystals and in-between neighboring dye crystals. Dye single crystals, grown on a diacetylene-containing lipid monolayer, display a dendritic morphology. The top monolayer of this lipid possesses heterogeneous surface structure with granular morphology due to polymerization caused by visible light. Polymerization of diacetylene containing lipid alkyl chains increases the mechanical stability of the composite molecular films.

ACKNOWLEDGMENTS

The authors are very grateful to the following colleagues for supplying the samples for investigations, fruitful cooperation and helpful discussions: Dr. H. Bengs, and Prof. H. Ringsdorf (Mainz U.), discotic liquid crystals; Prof. M. Foster (U. Akron), X-ray data, Dr. A. Schmidt, Dr. W. Knoll, Dr. K. Mathauer, and Prof. G. Wegner (M.P.I., Mainz), polyglutamate; Prof. A Rusanov (Academy of Science, Russia), and Dr. Bliznyuk (Academy of Science, Ukraine), ladder polymers; Dr. V. N. Bliznyuk, Dr. S. Kirstein, and Prof. H. Mühwald (Mainz U.), dye-containing films; Prof. W. Brittain (U. of Akron), silanes. Our thanks for the discussions of various parts of this work go to Prof. A. Keller (Bristol U.), Prof. P. Pershan (Harvard U.), Prof. E. Samulski (North Carolina U.), DSc. O. Lavrentovich (LCI, Kent U.), and Dr. L. F. Chi (B.A.S.F.). We appreciate financial support received from the U.S. Army, the National Science Foundation Center for Molecular and Microstructure Composites, and E.P.I.C., the Edison Polymer Innovation Corporation of the State of Ohio.

REFERENCES

1. M.C. Petty, Possible applications of Langmuir-Blodgett films, *Thin Solid Films*, 210/211:417-430 (1992).
2. A. Ulman, "Introduction to Ultrathin Organic Films", Academic Press, Boston (1991).
3. D. Sarid, "Scanning Force Microscopy," Oxford University Press, New York (1991).
4. J. Rabe and H.J. Butt, Molecules at interfaces: STM in materials and life sciences, *Ultramicroscopy*, 42-44A:41-54 (1992).
5. D.H. Reneker, R. Patil, S.-J. Kim, V.V. Tsukruk, Observation of oligomers, polymers and their crystals by scanning tunneling and atomic force microscopy, *in:*"Polymer Characterization," M. Dosiere, ed., NATO ASI Series, Kluwer Acad. Press, London C405:357-400 (1993).

6. J. Frommer, Scanning tunneling and atomic force microscopy in organic chemistry, *Angew. Chem. Int. Ed. Engl.*, 31:1298-1318 (1992).
7. L.F. Chi, L. M. Eng, K. Graf, H. Fuchs, Structure and stability of Langmuir-Blodgett films investigated by scanning force microscopy, *Langmuir*, 8:2255-2261 (1992).
8. D.K. Schwarz, J. Garnaes, R. Viswanathan, J.A. Zasadzinski, Commensurate defect superstructures in Langmuir-Blodgett films, *Science*, 257:508-511 (1992).
9. R. Viswanathan, D.K. Schwartz, J. Garnaes, J.A. Zasadzinski, Atomic force microscopy images of substrates and PH effects on Langmuir-Blodgett films detected with atomic force microscopy, *Langmuir*, 8:1603-1607 (1992).
10. L. Bourdieu, P. Silberzan, D. Chatenay, Langmuir-Blodgett films: from micron to angstrom, *Phys. Rev. Lett.*, 67:2029-2032 (1991).
11. J.P.K. Peltonen, P. He, J.B. Rosenholm, Order and defects of Langmuir-Blodgett films detected with atomic force microscopy, *J. Am. Chem. Soc.*, 114:7637-7642 (1992).
12. J.M.Tidswell, T.A. Rabedeau, P.S. Pershan, J. Folkers, M. Baker, G.M.Whitesides, X-ray specular reflection studies of silicon coated by organic monolayers, *Phys. Rev.*, B, 44:10869-10878 (1991).
13. L.F. Chi, M. Anders, H. Fuchs, R.R. Johnston, H. Ringsdorf, Domain structures in Langmuir-Blodgett films investigated by atomic force microscopy, *Science*, 259:213-216 (1993).
14. W. Brittain, unpublished results.
15. M. Möller, V.V. Tsukruk, J.H. Wendorff, H. Bengs, H. Ringsdorf, Discotic charge transfer twins: structure and mesomorphic behaviour, *Liquid Crystals*, 12:17-36 (1992).
16. A. Mathy, K. Mathauer, G. Wegner, C. Bubeck, Preparation and waveguide properties of polyglutamate LB films, *Thin Solid Films,* 215:98-102 (1992).
17. V.N. Bliznyuk, O. Lokhonya, A.L. Rusanov, I.I. Ponomarev, V.V. Shilov, Structure of conjugated heterocyclic polymers, *Polymer. Sci.USSR,* 34A, 120-127 (1992).
18. S. Kirstein, H. Mühwald, Structure and optical properties of monolayer single crystal of a cyanine dye, *Chem. Phys. Lett.,* 189:408 (1992).
19. V.V. Tsukruk, D.H. Reneker, H. Bengs, H. Ringsdorf, Atomic force microscopy of ordered monolayer films from discoids liquid crystals, *Langmuir*, 9:2141-2144 (1993).
20. V.V. Tsukruk, M. Foster, D.H. Reneker, A. Schmidt, W. Knoll, Morphology of Langmuir-Blodgett films from polyglutamates observed by atomic force microscopy, *Langmuir*, 9:3538-3547 (1993).
21. V.V. Tsukruk, V.N. Bliznyuk, D.H. Reneker, S. Kirstein, H. Mühwald, Surface morphology of cyanine dye single crystal sheets grown on lipid monolayer, *Thin Solid Films*, in press.
22. V.N. Bliznyuk, S. Kirstein, H. Mühwald, Structural control of optical spectra of two-dimensional mixture of dye crystals, *J. Phys. Chem.*, in press.
23. P.G. de Gennes, Wetting: statics and dynamics, *Rev. Modern Phys.*, 57: 827-878 (1985).
24. S. Garoff, E.B. Sirota, S.K. Sinha, H.B. Stanley, The effect of substrate roughness on ultrathin water films, *J. Chem. Phys.*, 90:7505-7513 (1989).
25. O. Karthaus, H. Ringsdorf, V.V. Tsukruk, J.H. Wendorff, Columnar ordering of LC discotics in LB films, *Langmuir*, 8:2279-2283 (1992).
26. M. Auweraer, C. Catry, L. Chi, O. Karthaus, W. Knoll, H. Ringsdorf, M. Sawodny, C. Urban, Mono and multilayers made from discotic liquid crystals, *Thin Solid Films*, 210/211:39-43 (1992).
27. D. Adam, F. Closs, T. Frey, D. Finhoft, D. Haarer, H. Ringsdorf, P. Schumacher, K. Siemensmeyer, Transient photoconductivity in discotic liquid crystals, *Phys. Rev. Lett.*, 70:457-460 (1993).
28. J. Watanabe, H. Ono, I. Uematsu, A. Abe, Molecular packing and thermotropic behaviour of polyglutamates with long alkyl side chains, *Macromolecules*, 18:2141-2150 (1985).
29. Y. Tsujita, R. Ojika, A. Takizawa, T. Kinoshita, Thermal and transport properties of copolyglutamates, *J. Polym. Sci. Part A, Polym. Chem.*, 28:1341-1353 (1990).
30. G.F. Meyers, B. M. DeKoven, J.T. Seitz, Is the molecular surface of polystyrene really glass? *Langmuir*, 8:2330-2337 (1992).
31. D.R. Ulrich, Multifunctional macromolecular ultrastructures, *Polymer*, 28:533-545 (1987).
32. V.V. Tsukruk, V.N. Bliznyuk, D.H. Reneker, Morphology and molecular ordering in Langmuir-Blodgett films from ladder polymers, *Thin Solid Films*, in press.
33. T.H. James, "The Theory of Photographic Process," Macmillan, NY (1977).
34. C. Bubeck, B. Ticke, G. Wegner, Formation of lipid monolayers, *Ber. Bunseng. Phys. Chem.*, 86:499-507 (1982).

PT/IR NANOTIPS--A MECHANICAL TECHNIQUE FOR SHARPENING TIPS UTILIZED BY A SCANNING TUNNELING MICROSCOPE

Ernest C. Hammond, Jr. and Brenda Queen

Department of Physics
Morgan State University
Baltimore, Maryland 21239

Abstract: The image quality of a scanning tunneling microscope (STM) depends on the tip sharpness. The literature is replete with examples of etching techniques to construct tip thickness within a few atomic dimensions, thus enhancing the STM image. This research team has developed a mechanical technique using a Dremel Motor tool and a diamond wheel stone to produce similar atomic dimensions. The technique is very cost effective because one's tip lifetime is extended by a factor of four. Moreover, this paper will compare and contrast scanning electron micrographs produced from the conventional techniques with the protocol developed by this research team. Further, our technique will ensure reasonable tip sharpness and reproducibility.

INTRODUCTION

The ultimate problem in doing scanning tunneling microscopy involves the manipulation of the tips of the scanning tunneling microscope to form a point whose dimensions involve multiples of the atomic width.[1,2,3] Many researchers have attempted to use etching techniques to draw or etch the needle points within several atomic dimensions.

Our earliest involvement in scanning tunneling microscopy produced the idea that any wire of the appropriate diameter could be used as a tip for detecting tunneling currents from various conductive samples. Many of the major laboratories and manufacturers of scanning tunneling microscopes employ only a pair of scissors to cut the wire and produce the appropriate multi-atomic dimensions necessary for the tip.

At the IBM Thomas J. Watson Research Center, tunneling has been observed by using ordinary wire and simply cutting it with scissors, while other groups have attempted to construct multiple atomic dimensional tips by using etching processes including elaborate coating and techniques. This research group, therefore, initiated a research project that would utilize mechanical techniques for sharpening iridium/platinum tips for immediate reuse and improved tip sharpness and enhanced image quality.

MATERIALS AND METHODS

Scanning tunneling microscope tungsten tips used in field ion microscopy produced by electrochemical etchings have revealed protrusions of oxidized tungsten with low desorption field strength (between 10 and 30 V/nm). These protrusions are considered to be operational as electron donors and acceptors in the STM.

There have been a number of papers discussing the theory of tunneling involving metal tips. These tips consist of a single tungsten atom that is absorbed at a particular surface that have been modelled using supercomputer and computer visualization techniques. The theory is used to investigate the barrier height of the surface.

Uzi Landman, Regents Professor at the Georgia Institute of Technology, spoke at the Atomic and Nanoscale Modification of Materials--a conference sponsored by the Engineering Foundation. He described and demonstrated computer modelling of tip interactions under various physical conditions. These interactions were created using idealized physical conditions.

Salmeron, et al., have also demonstrated that STM tips undergo compression or repulsive forces producing permanent damage both in elastic and plastic deformation when analyzing the sample between current and tip. They have observed and we have demonstrated the shearing forces where the tip interacts with the sample.

An analysis of the broken tips indicates that during the scanning process as the tip is moved in the Z direction, a major number of the tips crash into the sample thus destroying the point on the tip and the scanning image resolution. At the American Physical Society meeting, researchers have observed that certain tips actually penetrate the surface of the atom and have obtained tunneling effects while the tip remains within the surface. This research team has never observed this phenomenon using a pyrolytic carbon sample. An analysis, using the scanning electron microscope, has confirmed the

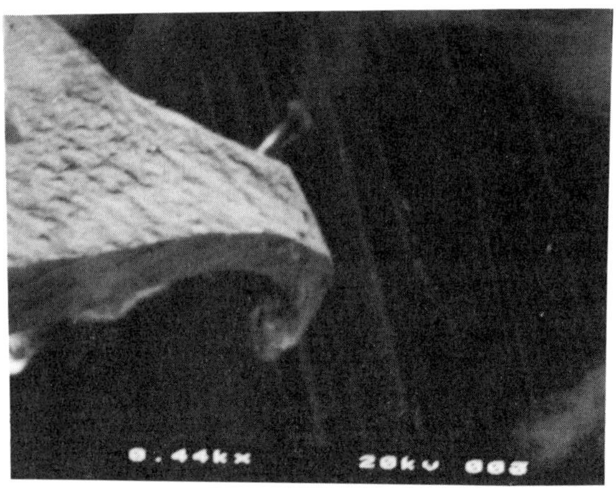

Figure 1. This is a typical micrograph of the bending of the tip at approximately 44 x magnification.

destruction action after the tip has crashed into the sample (see Figures 1 and 2). Also, the electron micrographs clearly reproduce a visualization of the shearing effects associated with using scissors to produce the direction associated with the shearing motion of the scissors (Figure 3).

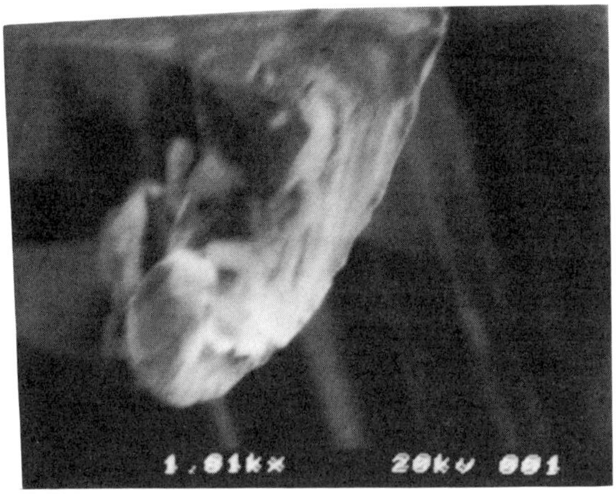

Figure 2. This is a closeup picture of the rounded tip of the Ir/Pt tip indicating forced distortion in the tip.

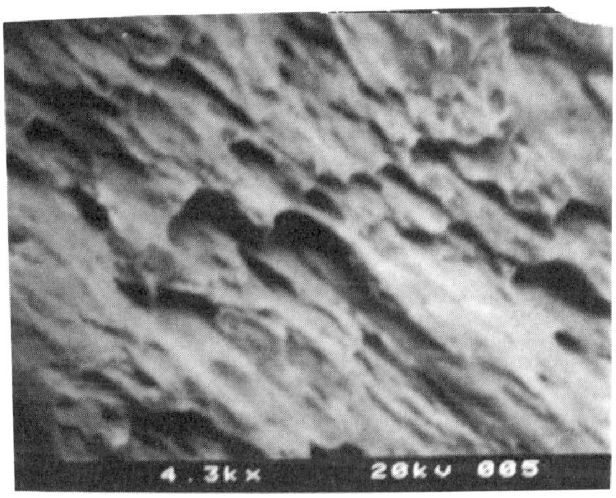

Figure 3. This is a micrograph at about 4,000 magnification showing the shearing effects due to the commercially produced scissors.

An attempt was made to sharpen the tips using the hand-held method to produce a sharp point. This method works within limits, but the disadvantage in using the hand-held method of sharpening the tip produces numerous metallic spurs. These spurs offer multi-positional tunneling, which can reduce the tunneling signal from a particular point within the limits of a few atomic dimensions. Special handling devices were devised to move the tip forward to the grindstone and device for holding the Ir/Pt tip. There is still difficulty in observing exactly how much material is being ground. Currently, we are searching for grinding materials with grain size is smaller than that of the grain size of the commercially available material.

The solution to this problem seems to involve approaching the lapidary industry to see the type of grinding adhesives available to them. Useful information has been obtained from representatives of the lapidary industry that a fine grain of polishing grit may prove to be beneficial in obtaining a multiple atomic dimension at the tip. We have attempted to devise a grinding protocol device that ultimately gives the tips a point within those desired dimensions. The device should include a hardened rouge material whose diameter is no more than 1/8th of an inch thick. Its conical opening in the middle would permit the insertion of the tip, while this rouge grinding stone is spinning at very high rotational rates. Essentially, we refer to it as a Nanotip Pencil Sharpening Device.

The real problem after grinding the tip is to remove all excess metallic spurs, as seen in Figures 4 and 5. These spurs tend to act as tunneling points above the surface of the sample, which can reduce the sensitivity of the tunneling device. This above device is currently being constructed and will be tested. It should be noted that there is substantial improvement in the images even when utilizing the hand-held techniques of sharpening the tips. This implies a four-to-five times increase in the lifetime of the tips where currently the tips are being thrown away after a one-time use.

When the marred surface of propylic carbon is visualized in the two-dimensional mode, the features showing the effect of the continuous scan are clearly seen (Figure 6). These same surface irregularities can be seen more clearly in a three-dimensional rendition (Figure 7).

The mechanical Dremel (Figure 8) was used to obtain the appropriate tip sharpness for this study.

Figure 4. This is a 500 x magnification micrograph of a hand-ground tip of Ir/Pt. Note the spurs that are attached to the tip.

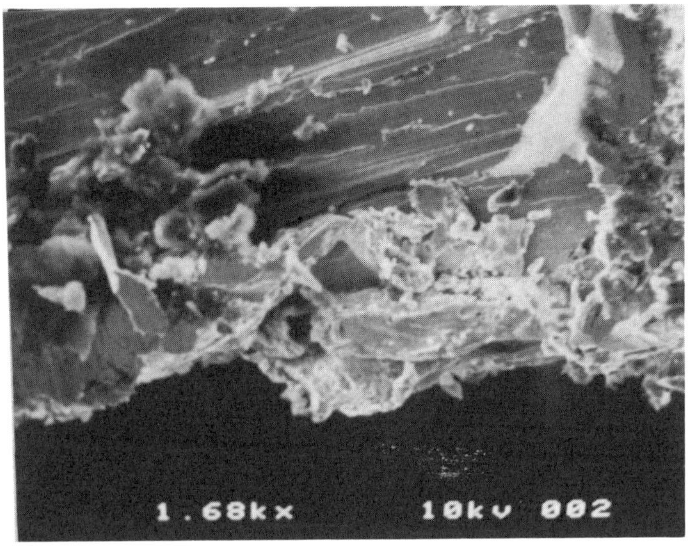

Figure 5. This is a 1.68 kx magnification for the closeup of the tip showing the number of centers, spurs and points where multiple tunneling may occur.

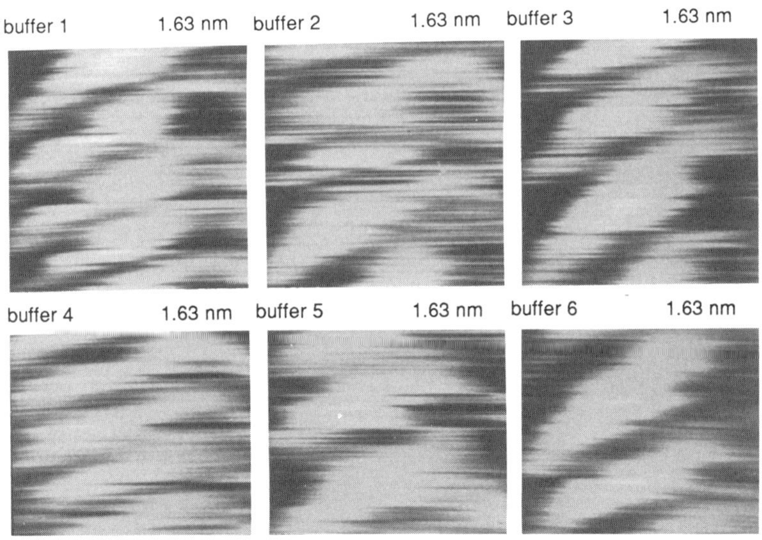

Figure 6. This is a micrograph showing the continuous scan using the Ir/Pt tip to look at a marred surface of pyrolytic carbon in the two-dimensional mode.

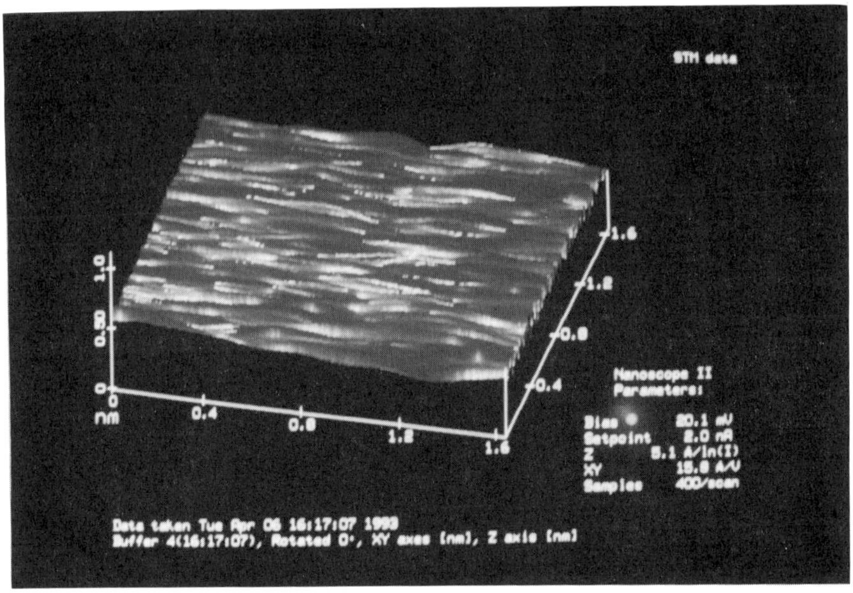

Figure 7. This is a micrograph showing a three-dimensional using the hand held sharpening technique.

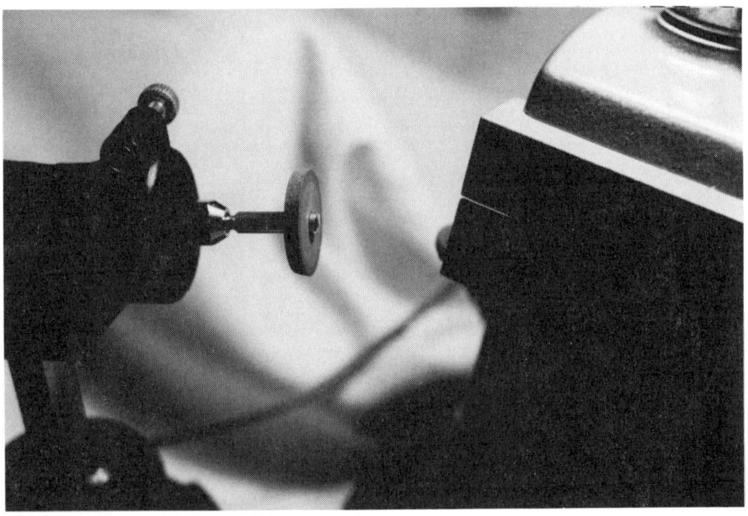

Figure 8. This represents the mechanical Dremel set up to gain the appropriate sharpness at the tip.

CONCLUSIONS

With nanotechniques and nanomanipulative devices as proposed above, the resolution of the scanning tunneling microscope images can be improved as well as the longevity and usage of the tip. We are currently experimenting with various rouge and polishing grits to improve the sharpness of the tip. Careful examination of these powders and grains will be investigated using the scanning electron microscope at very high resolution, examining spurs and other protrusions. This technique has permitted us therefore to increase the functioning lifetime of the Ir/Pt tips, even in the hand-held mode for sharpening and without any manipulating tools.

REFERENCES

1. U. Schmidt, Characterization of STM tungsten tips by FIM with an organic image gas, *Surf. Sci.* 266:249-52 (1992).
2. G. Doyen, Theory of tunneling from transition metal tips, *NATO ASI Ser.*, Ser. E. 184:97-111 (1990).
3. M. Salmeron, Tip-surface forces during imaging by scanning tunneling microscopy, *J. Vac. Sci. Technol.*, B. 9:1347 (1990).

ATOMIC FORCE MICROSCOPY STUDY OF ELECTRON BEAM PATTERNED SIO_2

John F. Kimball,[1] Patricia E. Allen,[1] Dieter P. Griffis,[2]
Zbigniew J. Radzimski,[2] and Phillip E. Russell[2]

[1]Appalachian State University
Physics & Astronomy Department
Boone, N.C. 28608
[2]North Carolina State University
Analytical Instrumentation Facility
Raleigh, N.C. 27695

Abstract: An atomic force microscope (AFM) is used to study the patterns obtained from E-beam patterning of SiO_2. The SiO_2 is patterned using an E-beam, a plasma etcher and a chemical P-etch. Once patterned, the sample is studied with the AFM to determine the topography. Various patterns are made to determine the feasibility of this patterning technique in designing optical devices. Ideal patterning parameters are also explored.

INTRODUCTION

A process for making maskless patterned semiconductor devices, which could eliminate organic resists, would greatly reduce the number of processing steps necessary in the fabrication[1] of such devices. Due to the abundance of SiO_2 in the semiconductor industry, the use of SiO_2 in the patterning process is easily justified. This process could lead to a totally dry patterning procedure based on the use of SiO_2 thus allowing the integration of this procedure into single wafer processing module clusters.[2]

Patterning of SiO_2 has been performed with various techniques including optical, electron beam, X-ray and ion beam lithography.[3] Present manufacturing techniques in the integrated circuit (IC) market predominantly utilize optical lithography, with a resolution of about 1 μm. However, in certain IC applications electron beam lithography has been used with a resolution of about 0.5 μm. The ability to pattern SiO_2 using electron irradiation has been demonstrated using low[4] and high[5] energy electron beams. In 1967, the ability to do resistless fabrication of planar transistors using low energy (1-15 keV) electron beam exposure of SiO_2 followed by "P"

Atomic Force Microscopy/Scanning Tunneling Microscopy
Edited by S.H. Cohen *et al.*, Plenum Press, New York, 1994

etch was demonstrated.[4] Recently, 20 nm line widths were fabricated using high energy (300 keV) electrons.[5] These results indicate that it may be possible to develop a practical resistless lithography procedure, capable of <0.1 μm line widths.

The work that follows describes low energy (<4.0 keV) electron beam patterning of SiO_2[6] samples including: thermally grown in dry O_2 gas ("dry oxide"), thermally grown in wet O_2 gas ("steam oxide"), remote plasma enhanced chemical vapor deposition (RPECVD)[7] and low pressure chemical vapor deposition (LPCVD). Monte Carlo electron beam trajectory calculations were used to select the optimum beam energy for each oxide exposure. The contamination formed during electron beam irradiation was removed[8] during oxygen plasma "ashing".[9] All samples were then P etched. Atomic Force Microscope (AFM) images of the SiO_2 after plasma ashing and after P etching were made. The average height variation in the exposed SiO_2 was determined from the AFM images.

EXPERIMENT

The effect of *e*-beam exposure of the four types of SiO_2 was determined from images obtained with the AFM. Two thicknesses of SiO_2 were used. For each type and thickness of SiO_2, Monte Carlo electron trajectory calculations[10] were performed to determine the optimum electron beam energy to be used. The simulation program, which was applied to multilayer/multielement structures, used Mott cross section for scattering and Love energy loss formula for the energy decrease along the electron trajectories. Once the optimum energy was determined for each thickness, each type and thickness of oxide was exposed over a range of electron energy levels, centered around the predicted value, to determine the validity of the Monte Carlo calculations.

The patterns formed in the SiO_2 were made be flooding the SiO_2 films with electrons through a transmission electron microscope (TEM) grid. The exposure patterns formed, shown in Figure 1, consist of a collection of exposed squares surrounded by unexposed regions. Several exposure doses were used to determine their effect on the etch depth. Doses for the 100 nm films were 0.05, 0.10, 0.25 and 0.50 C/cm^2 at electron energies of 2.5, 2.8, 3.0 and 4.0 keV. Doses for the 50 nm films were 0.05, 0.15 and 0.25 C/cm^2 at electron energies of 1.5, 2.0, 2.5 and 3.0 keV. All patterns were exposed with a Hitachi S-570 scanning electron microscope (SEM). The SEM was adjusted to produce the maximum beam current in order to reduce the exposure time.

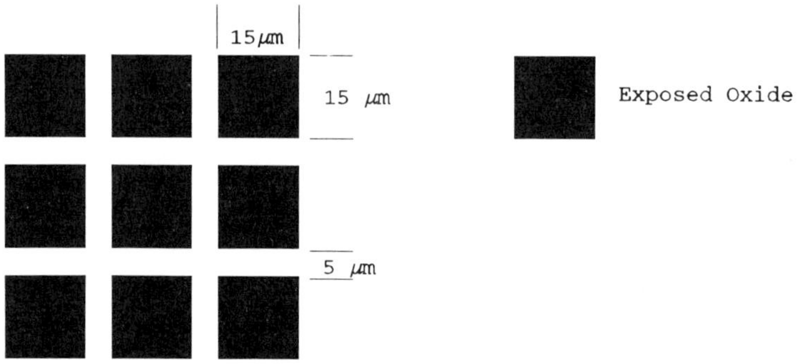

Figure 1. Exposure pattern of SiO_2 due to TEM grid.

Once the samples were patterned for the desired dose at the desired electron energy, each sample was placed in an oxygen plasma (plasma ashing)[8] for 30 minutes. The samples were then chemical etched using P etchant[4,5] (15 parts 49% HF, 10 parts 70% HNO_3 and 300 parts H_2O). Etching times were 3.0 minutes for the 100 nm dry and steam oxides, 1.5 minutes for the 50 nm dry and steam, 30 seconds for the 100 nm LPCVD and RPECVD oxides. Etching depth measurements of the exposed regions were made using a Digital Instruments Nanoscope III Atomic Force Microscope.

RESULTS

Although, it has been shown that electron beam exposure of SiO_2 induces etching selectivity,[4] doses of 1 C/cm^2 were required for saturation of the etching enhancement. Reduction of the required dose through optimization of energy deposition within the desired layer would provide a reduction in the required exposure time. Also, the beam damage to the underlying substrate or other layers can be avoided or reduced by utilizing low beam energies. Monte Carlo electron trajectory calculations[9] were performed for both SiO_2 layer thicknesses on a Si substrate in order to predict the electron beam energy that would provide the optimum energy deposition within the SiO_2 layer. Sample electron trajectories are shown in Figure 2(a) for a SiO_2 film thickness of 100 nm and Figure 2(b) for a SiO_2 film thickness of 50 nm. For each thickness there are electron energies that can not penetrate the SiO_2 layer completely, thus the film is only partially exposed. Also for each thickness there are energies which are too high and penetrate into the Si substrate, thus overexposing the film. The ideal electron energy that exposes the film completely without damaging the Si substrate can be determined by the Monte Carlo trajectory calculations. For the 100 nm and 50 nm films, 3.0 keV and 2.0 keV, respectively, are the ideal energies.

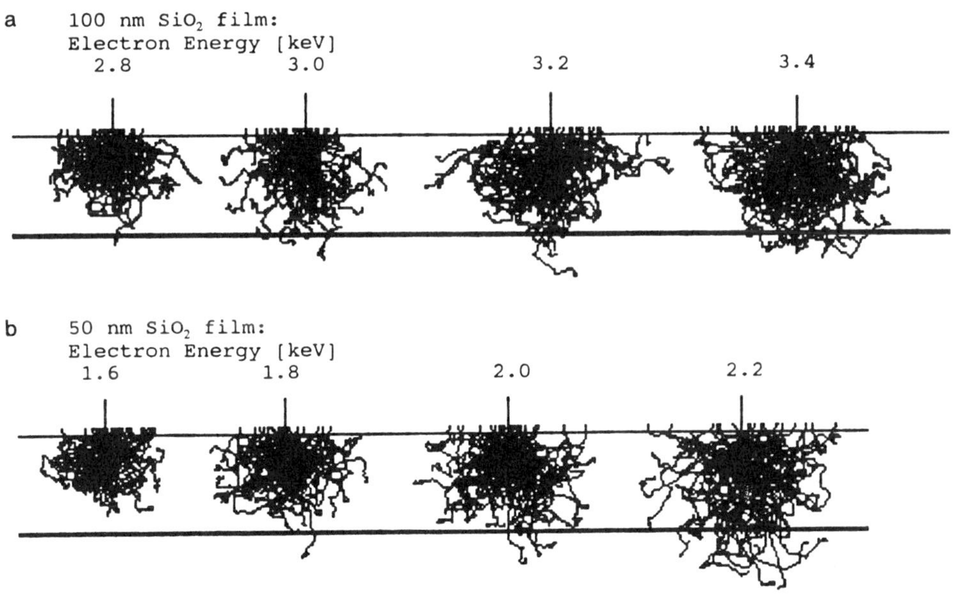

Figure 2. The electron trajectories simulated by Monte Carlo method in SiO_2/Si structure. Number of trajectories = 100.

By looking at the total energy deposited in the SiO_2 layer the proper beam energy for each thickness can be determined, as seen in Figure 3. The ideal energy will be the energy which deposits the majority of its energy in the oxide layer.

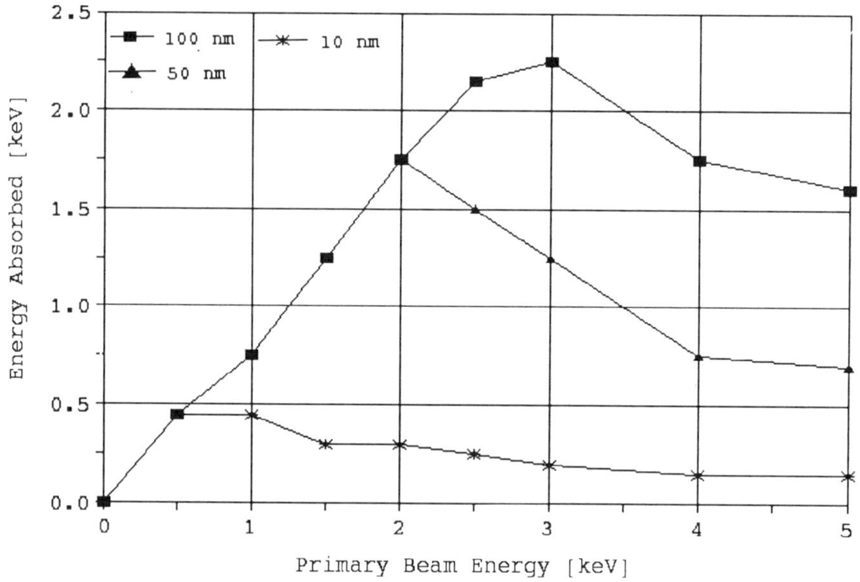

Figure 3. Total energy deposited in SiO_2 layer as a function of primary beam energy for various thicknesses of SiO_2 layer.[6]

To verify the Monte Carlo predictions, the four types of SiO_2 were exposed over energies centered on the predicted values. These samples were plasma ashed, P etched and examined with the AFM. Height information was determined from AFM images by analyzing vertical cross sections across the boundaries of exposed and unexposed regions. The average height was determined for each thickness, dose and type examined and the results are given in Table 1. A negative height value represents an exposed region which is lower than the unexposed region. Examples of AFM images used to determine the difference in height between exposed and unexposed regions can be seen in Figures 4 through 7.

Figure 5 is an AFM image of 100 nm dry SiO_2 which was *e*-beam exposed at a beam energy of 3 keV. The darker region in the lower left portion of the image corresponds to the exposed area, while the brighter bands along the top and left are the unexposed regions. The unexposed bands are narrower than the TEM mask used (refer to Figure 1) due to over-etching of the exposed SiO_2 which resulted from the 0.50 C/cm^2 dose used for the exposure. The AFM image of 100 nm steam oxide exposed at 3 keV, as shown in Figure 6, illustrates that for a lower dose of 0.05 C/cm^2 the unexposed band more closely mirrors the TEM grid. From previous work[6] it was shown that the ideal dose for 100 nm thermal oxides is 0.25 C/cm^2.

Figure 7 is a three dimensional AFM image of 100 nm LPCVD SiO_2 exposed at 3 keV. The upper and lower right regions correspond to the exposed oxide. Figure 8 is a three dimensional AFM image of 50 nm Dry SiO_2 exposed at 2 keV. The upper region in this image corresponds to the unexposed oxide. As can be seen in Figures 7 and 8, the LPCVD is serving as a negative resist while the dry oxide is serving as a positive resist. A negative resist is defined

Table 1. Average height difference between exposed and unexposed SiO_2 after P etch. (a) Height information not obtainable from AFM images. (b) LPCVD and RPECVD 50 nm samples were unavailable at time of press.

(a) 100 nm SiO_2 Samples: Electron Energy = 3.0 keV

Dose \ Type	Dry	Steam	LPCVD	RPECVD
0.50 C/cm^2	- 63 nm	- 56 nm	+ 11 nm	+ 91 nm
0.25	- 30 nm	- 39 nm	a	+ 94 nm
0.10	- 17 nm	- 34 nm	+ 5 nm	+ 84 nm
0.05	- 12 nm	- 27 nm	+ 8 nm	a

(b) 0 nm SiO_2 Samples: Electron Energy = 2.0 keV

Dose \ Type	Dry	Steam	LPCVD	RPECVD
0.25 C/cm^2	- 23 nm	- 14 nm	b	b
0.15	- 18 nm	- 11 nm	b	b
0.05	- 8 nm	- 2 nm	b	b

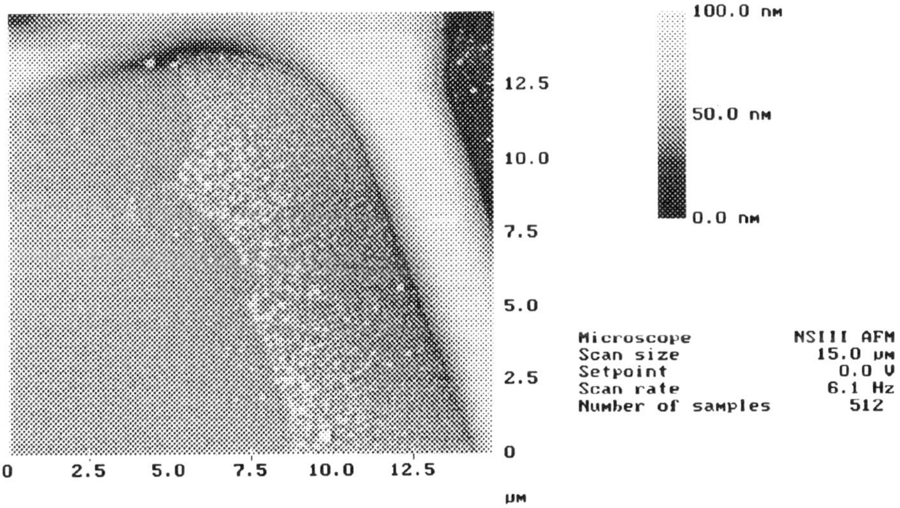

Figure 4. AFM top view image of 100 nm Dry SiO_2 at 3 keV, 0.50 C/cm^2.

as material which remains after exposure and etching, while for a positive resist the exposed regions are removed after etching. From the height values in Table 1, the negative values associated with the thermal oxides indicate positive resist behavior, while the positive values for the CVD films indicate negative resist behavior.

DISCUSSION

The etch rate of SiO_2 in P etch is known to depend on a variety of factors including stoichiometry, density, and impurities.[11,12] It is believed that the major factor influencing the non-*e*-beam exposed oxide P etch rates is oxide density which varies with the method of oxide growth. High density thermal oxides such as steam and dry oxides have reported etch rates of 0.2 nm/s compared with the 1-2 nm/s etch rates for lower density CVD oxides[13,14] or with 7 nm/s etch rate for very low density, porous *e*-beam evaporated oxide.[10] The above literature data indicates that unannealed CVD films etch faster than thermal steam oxides due to structural defects[15] and a more atomically open structure.[16] It is also known that annealing of CVD oxides[15] results in an increase in oxide density with an accompanying decrease in the oxide etch rate.

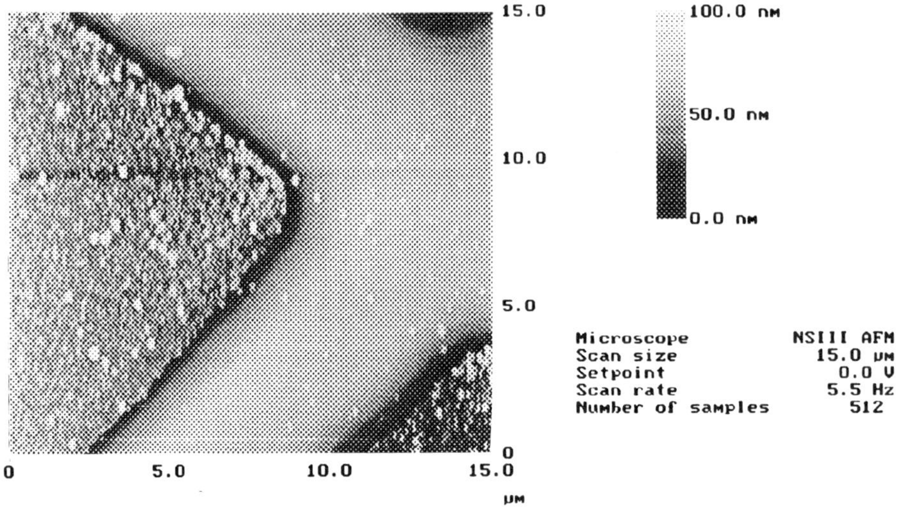

Figure 5. AFM top view image of 100 nm steam SiO_2 at 3 keV, 0.05 C/cm^2.

The results of P etching of the different oxide films are summarized in Table 1. In the case of the steam and dry SiO_2 the increase in the etch rate after *e*-beam exposure appears to be due to a decrease in density.[6] This is evident by the swelling of the exposed layer after plasma ashing, presumably caused by *e*-beam induced damage to the originally very dense thermal and dry oxides. This swelling was observed from AFM images taken after plasma ashing. This decrease in density for the steam and dry oxides is in direct contrast to the behavior of the

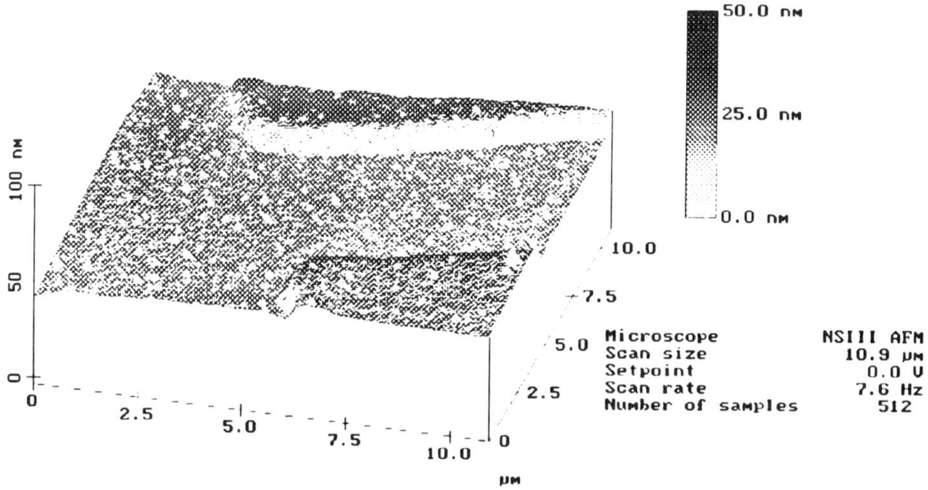

Figure 6. AFM 3-D image of 100 nm LPCVD SiO_2 at 3 keV, 0.50 C/cm^2.

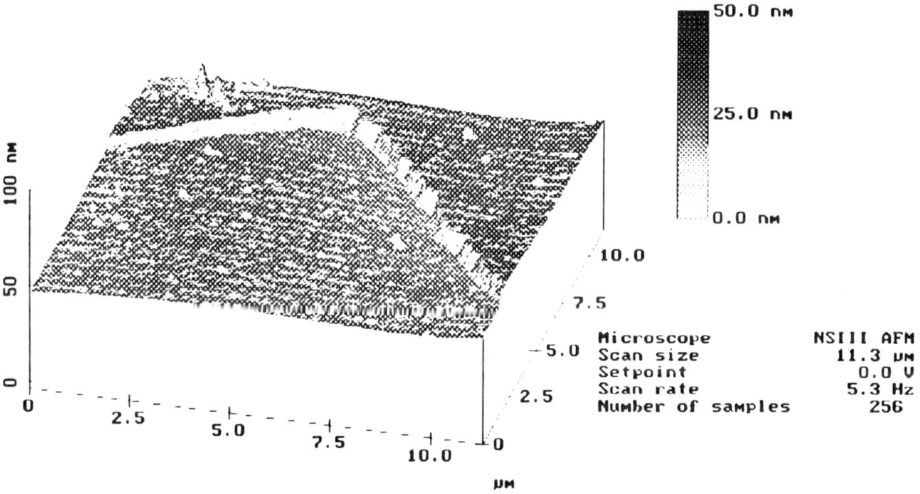

Figure 7. AFM 3-D image of 50 nm Dry SiO_2 at 2 keV, 0.05 C/cm^2.

originally low density RPECVD and LPCVD oxides, which show an increase in density after *e*-beam exposure. This increase in density was due to a reduction of the exposed layer as observed from AFM images acquired after plasma ashing.

The importance of the dose is evident in the etch variations shown in Figures 5 and 6 and Table 1. More work is necessary to determine the ideal dose for each thickness and type of oxide. In addition, the effect of P etch time on the pattern formation needs to be explored.

Complete explanation of the mechanisms of the transformation of the oxides is beyond the scope of this study. IR spectroscopy and ellipsometry studies of these oxides as a function of *e*-beam dose should help to explain the transformations taking place within these films. Further work is necessary to determine whether P etch or another wet etchant provides the best etching selectivity with the oxides types considered. Finally, if a dry etching process such as reactive ion etching (RIE) can be developed which gives similar or perhaps improved etching selectivity compared to that demonstrated with P etch, the work of integration of this promising lithography process into single wafer processing cluster technology can begin.

CONCLUSION

Electron bombardment of the SiO_2 layers results in an enhancement of the etch rates as compared to the native layers. The etching selectivity as well as the regions, exposed or unexposed, which etched preferentially depend on the formation process of the SiO_2 films. Resulting increases or decreases in etch rate appear to be related to layer density after *e*-beam exposure. This study shows that *e*-beam patterning of SiO_2 films is a promising path toward development of a resistless *e*-beam lithography technique. The AFM allows a quantitative method of determining the alteration due to the patterning processes. Further study is required to understand the structure of *e*-beam exposure of oxides versus the native films, and to determine whether the etching selectivity induced can be extended to dry etching techniques.

ACKNOWLEDGMENTS

This work was supported by the NSF Grant No. ECS9110746, ASU Graduate Student Association, the NCSU Center for Advanced Materials Processing NSF Grant No. CDR87211505, the NCSU Precision Engineering Center, ONR531234, the NSF PYI Program No. DMR8657813, and the NCSU Analytical Instrumentation Facility. Special thanks to Dr. Ruth Dewell of ASU for the use of the SEM. Thanks to G. G. Fountain of Research Triangle Institute for providing the RPECVD samples and to J. O'Sullivan of NCSU for providing the dry, steam and LPCVD samples.

REFERENCES

1. G.W. Rubloff, Maskless selected area processing, *J. Vac. Sci. Technol.* B7:1454-1461 (1989).
2. M. Taneya, Y. Sugimoto, H. Hikeda, and K. Akita, Electron-Beam Induced C_{12} Etching of GaAs, *Jpn. J. Appl. Phys.* 28:L515-L517 (1989).
3. S.M. Sze, "Semiconductor Devices," John Wiley & Sons, New York, 429 (1985).
4. T.W. O'Keeffe and R.M. Handy, Fabrication of Planar Silicon Transistors without Photoresist, *Solid State Electron.*11, 261-266 (1989).
5. D.R. Allee and A.N. Broers, Direct nanometer scale patterning of SiO_2 with electron beam irradiation through a sacrificial layer, *Appl. Phys. Lett.* 57: 2271-2273 (1990).
6. P.E. Allen, D.P. Griffis, Z.J. Radzimski, and P.E. Russell, Electron beam patterning of SiO_2, *J. Vac. Sci. Technol.* A 4: 965-969 (1992).

7. P.D. Richard, R.J. Markunis, G. Lucovsky, G. G. Fountain, A.N. Mansour, and D.V. Tsu, Remote plasma enhanced CVD deposition, *J. Vac. Sci. Technol.* A 3:867 (1985).
8. D.C. Joy, "Principles of Analytical Electron Microscopy," Plenum, New York, 353 (1986).
9. R.F. Reichelderfer, J.M. Welty, and J.F. Battey, J. Electrochem. Soc., 124:1926 (1977).
10. J.C. Russ, Z.J. Radzimski, A. Buczkowski, and L. Maynard, Monte Carlo Modeling of Electron Signals from Heterogeneous Specimens with Nonplanar Surfaces, *J. Comput. Assist. Microsc.* 2: 59-89 (1990).
11. W.A. Pliskin, and H.S. Lehman, *J. Electrochem. Soc.* 112:1013 (1965).
12. W. Kern and C.A. Deckert, "Thin Film Processes," J. L. Vossen and W. Kern, eds. Academic Press, New York, 415 (1978).
13. S. Rojas A. Modelli, W.S. Wu, A. Borghesi, and B. Pivac, Properties of silicon dioxide films by low-pressure chemical vapor deposition from tetraethlorthosilicate, *J. Vac. Sci. Technol.* B 8: 1177-1184 (1990).
14. C. Falcony. A. Ortiz, S. Lopez, J.C. Alonso, and S. Muhl, Low temperature SiO_2 films, *Thin Solid Films* 193/194:638-647 (1990).
15. J. Oroshnik and J. Kraitchman, Pyrolytic Deposition of Silicon Dioxide in an Evacuated System, *J. Electrochem. Soc.* 115:649-652 (1968).
16. G. Lucovsky, S.S. Kim, and J.T. Fitch, Formation of device quality Si/SiO_2 interfaces at low substrate temperatures by remote plasma enhanced chemical vapor deposition of SiO_2, *J. Vac. Sci. Technol.* B 8:822-831 (1990).

MEASURING THE MECHANICAL PROPERTIES OF PREFORMED, NANOMETER-SIZE GOLD CLUSTERS WITH THE ATOMIC FORCE MICROSCOPE

D.M. Schaefer,[1] A. Patil,[2] R.P. Andres,[2] and R. Reifenberger[3]

[1]Naval Research Laboratory
Surface Chemistry Branch
Washington, DC 20375-5000
[2]Purdue University
School of Chemical Engineering
W. Lafayette IN 47907
[3]Purdue University
Department of Physics
W. Lafayette IN 47907

Abstract: An atomic force microscope has been used in the attractive (noncontact) force mode to image individual nanometer-size Au clusters preformed in the gas phase and deposited on a wide variety of atomically flat substrates. Contact imaging modes are shown to provide sufficient lateral forces to dislodge the clusters. Using a noncontact technique, it is possible to reliably image preformed clusters in their as-deposited positions. This capability allows nanoindentation studies to measure the mechanical properties of individual nanometer-size Au clusters supported on highly oriented pyrolitic graphite and permits a measure of the deformation of a nanometer-size cluster as a function of the applied load.

INTRODUCTION

Information on preformed nanometer-size clusters is useful for new advances in such diverse areas as catalysis, nanophase materials and nanoelectronics. A primary requisite for studying a preformed, supported cluster on a substrate is a reliable method for producing well characterized, nanometer-size clusters. In this regard, Andres and coworkers[1,2] have shown that using a multiple expansion gas aggregation source, clusters with a controlled mean size and narrow size distribution can be produced and subsequently deposited onto a wide variety of substrates. Scanning probe microscopy has provided a technique to probe the properties of nanometer-size supported clusters. A generic problem with previous scanning probe studies of supported clusters is the difficulty encountered due to the high cluster mobility on substrates. In a recent paper, this long-standing problem has been solved and a reliable way to image preformed nanometer-size metallic clusters on atomically flat substrates has been reported.[3] Schaefer et al. have demonstrated that reliable images of deposited clusters on a wide variety of atomically flat substrates can be achieved. This technique has the distinct advantage of revealing clusters in their as-deposited condition.

With the ability to locate clusters reliably and accurately on atomically flat substrates, it becomes possible to perform nanoindentation experiments on individual nanometer-size clusters. Such an experiment is an extension of the established techniques, which use indentations to characterize the material properties of a substrate. The ability of the AFM as an indentation apparatus has been demonstrated by Burnham et al.[4] whereby the elastic modulus for bulk highly oriented pyrolitic graphite (HOPG) and an elastometer was determined. The hardness was also determined for bulk gold. These experiments involved large loading forces and did not require a precise placement of the AFM tip on the substrate of interest.

In this experiment, we utilize the AFM's unique capability to resolve forces and displacements, along with its ability to image the topography of a surface. As described below, the AFM was used to scan the surface and locate individual clusters. By positioning the tip over a preselected cluster, it is demonstrated that the nanometer-size cluster can be moved to a selected site. Nanoindentation tests were also performed on selected Au clusters using loading forces of ~ 4 nN.

CLUSTER SOURCE

A description of the multiple expansion cluster source (MECS) used in this study can be found in the literature.[1,2,5] It is a gas aggregation source, in which both cluster growth via accretion of single atoms and via cluster-cluster aggregation can be promoted. In the present experiments, Au clusters with diameters as large as 20 nm were produced by aggregation of primary clusters with diameters in the 1 nm to 1.5 nm range. As in previous studies,[6] a small portion of the cluster aerosol produced in the MECS is expanded through a capillary nozzle into a vacuum to produce a cluster beam, is decelerated in a gas scattering cell, and is deposited on clean substrate.

For the experiments described below, the cluster aerosol emerging from the MECS was directed through a tubular annealing reactor before being expanded into vacuum.[7] The mean residence time of a cluster in this annealing reactor is approxirnately one second. With this arrangement, clusters exiting the growth region of the MECS with a temperature near 300K could be heated to a predetermined temperature and then cooled again. After annealing, the clusters pass into a vacuum chamber with a base pressure of 1×10^{-6} torr and are soft landed on both carbon coated TEM grids (carbon film thickness 5 nm) for TEM analysis or HOPG substrates for future scanning probe studies. Without annealing, it was found that the clusters are multiply twinned polycrystals with fcc domains between 1 and 2 nm in diameter. However, when the clusters were annealed at 1400K, all clusters oriented in such a way that clear lattice fringes could be observed in a HRTEM were single crystals with an unambiguous fcc structure.[7]

Based on this previous work, the Au clusters investigated in the present study were annealed at 1400K and are expected to be single fcc crystals having roughly the shape of a truncated octahedron. Furthermore, electron microdiffraction from clusters supported on amorphous SiO_2 and carbon films reveals that preformed Au clusters deposited on these substrates have a preferred orientation, with their (111) planes parallel to the substrate surface.

IMAGING PREFORMED CLUSTERS WITH THE AFM

The atomic force microscope (AFM) used in this work is a custom-built instrument, shown schematically in Figure 1. The design of this instrument is similar to other atomic force microscope designs. Some of the key features of this instrument include a) a laser deflection system for measuring tip motion, b) a second piezoelectric tube, used to oscillate the tip, and c) a hysteresis correction system to linearize the sample motion in the z direction. A 68030 CPU-based computer is used for all data acquisition and system control. This instrument is presently mounted onto an 8-inch flange and inserted into a stainless steel vacuum chamber. The instrument is, therefore, capable of operating in an ambient, moderate vacuum (10^{-1}--10^{-2} torr), or an inert gas environment. All data presented here were taken under ambient conditions.

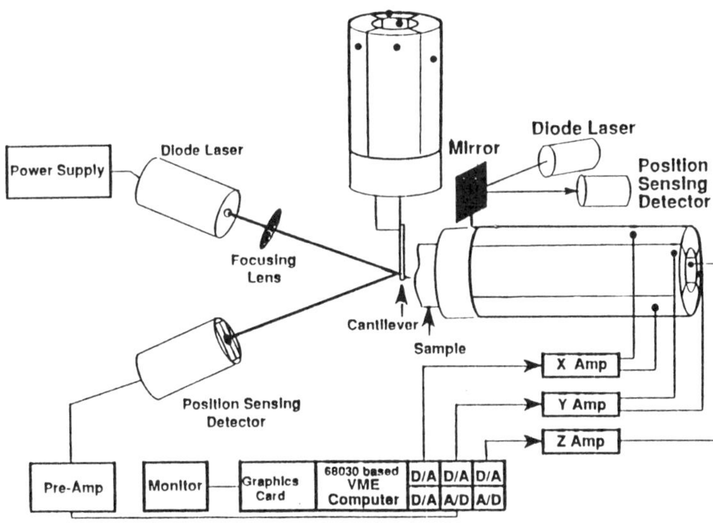

Figure 1. Schematic diagram of atomic force microscope.

With the addition of a second piezotube, this AFM is capable of operating in the noncontact mode. All noncontact mode scans were performed using an amplitude modulation technique. Figure 2 shows a typical noncontact mode AFM image of annealed Au clusters on the surface of highly oriented pyrolitic graphite (HOPG). The large steps evident from this image were introduced into the substrate by heating the HOPG in a moist oxygen atmosphere as described elsewhere.[3] The noncontact AFM images show a large number of supported clusters. Furthermore, succeeding scans proved to be quite reproducible, leading to the conclusion that we are imaging preformed, nanometer-size clusters in their as-deposited positions.

As a demonstration of the need to use noncontact techniques for reliable imaging of preformed clusters, an image of the sample was first obtained in the noncontact mode (see

Figure 3 (a)). The scan size was approximately 1.2 µm x 1.2 µm. This same area was scanned in the noncontact mode several times, demonstrating that the scans proved to be very reproducible. The tip oscillation was then stopped, and a scan in the contact (constant

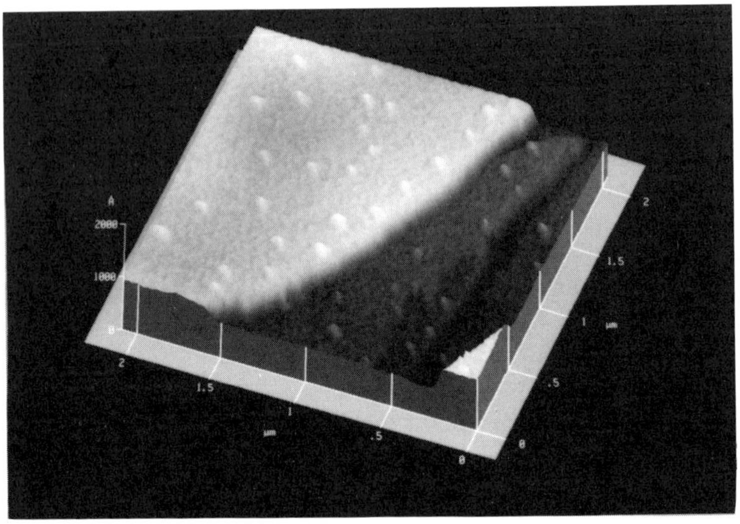

Figure 2. Noncontact AFM scan of Au clusters supported near a step edge on an HOPG substrate. The height of the step is roughly 50 nm. The x,y scan range is 2.09 µm.

force) mode was taken. The scan size was 0.5 µm x 0.5 µm with a normal force of ~10^{-8} N. The image taken in contact mode showed only the bare substrate with no clusters. Finally, another noncontact mode scan was taken, identical to the first (see Figure 3(b)). From this image we see that no clusters are present over the region that the central region scanned in the contact mode. An outline of the contact mode scan is shown by the dashed overlayed lines. This shows direct evidence that the AFM tip dislodges the clusters during the contact mode image.

The weak binding of the cluster to the substrate can prove advantageous. By using the AFM tip, it is possible to move the clusters to preselected positions on the sample. This is illustrated in Figure 4. Figure 4 (a) shows the initial image of the sample. The arrow indicates the cluster selected for displacement. After disconnecting the tip oscillation voltage, the AFM probe was lowered to contact with the surface and used to move the individual cluster across the surface. Subsequently, another noncontact scan was taken and revealed that the preselected cluster had been moved (see arrow in Figure 4 (b)). The ability to vector clusters across smooth surfaces demonstrated here opens up the possibility of fabricating prototypical two-dimensional devices fabricated from an assembled array of nanometer-size clusters of controlled size and composition.

INDENTATION TESTS: BULK MATERIALS

With the ability to reliably image the preformed clusters, it becomes possible to perform mechanical studies using indentation techniques. One difficulty in using the AFM to perform indentation experiments is the inherent hysteresis present in the piezotubes. The nonlinear motion in the z direction distorts the indentation curves, leading to large errors and irreproducible results. It was determined that for the piezotubes used in these experiments, the nonlinearity becomes significant when tube displacements exceed ~30nm. To overcome this problem, a correction system was installed, which provided an independent measure of the sample displacement as a function of the z-voltage applied to

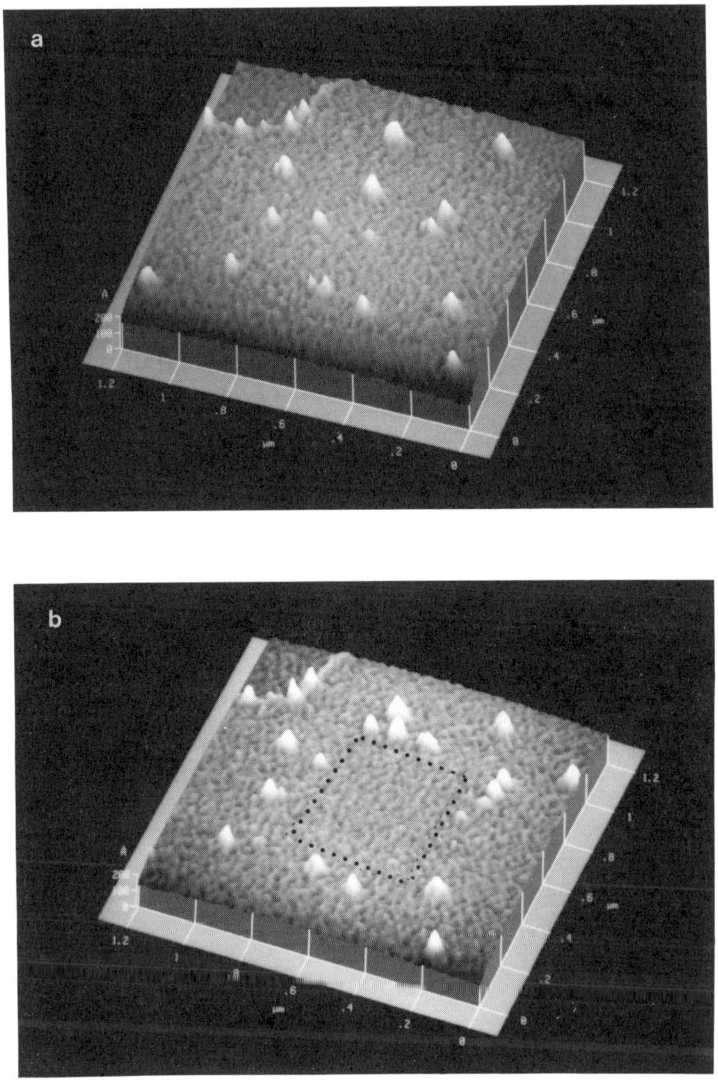

Figure 3. (a) AC mode scan of 1.2 µm x 1.2 µm area showing Au clusters on HOPG. (b) AC mode scan of same area after a contact mode scan had been performed.

the piezotube. This correction system consisted of a small mirror mounted onto the piezotube near the sample mount. A laser photo-diode was used to reflect a beam off the mirror, into a position sensitive detector. The output was then used to linearize the sample motion during post acquisition data analysis.

A number of initial experiments were performed to confirm the ability of the AFM as an indentation device. Indentation experiments were performed on bulk samples including an elastometer, HOPG, and bulk gold. Experiments were performed with large loading forces (~ 200 nN) and small loading forces (~4 nN) with similar results.

The elastometer was formed by cutting an ordinary rubber band into a small strip. No special cleaning or sample preparation was performed. Figure 5 (a) demonstrates a force

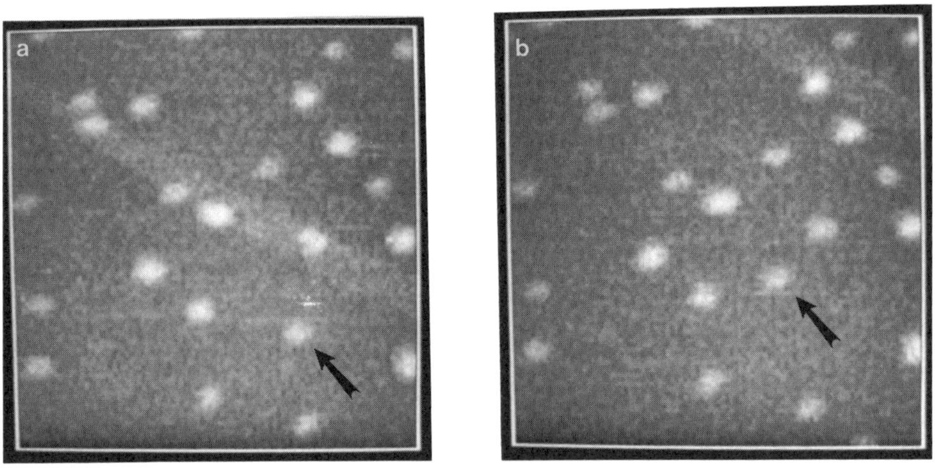

Figure 4. (a) AC mode scan of Au clusters on an HOPG substrate. (b) AC mode scan after the AFM tip was used to move an individual cluster to a selected position on the sample.

curve on the elastometer with large loading forces. Here we see a retrace of the loading and unloading region of the force curve, a characteristic of elastic behavior. This retrace is also seen in the plot of loading force vs. penetration depth, shown in Figure 5 (b). This plot shows both loading and unloading curves. No plastic deformation is observed. The elastic modulus, E, was determined using the equation for an axisymmetric indenter

$$E = \frac{(1 - \nu^2)F}{2rh} \tag{1}$$

where ν is the Poisson ratio, F is the applied force, r is the contact radius, and h is the penetration depth. The tips used were commercially bought ultralevers with known aspect ratios of 2:1. Knowing the penetration depth, the contact radius was determined to be ~43 nm. Therefore, an elastic modulus was determined to be 8.2 MPa, which is in excellent agreement with previous reports.[4]

An example of force curves taken with small loading forces is shown in Figure 6 using bulk gold. The sample was made from 99.999% pure gold wire, which was cleaned by alternately rinsing in acetone and methonal several times, and then placed into a 5% nitric acid solution for 1.5 hours. The sample was then rinsed in deionized water, blown off with dry nitrogen gas, and immediately placed into the AFM. Figure 6 (a) is the force vs. sample displacement curve, with Figure 6 (b) showing the loading force vs. tip penetration depth. For clarity, only the unloading portion of the curve is shown in the force vs. penetration plot. The elastic modulus of the gold was determined using Eq. 1 to be 79.2 GPa. This is in excellent agreement with the bulk value of 80 GPa for indentation into a randomly oriented gold sample. The hardness, defined as the force divided by the contact area, was also determined. The measured value of 0.67 GPa is also in good agreement with the literature value of 0.50 GPa. These experiments confirm the ability of the AFM to perform indentation experiments, even for small loading forces.

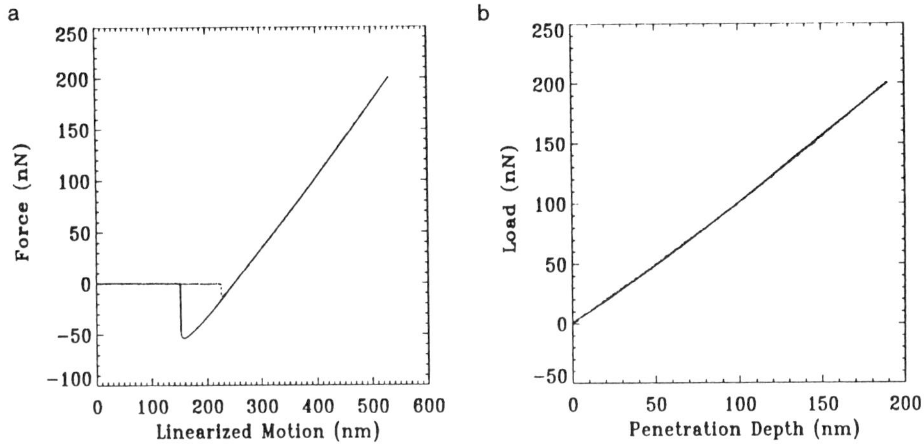

Figure 5. Indentation test performed on an elastometer using large loading forces. (a) The applied force vs. sample displacement curve. (b) The applied load vs. tip penetration depth.

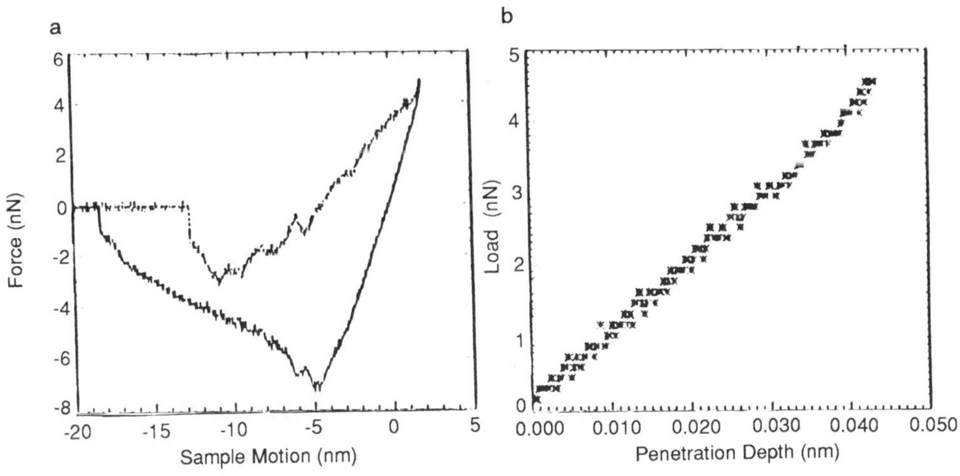

Figure 6. Indentation test performed on bulk gold using small loading forces. (a) The applied force vs. sample displacement curve. (b) The applied load vs. tip penetration depth.

NANO-INDENTATION OF A SUPPORTED CLUSTER AT SMALL LOADS

The elastic properties of a supported cluster were measured by (i) locating a cluster using the noncontact mode of operation, (ii) removing the oscillatory motion of the cantilever, (iii) positioning the tip over the cluster, (iv) moving the sample toward the Si tip on the cantilever until a specified indentation force was reached, and (v) withdrawing the sample from the cantilever.

Nano-indentation data for a 3.1 nm high Au cluster supported on an HOPG substrate are shown in Figure 7. Before measuring the deformation of the cluster under loading (see Figure 7 (a)), a nano-indentation curve (see Figure 7 (b)) was obtained for the HOPG substrate. This served several purposes. First, it was used to distinguish between a successful and unsuccessful indentation on a cluster. If the tip misses the cluster, the resulting nano-indentation data are characteristic of the substrate. Secondly, the substrate also acts as a calibration of the tip, since HOPG has known elastic properties. These data are replotted in Figure 7 (c) to compare the deformation of the cluster and substrate as a function of applied load.

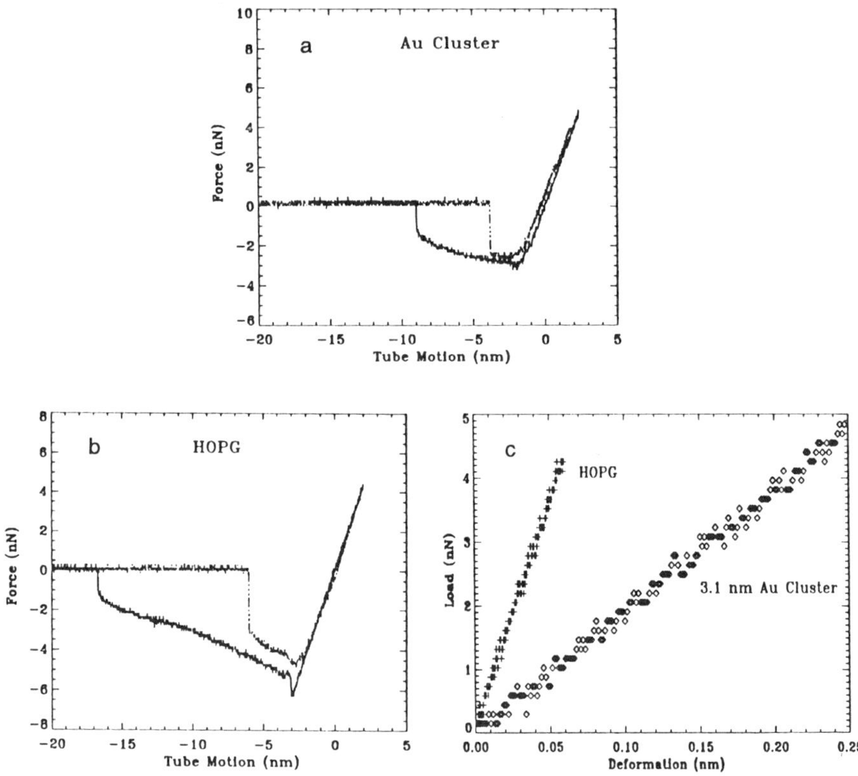

Figure 7. (a) A plot of the force vs. sample motion obtained with the tip positioned over a 3.1 nm high Au cluster supported on HOPG. (b) The same data except now the tip is positioned over the HOPG substrate. (c) The nano-indentation data in (a) and (b) are plotted as a function of the penetration depth of the tip for a given applied load. The slopes determined from these data are used to calculate the elastic modulus of the cluster.

When interpreting such data, it is important to determine whether the tip area is larger or smaller than the cluster face. The nano-indentation data into bare HOPG was therefore analyzed to determine the contact area of the tip. The analysis is performed using the standard equation for an axisymmetric indenter

$$S_{HOPG} = \beta E^{*}_{HOPG}\sqrt{A_{tip}} \tag{2}$$

where S_{HOPG} is the slope of the HOPG unloading curve, E_{HOPG} is the effective elastic modulus for HOPG given by

$$\frac{1}{E^{*}_{HOPG}} = \frac{(1-\nu^{2}_{HOPG})}{E_{HOPG}} + \frac{(1-\nu^{2}_{tip})}{E_{tip}} \tag{3}$$

where ν_{HOPG} and ν_{tip} are the Poisson ratios and E_{HOPG} and E_{tip} are the Young's moduli for HOPG and the tip material (SiO_2) respectively. β is a constant which depends weakly on the geometry of the tip. From Eq. 1, a constant slope implies a constant contact area. This behavior is observed in Figure 7 (c). We, therefore, model the tip as an axisymmetric indentor with cylindrical symmetry. Using $\beta = 1.129$ (circular contact area), $E_{HOPG} = 17$ GPa,[8] $E_{tip} = 60$ GPa (SiO_2),[9] and $\nu_{HOPG} = \nu_{Tip} = 0.3$, the contact area was experimentally determined to be 19.8 nm^2.

As a check of this contact area, TEM was used to determine the tip profile after the nanoindentation experiment was completed. From TEM micrographs, a spherical tip radius, R, of 37 nm was determined. Using a Hertzian model for a spherical tip elastically contacting a flat surface[10], the contact area, given by

$$A_{tip} + \pi\left(\frac{3FR}{4E^{*}}\right)^{\frac{2}{3}} \tag{4}$$

was determined to be 12.7 nm^2. We conclude that the contact area of the tip is on the order of 10 to 20 nm^2.

Microdiffraction TEM studies have shown that annealed Au clusters have a strong tendency to orient on a substrate with the [111] direction normal to the surface of the substrate.[7] Thus, it becomes necessary to estimate the area of the (111) facet of the cluster to complete the analysis of the nano-indentation experiment. To do this, the area of the (111) face of a perfect truncated octahedral gold cluster as a function of cluster height h_{cl} is required. As shown in Figure 8, it is easy to calculate that 1289 atoms arranged in a truncated octahedral geometry has a height of 2.8 nm and a face area of 4.40 nm^2. Likewise, a Au cluster with 2406 atoms has a height of 3.5 nm and a face area of 6.55 nm^2. Thus, for the 3.1 nm cluster under study, it is reasonable to assume that $4.4nm^2 \leq A_{Cl} \leq 6.55nm^2$. An important conclusion that follows from this discussion is that the contact area of the tip is greater than the area of the cluster face under the conditions of the experiment.

The measured deformation of the cluster is assumed to consist of three contributions: (i) the deformation of the cluster-tip interface, (ii) the compression of the cluster itself along its [111] direction, and (iii) deformation of the cluster-substrate interface. The details of the analysis are presented elsewhere.[3]

Figure 9 shows how the modulus of the cluster, E_{cl}, varies as a function of the slope S of the unloading curve measured from Figure 7(c). Since the exact area of the (111) face is uncertain, results assuming perfect truncated octahedral structures for Au_{1289} and Au_{2406}

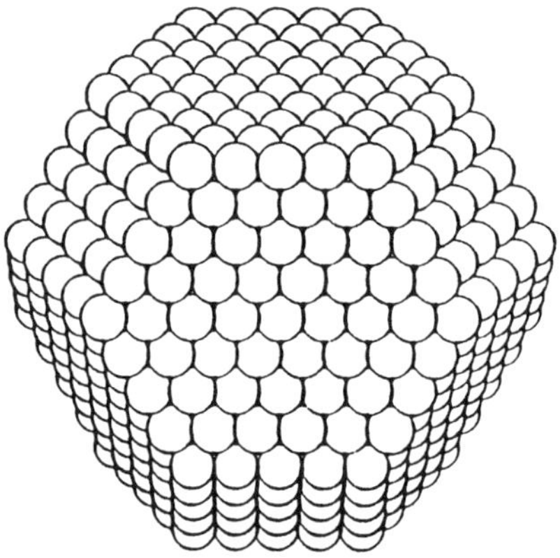

Figure 8. Truncated octahedral geometry of a 1289 atom cluster.

are presented as limiting cases. The experimental slope S is also uncertain, due largely to uncertainties in calibrating the tip motion and two limiting values for the slope are presented. Our best estimate for S of 20.3 N/nm is also indicated. Based on this analysis, we conclude that the elastic modulus of the 3.1 nm high Au cluster is 56±39 GPa.[11] This should be compared to a bulk value of 80 GPa for polycrystalline Au,[12] and 120 GPa for crystalline Au compressed along the [111].[12]

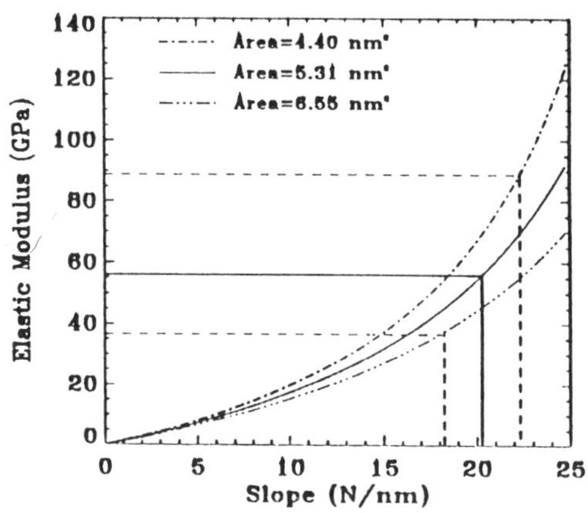

Figure 9. The elastic modulus of a 3.1 nm high cluster as a function of the slope measured from a nano-indentation experiment. These cases are presented assuming different contact areas between the tip and cluster. The uncertainty in the measured slope is indicated by the vertical dashed lines. The corresponding uncertainty in the elastic modulus is shown by the dashed horizontal lines.

CONCLUSIONS

It has been demonstrated that using an atomic force microscope in the noncontact mode, preformed nanometer-size metallic clusters deposited onto atomically flat substrates can be reliably imaged in their as-deposited positions. The ability to move clusters across smooth surfaces was also demonstrated, opening up the possibility of fabricating proto-typical 2-dimensional devices fabricated from an assembled array of nanometer-size clusters of controlled size and composition. Using the AFM as a nano-indentation device, the mechanical properties of individual nanometer-size clusters can also be determined. Thus, it now becomes possible to measure the size-dependence of the elastic properties of nanometer-scale objects.

ACKNOWLEDGMENTS

This work was partially funded by the National Science Foundation under contract ECS9117691.

REFERENCES

1. Seung Bin Park, Ph.D thesis, Purdue University, 1988.
2. E. Choi and R.P. Andres, *in*: Physics and Chemistry of Small Clusters, P. Jena, B.K. Rao, and S.N. Ihanna, eds., Plenum Press, New York, 61 (1987).
3. D.M. Schaefer, A. Ramachandra, R.P. Andres, and R. Reifenberger, Imaging nanometer-size metallic clusters with the atomic force ficroscope, *Z. Phys.* D, 26:249-252 (1993).
4. N.A. Burnham and R.J. Colton, *J. Vac. Sci. Technol.* A7: 2906 (1989).
5. T. Castro, Y.Z. Li, R. Reifenberger, E. Choi, S.B. Park, and R.P. Andres, *J. Vac. Sci. Technol.* A7: 2845 (1989).
6. T. Castro, R. Reifenberger, E. Choi, and R.P. Andres, *Surf. Sci.* 234:43 (1990).
7. A.N. Patil, D.Y. Paithankar, N. Otsulka, and R.P. Andres, The minimum energy structure of nanometer-size gold clusters, *Z. Phys.* D, 26:135-138 (1993).
8. J. A. Harrison, R. J. Colton, C. T. White, and D. W. Brenner, *Surf. Sci.* 271: 57 (1992).
9. Dror Sarid, "Scanning Force Microscopy," Oxford University Press (1991).
10. J.B. Pethica and W.C. Oliver, *Physica Scripta* T19, 61 (1987).
11. D.M. Schaeffer, A.N. Patil, R.P. Andres, R.R. Riefenberger, Nanoindentation of a supported Au cluster, *Appl. Phys. Lett.*, 63:11, 1492-1496 (1993).
12. Lawrence E. I. Van Vlack, "Material Science for Engineers," Addison-Wesley Publishing Company (1971).

ANALYSIS AND INTERPRETATION OF SCANNING TUNNELING MICROSCOPY IMAGES IN AN ELECTROCHEMICAL ENVIRONMENT: COPPER ON AU(111)

W. Haiss, J.K. Sass, D. Lackey, and M. van Heel

Fritz-Haber-Institut der MPG
Faradayweg 4-6, D-14195 Berlin, Germany

Abstract: The electrochemical deposition of a copper monolayer on Au(111) has been studied by STM in H_2SO_4 and Na_2SO_4 solutions. Three different superstructures, which are attributed to the adsorption of either HSO_4^-, SO_4^{2-} or both anions, have been observed for the full monolayer. Evidence for the expected epitaxial (1 x 1) overlayer was also obtained at very negative potentials, by avoiding bulk copper growth. A novel STM data analysis by multivariate statistical procedures is shown to provide detailed information, which is not evident in the raw data. The need for a better understanding of the tunneling and imaging mechanisms in an electrochemical environment is emphasized.

INTRODUCTION

Compared to scanning tunneling microscopy under ultrahigh vacuum conditions the transfer of electrons from the tip to the sample is obviously a much more complicated physical process when an electrolyte solution is present in the STM gap.[1] In electrochemical investigations the electrode potential of the sample and of the tip are externally controlled by the usual potentiostatic arrangement. Therefore the formation of electrical double layers on both surfaces must be considered, with the amount of excess ions and the orientation of the water molecules at the interfaces depending on the applied potential.[2]

As a result it is to be expected that the tunneling electron trajectories in the gap have a completely different character in the electrochemical situation, in particular since recent molecular dynamics calculations indicate that the water molecule is a strong backscatterer of electrons.[3] Instead of a smooth, one-step tunneling process, as in the vacuum, it is more likely that the electrons suffer intermediate collisions and that their path is of a more diffuse, three-dimensional nature.[1] In addition, there exists the possibility that the orientational dielectric behavior of water is sufficiently fast to respond to the traversing electrons, which in turn would lower the barrier and could further increase the tunneling time.[1,4] This concept has given rise to considerable controversy,[5] however, and requires additional study.

It may be useful, in view of the multidisciplinary nature of this conference, to provide an illustration of the particular environment in which electrochemical STM takes place and in Figure 1 this has been attempted in a qualitative manner. That figure depicts only the salient features of the multicomponent solution environment. Of the water molecules and ions residing in the gap, only a few are actually shown. In the double layers on both surfaces there may be ions, for example Na^+, which have essentially retained their bulk hydration shell ("nonspecific" adsorption[2]), whereas others, such as Br^-, tend to be in direct contact with the metal electrode surface ("specific" adsorption) and have only a more loosely bound surface hydration shell.[2,6] In-between the ions, hydrogen-bonded water patches (only one molecule depicted) with preferential dipole orientation, induced by the metal surface, have been shown to prevail.[6,7]

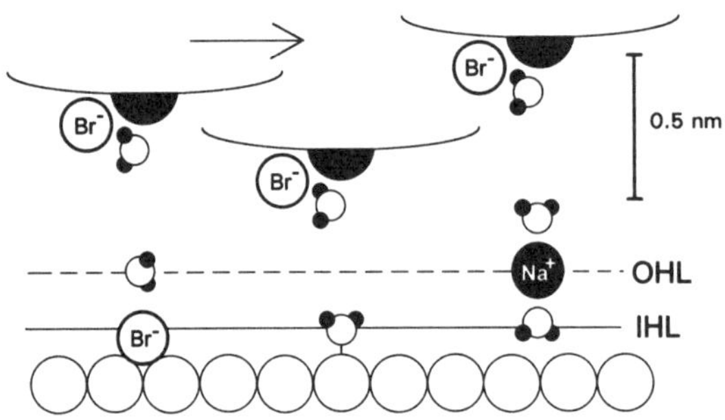

Figure 1. Schematic representation of tunneling in an electrochemical environment showing the associated double layers of the tip and the electrode surface.

As indicated in Figure 1, the extension of one double layer is in the range 0.3-0.5 nm, which implies a minimum gap distance of probably 1 nm, if the tip is to scan along the surface without the two double layers interfering with each other. However, to achieve atomic resolution of the sample surface, such distances are prohibitively large and a compromise between the local perturbance of the electrode potential and the image quality must be sought. We should like to emphasize that these important aspects have not yet received sufficient attention in electrochemical STM studies and that a detailed analysis of the distortion of the electrode potential by the tip is required.

The electrochemical deposition of foreign metal overlayers on single-crystal metal substrates, particularly in the monolayer range, has served in recent years as a model system for studying the *in situ* applicability of STM.[8] Typically, such metallic adsorbates exhibit a submonolayer superstructure with significant corrugation and spatial stability and can therefore be more easily resolved than a close-packed metallic surface. One such system, which has been studied in considerable detail, is the formation of copper adlayers on the Au(111) surface in sulphate or bisulphate containing solutions.[9-11] In accordance with the electrochemical charge measurements of 1/3 of a monolayer, a well-defined $(\sqrt{3} \times \sqrt{3})R30°$ pattern is observed as the initial stage of the first layer build-up. In a second step an additional charge, corresponding to two thirds of a monolayer, is registered. This indicates quite clearly that a full monolayer is formed prior to the bulk growth of copper on this surface. However, in the previous STM-studies there are no reports that any structure (for example a (1x1)) has been seen in this coverage range, which seems quite peculiar.

We have therefore performed a detailed investigation of this copper monolayer on Au(111) in both H_2SO_4 and Na_2SO_4 solutions. Our results indicate that characteristic ordering phenomena, different from the (1 x 1) structure and induced by the anions (HSO_4^-/SO_4^{2-}), occur for the monolayer. In the interpretation of our STM-images the crucial question arises whether the ordered structures reflect the arrangement of the copper atoms or of the adsorbed anions. In an attempt to answer this question we have chosen a novel approach to the analysis of STM-data, which is based on a multivariate statistical image processing procedure and has been successfully applied to electron microscopy images of biological samples for many years.[12-14] The basic idea in this analysis is that the repetitive information content in STM images can be used for selectively averaging specific structures occurring in the data and thus significantly increasing the signal-to-noise ratio. Our preliminary experience indicates that this type of image processing is of quite general value and may evolve into a standard tool in STM studies.

EXPERIMENTAL PROCEDURES

We combined a commercial, beetle-type[15] STM instrument (Delta-Phi-Electronics), which had originally been designed for UHV studies, with an electrochemical cell configuration, providing potentiostatic control of the metal sample and facilitating deaeration of the electrolyte solution prior to the experiments. Evaporated Au(111) films were prepared by flame-annealing and quenching in ultra pure water, as described previously.[16,17] Electrolyte solutions were prepared from doubly distilled water, H_2SO_4 or Na_2SO_4 (Merck, suprapur) and $CuSO_4$ (Merck, p.a.). A copper wire, inserted into the solution, served as the reference electrode and potentials are subsequently quoted with respect to the Cu/Cu^{2+} couple. Iridium tips, obtained by mechanical polishing, were coated with Apiezon wax, such that only a few tens of microns at the end of the tip were exposed to the electrolyte and the faradaic current on the tip was minimized. Tunneling voltages are quoted as positive when the tip is positive relative to the sample. All STM images were obtained in the topographic constant current mode. Unless otherwise stated the grey values in the images represent the derivative of the z-piezo motion with respect to time (i.e. with respect to position on the surface). Because of the limited range of our 8-bit AD converter, this operational mode was used to suppress slight tilts of the sample which invariably occur.

RESULTS

The current flow in the electrochemical cell in response to a cyclic change, linear in time, of the electrode potential is termed a voltammogram.[2] In Figure 2 three voltammograms obtained with the Au(111) films in different representative electrolyte solutions are shown. Typically a scan of the potential was started at 400 mV, where there is no copper on the surface, and the direction of the scan reversed at 0 mV, to avoid bulk deposition. This yields the lower and upper current traces, respectively, which are seen in Figure 2. The sharp peaks at 230-240 mV, which occur in all solutions, are associated with a charge of 1/3 of a monolayer of copper adsorption/desorption as already mentioned in the Introduction. Significant differences between the solutions containing either HSO_4^- (Figure

2 (a)) or SO_4^{2-} (Figure 2 (c)) are observed at lower potentials where the monolayer is completed. In the sulphate solution the adsorption/desorption peaks appear at more positive potentials and are more widely separated (i.e. less reversible).

The dissociation of the second proton in H_2SO_4 solutions is characterized by a pK-value of 1.92, implying that in this pH-range HSO_4^- and SO_4^{2-} species coexist. In Figure 2 (b) a voltammogram is shown for a dilute acid solution, in which this coexistence is apparently manifested by two separate desorption peaks in the potential range where the transition from full monolayer to 1/3 of a monolayer occurs. The corresponding features in the adsorption current are not as well separated, but a shoulder preceding the main peak may be distinguished. For this 4 mM solution of H_2SO_4 the anion ratio HSO_4^-/SO_4^{2-} is approximately 1/2.

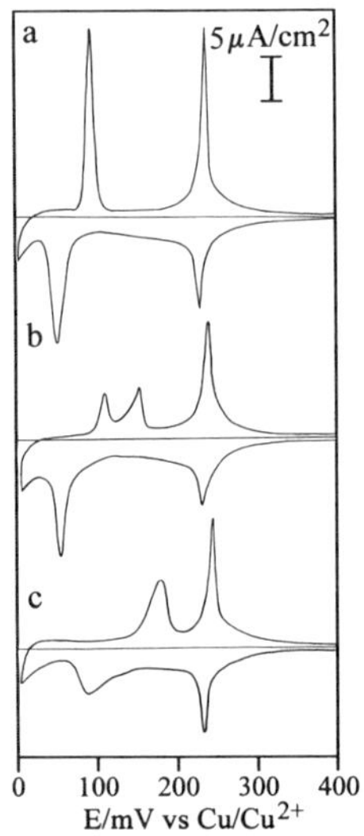

Figure 2. Cyclic voltammograms of copper monolayer formation on Au(111) films recorded at 2 mV s^{-1} in H_2SO_4 and Na_2SO_4 solutions; (a) 0.1 M H_2SO_4 + 1 mM $CuSO_4$; (b) 4 mM H_2SO_4 + 1 mM $CuSO_4$; (c) 50 mM Na_2SO_4 + 5 mM $CuSO_4$.

The voltammograms shown in Figure 2 provide guidance for selecting the relevant electrode potentials at which the STM observations are performed. In Figure 3 we first of all show, as a basis for the subsequent images, data for the Au(111) surface in the absence of copper adsorption. At the metal-electrolyte interface both the reconstructed (22 x √3)[18] and the unreconstructed (1 x 1) surface phases can be formed, by applying the appropriate potential.[19,20] In Figure 3 (a) the characteristic twinned lines with a distance of 25 Å, a corrugation of 0.15 Å and repeating every 63 Å, which are due to the uniaxial stress in the <110> direction, are seen. Indications for the development of the chevron or herringbone structure,[18] which results from a domain rotation by 120° at regular intervals when fully developed, are also visible. The unreconstructed (1 x 1) surface with a corrugation of 0.18 Å, which can be obtained at more positive potentials, is shown in Figure 3 (b) with atomic resolution. This image is characterized by a low noise level and demonstrates the capabilities of our electrochemical STM apparatus.

As mentioned above the initial stage of copper deposition occurs when the potential is moved to below the first adsorption peak in the voltammograms of Figure 2. For all three solutions one then observes the ($\sqrt{3}$ x $\sqrt{3}$)R30° pattern with the STM and an example of this structure is shown in Figure 4. It was possible to measure this image without differentiation (cf. Experimental Procedure) because of the low tilt angle. The measured corrugation for this pattern is 0.5 Å, which may be attributed to the increase in nearest neighbor distance compared to the gold substrate (4.95 Å vs. 2.89 Å).

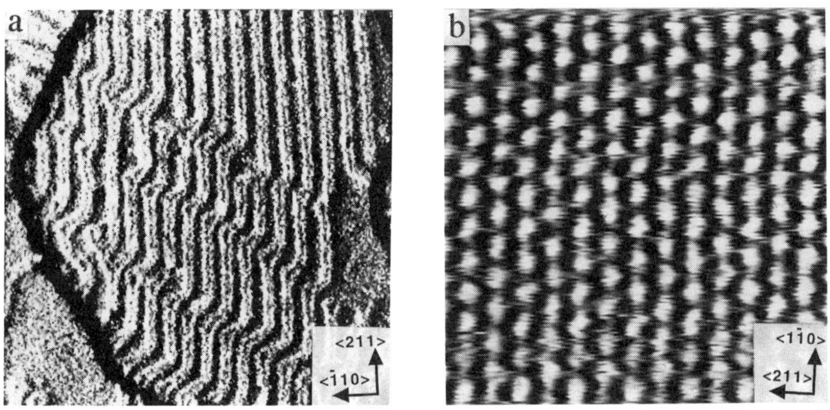

Figure 3. STM images taken in 0.1 M H_2SO_4 showing the Au(111) substrate. (a) In this large-scale scan the characteristic twinned lines of the reconstructed Au(111) surface can be seen at E = + 300 mV vs. Cu/Cu^{2+} (1300 x 1300 $Å^2$; V_t = +190 mV, I_t = 30 nA). (b) At more positive potentials (E = +600 mV vs. Cu/Cu^{2+}) the unreconstructed Au (111) surface is formed and is imaged here with atomic resolution (35 x 35 $Å^2$; V_t = -10 mV, I_t = 30 nA).

Our main interest in this study was directed at the potential range beyond the second adsorption peak in the voltammograms of Figure 2, where the charge evaluation indicates that a full monolayer has been deposited and one would expect the formation of a (1 x 1) overlayer.

In Figure 5 (a) it is shown that in 0.1 M H_2SO_4 at a potential of 0 mV there is no indication of this expected structure and instead a very diffuse STM image is observed (the white diagonal band is a monoatomic step). When merely inspecting the raw data no long-range order can be detected. Only after performing a Fourier transform of the image does one clearly find that a periodic variation of the height is present along the <211>

direction. In Figure 5 (b) this is demonstrated by superimposing the original data and the Fourier-transform filtered image. Somewhat irregular rows extending along the <110> direction, with a separation corresponding to the distance (2.50 Å) between the rows of the gold substrate, are now visible. In addition there are patch-like variations of the tip height occurring on a larger scale, but without any obvious periodic arrangement.

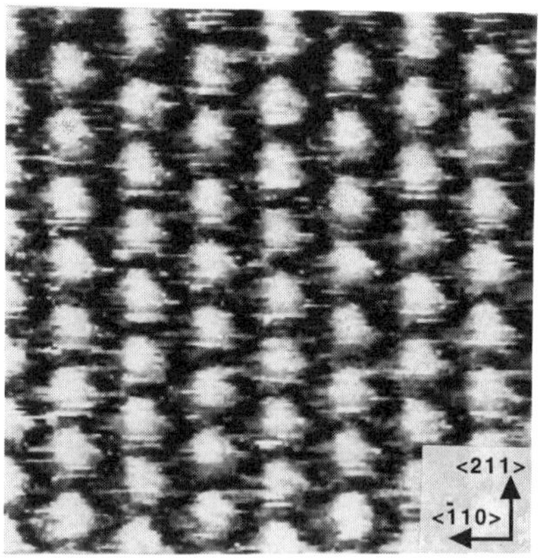

Figure 4. An STM image of a Au(111) film at E = +140 mV vs. Cu/Cu^{2+} in 50 mM Na_2SO_4 + 5 mM $CuSO_4$ showing the ($\sqrt{3}$ x $\sqrt{3}$)R30° structure (35 x 35 $Å^2$; V_t = +180 mV, I_t = 30 nA).

In the 0.05 M Na_2SO_4 solution, which contains SO_4^{2-} instead of HSO_4^- ions, the observations in the same low potential region are very different, as seen in Figure 5 (c). This high resolution image shows a well-ordered p(2 x 2) superstructure, which persists up to the peak corresponding to desorption of the full monolayer (cf. Figure 2 (c)). The growth of this phase in island form on the ($\sqrt{3}$ x $\sqrt{3}$)R30° structure can also be followed by setting the potential to +100 mV (initial rise of the adsorption peak associated with the full monolayer formation). An example of this coexistence of the two phases is shown in Figure 5 (d). Here the p(2 x 2) superstructure in the islands is not as well ordered as when it extends over the whole surface. Its height above the coexisting phase is 0.2-0.3 Å.

The striking differences in the STM observations between the HSO_4^- and SO_4^{2-} containing solutions (cf. Figures 5(a) and (c)), at potentials where a full monolayer is deposited, immediately suggest that adsorption of the anions on top of the copper layer is the origin of the failure to observe the expected (1 x 1) epitaxial layer. Such adsorption was also inferred from radio tracer experiments for copper adsorption on a polycrystalline gold surface.[21] This anion adsorption is expected to decrease with lower potentials, where the surface becomes more negatively charged. However, because of the copper bulk deposition below 0 mV this effect cannot be exploited when the usual $CuSO_4$ concentration (1 mM) is used. We therefore gradually reduced the copper content of the solution, such that its total amount in the cell volume equalled only a few monolayers. In the region directly below the tip there is an additional shielding effect, which hinders the diffusion of copper ions, such that bulk growth is further prevented. Under these conditions it was possible to study the full monolayer at more negative potentials, with the hope that the anion desorption is significantly decreased.

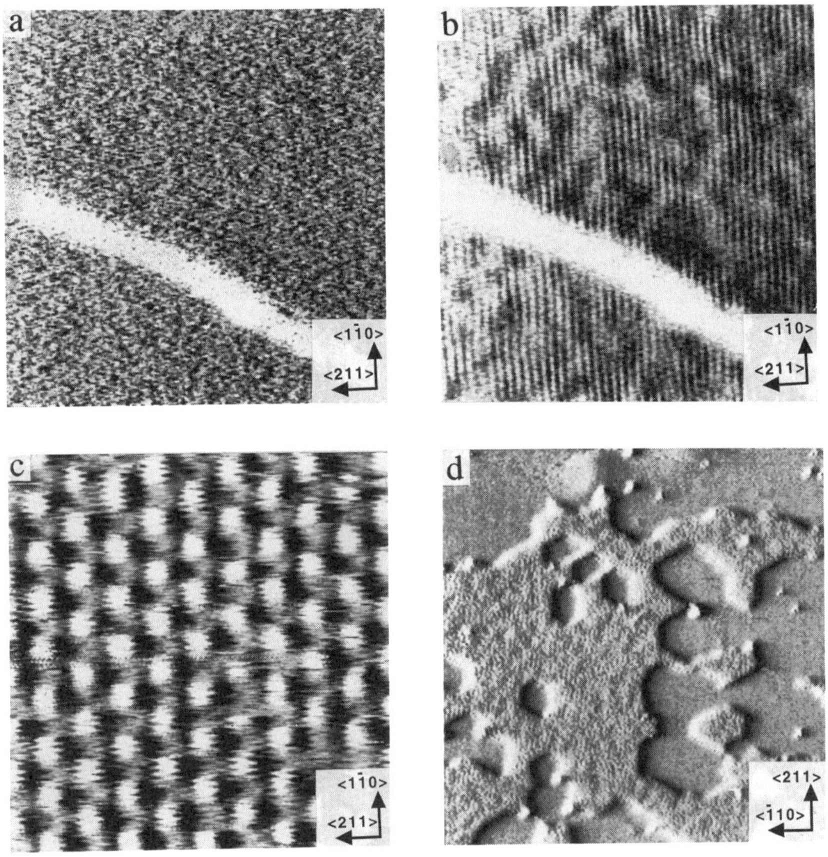

Figure 5. STM images comparing the structures formed in the presence of HSO_4^- and SO_4^{2-} in the low potential region. (a) Unresolved image at E = 0 mV vs. Cu/Cu^{2+} in 0.1 M H_2SO_4 + 1 mM $CuSO_4$ (143 x 143 $Å^2$; V_t = +20 mV, I_t = 30 nA). (b) A Fourier transform of the image in (a) superimposed on the raw data revealing a structure with irregular rows along the <110> direction. (c) An image taken in 50 mM Na_2SO_4 + 5 mM $CuSO_4$ at E = +20 mV vs. Cu/Cu^{2+} showing the p(2 x 2) structure (48 x 48 $Å^2$; V_t = +40 mV, I_t = 30 nA). (d) Image taken in 50 mM Na_2SO_4 + 5 mM $CuSO_4$ at the onset of the second adsorption peak (E = +100 mV vs. Cu/Cu^{2+}) showing island growth of the p(2 x 2) structure on the (√3 x √3)R30° structure (1000 x 1000 $Å^2$; V_t = +140 mV, I_t = 30 nA).

In Figure 6(a) the raw STM data obtained in 0.1 M H_2SO_4 and 0.01 mM $CuSO_4$ at a potential of -200 mV are shown. Compared to Figure 5 (a), the image is qualitatively quite different now, although long-range ordering is also not immediately obvious. However, when again Fourier-filtering the image and superimposing the result on the original data a definite (1 x 1) pattern is obtained, which is shown in Figure 6 (b). This clearly confirms that in the absence of the anions on the surface the first copper adlayer tends to be arranged in epitaxial registry with the Au(111) substrate.

We also performed STM-investigations in the 4 mM H_2SO_4 solution to assess the influence of the combined presence of both HSO_4^- and SO_4^{2-}, which in the voltammogram

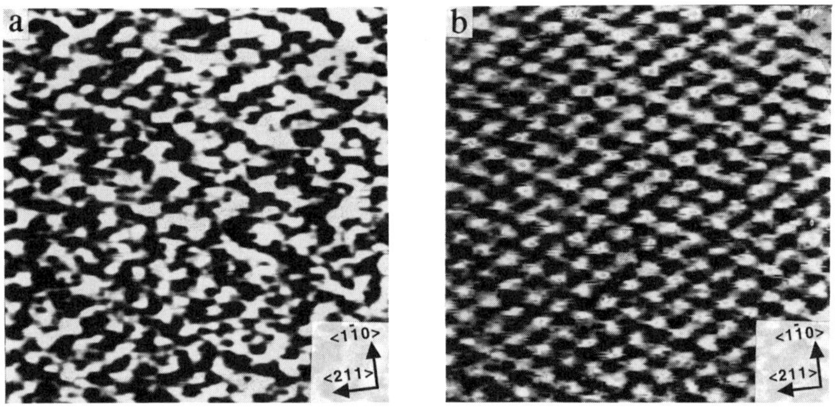

Figure 6. STM image obtained in a solution with a low concentration Cu^{2+} ions to hinder bulk deposition at negative potentials. (a) Raw image obtained at E = 200 mV vs. Cu/Cu^{2+} in 0.1 M H_2SO_4 + 0.01 mM $CuSO_4$ (43 x 43 Å^2 V_t = + 20 mV, I_t = 30 nA). (b) Fourier transform of the image in (a) superimposed on the raw data revealing the (1 x 1) copper adlayer at a reduced coverage of adsorbed anions.

(cf. Figure 2 (b)) manifests itself by two distinct desorption peaks of the monolayer. This raises the question of whether the structural features associated with the two anions (cf. Figures. 5 (a) and (c) are also found to coexist in this solution.

The image shown in Figure 7 (a) indicates that indeed there are different structural phases present on the surface. However, only the island in the upper part of the image is similar to Figure 5 (a), whereas the other area is of a new type of structure and not the p(2 x 2), as may perhaps have been expected. Upon merely visual inspection this new structure, which is shown in Figure 7 (b) at higher resolution, is characterized by rows along the <110> direction with a separation of twice the substrate row distance.

The Fourier transform of this image provides further structural information, because in addition to the bright spots resulting from the obvious periodicity along <211> there are weaker spots in the same positions which are seen for the p(2 x 2) superstructure shown in Figure 5 (c). This seemingly new structure is thus quite similar to the pattern observed in the Na_2SO_4 solution, although there is a distinct difference. If the two structures seen in Figure 7(a) are associated with the different anions, which in turn are reflected in the double-peak, sequential desorption of copper, then it may be expected that the more reversible HSO_4--induced structure can be removed by changing the potential from the full monolayer region to a value in-between the two desorption peaks (cf. Figure 2(b)).

In Figure 7(c) a typical example of the STM-observations, resulting from this procedure, is shown. Only one of the structures, associated with the full monolayer (cf. Figure 7(a)), is now present on the surface and coexists with ($\sqrt{3}$ x $\sqrt{3}$)R30° areas where the HSO_4^--induced pattern (cf. Figure 5(a)) has indeed been removed.

Summarizing the results presented in this section, it has been shown that the two anions HSO_4^- and SO_4^{2-} interact strongly with the full monolayer of copper on Au(111) and give rise to distinctly different STM-images in the potential region preceding copper bulk deposition. In a solution, which contains both anions, only the HSO_4^--induced structure is observed, whereas the p(2 x 2) pattern seen in pure sulphate solution is significantly modified. The copper associated with the HSO_4^- is more weakly bound and the STM structure associated with it can be selectively removed by adjusting the potential positive of the first desorption peak in the voltammogram of Figure 2(b).

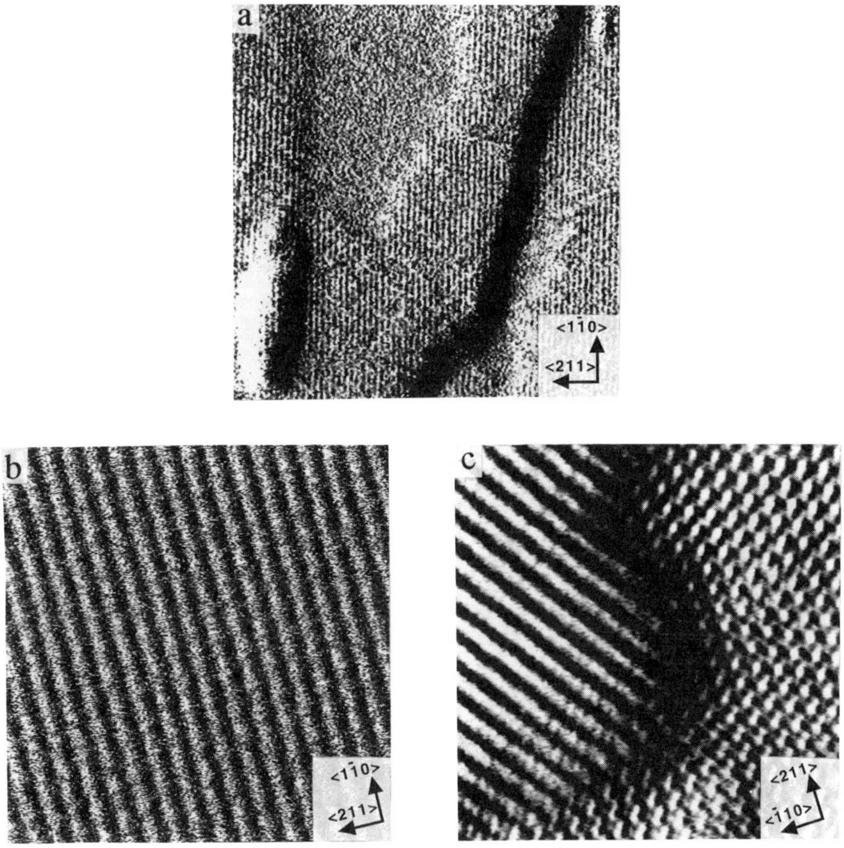

Figure 7. STM images in a solution containing both HSO_4^- and SO_4^{2-} anions (4 mM H_2SO_4 + 1 mM $CuSO_4$). (a) Image taken at E = 0 mV vs. Cu/Cu^{2+} showing a joint domain structure; the unresolved island is the same structure as observed in the HSO_4^- solution at low potential (cf. Figures 5(a) and (b)), the second structure (rows in the <110> direction) is a new structure not previously observed in either HSO_4^- or SO_4^{2-} containing solutions (340 x 340 Å^2; V_t = + 30 mV, I_t = 30 nA). (b) A high resolution image of the new structure observed in (a) revealing rows along the <110> direction with a separation of twice the substrate row spacing (85 x 85 Å^2; V_t = + 20 mV, I_t = 30 nA, E = +2 mV vs. Cu/Cu^{2+}). (c) Image taken at a potential between the two "monolayer" desorption peaks in Figure 2(b) (E = + 130 mV vs. Cu/Cu^{2+}) showing the coexistence of only one of the "monolayer" structures (cf. Figure 7(b)) with the ($\sqrt{3}$ x $\sqrt{3}$)R30° (85 x 85 Å^2; V_t = + 20 mV, I_t = 30 nA).

MULTIVARIATE STATISTICAL ANALYSIS OF STM IMAGES

In scanning tunneling microscopy of solid surfaces, either without or with an adsorbate, a periodic arrangement of the surface atoms is investigated in the majority of studies. Even in those cases where the arrangement is not periodic, with biological surface species for example, the object of interest is also present many times in the image. Therefore STM images can be expected to have an information content which is considerably larger than what can be exploited by the standard procedure of visual inspection.

A very similar situation, in which many images of the same object are collected, is encountered in electron-microscopy studies of biological macromolecules. Such studies have, of course, been performed for a much longer time than STM and have been accompanied by extensive computational efforts to extract the significant information from a large set of noisy images. One such image-processing technique, which has been developed during the last decade and has been particularly successful for this purpose, is based on multivariate statistical analysis (MSA)[12-14] and is commercially available ("Imagic 5", Image Science).

To our knowledge this type of elaborate image processing has not yet been applied to STM data and we present now our initial efforts in this direction. The computational details of the "Imagic 5" software have been described in detail elsewhere.[12-14] Therefore it may suffice here to merely outline the basic features of the MSA analysis.

In the first step the data are compressed with the help of an appropriate transform of the original coordinate system in which the data are described. The new coordinate system is spanned by the first few eigenvectors (eigenimages) of the data set (associated with the largest eigenvalues), which together describe a large portion of the variance present in the images. In the resulting compact factor space every image is described by its coordinates (typically 10-20) relative to the main eigenimages rather than by its original number of pixel grey values (512 x 512 in our images, for example). Thus it becomes much easier to divide the elements of the data cloud into groups (subclouds) exhibiting similar behavior. Efficient automatic clustering procedures have been developed for this purpose. This type of analysis has been applied to the p(2 x 2) structure (cf. Figure 5(c)), which is seen in the Na_2SO_4 solution.

Four STM images measured on the same area of the surface and containing a total of approximately 1200 bright spots, which give rise to the superstructure, were analyzed. The classes of significantly different structure, which resulted from the image processing, are shown in Figure 8. The image in the upper left-hand corner is the average of the 14 different classes shown subsequently, to which we shall refer by number from left to right (1st row: classes 1-4, 2d row: classes 5-9, etc.). The average image is merely the result of the choice of the dimensions, in which the similarities have been analyzed. For example, class 7 contains one half of the p(2 x 2) unit cell and class 8 the other half, which when overlapped give the rectangular pattern of the average of classes.

The individual classes exhibit statistically significant molecular details, which occur many times in the large data set. Quite generally, there is some systematic structure in addition to the bright p(2 x 2) spots and a tendency of pairing between spots of higher and lower intensity (see, for example, classes 5 and 6). In some classes there are also strong deviations from the equilateral triangular arrangement of the more intense p(2 x 2) spots, such that they are closer to each other (9 and 14) or appear to merge into a larger spot (classes 3 and 5, for example). It is interesting that these details, which the MSA image processing yields, can indeed be detected with some effort by inspecting a large amount of raw data, but only in retrospect with the knowledge that Figure 8 provides. We should like to add a note of caution, however, at this early stage of applying MSA to STM images. In the area of electron microscopy there exists detailed experience for many years in choosing the appropriate parameters for the data analysis and obtaining a similar confidence in the area of STM will require further studies of other adsorbate systems and in different environments.[22]

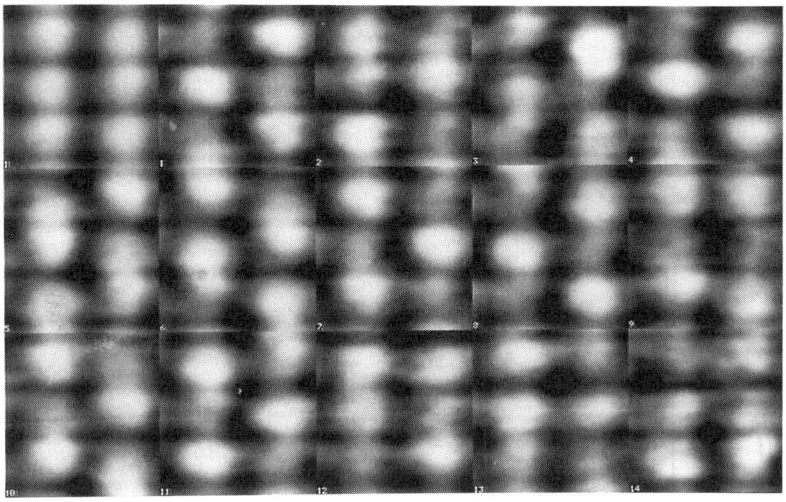

Figure 8. Result of the MSA analysis of the p(2 x 2) superstructure, using approximately 1200 bright spots (cf. Figure 5(c)). There are 14 classes, into which similarities in the date set were categorized. The image in the upper left-hand corner is the sum of all classes.

DISCUSSION

The interpretation of the results presented in this paper is closely connected to an understanding of the tunneling mechanisms in the electrochemical environment (cf. Introduction) and to the question which adsorbed solution species can be imaged in STM. For example, it is not immediately obvious how at constant tunneling current the tip does react when water molecules in the gap region are replaced by anions such as HSO_4^- or SO_4^{2-}. It seems, however, that an ordered anion phase on Au(111) in pure 0.1 M H_2SO_4 has recently been observed by STM[23] and this may be quite a general phenomenon.

The basic question, which we are therefore faced with, is whether the observed superstructures associated with the full copper monolayer reflect the arrangement of the adlayer metal atoms or of the anions, which apparently adsorb on top of the copper[21]. A useful starting point for this distinction is a comparison of the periodic dimensions of the patterns with the diameter of the fairly large anions (4.0-4.5 Å). The separation between the rows seen in Figure 5(b) equals the distance between the rows of the substrate lattice (2.50 Å) and at least in this system a periodic array of the anions cannot be the origin of the observed structure. However, if we wish to attribute the observed pattern in 0.1 M H_2SO_4 to the arrangement of the copper atoms, the (1 x 1) pattern seen in Figure 6 indicates that this arrangement depends on the amount of adsorbed anions and thus varies with applied potential. When there is a larger anion coverage the positions of the copper atoms apparently deviate from the simple epitaxial arrangement and more complicated structures evolve.

When applying this concept to the p(2 x 2) superstructure, the details of the MSA image processing would be attributed to a considerable mobility of the copper atoms in the adlayer, such that a significant population of adsorption sites outside the p(2 x 2) array takes place. This may involve the formation of copper dimers, although it is difficult to understand that the orientation of these dimers should be at a substantial angle to the surface plane as Figure 8 seems to indicate. There is also the problem of accounting for the coverage of one full monolayer in this interpretation, since the p(2 x 2) pattern corresponds to only 1/4 and even when counting the lower intensity features in Figure 8 only 1/2 of a monolayer is visible.

In summary, our work clearly shows the strong influence of the ionic solutions species on the structures observed in electrochemical STM of seemingly simple overlayers, such as a full monolayer. This points to the need for more careful considerations of the basic factors which control the tunneling process and hence the imaging in electrochemical environments. To this extent the additional information that can be gained by the image processing technique described above may make a significant contribution towards achieving this goal.

ACKNOWLEDGMENTS

We would like to thank H. Gerischer, J. K. Gimzewski, P. Dube and R. Nichols for valuable discussions. We are also grateful to J. H. Block for his continuing support and encouragement of this work.

REFERENCES

1. J. K. Gimzewski and J. K. Sass, Fundamental aspects of scanning tunneling microscopy in polar solvents, *in:* "Proc. ICTP Conf. Cond. Matt. Phys. Aspects of Electrochem.," M. P. Tosi and A. A. Kornyshev, eds., World Scientific, New Jersey, 303-326 (1991).
2. See for example: J. O'M. Bockris and A. K. N. Reddy,"Modern Electrochemistry," Plenum Press, New York (1977).
3. R. N. Barnett, U. Landmann and A. Nitzan, Primary events following electron injection into water and adsorbed water layers, *J. Chem. Phys.* 93:6535-6542 (1990).
4. J. K. Sass and J. K. Gimzewski, Solvent dynamical effects in scanning tunneling microscopy with a polar liquid in the gap, *J. Electroanal. Chem.* 308:333-337 (1991).
5. K. L. Sebastian and G. Doyen, Electrochemical scanning tunneling microscopy: Does the orientational polarisation of the liquid play any role?, *Surf. Sci.*, in press.
6. K. Bange, B. Straehler, J. K. Sass and R. Parsons, The interaction of Br with Ag(110): Comparison of electrochemical and gas-phase adsorption measurements, *J. Electroanal. Chem.* 229:87-98 (1987).
7. J. K. Sass, J. Schott and D. Lackey, Prospects for the direct determination of the dielectric constant of water in the double layer using model adsorption experiments in UHV: Coadsorption of Cs and H_2O on Cu(110), *J. Electroanal. Chem.* 283, 441-448 (1990) .
8. D. M. Kolb, The initial stages of metal deposition: An atomistic view, *Schering Lecture Publications*, 2:1-35, Schering Research Foundation, Berlin (1991).
9. O. M. Magnussen, J. Hotlos, R. J. Nichols, D. M. Kolb and R. J. Behm, Atomic structure of Cu adlayers on Au(100) and Au(111) electrodes observed by in situ scanning tunneling microscopy, *Phys. Rev. Lett.* 64:2929-2932 (1990).
10. T. Hachiya, H. Honbo and K. Itaya, Detailed underpotential deposition of copper on gold(111) in aqueous solutions, *J. Electroanal. Chem.* 315:275-291 (1991).
11. S. Manne, P. K. Hansma, J. Massie, V. Elings and A. A. Gerwith, Atomic resolution electrochemistry with the atomic force microscope: Copper deposition on gold surfaces, *Science* 251:183-186 (1991).
12. M. van Heel, Classification of very large electron microscopical image data sets, *Optik* 82:14-126 (1989).
13. E. J. Boekema and M. van Heel, Molecular shape of lumbricus terrestris erythrocruorin studied by electron microscopy and image analysis, *Biochim. Biophys. Acta* 957:370-379 (1989).
14. L. Borland and M. van Heel, Classification of image data in conjugate representation spaces, *J. Opt. Soc. Am.* A 7: 601-610 (1990).

15. K. H. Besocke, An easily operable scanning tunneling microscope, *Surf. Sci.* 181:145-153 (1987).
16. W. Haiss, D. Lackey, J. K. Sass and K. H. Besocke, Atomic resolution scanning tunneling microscopy images of Au(111) surfaces in air and polar organic solvents, *J. Chem. Phys.* 95: 2193-2196 (1991).
17. W. Haiss, D. Lackey, J. K. Sass, H. Meyer and R. J. Nichols, A determination of copper overlayer structures on Au(111) in the presence of electrolyte additives, *Chem. Phys. Lett.* 200: 343-349 (1992).
18. J. V. Barth, H. Brune, G. Ertl and R. J. Behm, Scanning tunneling microscopy observations on the reconstructed Au(111) surface: Atomic structure, long-range superstructure, rotational domains, and surface defects, *Phys. Rev.* B 42, 9307-9318 (1990).
19. X. Gao, A. Hamelin and M. J. Weaver, Atomic relaxation at ordered electrode surfaces probed by scanning tunneling microscopy: Au(111) in aqueous solution compared to ultrahigh-vacuum environments, *J. Chem. Phys.* 95:6993-6996 (1991).
20. N. J. Tao and S. M. Lindsay, Observations of the 22 x $\sqrt{3}$ reconstruction of Au(111) under aqueous solutions using scanning tunneling microscopy, *J. Appl. Phys.* 70:5141-5143 (1991).
21. P. Zelenay, L. M. Rice-Jackson, A. Wieckowski and J. Gawlowski, Radioactive labeling study of isulfate adsorption on copper adatoms deposited on the gold electrode in neutral media, *Surf. Sci.* 256:253-263 (1991).
22. STM-images of C_{60} molecules adsorbed in UHV on gold and copper surfaces are currently being analyzed by the MSA method and the preliminary results are very encouraging; J. K. Gimzewski, R. Schlittler, M. van Heel and J. K. Sass, to be published.
23. O. M. Magnussen, J. Hageböck, J. Hotlos and R. J. Behm, In situ scanning tunnelling microscopy observations of a disorder-order transition in hydrogensulfate adlayers on Au(111), *Faraday Discuss.*, 94:154-163 (1992).

A CLOSED-LOOP OPTICAL SCAN CORRECTION SYSTEM

FOR SCANNING PROBE MICROSCOPES

Daniel R. Marshall,[1] Eric M. Fray,[2] James D. Mueller,[2]
L. Martin Courtney, III,[2] John C. Podlesny,[2] John B. Hayes,[2]
Tami L. Balter,[2] and Jay Jahanmir[2]

[1]Hewlett Packard
Boise, ID 83714
[2]WYKO Corporation
Tucson, AZ 85706

Abstract: Piezoelectric actuators are widely used for scanning probe microscopes (SPMs) because they are capable of motions that can be continuously subdivided down to atomic dimensions and smaller. These actuators have three serious drawbacks -- drift, hysteresis, and nonlinearity of displacement versus driving voltage -- that have limited the usefulness of scanning probe microscopes as metrology instruments capable of accurate dimensional measurements down to the nanometer level. This paper describes a novel optical method for measuring and actively controlling multiple translational degrees-of-freedom at a single location in a scanning probe microscope using a folded retroreflector. Shot noise-limited resolution is shown to be less than 0.1 nm. This scanner was incorporated in a scanning probe microscope and its performance was characterized. Measured distortion is less than 1% over a 100 μm x 100 μm scan range. Lateral thermal drift is less than 0.45 μm per hour.

INTRODUCTION

Scanning probe microscope (SPM) technology has depended on piezoelectric actuators to generate motion with sub-Angstrom control.[1] The virtues of these actuators include: excellent resolution, good stiffness, low power consumption, and reasonable scan range. Their vices include: nonlinearity of position versus driving voltage (Figure 1); hysteresis (position for a given driving voltage is dependent on the past history of movement); thermal drift; and creep (position slowly drifts in the direction of the most recent movement after a rapid change in voltage). These vices, if left uncorrected, induce significant distortion and inaccuracies in the scan data.

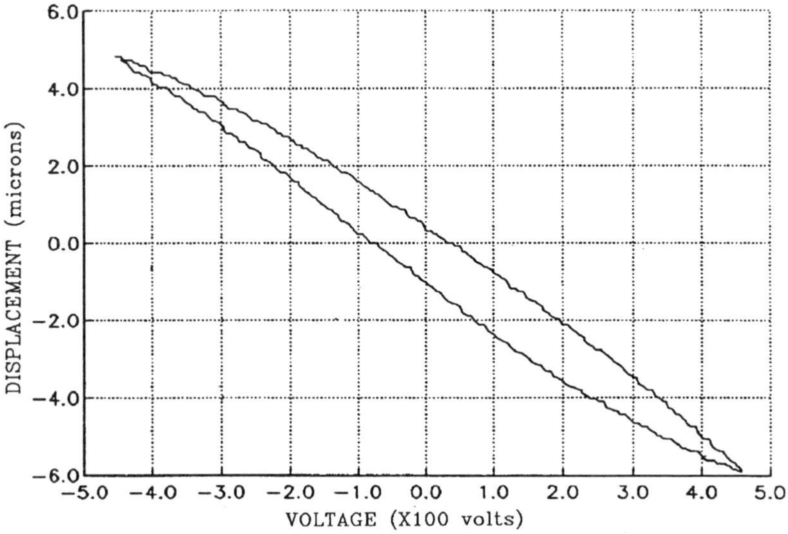

Figure 1. Displacement versus applied voltage for a tube-shaped piezoelectric ceramic. Data were taken with a Michelson interferometer. The motion of the ceramic is nonlinear and has hysteresis.

Two general correction methods have been used to compensate for the nonideal performance of piezoelectric actuators: open loop and closed loop. In open loop methods, the nonlinearity can be removed either by postprocessing of the image[2] or by applying appropriate nonlinear voltages to the ceramic. The effects of hysteresis on the image can be eliminated by taking data in the same scan direction. The hysteresis and creep can also be reduced by controlling the applied charge to the ceramic[3] or by placing a small capacitor in series with the ceramic.[4] Distortion caused by nonlinearity can usually be removed to less than 5% of full scale using a combination of these methods. Because these techniques do not control the actual position of the actuator in a closed loop fashion, it is not possible to correct for environmentally related errors such as those caused by shock, vibration, and temperature change. Some of the other parameters that can affect the accuracy of the calibration in open loop methods include scan speed, scan range, and the location of the scan relative to the maximum scan range.

A correction can also be obtained by a secondary measurement of displacement of the ceramic. The scanner position is monitored as the image is taken and used to correct the distortion induced by the piezoelectric ceramic. Yamada et al.[5] used an optical interferometer to monitor the scans with nanometer accuracy. Griffith et al.[6] have developed a simple capacitance-based position monitor. Even though both these techniques can be adapted to operate in closed loop fashion, they are difficult to align and manufacture.

Closed loop methods measure the actual position of the scanning element and make real-time corrections. Closed loop control techniques potentially offer the following advantages over open loop methods: improved scanning speed, direct correction of nonlinearity, hysteresis and creep, improved shock and vibration rejection, and reduced drift. A simple closed loop scanner was demonstrated by Barrett and Quate.[7] The authors used a simple two-axis optical beam displacement sensor to measure the x and y position of the piezoelectric tube scanner. Using this technique, the authors reported reduction of both linear and quadratic scan distortion to less than 2% of the image size. This design is simple and inexpensive but has three main limitations: (1) the signal-to-noise ratio is limited by the small movement of the illuminated spot on the detector and by the limited radiance of the source; (2) Abbe offset errors caused by the necessity of locating position flags for different coordinate axes in different locations create offset errors that can be relatively difficult to remove; and (3) the method offers limited flexibility in choosing a range/resolution tradeoff because of the small motion of the aperture.

In this paper, we describe a new optically based position sensing method that removes the above limitations. This is followed by a summary of results obtained with a commercially available SPM incorporating this closed-loop position encoder.

SCANNER DESIGN AND PERFORMANCE

Figure 2 is a schematic of the optical position sensor. The position sensor uses a high radiance source that is focused to a small point image on a position sensor by way of a folded path through a corner cube (retroreflector). Because of the imaging properties of the corner cube, a moving image of the stationary source replicates the motion of the vertex of the cube in all three translational degrees of freedom. The corner cube is attached to the piezoelectric actuator near the scanned sample. A lateral position effect detector measures the motion in the plane of the face of the corner cube.

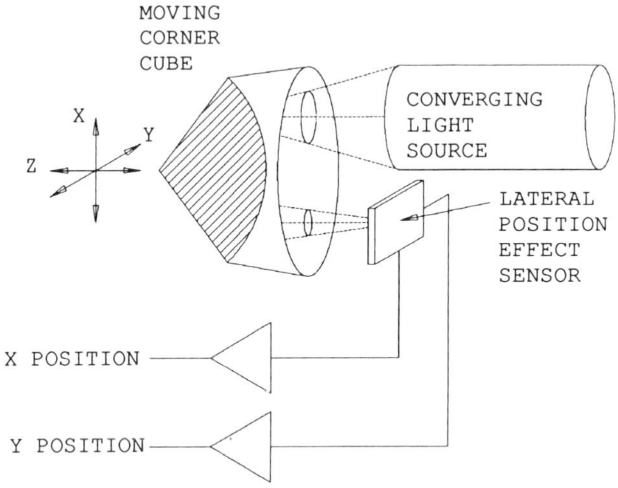

Figure 2. Schematic representation of the optical position sensor. The position sensor uses a high radiance source that is focused to a small point image on a position sensor by way of a folded path through a corner cube.

Figure 3 is a schematic drawing of the optical path of a position encoder incorporating a corner cube position sensor. The only moving optical part in this design is the corner cube, which is attached to a piezoelectric scanner with its vertex immediately adjacent to the sample being measured. Since the sensor is attached directly to the sample plane, displacement-induced errors are minimized. Much of the flexibility of this encoder design results from the fact that the moving spot of light is reimaged by a magnification module. The magnification module may be chosen to suit the range/resolution requirements of the measurement, and can provide virtually noise-free "amplification" of the position error signal. Variable magnification is possible because the optical path preserves a point image of the source. That is, any magnification of the motion of a point image will similarly magnify the slope of the error signal curve. While this sort of magnification is possible with the aperture-and-split-detector method, it is cumbersome, especially for a wide range of magnification values.

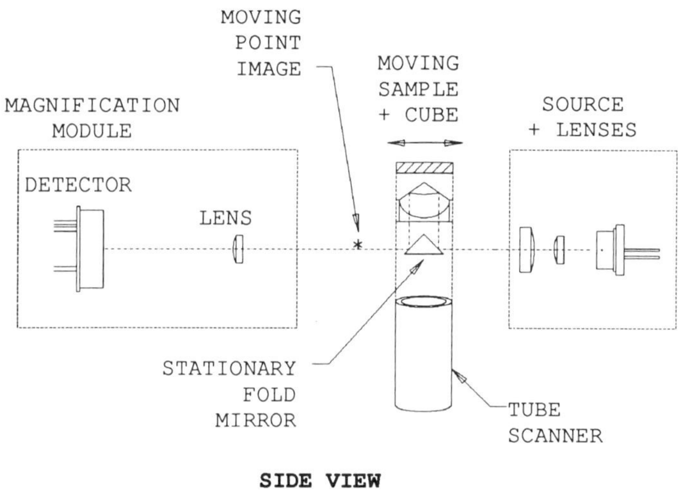

Figure 3. Schematic representation of the optical path of the position encoder incorporating a corner cube. The only moving part is the corner cube, which is attached to the piezoelectric scanner near the sample position.

The position sensor is a lateral effect photodiode, chosen for excellent linearity and high signal-to-noise ratio. The dominant noise sources are preamplifier noise and shot noise resulting from the detection process in the photodetector. A representative noise figure for high quality preamplifiers is about 0.15 $\mu V/\sqrt{Hz}$ rms at the input, or 1.5 $pA/\sqrt{Hz}$ at the input with 100 kΩ transimpedance gain. The shot noise limit is approximately 4.4 $pA/\sqrt{Hz}$, and the combined contribution of amplifier noise and shot noise is less than 5 $pA/\sqrt{Hz}$.

The optical magnification of the microscope in this example is chosen to be 10X. An additional factor of 2 results from the imaging properties of the corner cube, for a net magnification of 20X. This magnification is chosen so that a deflection of $\pm$100 μm x $\pm$100 μm at the sample generates maximum output from a 4 mm square detector. Based upon the values of optical power and responsivity used above, the net position sensor gain would be 60 μA/100 μm, or 60 pA/Å. Therefore the shot-noise-limited resolution of the detector in this design would be 0.08 $Å/\sqrt{Hz}$. If the bandwidth is chosen to be, for example 100 Hz, the shot-noise-limited resolution of the position sensor would be less than 1 Å. Actual measurements may be limited by other noise sources, especially environmental vibration.

Figure 4 shows a time domain trace of the output signal of the position encoder taken with a stationary corner cube in open loop configuration. Figure 5 shows the power spectrum of the same data. The 60 Hz peak is a measurement artifact resulting from pickup of radiated power line energy by the measuring instrument. This measurement includes noise and drift of the power amplifier that drives the piezoelectric scanning tube, thermal drift, and mechanical vibration. In closed loop operation, some of these contributions to noise and drift can be reduced by the servo loop by as much as two orders of magnitude, depending on their frequency spectrum.

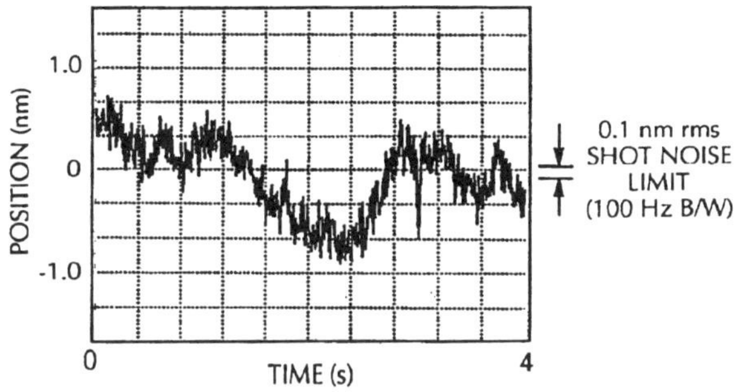

Figure 4. Oscilloscope trace of output signal from the position sensor with a stationary corner cube in an open loop mode.

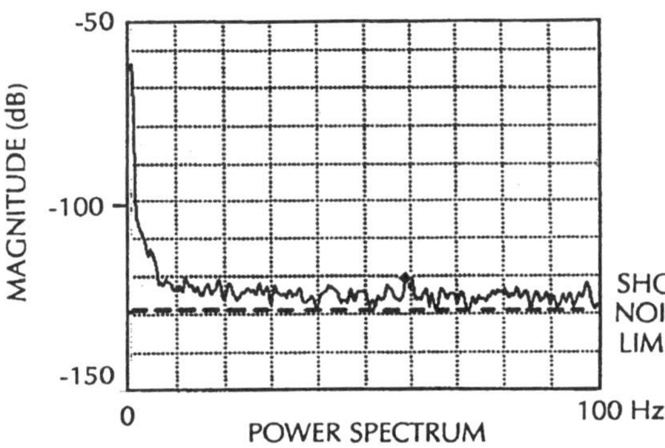

Figure 5. The power spectrum of the data in Figure 4. The 60 Hz peak is a measurement artifact resulting from pickup of power line energy by the measuring instrument.

The major advantages of this design are flexibility (adaptability to different scan ranges and geometries), high signal-to-noise ratio (which may be as much as 20 dB higher than a low radiance source and aperture type), and reduced offset between the scanned sample and the sensing element.

SPM MEASUREMENTS

The closed loop scanner was incorporated in a commercially available SPM (WYKO MicroProbe 3D). Measurements were performed in air using an atomic force microscope probe in repulsive mode. A typical scan speed of 5 Hz was used for most measurements. A 100 μm x 100 μm image of a two-dimensional grating target, made without (left) and with (right) closed loop correction of position is shown in Figure 6. Notice the characteristic distortion of the image caused by nonlinearity in the piezoelectric actuator when the correction loop is open. This distortion is corrected when the position correction loop is closed. The measured distortion is less than 1% of the total scan range.

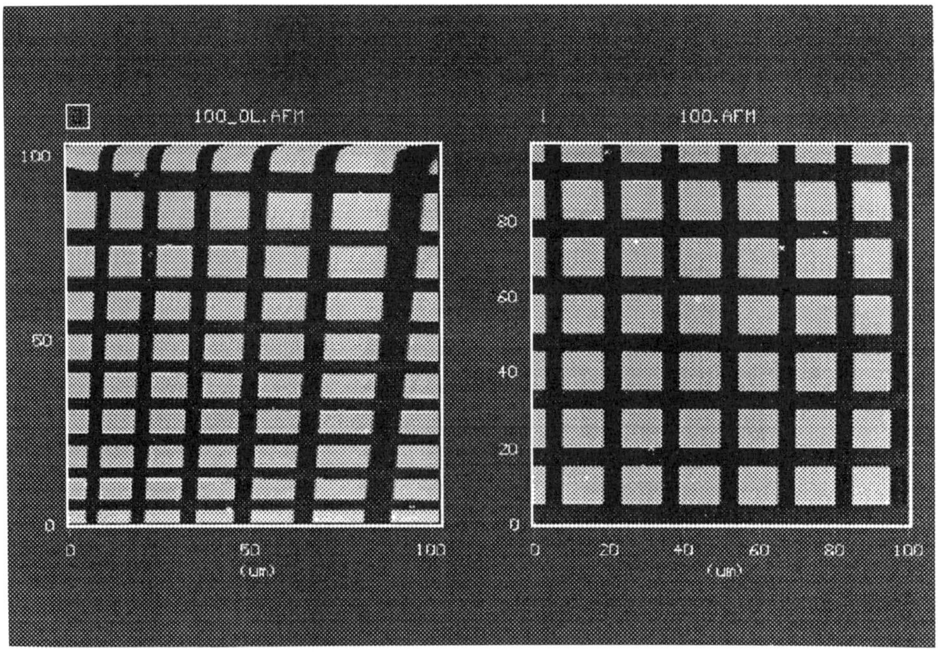

Figure 6. A 100 μm x 100 μm image of a two-dimensional grating taken with SPM using the position encoder. The image on the left-hand side was taken with the feedback loop open. The measured distortion in the image on the right-hand side, taken with the feedback loop closed, is less than 1%.

A major benefit of the closed loop scanner is the ability to accurately locate and scan a selected area. A series of scans taken on the grating target at different magnifications is shown in Figure 7. This series is a test of positioning accuracy after changes are made in magnification and scan location. The measured positioning error in changing the scan size in all cases is less than 1% of the size of the larger frame.

A series of measurements were also taken to test for thermal drift. Successive scans of a specific feature taken at half-hour intervals is shown in Figure 8. This series of scans tests the stability of the entire system, some elements of which are outside of the position control loop, and are, therefore, uncorrected. Average lateral drift was 0.45 μm/h.

Fast scan speeds can introduce significant servo positional errors and some distortion on one edge of the scan; for example, a positional offset of about 2 μm and distortion of about 2% at a scanning speed of 15 Hz over a 100 μm x 100 μm area (Figure 9, left). The largest scan is the worst case because the surface speed is the highest. Fast scanning is a very useful tool for quick surveys of an area to choose local areas of interest, and the errors decrease rapidly as the scanning speed or scan size is reduced. This is demonstrated in the image shown on the right-hand side of Figure 9. This is a 30 μm x 30 μm scan taken at 20 Hz.

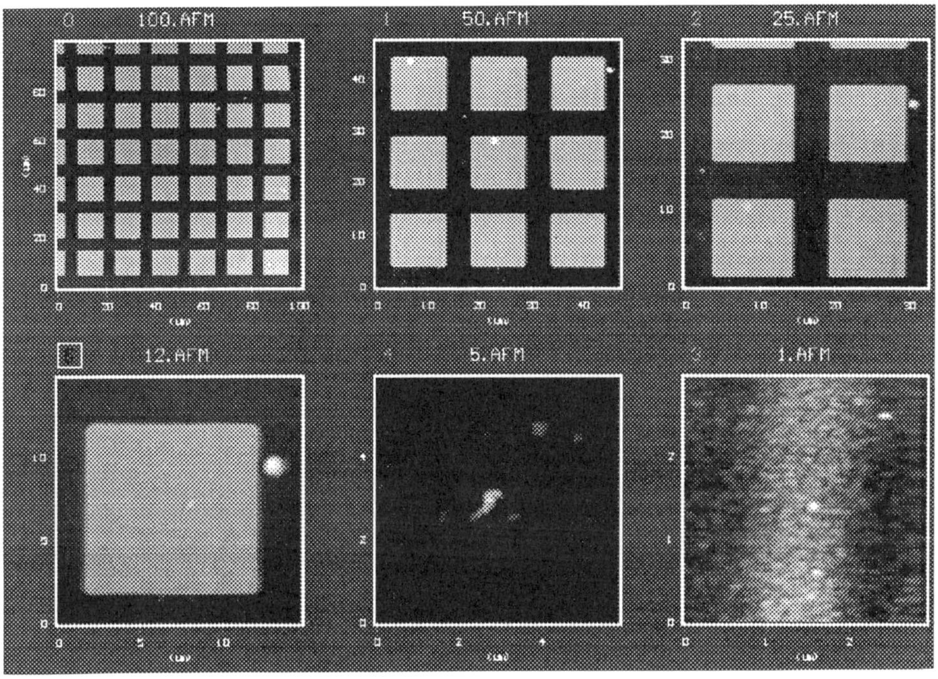

Figure 7. A series of scans taken on the two-dimensional grating at different magnifications. This series of images is a test of positional accuracy after changes are made in magnification and scan location. The measured positioning error in all cases is less than 1% of the size of the larger frame.

CONCLUSION

The closed loop corner cube optical position encoder provides an effective method for measuring the position of the scanned element *in situ*, that has the correct mass, simplicity, high frequency bandwidth, relative insensitivity to tilt and misalignment, low cost, and freedom from creep, nonlinearity, and hysteresis that will transform the scanning probe microscope from a research laboratory imaging tool to a production environment metrology instrument. This sensing method makes possible fundamental improvement in scanning microscopes through correction of the shortcomings of piezoelectric-like actuators through real-time servomechanical control of the moving element, thereby providing improved speed, robustness, and accuracy.

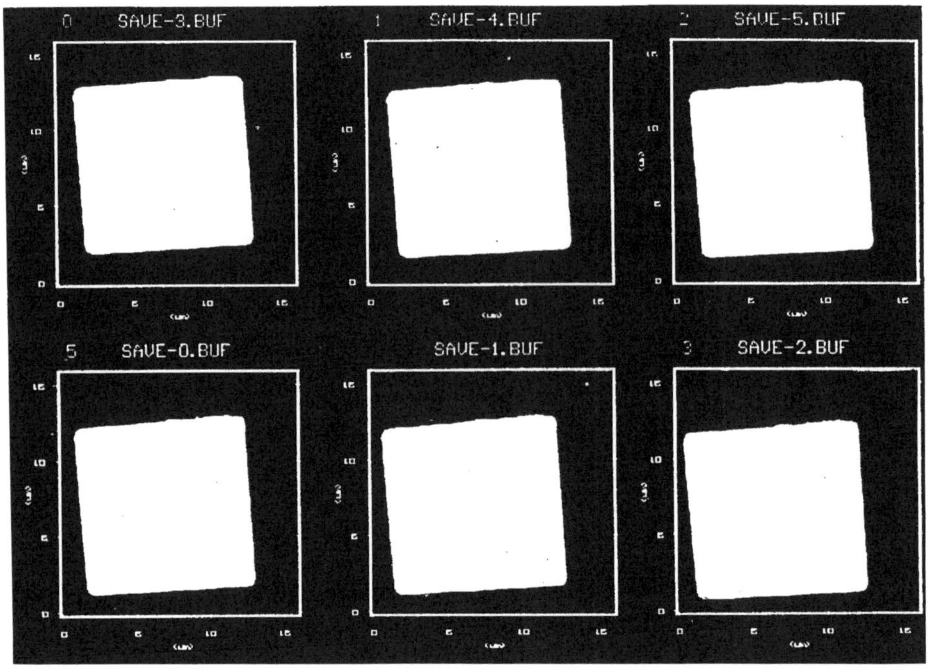

Figure 8. A series of measurements taken to test for thermal drift. Successive measurements were taken over the same area at half-hour intervals (clockwise from top left). The average lateral drift was 0.45 μm/h.

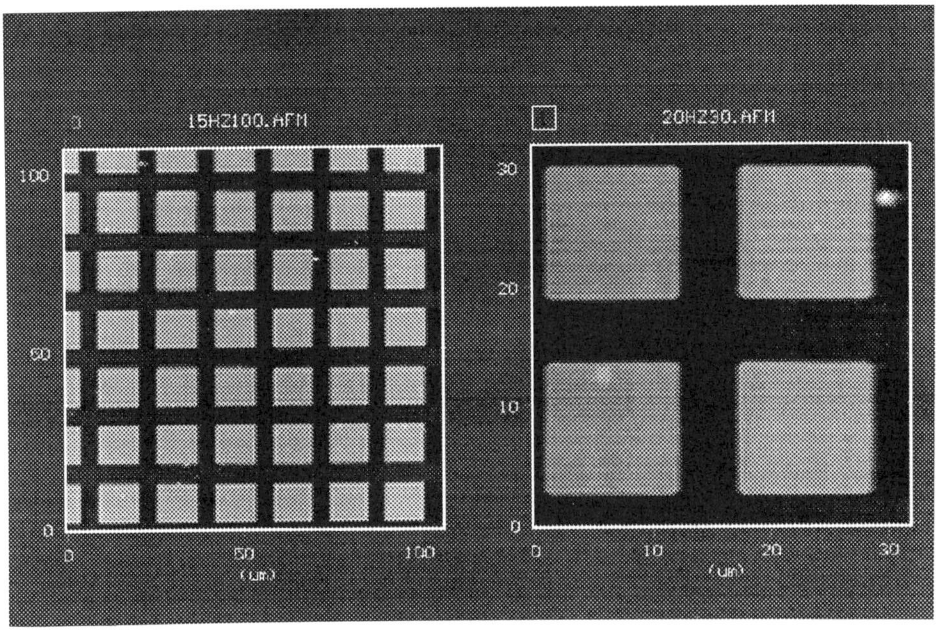

Figure 9. Measurements taken at high scan speeds. The image on the left-hand side is a 100 µm x 100 µm image taken at 15 Hz. A positional offset of 2 µm and distortion of about 2% is observed due to servo positional errors. The error decreases when either the scan speed or the scan size is reduced. This can be seen in the image on the right-hand side, which is a 30 µm image taken at 20 Hz.

ACKNOWLEDGMENT

We would like to thank L. Stahl for much helpful advice toward the preparation of this paper.

REFERENCES

1. J. Jahanmir, B.G. Haggar, and J.B. Hayes, The scanning probe microscope, *Scanning Microscopy*, 6:625-660 (1992).
2. M. Gehrtz, H. Strecker, and H. Grimm, 1988. Scanning tunneling microscopy of machined surfaces, *J. Vac. Sci. Technol.* A, 6.432-435 (1992)
3. C.V. Newcomb, and I. Flinn, Improving the linearity of piezoelectric ceramic actuators, *Electron. Lett.*, Vol. 18, pp.442-444 (1982).
4. H. Kaizuka, and B. Siu, A simple way to reduce hysteresis and creep when using piezoelectric actuators, Jpn. J. Appl. Phys., 27: L773-L776 (1988).
5. H. Yamada, T. Fujii, and K. Nakayama, Linewidth measurement by a new STM, Jpn. J. Appl. Phys., 28:2402-2404 (1989).
6. J.E. Griffith, G.L. Miller, C.A. Green, D.A. Grigg, and P.E. Russell, A scanning tunneling microscope with a capacitance-based position monitor, *J. Vac. Sci. Technol.* B, 8:2023-2027 (1990).
7. R.C. Barrett, and C.F. Quate, Optical scan-correction system applied to atomic force microscopy, *Rev. Sci. Instrum.*, 62:13931399 (1991).

VIBRATING PROBE (AC) METHODS IN ATOMIC FORCE MICROSCOPY

C.B. Prater

Digital Instruments
Santa Barbara CA 93103

Abstract: The atomic force microscope is expanding its capabilities through the use of vibrating probe (AC) techniques. One mode of operation, called Tapping Mode™ oscillates the probe so that it lightly "taps" on the surface at the bottom of each oscillation cycle. This method virtually eliminates frictional forces between the probe and surface yet maintains the high resolution and ease of operation of the contact mode technique. Tapping mode can also be combined with noncontact methods to measure long-range forces like those from magnetic and electric fields. This paper describes a two-step process called Lift Mode™ where a computer records the sample surface topography for each scan line and then initiates a noncontact measurement by flying the probe a small distance above the recorded topography. this provides simultaneous and yet separate measurements of both the sample topography and the long-range forces above the sample.

INTRODUCTION

The atomic force microscope (AFM)[1] has become an increasingly useful tool for profiling surfaces at high resolution. The AFM scans a stylus attached to a spring lever (or cantilever) over a sample surface. A detection system measures the motion of the cantilever in response to features on a sample surface. Images of the sample are created by plotting the motion of the cantilever as a function of the stylus position over the sample.

Most commercial AFM are operated in the "contact mode," where the AFM stylus is in constant contact with the sample surface. In this mode the AFM cantilever deflects directly in response to any topographic features on the sample. Usually a feedback loop is used to maintain a constant and small tracking force on the sample. While this style of AFM has been very successful, there are a number of disadvantages. Most of these disadvantages result from capillary forces that are present between the stylus and sample due to liquid contamination on the tip and sample.[2] This capillary attraction begins abruptly as soon as the liquid contamination layers on the stylus and on the sample come into contact. Typically, the capillary forces are of order 10^{-7} N and set the minimum forces at which a contact mode AFM can be operated under ambient conditions. There are also

Atomic Force Microscopy/Scanning Tunneling Microscopy
Edited by S.H. Cohen *et al.*, Plenum Press, New York, 1994

other forces, like electrostatic and chemical interactions which can contribute to a large attractive force between the tip and sample. These attractive forces also produce large frictional forces between the tip and sample. In practice frictional forces are much more destructive to both the stylus and the sample than normal forces and can also create false contrast in contact mode AFM images.

It has been possible to reduce or even negate attractive forces present in contact mode AFM by operating with the stylus and sample under a fluid.[2] This type of operation is very appealing for the study of biological samples, for example, under physiological conditions. For many samples, however, it is not practical or desirable to image the sample under a fluid.

As an alternative to contact mode AFM, a number of researchers have worked to develop noncontact mode AFMs.[3] These AFMs are based on the detection of the Van der Waals attraction between the stylus and sample, before the stylus actually contacts the sample. While this technique is very appealing in principle, it has a number of practical limitations. These limitations stem from the extremely short range of the Van der Waals force and also from the capillary forces between the stylus and sample. The Van der Waals attraction between the stylus and sample become large only at very small separations--typically on the order of 1 nm. The cantilever is then typically oscillated at a very small amplitude, around 1 nm so that it can stay in the range of the Van der Waals force gradient. Unfortunately, almost all samples in typical laboratory environments have a fluid contaminant layer of greater than 1 nm. The result is that small oscillation amplitudes at small separations are unstable--it is very easy for the stylus to get trapped by the fluid layer and the oscillation will be totally damped.

We have recently introduced an alternative approach, called Tapping Mode™. Tapping Mode[4] (Figure 1) is an AC technique where the cantilever is oscillated at a much higher amplitude, typically 10-100 nm. The cantilever is positioned so that the stylus "taps" the sample at the bottom of each oscillation cycle. Since this mode relies on direct contact forces, and it operates over a wide range of cantilever oscillation amplitudes (Figure 2), it is a more robust imaging method than noncontact mode. In addition, since the vibration amplitude is much greater than the thickness of typical fluid contaminant layers, the cantilever has sufficient energy to prevent the stylus from becoming trapped in the fluid layer. Also, because the stylus spends only a very brief time in contact with the sample, frictional damage and distortions are minimized. This technique has been used to successfully image a wide variety of samples that have been difficult to image with other AFM modes.

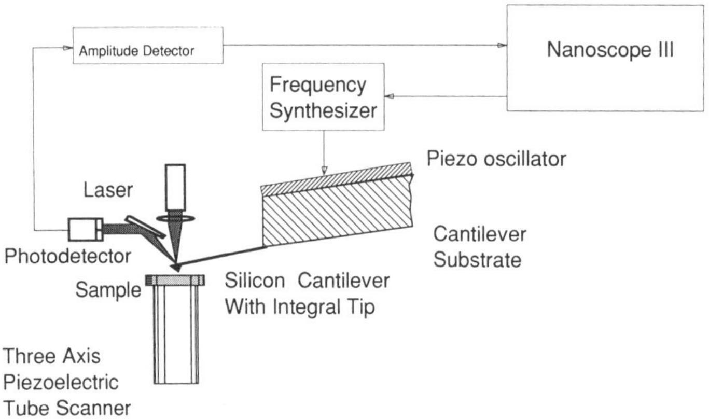

Figure 1. Schematic diagram of Tapping Mode AFM operation.

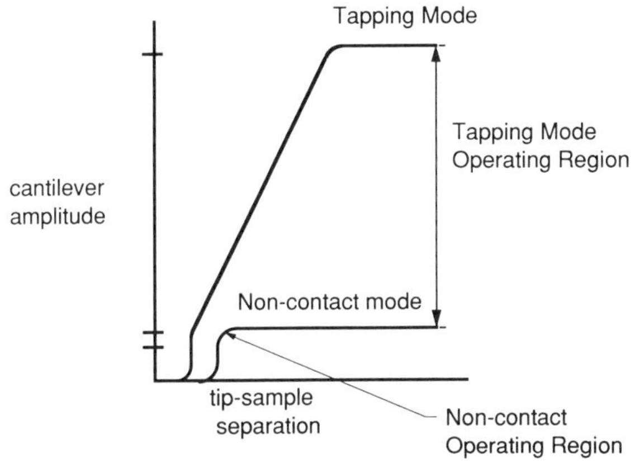

Figure 2. Comparison of Tapping Mode and noncontact mode operating ranges. Because the Tapping Mode AFM operates at much larger oscillation amplitude, there is a large, stable operating range where the cantilever stylus does not get trapped by the fluid layer on the sample.

In addition, we have recently developed a technique for integrating true noncontact AFM imaging of magnetic and electric fields with precise topographic measurements. This is accomplished in a two-step process we call "Lift Mode" (Figure 3). First the stylus is scanned over the sample to measure the local surface topography and this information is stored in memory. Next the stylus is lifted some distance above the surface (typically 10-200 nm) and the stylus is scanned above the surface, moving up and down to maintain a constant separation from the surface by using the recorded surface topography to calculate the path of the stylus.

During this second trace, alternative measurements of properties like magnetic or electric field strength can be made, by measuring changes in the cantilever's deflection, oscillation amplitude or phase, for example. A Lift Mode image of a magnetic sample is shown in Figure 4. This technique has been used successfully for both contact mode and Tapping Mode.

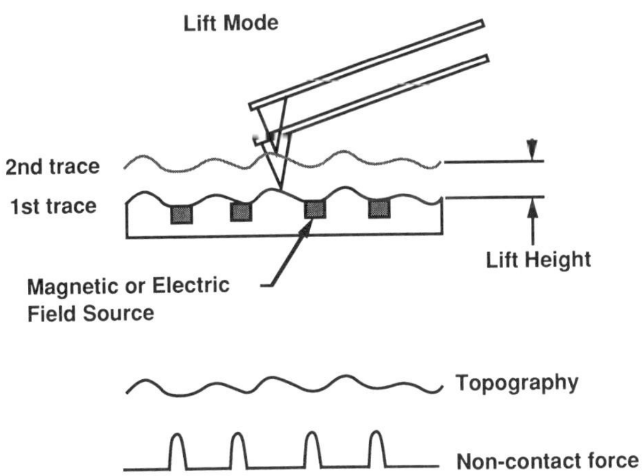

Figure 3. Schematic diagram of Lift Mode operation. After completing one trace to measure surface topography, the stylus is scanned above the surface maintaining a constant height above the local surface features. Other properties like electric or magnetic field strength are measured on this second trace.

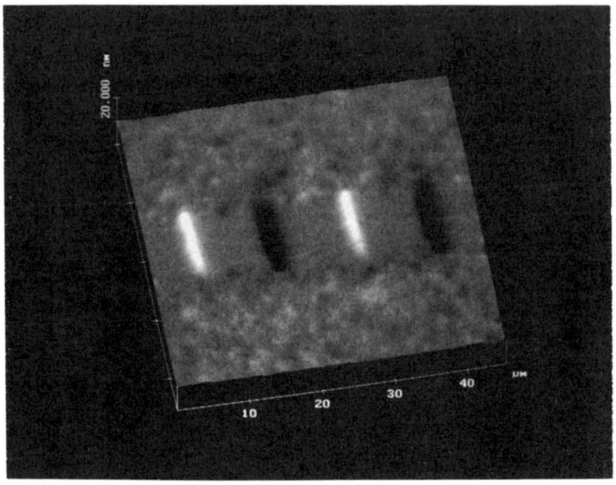

Figure 4. AFM image of magnetic bits on a computer hard disk, taken in Lift Mode.

REFERENCES

1. G. Binnig, C.F. Quate, and Ch. Gerber, Atomic force microscope. *Phys. Rev. Lett.* 12:930-933 (1986).
2. B. Drake, C.B. Prater, .L. Weisenhorn, S.A.C. Gould, T.R. Albrecht, C.F. Quate, D.S. Cannell, H.G. Hansma, and P.K. Hansma, Imaging crystals, polymers, and processes in water with the atomic force microscope. *Science* 243:1586-89 (1989).
3. Y. Martin, C.C. Williams, and H.K.Wickramasinghe, Atomic Force Microscope--Force Mapping and profiling on a sub 100 Å-scale, *J. Appl. Phys.* 61:4723-29 (1987).
4. Q. Zhong, D. Inniss, K. Kjoller, V.B. Elings, Fractured polymer/silica fiber surface studied by Tapping Mode atomic force microscopy. *Surface Science Letters*, 290:L688-92 (1993).

INDEX

The numbers following each entry refer to the first page of individual papers.